T0142091

Communications in Computer and Information Science 1633

More information about this series at https://link.springer.com/bookseries/7899

Gabriele Kotsis · A Min Tjoa · Ismail Khalil ·
Bernhard Moser · Alfred Taudes ·
Atif Mashkoor · Johannes Sametinger ·
Jorge Martinez-Gil · Florian Sobieczky ·
Lukas Fischer · Rudolf Ramler · Maqbool Khan ·
Gerald Czech (Eds.)

Database and Expert Systems Applications - DEXA 2022 Workshops

33rd International Conference, DEXA 2022
Vienna, Austria, August 22–24, 2022
Proceedings

Editors
Gabriele Kotsis
Johannes Kepler University of Linz
Linz, Oberösterreich, Austria

Ismail Khalil
Johannes Kepler University of Linz
Linz, Austria

Alfred Taudes
(WU) Vienna University of Economics and Business
Vienna, Austria

Johannes Sametinger
Johannes Kepler University of Linz
Linz, Austria

Florian Sobieczky
Software Competence Center Hagenberg
Hagenberg, Austria

Rudolf Ramler
Software Competence Center Hagenberg
Hagenberg, Austria

Gerald Czech
Software Competence Center Hagenberg
Hagenberg, Austria

A Min Tjoa
Technical University of Vienna
Vienna, Austria

Bernhard Moser
Software Competence Center Hagenberg
Hagenberg, Austria

Atif Mashkoor
Johannes Kepler University of Linz
Linz, Austria

Jorge Martinez-Gil
Software Competence Center Hagenberg
Hagenberg, Austria

Lukas Fischer
Software Competence Center Hagenberg
Hagenberg, Austria

Maqbool Khan
Pak-Austria Fachhochschule - Institute of Applied Sciences and Technology (PAF-IAST)
Haripur, Pakistan

ISSN 1865-0929 ISSN 1865-0937 (electronic)
Communications in Computer and Information Science
ISBN 978-3-031-14342-7 ISBN 978-3-031-14343-4 (eBook)
https://doi.org/10.1007/978-3-031-14343-4

This Springer imprint is published by the registered company Springer Nature Switzerland AG
The registered company address is: Gewerbestrasse 11, 6330 Cham, Switzerland

Preface

After two years of virtual events, the 33rd International Conference on Database and Expert Systems Applications (DEXA 2022) and the related conferences and workshops were back in person. This year, the DEXA events were held at the Vienna University of Economics and Business (WU) during August 22–24, 2022.

DEXA workshops are a platform for the exchange of ideas, experiences, and opinions among scientists and practitioners – those who are defining the requirements for future systems in the areas of database and artificial intelligence technologies.

This year DEXA featured six international workshops:

- The 6th International Workshop on Cyber-Security and Functional Safety in Cyber-Physical Systems (IWCFS 2022)
- The 4th International Workshop on Machine Learning and Knowledge Graphs (MLKgraphs 2022)
- The 2nd International Workshop on Time Ordered Data (ProTime 2022)
- The 2nd International Workshop on AI System Engineering: Math, Modelling and Software (AISys 2022)
- The 1st International Workshop on Distributed Ledgers and Related Technologies (DLRT 2022)
- The 1st International Workshop on Applied Research, Technology Transfer and Knowledge Exchange in Software and Data Science (ARTE 2022)

This proceedings includes papers that focus mainly on very specialized topics linked to applications of database and expert systems technologies, which were selected for presentation at the DEXA 2022 workshops. There were 62 submissions sent for peer-review. Out of these, 40 full papers were accepted after a mixed review process, which included single-blind and double-blind reviews. Each paper underwent approximately 3 reviews.

We would like to thank all workshop chairs and Program Committee members for their excellent work, namely, Atif Mashkoor and Johannes Sametinger, the chairs of IWCFS 2022; Jorge Martinez-Gil, the chair of MLKgraphs 2022; Siegfried Hörmann, Hans Manner, and Florian Sobieczky, the chairs of ProTime 2022; Paolo Meloni, Maqbool Khan, Gerald Czech, Thomas Hoch, and Bernhard Moser, the chairs of AISys 2022; Alfred Taudes, Edgar Weippl, and Bernhard Haslhofer, the chairs of DLRT 2022; and Lukas Fischer and Rudolf Ramler, the chairs of ARTE 2022.

Last but not least, we would like to express our thanks to all institutions actively supporting this event:

- Johannes Kepler University Linz (JKU)
- Software Competence Center Hagenberg (SCCH)

- International Organization for Information Integration and Web-based Applications and Services (@WAS)
- Vienna University of Economics and Business (WU)

Finally, we hope that all the participants of the DEXA 2022 workshops enjoyed the program we put together.

August 2022

Gabriele Kotsis
A Min Tjoa
Ismail Khalil

Organization

Steering Committee

Gabriele Kotsis	Johannes Kepler University Linz, Austria
A Min Tjoa	Technical University of Vienna, Austria
Robert Wille	Software Competence Center Hagenberg, Austria
Bernhard Moser	Software Competence Center Hagenberg, Austria
Alfred Taudes	Vienna University of Economics and Business and Austrian, Blockchain Center, Austria
Ismail Khalil	Johannes Kepler University Linz, Austria

AISys 2022 Chairs

Paolo Meloni	University of Cagliari, Italy
Maqbool Khan	PAF IAST, Pakistan
Gerald Czech	Upper Austrian Fire Brigade Association, Austria
Thomas Hoch	Software Competence Center Hagenberg, Austria
Bernhard Moser	Software Competence Center Hagenberg, Austria

AISys 2022 Program Committee

Jan Bosch	Chalmers University of Technology, Sweden
Gabriele Gianini	University of Milan, Italy
Philipp Haindl	Software Competence Center Hagenberg, Austria
Mihhail Matskin	KTH Royal Institute of Technology, Sweden
Nazeer Muhammad	PAF IAST, Pakistan
Helena Holmström Olsson	Malmö University, Sweden
Pierre-Edouard Portier	INSA Lyon, France
Wajid Rafique	University of Montreal, Canada
Muhammad Habib ur Rehman	King's College London, UK
Dou Wanchun	Nanjing University, China
Xiaolong Xu	NUIST, China
Shui Yu	University of Technology Sydney, Australia

ARTE 2022 Chairs

Lukas Fischer	Software Competence Center Hagenberg, Austria
Rudolf Ramler	Software Competence Center Hagenberg, Austria

ARTE 2022 Program Committee

Markus Brillinger	Pro2Future GmbH, Austria
Katja Bühler	VRVis Zentrum für Virtual Reality und Visualisierung Forschungs-GmbH, Austria
Frank Elberzhager	Fraunhofer Institute for Experimental Software Engineering, Germany
Gabriel Gonzalez-Castañé	Insight SFI Research Centre for Data Analytics and University College of Cork, Ireland
Michael Granitzer	University of Passau, Germany
Eckehard Hermann	University of Applied Sciences Upper Austria, Austria
Andrea Janes	Free University of Bozen-Bolzano, Italy
Ossi Kotavaara	University of Oulu, Kerttu Saalasti Institute, Finland
Stefanie Kritzinger	RISC Software GmbH, Austria
Harald Lampesberger	University of Applied Sciences Upper Austria, Austria
Martin Leucker	UniTransferKlinik Lübeck GmbH and University of Lübeck, Germany
Nikos Makris	Core Innovation and Technology Center, Greece
Silverio Martínez-Fernández	Polytechnic University of Catalonia - BarcelonaTech, Spain
Ignacio Montero Castro	AIMEN Technology Centre, Spain
Matti Muhos	University of Oulu, Kerttu Saalasti Institute, Finland
Bernhard Nessler	Software Competence Center Hagenberg, Austria
Antonio Padovano	University of Calabria, Italy
Mario Pichler	Software Competence Center Hagenberg, Austria
Dietmar Pfahl	University of Tartu, Estonia
Christian Rankl	RECENDT Research Center for Non-Destructive Testing GmbH, Austria
Malin Rosqvist	RISE, Sweden
Stefan Sauer	Software Innovation Campus Paderborn, Paderborn University, Germany
Georg Weichhart	PROFACTOR GmbH, Austria
Edgar Weippl	SBA Research and University of Vienna, Austria
Dietmar Winkler	TU Wien, Austria

DLRT 2022 Chairs

Alfred Taudes	Vienna University of Economics and Business and Austrian Blockchain Center, Austria
Edgar Weippl	University of Vienna, Austria
Bernhard Haslhofer	AIT Austrian Institute of Technology, Austria

DLRT 2022 Program Committee

Susanne Kalss	WU Wien, Austria
Klaus Hirschler	WU Wien, Austria
Thomas Moser	St. Pölten University of Applied Science, Austria
Alexander Eisl	Austrian Blockchain Center, Austria
Stefan Craß	Austrian Blockchain Center, Austria
Fabian Schär	Basel University, Switzerland
Phillip Sander	Frankfurt School of Finance & Management, Germany
Soulla Louca	University of Nicosia, Cyprus
Rainer Böhme	Innsbruck University, Austria
Hitoshi Yamamoto	Rissho University, Japan
Isamu Okada	Soka University, Japan
Alexander Norta	Tallinn University of Technology, Estonia
William Knottenbelt	Imperial College London, UK
Lam Kwok Yan	Nanyang Technological University, Singapore
Tong Cao	University of Luxembourg, Luxembourg
Walter Blocher	Kassel University, Germany

IWCFS 2022 Chairs

Atif Mashkoor	Johannes Kepler University Linz, Austria
Johannes Sametinger	Johannes Kepler University Linz, Austria

IWCFS 2022 Program Committee

Yamine Ait Ameur	IRIT/INPT-ENSEEIHT, France
Paolo Arcaini	National Institute of Informatics, Japan
Richard Banach	University of Manchester, UK
Ladjel Bellatreche	ENSMA, France
Silvia Bonfanti	University of Bergamo, Italy
Jorge Cuellar	University of Passau, Germany
Angelo Gargantini	University of Bergamo, Italy
Irum Inayat	National University of Computer and Emerging Sciences, Pakistan

Jean-Pierre Jacquot — University of Lorraine, France
Muhammad Khan — University of Greenwich, UK
Christophe Ponsard — CETIC, Belgium
Rudolf Ramler — Software Competence Center Hagenberg, Austria
Neeraj Singh — INPT-ENSEEIHT/IRIT, France
Michael Vierhauser — Johannes Kepler University Linz, Austria
Edgar Weippl — University of Vienna, Austria

MLKgraphs 2022 Chair

Jorge Martinez-Gil — Software Competence Center Hagenberg, Austria

MLKgraphs 2022 Program Committee

Anastasia Dimou — Ghent University, Belgium
Lisa Ehrlinger — Johannes Kepler University Linz and Software Competence Center Hagenberg, Austria
Isaac Lera — University of the Balearic Islands, Spain
Femke Ongenae — Ghent University, Belgium
Mario Pichler — Software Competence Center Hagenberg, Austria
Artem Revenko — Semantic Web Company GmbH, Austria
Marta Sabou — Vienna University of Technology, Austria
Iztok Savnik — University of Primorska, Slovenia
Sanju Mishra Tiwari — Universidad Autonoma de Tamaulipas, Mexico
Marina Tropmann-Frick — Hamburg University of Applied Sciences, Germany

ProTime 2022 Chairs

Siegfried Hörmann — TU Graz, Austria
Hans Manner — University of Graz, Austria
Florian Sobieczky — Software Competence Center Hagenberg, Austria

ProTime 2022 Program Committee

David Gabauer — Software Competence Center Hagenberg, Austria
Sebastian Müller — Aix-Marseille University, France
Ivo Bukovsky — University of South Bohemia, Czech Republic
Anna-Christina Glock — Software Competence Center Hagenberg, Austria
Michal Lewandowsky — Software Competence Center Hagenberg, Austria

Organizers

scch { }
software competence
center hagenberg

JKU
Institute for
Telecooperation

Web Applications Society
@WAS
www.iiwas.org

Contents

AI System Engineering: Math, Modelling and Software

Unboundedness of Linear Regions of Deep ReLU Neural Networks

Anton Ponomarchuk[1(✉)], Christoph Koutschan[1], and Bernhard Moser[2]

[1] Johann Radon Institute for Computational and Applied Mathematics (RICAM), ÖAW, Linz, Austria
anton.ponomarchuk@ricam.oeaw.ac.at
[2] Software Competence Center Hagenberg (SCCH), Hagenberg, Austria

Abstract. Recent work concerning adversarial attacks on ReLU neural networks has shown that unbounded regions and regions with a sufficiently large volume can be prone to containing adversarial samples. Finding the representation of linear regions and identifying their properties are challenging tasks. In practice, one works with deep neural networks and high-dimensional input data that leads to polytopes represented by an extensive number of inequalities, and hence demanding high computational resources. The approach should be scalable, feasible and numerically stable. We discuss an algorithm that finds the H-representation of each region of a neural network and identifies if the region is bounded or not.

Keywords: Neural network · Unbounded polytope · Linear programming · ReLU activation function

1 Introduction

In recent years, neural networks have become the dominant approach to solving tasks in domains like speech recognition, object detection, image generation, and classification. Despite their high prediction performance, there still exist undesirable network properties that are not fully understood. We investigate phenomena related to unbounded regions in the network's input space: the network should not provide high confidence predictions for data far away from training data. Concrete examples show how unbounded regions can be used to produce fooling images or out-of-distribution images that lead to misclassification of the network [4,7,8,10,13]. Robustness against such adversarial attacks is of utmost importance in critical applications like autonomous driving or medical diagnosis. There are still open questions concerning the correct processing and the meaning of unbounded linear regions for neural network accuracy.

A neural network F can be viewed as a function that maps an input vector $\mathbf{x} \in \mathbb{R}^{n_0}$ to an output vector in $\mathbb{R}^{n_L}$, by propagating it through the L hidden layers of the network. Each layer performs an affine-linear transformation, followed by a nonlinear activation

The research reported in this paper has been partly funded by BMK, BMDW, and the Province of Upper Austria in the frame of the COMET Programme managed by FFG in the COMET Module S3AI.

G. Kotsis et al. (Eds.): DEXA 2022 Workshops, CCIS 1633, pp. 3–10, 2022.
https://doi.org/10.1007/978-3-031-14343-4_1

function. Here we use the ReLU-activation function $\sigma(x) := \max(0, x)$, and therefore $F\colon \mathbb{R}^{n_0} \to \mathbb{R}^{n_L}$ is a continuous piecewise linear function. It splits the input space $\mathbb{R}^{n_0}$ into a finite set of linear regions (polytopes), on each of which the function F is linear, i.e., it can be described as $\mathbf{x} \mapsto \mathbf{A} \cdot \mathbf{x} + \mathbf{b}$ for some $A \in \mathbb{R}^{n_L \times n_0}$, $\mathbf{b} \in \mathbb{R}^{n_L}$.

It has been shown [6] that neural networks with ReLU activation function and softmax as output (the network is then called a classifier) achieve overconfident predictions in all unbounded linear regions. Moreover, it is not clear what impact unbounded regions have on the neural network's calibration [12]. It is beneficial to know that the trained model provides accurate predictions and the ones that describe the confidence in the output class correctness. This leads us to the task of deciding which linear regions are bounded and which are unbounded. For the bounded regions the next question is if their volume is sufficiently large to contain adversarial examples. Determining the volume of a convex polytope is a polynomial problem. It can be done either by triangulation [1] or by volume approximation using Monte Carlo sampling or random walk methods [3, 11]. The recent approximation approaches achieve the complexity of $\mathcal{O}(Nn_0^4)$ steps, where n_0 is the dimension of the input space and N is the number of inequalities defining the given polytope. As a result, it is challenging to apply them in practice for high-dimensional input spaces and deep neural networks.

We discuss an algorithm for checking the boundedness of a polytope $\mathbf{H} \subset \mathbb{R}^{n_0}$. In order to do so, we first revise the algorithm from [16] that for any point $\mathbf{x} \in \mathbb{R}^{n_0}$ calculates a maximal linear region $\mathbf{H}$ such that $\mathbf{x} \in \mathbf{H}$. Then by using the H-representation of the polytope $\mathbf{H}$ we provide an algorithm to check whether $\mathbf{H}$ is bounded or not.

2 Preliminaries

A function $F\colon \mathbb{R}^{n_0} \to \mathbb{R}^{n_L}$ defined by a neural network propagates an input vector $\mathbf{x} \in \mathbb{R}^{n_0}$ through L hidden layers to some output vector in $\mathbb{R}^{n_L}$, where n_i denotes the number of neurons in the i-th layer. More precisely, the i-th hidden layer, $i \in \{1, \ldots, L\}$, performs the mapping $\mathbf{a}_i$, which is described by the following composition of functions:

$$\mathbf{a}_i(\mathbf{x}) := \sigma \circ f_i(\mathbf{x}),$$

where $f_i\colon \mathbb{R}^{n_{i-1}} \to \mathbb{R}^{n_i}$ is an affine mapping and $\sigma\colon \mathbb{R} \to \mathbb{R}$ is a non-linear function that acts componentwise on vectors. Each affine mapping f_i is represented by a linear function $f_i(\mathbf{x}) := \mathbf{A}_i\mathbf{x} + \mathbf{b}_i$, where $\mathbf{A}_i \in \mathbb{R}^{n_i \times n_{i-1}}$ and $\mathbf{b}_i \in \mathbb{R}^{n_i}$, and where the non-linear part is the ReLU activation function $\sigma(x) := \max(x, 0)$. Hence, a neural network $F\colon \mathbb{R}^{n_0} \to \mathbb{R}^{n_L}$ has the form

$$F(\mathbf{x}) := f_L \circ \sigma \circ f_{L-1} \circ \ldots \circ \sigma \circ f_1(\mathbf{x}).$$

Since $F(\mathbf{x})$ is a composition of affine and ReLU activation functions, it follows that it is piecewise linear. This property of $F(\mathbf{x})$ implies that the input space $\mathbb{R}^{n_0}$ is split into a finite set of linear regions $\{\mathbf{H}_i\}_{i=1}^{J}$, such that the function $F(\mathbf{x})$ is linear in each $\mathbf{H}_i$ and $\mathbb{R}^{n_0} = \bigcup_{i=1}^{J} \mathbf{H}_i$. It can been shown [16] that all the regions are polytopes and for each one its so-called H-representation can be determined: every polytope $\mathbf{H} \subseteq \mathbb{R}^{n_0}$

can be represented as a finite intersection of halfspaces, i.e., as the set of all points that satisfy a finite list of linear inequalities:

$$\mathbf{H} := \{\mathbf{x} \in \mathbb{R}^{n_0} \mid \mathbf{W}\mathbf{x} \leq \mathbf{v}\},$$

where $\mathbf{W} \in \mathbb{R}^{N \times n_0}$, $\mathbf{v} \in \mathbb{R}^N$, and $N = n_1 + \cdots + n_L$ denotes the total number of neurons of the network F. Moreover, an algorithm was proposed in [16] that computes for a given point $\mathbf{x} \in \mathbb{R}^{n_0}$ the corresponding polytope $\mathbf{H}(\mathbf{x}) \subseteq \mathbb{R}^{n_0}$ (see Appendix A).

3 Representation and Analysis via Code Space

With the preparations introduced above, we can now present an algorithm that checks whether the polytope $\mathbf{H}(\mathbf{x})$ is bounded or not, for a given point $\mathbf{x} \in \mathbb{R}^{n_0}$. Without loss of generality, we can assume that the H-representation of $\mathbf{H}(\mathbf{x})$ does not have any duplicated inequalities, i.e., the augmented matrix $(\mathbf{W}|\mathbf{v}) \in \mathbb{R}^{N \times (n_0+1)}$ does not have any repeated rows. Let us denote by $\ker(\mathbf{W})$ the null space of the matrix $\mathbf{W}$. The following lemma enables us to check the boundedness of $\mathbf{H}(\mathbf{x})$.

Lemma 1. *A non-empty polytope* $\mathbf{H} = \{\mathbf{x} \in \mathbb{R}^{n_0} \mid \mathbf{W}\mathbf{x} \leq \mathbf{b}\}$ *is bounded if and only if* $\ker(\mathbf{W}) = \{\mathbf{0}\}$ *and the following linear program admits a feasible solution:*

$$\min \|\mathbf{y}\|_1 \quad \textit{subject to} \quad \mathbf{W}^T\mathbf{y} = \mathbf{0} \textit{ and } \mathbf{y} \geq \mathbf{1}. \tag{1}$$

Proof. ("$\Rightarrow$") Assume, that $\mathbf{H}$ is bounded and non-empty. Then it follows that the null space of $\mathbf{W}$ is trivial, because otherwise, there would exist a non-zero vector $\mathbf{v} \in \ker(\mathbf{W})$ such that for all scalars $\lambda \in \mathbb{R}$ and any point $\mathbf{x} \in \mathbf{H}$ the following holds:

$$\mathbf{W}(\mathbf{x} + \lambda\mathbf{v}) \leq \mathbf{b} \Rightarrow \mathbf{x} + \lambda\mathbf{v} \in \mathbf{H}.$$

As a result, the polytope $\mathbf{H}$ is unbounded, contradicting our assumption.

Now, let us assume that the linear program (1) does not have a solution. It can only be the case when the set of feasible solutions that is described by the restrictions $\mathbf{W}^T\mathbf{y} = \mathbf{0}$ and $\mathbf{y} \geq \mathbf{1}$ is empty. By Stiemke's Lemma, see Appendix C, it follows that there exists a vector $\mathbf{v} \in \mathbb{R}^{n_0}$ such that $\mathbf{W}\mathbf{v} > \mathbf{0}$. Thereby, for all vectors $\mathbf{x} \in \mathbf{H}$ and for all non-positive scalars $\lambda \in \mathbb{R}$ we have:

$$\mathbf{W}(\mathbf{x} + \lambda\mathbf{v}) = \mathbf{W}\mathbf{x} + \lambda\mathbf{W}\mathbf{v} \leq \mathbf{b} \Rightarrow \mathbf{x} + \lambda\mathbf{v} \in \mathbf{H}.$$

As a result, the polytope $\mathbf{H}$ is unbounded, that is contradiction. Thus, if the given polytope $\mathbf{H}$ is bounded then the given linear program has a solution and the null space of the matrix $\mathbf{W}$ is trivial.

("$\Leftarrow$") Assume that the null space of the matrix $\mathbf{W}$ is trivial and the linear program (1) has a solution. If the given (non-empty) polytope $\mathbf{H}$ was unbounded, then for all vectors $\mathbf{x} \in \mathbf{H}$ there would exist a nonzero vector $\mathbf{v} \in \mathbb{R}^{n_0}$, such that for all non-negative $\lambda \in \mathbb{R}$ holds $\mathbf{x} + \lambda\mathbf{v} \in \mathbf{H}$. It follows that

$$\lambda\mathbf{W}\mathbf{v} \leq \mathbf{b} - \mathbf{W}\mathbf{x}, \tag{2}$$

where $\mathbf{b} - \mathbf{W}\mathbf{x} \geq \mathbf{0}$. If at least one of the entries of $\mathbf{W}\mathbf{v}$ is positive, then there exists $\lambda' > 0$ such that the inequality (2) breaks. Thus $\mathbf{W}\mathbf{v} \leq \mathbf{0}$, and Stiemke's Lemma, together with the trivial null space of $\mathbf{W}$, leads to a contradiction. $\square$

Concerning the complexity, note that in the worst case, it is necessary to compute the null space of the matrix $\mathbf{W}$ and to solve the linear program described in Lemma 1. The basic algorithm to solve the linear programming problem is an interior-point algorithm [15]. It has the worst-case complexity $\mathcal{O}(\sqrt{n_0}L)$, where L is the length of the binary encoding of the input data:

$$L := \sum_{i=0}^{N}\sum_{j=0}^{n_0}\lceil \log_2(|w_{ij}|+1)+1\rceil,$$

where $w_{i0} = b_i$ and $w_{0j} = 1$. The complexity of the solver depends on the input dimension n_0 and the number N of inequalities. The HiGHS [9] implementation of the interior-point method finds an optimal solution even faster in practice.

On the other hand, the complexity for computing the null space of the matrix $\mathbf{W} \in \mathbb{R}^{n_0 \times N}$ is at most $\mathcal{O}(n_0 N \min(n_0, N))$. Hence, the total-worst case time complexity for the given approach is at most $\mathcal{O}(n_0 N \min(n_0, N) + \sqrt{n_0}L)$. In practice, modern implementations interior-point algorithms, like HiGHS, use techniques for speeding up computation, for instance, parallelization.

We implemented Lemma 1 and the polytope computation algorithm, see Appendix A, in Python with libraries for scientific and deep learning computing: SciPy [18], NumPy [5], PyTorch [14]. For time execution measurements we used the basic Python library timeit. For creating the graphics we used the matplotlib library [2].

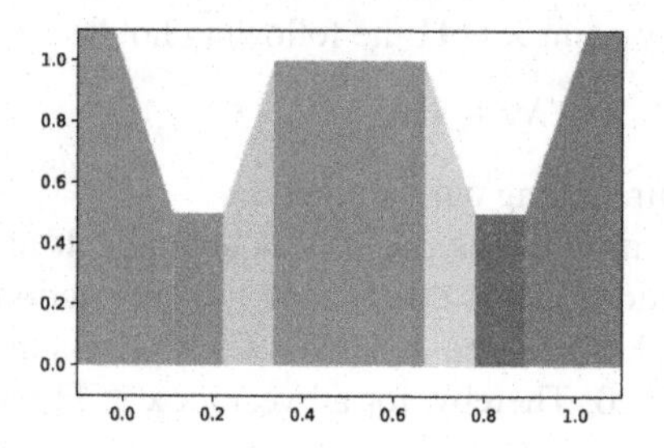

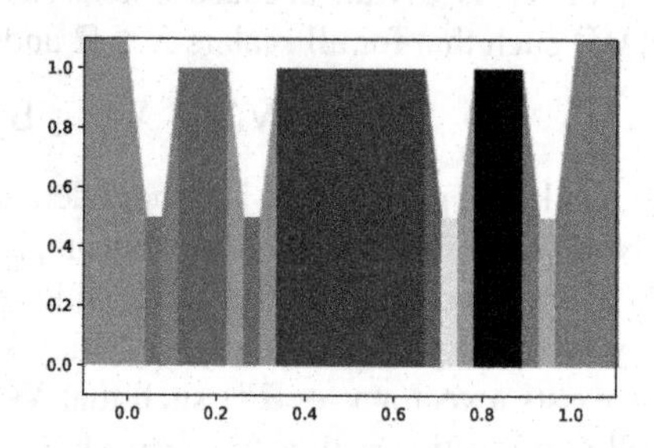

Fig. 1. Decision surfaces splitted into polytopes for depths $L = 2, 3$. In each polytope a neural network is linear. The leftmost and rightmost regions are unbounded.

4 Experiments

In this section we are going to discuss some preliminary experiments that we carried out with our implementation.

The first experiment is based on a neural network that is defined recursively. On the unit interval $[0, 1]$ we define the function $F(x) := \max\{-3x+1, 0, 3x-2\}$. Its nested

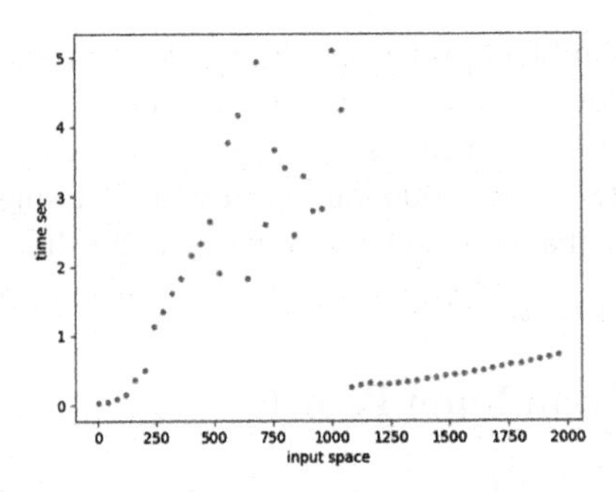

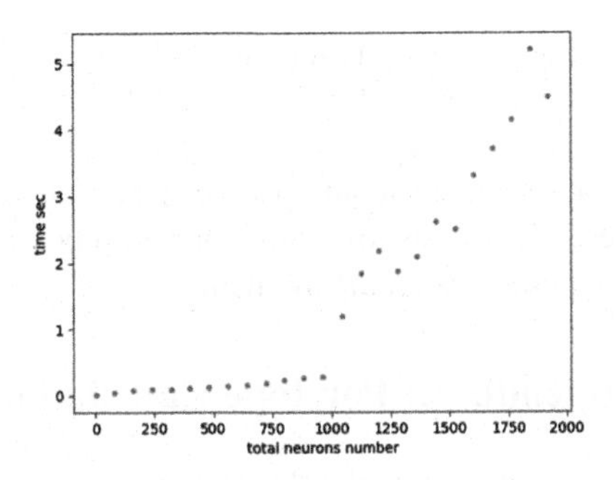

Fig. 2. Time needed for checking whether the calculated polytope is bounded. The left plot shows timings for different input spaces and a fixed neural network with $N = 1074$ neurons. In the right plot, the input space is fixed ($n_0 = 1024$) and the total number of neurons varies.

composition is the following: $F^{l+1}(x) := F\big(F^{l}(x)\big)$ and $F^1(x) := F(x)$, see Fig. 1. The decision boundary is defined as the upper border of the regions:

$$\mathbf{H}_L := \big\{(x, y) \in [0,1]^2 \mid y \leq \tfrac{1}{2}(F_L(x) + 1)\big\}.$$

The network $F^L(x)$ is used to test the algorithm. The reason is that this network could generate any number of explicit linear regions. The greater recursion L, the greater is the number of linear regions the network generates. While experimenting with this network, we found several problems that could occur during the execution of the algorithm. Firstly, after computing the H-representation of a polytope, all identical rows should be removed from the corresponding matrix $\mathbf{W}$. If the matrix contains identical rows, it can provide incorrect results. Secondly, not all linear problem solvers can handle problems with floating point numbers in practice. For example, the basic interior-point linear program solver in SciPy failed to solve such problems due to numerical problems. On the other hand, the HiGHS implementation succeeded.

The second experiment evaluates the time dependency on the dimension n_0 of the input space and the number N of neurons. For this experiment, fully-connected neural networks were generated with random weights. Suppose that the number of neurons is smaller than the dimension of the input space $N < n_0$. In this case, any region $\mathbf{H}$ is unbounded because the corresponding matrix $\mathbf{W}$ will have a non-trivial null space. As a result, there is no need for solving the linear program task in such a situation. Also, one can see the "drop" in the time measurements in the left plot of Fig. 2 because of it. Otherwise, if $n_0 \leq N$, then there are cases where one needs to compute the null space and the linear program for the given $\mathbf{W}$. As a result, in the right plot, the jump at $N = 1024$ is explained because not all null spaces are non-trivial beyond this point.

5 Outlook and Future Work

The algorithms presented in the paper are the first step in further understanding the properties of the linear regions that correspond to a neural network. There is an open question of detecting linear regions prone to containing adversarial examples. Is there a relation between the region's volume and this problem? What should one do to bypass the problem in such networks? Also, the same questions relate to unbounded regions.

Furthermore, there is an open question of how geometric knowledge about the tessellation of the input space can help us create better validation and test sets? Using these algorithms, one can work with polytopes that correspond to training and validation points and compare them.

Appendix A: Polytope Calculation for an Input Point x

Let us remind that the ReLU neural network $F(\mathbf{x}) = f_L \circ \sigma \circ f_{L-1} \circ \ldots \circ \sigma \circ f_1(\mathbf{x})$ is a composition of L affine functions $f_i(\mathbf{x}) = \mathbf{A}_i\mathbf{x} + \mathbf{b}_i$, where $\mathbf{A}_i \in \mathbb{R}^{n_i \times n_{i-1}}$ and $\mathbf{b}_i \in \mathbb{R}^{n_i}$ for all $i \in \{1, \ldots, L\}$, with a point-wise non-linear function $\sigma(x) = \max(x, 0)$. We denote the i-th hidden layer of the network $F(\mathbf{x})$ by $\mathbf{a}_i(\mathbf{x}) = \sigma \circ f_i(\mathbf{x})$.

A **binary activation state** for an input vector $\mathbf{x} \in \mathbb{R}^{n_0}$ is the function

$$\beta_k^{i_k}(\mathbf{x}) := \begin{cases} 1, & a_k^{i_k}(\mathbf{x}) > 0, \\ 0, & \text{otherwise}, \end{cases}$$

where $a_k^{i_k}(\mathbf{x})$ is the i_k-th output of the k-th hidden layer $\mathbf{a}_k$, for all $k \in \{1, \ldots, L\}$ and $i_k \in \{1, \ldots, n_k\}$.

A **polar activation state** for an input vector $\mathbf{x} \in \mathbb{R}^{n_0}$ is the function

$$\pi_k^{i_k}(\mathbf{x}) := 2\beta_k^{i_k}(\mathbf{x}) - 1,$$

for all $k \in \{1, \ldots, L\}$ and $i_k \in \{1, \ldots, n_k\}$. Note that we defined two binary functions which have the sets $\{0, 1\}$ and $\{-1, 1\}$ as codomains, respectively. By using $\beta_k^{i_k}(\mathbf{x})$ and $\pi_k^{i_k}(\mathbf{x})$, we now collect all states of a layer into a diagonal matrix form:

$$\mathbf{Q}_k^{\pi}(\mathbf{x}) := \operatorname{diag}\left(\pi_k^1(\mathbf{x}), \ldots, \pi_k^{n_k}(\mathbf{x})\right),$$
$$\mathbf{Q}_k^{\beta}(\mathbf{x}) := \operatorname{diag}\left(\beta_k^1(\mathbf{x}), \ldots, \beta_k^{n_k}(\mathbf{x})\right),$$

where $k \in \{1, \ldots, L\}$. We will use the matrix $\mathbf{Q}_k^{\beta}(\mathbf{x})$ to model the behavior of the activation function in the k-th layer. For each input vector $\mathbf{x} \in \mathbb{R}^{n_0}$, the matrices $\mathbf{Q}_k^{\pi}(\mathbf{x})$ and $\mathbf{Q}_k^{\beta}(\mathbf{x})$ allow us to derive an H-representation of the corresponding polytope $\mathbf{H}(\mathbf{x}) \in \{\mathbf{H}_i\}_{i=1}^{J}$ in explicit form. More precisely, the H-representation is given as a set of inequalities in the following way:

$$\mathbf{H}(\mathbf{x}) := \{\mathbf{x}' \in \mathbb{R}^{n_0} \mid \mathbf{W}_k(\mathbf{x}) \cdot \mathbf{x}' + \mathbf{v}_k(\mathbf{x}) \geq \mathbf{0},\ k \in \{1, \ldots, L\}\}, \tag{3}$$

where

$$\mathbf{W}_k(\mathbf{x}) := \mathbf{Q}_k^{\pi}(\mathbf{x})\mathbf{A}_k \prod_{j=1}^{k=1} \mathbf{Q}_{k-j}^{\beta}(\mathbf{x})\mathbf{A}_{k-j}, \tag{4}$$

$$\mathbf{v}_k(\mathbf{x}) := \mathbf{Q}_k^{\pi}(\mathbf{x}) \sum_{i=1}^{k} \left(\prod_{j=1}^{k-i} \mathbf{A}_{k-j+1}\mathbf{Q}_{k-j}^{\beta}(\mathbf{x}) \right) \mathbf{b}_i, \tag{5}$$

such that $\mathbf{W}_k(\mathbf{x}) \in \mathbb{R}^{n_k \times n_0}$ and $\mathbf{v}_k(\mathbf{x}) \in \mathbb{R}^{n_k}$. According to (3), the polytope $\mathbf{H}(\mathbf{x})$ is defined by exactly $N = n_1 + \ldots + n_L$ inequalities. However, in practice, the number of half-spaces whose intersection yields the polytope $\mathbf{H}(\mathbf{x})$ is typically smaller than N, so that the above representation is not minimal in general.

Appendix B: Unbounded Linear Region Problem

As mentioned in Appendix A, a ReLU neural network F splits the input space $\mathbb{R}^{n_0}$ into a set of linear regions $\mathbb{R}^{n_0} = \bigcup_{i=1}^{J} \mathbf{H}_i$. On each such linear region the network realizes some affine function

$$F_{\mathbf{H}_i}(\mathbf{x}) := \mathbf{A}_i \mathbf{x} + \mathbf{b}_i,$$

where $\mathbf{A}_i \in \mathbb{R}^{n_L \times n_0}$, $\mathbf{b} \in \mathbb{R}^{n_L}$, $\mathbf{x}_i \in \mathbf{H}_i$ and $i \in \{1, \ldots, J\}$. So the given network F is represented by a set of affine functions $F_{\mathbf{H}_i}$, each of which corresponds to some linear region $\mathbf{H}_i$ for all $i \in \{1, \ldots, J\}$.

Assume that $\mathbf{A}_i$ does not contain identical rows for all $i \in \{1, \ldots, J\}$, then for almost all $\mathbf{x} \in \mathbb{R}^{n_0}$ and $\varepsilon > 0$, there exists an $\alpha > 0$ and a class $k \in \{1, \ldots, n_L\}$ such that for $\mathbf{z} = \alpha \mathbf{x}$ the following holds:

$$\frac{\exp(F_k(\mathbf{z}))}{\sum_{j=1}^{n_L} \exp(F_j(\mathbf{z}))} \geq 1 - \varepsilon. \tag{6}$$

Inequality (6) shows that almost any point from the input space $\mathbb{R}^{n_0}$ can be scaled such that the transformed input value will get an overconfident output for some class $k \in \{1, \ldots, n_L\}$. See reference [6] for a proof of the above statement.

Appendix C: Stiemke's Lemma

Stiemke's Lemma states the following:

Lemma 2. *Let* $\mathbf{W} \in \mathbb{R}^{m \times n}$ *and* $\mathbf{W}\mathbf{x} = 0$ *be a homogeneous system of linear inequalities. Then only one of the following is true:*

1. *There exists a vector* $\mathbf{v} \in \mathbb{R}^m$ *such that the vector* $\mathbf{W}^T \mathbf{v} \geq 0$ *with at least one non-zero element.*
2. *There exists a vector* $\mathbf{x} \in \mathbb{R}^n$ *such that* $\mathbf{x} > 0$ *and* $\mathbf{W}\mathbf{x} = 0$.

This lemma has variants with different sign constraints that can be found in [17].

References

1. Büeler, B., Enge, A., Fukuda, K.: Exact volume computation for polytopes: a practical study. In: Polytopes-Combinatorics and Computation, pp. 131–154. Springer, Heidelberg (2000). https://doi.org/10.1007/978-3-0348-8438-9_6
2. Caswell, T.A., et al.: matplotlib/matplotlib: Rel: v3.5.1 (2021)

3. Emiris, I.Z., Fisikopoulos, V.: Efficient random-walk methods for approximating polytope volume. In: Proceedings of the Thirtieth Annual Symposium on Computational Geometry, pp. 318–327 (2014)
4. Guo, C., Pleiss, G., Sun, Y., Weinberger, K.Q.: On calibration of modern neural networks. In: International Conference on Machine Learning, pp. 1321–1330. PMLR (2017)
5. Harris, C.R., et al.: Array programming with NumPy. Nature **585**(7825), 357–362 (2020)
6. Hein, M., Andriushchenko, M., Bitterwolf, J.: Why ReLU networks yield high-confidence predictions far away from the training data and how to mitigate the problem. In: Proceedings of the IEEE/CVF Conference on Computer Vision and Pattern Recognition, pp. 41–50 (2019)
7. Hendrycks, D., Gimpel, K.: A baseline for detecting misclassified and out-of-distribution examples in neural networks. arXiv preprint arXiv:1610.02136 (2016)
8. Hsu, Y.C., Shen, Y., Jin, H., Kira, Z.: Generalized ODIN: detecting out-of-distribution image without learning from out-of-distribution data. In: Proceedings of the IEEE/CVF Conference on Computer Vision and Pattern Recognition (CVPR) (2020)
9. Huangfu, Q., Hall, J.A.J.: Parallelizing the dual revised simplex method. Math. Program. Comput. **10**(1), 119–142 (2017). https://doi.org/10.1007/s12532-017-0130-5
10. Leibig, C., Allken, V., Ayhan, M.S., Berens, P., Wahl, S.: Leveraging uncertainty information from deep neural networks for disease detection. Sci. Rep. **7**(1), 1–14 (2017)
11. Mangoubi, O., Vishnoi, N.K.: Faster polytope rounding, sampling, and volume computation via a sub-linear ball walk. In: 2019 IEEE 60th Annual Symposium on Foundations of Computer Science (FOCS), pp. 1338–1357. IEEE (2019)
12. Minderer, M., et al.: Revisiting the calibration of modern neural networks. Adv. Neural Inf. Process. Syst. **34**, 1–13 (2021)
13. Nguyen, A., Yosinski, J., Clune, J.: Deep neural networks are easily fooled: high confidence predictions for unrecognizable images. In: Proceedings of the IEEE Conference on Computer Vision and Pattern Recognition (CVPR) (2015)
14. Paszke, A., et al.: Pytorch: an imperative style, high-performance deep learning library. Adv. Neural Inf. Process. Syst. **32**, 8024–8035 (2019)
15. Potra, F.A., Wright, S.J.: Interior-point methods. J. Comput. Appl. Math. **124**(1–2), 281–302 (2000)
16. Shepeleva, N., Zellinger, W., Lewandowski, M., Moser, B.: ReLU code space: a basis for rating network quality besides accuracy. In: ICLR 2020 Workshop on Neural Architecture Search (NAS 2020) (2020)
17. Stiemke, E.: Über positive Lösungen homogener linearer Gleichungen. Mathematische Annalen **76**(2), 340–342 (1915)
18. Virtanen, P., et al.: Scipy 1.0: fundamental algorithms for scientific computing in python. Nature methods **17**(3), 261–272 (2020)

Applying Time-Inhomogeneous Markov Chains to Math Performance Rating

Eva-Maria Infanger[1,2](✉), Gerald Infanger[1], Zsolt Lavicza[2], and Florian Sobieczky[3]

[1] MatheArena GmbH, Engersdorf 30, 4921 Hohenzell, Austria
research@mathearena.com

[2] Johannes Kepler University, Altenberger Street 68, 4040 Linz, Austria
soe@jku.at

[3] Software Competence Center Hagenberg (SCCH), Hagenberg, Austria
florian.sobieczky@scch.at

https://www.mathearena.com, https://www.jku.at/linz-school-of-education

Abstract. In this paper, we present a case study in collaboration with mathematics education and probability theory, with one providing the use case and the other providing tests and data. The application deals with the reason and execution of automation of difficulty classification of mathematical tasks and their users' skills based on the Elo rating system. The basic method is to be extended to achieve numerically fast converging ranks as opposed to the usual weak convergence of Elo numbers. The advantage over comparable state-of-the-art ranking methods is demonstrated in this paper by rendering the system an inhomogeneous Markov Chain. The usual Elo ranking system, which for equal skills (Chess, Math, ...) defines an asymptotically stationary time-inhomogeneous Markov process with a weakly convergent probability law. Our main objective is to modify this process by using an optimally decreasing learning rate by experiment to achieve fast and reliable numerical convergence. The time scale on which these ranking numbers converge then may serve as the basis for enabling digital applicability of established theories of learning psychology such as spiral principal and Cognitive Load Theory. We argue that the so further developed and tested algorithm shall lay the foundation for easier and better digital assignment of tasks to the individual students and how it is to be researched and tested in more detail in future.

Keywords: Time-inhomogeneous Markov Chains · Elo rating system · Applications in mathematics education · Adaptive math training · Interactive learning environments

Supported by FFG Austria, project 41406789 called "Adaptive Lernhilfe".

G. Kotsis et al. (Eds.): DEXA 2022 Workshops, CCIS 1633, pp. 11–21, 2022.
https://doi.org/10.1007/978-3-031-14343-4_2

1 Introduction

Starting point of this whole project and research was the search for a math skill ranking method in the spirit of the Elo Chess ranking system [1] to develop our ideas, ways of teaching and explanations of mathematics in a way independent of time and place and with an automatic adaption to the individual needs of the learners [4,7]. In the following sections we will first shortly describe the origin of the use case and implementation to clarify the decisions made in the following chapters concerning our adaptive algorithm and data analysis. At the end, we will give an outlook on the refining steps, which shall improve the algorithm according to the underlying learning theories.

1.1 The Story Behind - Origins of the Use Case

Since explaining math to people in a tangible, motivating way is a well-known challenge for teachers and therefore a well-studied area (e.g.cognitive load [17], the influence of working memory components [10] or students and learner models [7]) we met with some success in fulfilling this task in an educational environment. In particular, due to the great demand, we elaborated our ideas into a math-student oriented smartphone application.

Focusing on the various needs of our target group - specifically the currently low achieving mathematics students - we are aware of their being digital natives and thus linking easier to smartphones than books. We observe that because of their many other duties during school time, they need an individual amount of learning time and support to fully grasp the corresponding math concepts. These are mostly unattached to the learner's mathematical school and study time, so that our team's aim is enhancing deep study for our students, as well as engaging them in training of mathematical concepts as much time as possible [4].

Hence, in order to optimize our own tool, we compared several mathematics learning platforms, but most of them lacked in one way or another our medial scenario (e.g. Moodle was not quite accessible via smartphone or tablet, learning card systems had difficulties with displaying math symbols, learning management systems have the drawback of being teacher focused and thus may enhance math anxiety, a circumstance that should be avoided at all costs [20]. Also, none of the tools had an adaptive feature). We created our own mobile adaptive math training application, working for a piece of technology

> "that supports some of the most effective learning principles identified during the last hundred years"

[18], involving the ideas of J. Bruner's "spiral principal" [2] and Cognitive Load Theory (CLT, [16]). Establishing the missing connection between much learning material and often overwhelmed and confused students by self-organizing their own learning is the main idea [4].

The Cognitive Load Theory is based on the assumption that the working memory necessary for learning is limited in its capacity. In order to be able to categorize and organize knowledge, it is necessary to challenge the learner's working memory and at the same time not to overload it. Ideally, this is achieved by optimizing the learning environment and by formulating the learning content clearly and without irrelevant information. In addition, the level of difficulty should be chosen in such a way that the learners are neither under nor overwhelmed. Hence, this learning process is always dependent on the previous history and current state of a learner, requires good prognosis about possible development and can therefore be represented particularly well by a Markov process. If this combination is successful, particularly effective learning is possible and the complexity of the selected tasks can be increased bit by bit - in addition to avoidance of frustration and gain in motivation [3].

In the following sections the results of these considerations (a work in process), the current algorithmic developments and their background theories are presented. Afterwards we will explain the program, test runs and the process of data collection, before we finish with analysis, conclusions and outlook.

1.2 An Adaptive Mobile App for Math Training

To counter the above mentioned issues in math education within the settings available to us, we developed short training of 10 mini-tasks, each divided in twenty different math topics. We implemented them in a mobile application, available in various online stores. In order to manage the cognitive load [3], specially developed tasks were created and classified in distinct difficulty levels to put the spiral principle [2] into practice. Also we get the possibility to assign them to the learners automatically [21], adapting to their individual skill level. Until now, this is achieved by expert assessment and a static algorithm designed by the team. As this is not practical, error prone and subjective ([7]), the next step - argued in this paper - will be the development of an automatically adapting algorithm, that simultaneously classifies users' abilities as well as task difficulties numerically.

2 Ranking Based on the Elo System

In the following two sections, the standard Elo ranking system is explained, and our extension producing numerical convergence is defined.

2.1 Automated Estimation of Skill and Difficulty as a Problem-Solving Match Between Learner and Exercise

Like other theories useful for estimation of skills (such as Item Response Theory in the form of so called Rasch Models [14]), the Elo rating system has the advantage of flexibility in modeling skills of individuals. With sufficient accuracy in identifying the parameters of the model, the model let's ranking scores evolve both according to a random and a systematic part. The systematic part expresses the convergence towards, while the random part expresses fluctuations which come from the step by step updating of the system according to match-scores (in Chess) - which will be replaced by solving attempts of math exercises by the training tool *MatheArena* ([15], Chap. 5).

Berg [1] has introduced an approach for adapting the Elo Chess Rating System, customized to the replacement of the Chess-match with the trial of strength between student and learning-item. A student meets an item and has either the chance to win or lose, depending on whether the solution/answer is right or wrong. Thus, the students' ability as well as the task difficulty gets a rating and therefore an automated classification, unattached to (possibly) biased expert assessment [7].

The Standard Elo Ranking System. Following Pelanek's suggestion, we not only assign Elo numbers to learners, but to exercises, as well [15]. The usual Elo rating implies the following definitions:

$$f_{x\beta} = \frac{1}{1 + 10^{-\beta \cdot x}} \tag{1}$$

The probability to solve a problem with Elo rank E_B by a person with Elo rank E_A is approximated by

$$E_A = F(v_i - v_j, \beta), \tag{2}$$

while the probability to fail in solving the problem is

$$E_B = F(v_j - v_i, \beta) = 1 - E_A. \tag{3}$$

After one round of problem-solving, the ranks of person and exercise are updated with these probabilities together with the information about the true current outcomes ($S_A, S_B \in \{0, 1\}$) by

$$R_A + k \cdot (S_A - E_A) \text{ and } R_B + k \cdot (S_B - E_B), \tag{4}$$

respectively [3].

Note that already the classical Elo system only allows rank-adaption by solving easy tasks to be low (due to a small difference of $S_A - E_A = 1 - E_A$ in (4)), while a high such adaption undergoes an unlikely win/solving of a task (due to a large corresponding difference between indicator S_A and winning probability). Also, in the classical Elo rating system, the learning rate k is fixed and the system with arbitrary initial ranks describes an asymptotically stationary Markov chain. This is equivalent to the convergence of the random variables R_A, R_B being'in distribution' [5]. It is the subject of the present case-study to determine an optimal **time-dependent learning rate**, to render the system an inhomogeneous Markov Chain with the ranks R_A, R_B converging to real numbers, almost surely ([19], Chap. 5.2).

The test program - implemented in the programming language R - implements a simulation of the discrete time Markov Chain defined by the Elo System, with a learner ('Emmy' - in memory of Emmy Noether) and several exercises to be solved by her during each time step. The code is designed to carry out experiments (see Sect. 3, below) with the aim of measuring the degree of convergence of a suitable mean under different learning-rate adaption strategies. For more detailed investigations, the basic program is available at any time under the following link: https://github.com/MatheArena/time-inhomogeneous-Markov-chains.

2.2 Time-inhomogeneous Markov Chains

In order to carry out the R program, we let the ranks and the learning rate depend on a discrete time variable $t \in \mathbb{N}$, expressed by $R_A(t), R_B(t)$, and $k(t)$, respectively. To make the random sequences of ranks $R_A(t), R_B(t)$ converge to real numbers, almost surely, we tested several methods to let the learning rate $k(t)$ converge to zero, making the rank-changes in (4) vanish, asymptotically. The certainty about numerical convergence of the considered Elo ranks is established by the following theorem:

Theorem By (4) every learning rate function $k : \mathbb{N} \to \mathbb{R}_+$ converging to zero for $t \to \infty$ induces a sequence of converging Elo Ranks.

Proof: The Elo ranks (here: R_A, same applies to ranks of Item levels R_B,) defined by (4) are bounded from below by zero, and from above by the sum of initial ranks (no additional players are entering during the game). By Bolzano-Weierstrass' Theorem, there is a converging sub-sequence. Any two converging subsequences will have the property, that their difference is bounded by $C \cdot k(t)$ with some $C > 0$. In particular, there are the sequences of the upper and lower accumulation points. Considering their difference,

$$\limsup_t R_A(t) - \liminf_t R_A(t) = \lim_{t\to\infty} (\sup_{r>t} R_A(r) - \inf_{s>t} R_A(s)) \leq C \cdot \lim_{t\to\infty} k(t) = 0,$$

shows their equality and therefore the existence of the limit $\lim_{t\to\infty} R_A(t)$. □

Remark: The convergence is not only'almost sure' (up to a set of measure zero), but true for every sequence of Elo ranks (with $k(t) \to 0$).

The variables are using Markov chains in order to estimate the applicability and problem areas of the selected algorithms. Therefore, we try various variants of given starting values and probabilities for solving tasks. To examine the possibilities on the given theoretical basis, we vary the following parameters:

The number m gives information about the number of different item families (further explanations about this term can be found in [9]) in the theoretically given item pool. We choose $m = 4$, one difficult (*ItemLevel_1*, 30% solving rate) and three low-level exercises (*ItemLevel_2-4*, 70% solving rate). The initial value of the rank is chosen equally for the person (to be 7), as well as the difficulty level of the tasks to show the correct convergence despite randomly chosen starting

Table 1. Varied parameters in the Markov process.

Parameters	Definition	Variants
R_A	Elo rank person	> 0
R_B	(Vector of) Elo rank problem	> 0
t	Discrete running time variable	$t \in \{1, \ldots, n\}$,
m	Number of item families	4
n	Number of solved tasks	80, 100, 1000
β	Steepness of sigmoid parameter	0.1
k	Time-dep. learning rate convergence	4 variants (given below)
l	range of running mean (RM)	0–25 of last times steps

ranks. The number of time steps n also represent the number of answers 'Emmy' is giving to each of the tasks, always randomly chosen out of the item pool above (In particular, no exercise is given, twice!)

We post-pone the task of determining the influence of the initial condition to another study. For $k(n)$ to converge we study the four different learning rates of Table 2, as characteristic representatives of different speeds of convergence:

Table 2. Four different types of time-dependent learning adaption rate $k(n)$:

Type:	Formula for $k(n)$	Parameters found through experimentation:
Constant	c	c
Arithmetic	$\frac{a}{1+b \cdot n}$	$a = 4, b = 1$ [15]
Exponential	$d \cdot \exp(-\lambda \cdot n)$	$d = 1, \lambda = 0.06$
Logarithmic	$\frac{e}{\log(1+f \cdot n)}$	$e = 2, f = 1$

While the arithmetic type is taken from [15], the logarithmic type is inspired by the theory of simulated annealing [13]. There, a so called 'cooling schedule' for Markov Chain Monte Carlo simulations [12] was chosen to be of the logarithmic type to ensure almost sure convergence towards of a time inhomogeneous Markov chain in an unbiased way (to its expected 'ranks') [8].

Finally, we also apply a running (arithmetic) mean in time with the range of the last l rank-values. The objective of this study is to determine by experiment an optimal pair of learning rate and running mean range to achieve fast convergence. The core statements of the learning psychology theories described at the beginning are used as a basis for decision-making.

3 Data and Experiment

So, if we now combine the learning psychology theories and the Markov processes, we get the following practical use case: We assume that a (simulated) user answers different (simulated) tasks of a certain (level) type with approximately the same difficulty n times (*time steps*). According to [11] (Sect. 2.2), this is a valid approach if the exercises difficulty level is evaluated with a parametric model. For example, in algebraic tasks it should be guaranteed that fractures do not coincide, zero division is excluded and the solving steps remain comparable. Note that we do not yet know the Elo ranks of these item levels.

We examine the 4 different learning rate types in our theoretical test environment to find the best possible approximation in few time steps for the assumption of the actually happening learning process [2,3].

With the aim of clarity and comprehensibility we limit the number of item families to 4 and choose only one learner ('Emmy' - black curve) for the test, since the results do not significantly differ with larger numbers. To check the validity of the ranking, we observe the convergence of the evaluations over time, and whether they eventually group in distinct characteristic clusters. Experimentation (see next section) yielded the optimal length of the running mean to be at a value of 10 backward time steps, independently of the speed of convergence of the learning rate $k(n)$. Figure 2 shows the onset of numerical convergence of the Elo ranks in the case of the logarithmic type learning rate as compared to the standard Elo ranking (weak convergence) shown in Fig. 1.

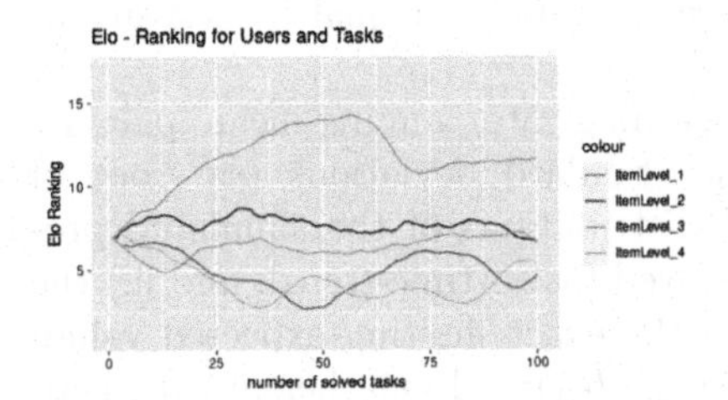

Fig. 1. Constant learning rate $k(t)$

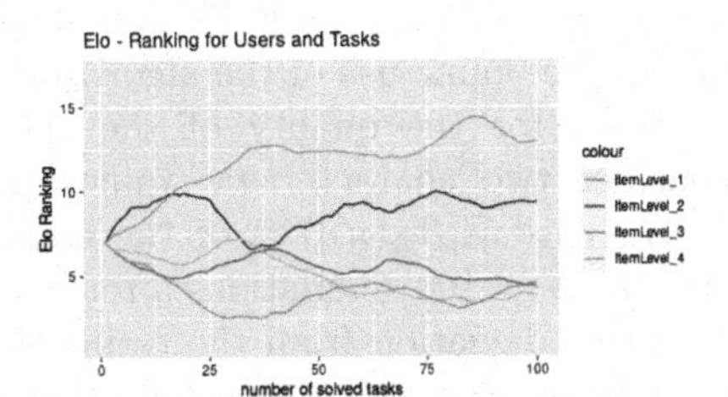

Fig. 2. Logarithmic convergence of $k(t)$.

4 Analysis and Results

After testing all relevant variations of convergence factors and the associated parameters, we are able to exclude the variants with constant and exponential factor before detailed analysis because there can too slow convergence be observed for the former (see Fig. 1) and too wide a range for the latter. Hence, we limit ourselves to compare the suggested convergence factors - arithmetic from [15] and logarithmic motivated by [1]. In the evaluation of the statistics after 100 runs of the mentioned random experiments it can be observed that the basic Elo rating system applied for education (Fig. 3) has a greater deviation and variance than the logarithmic convergence factor (Fig. 4). The same goes for the span. Hence, the second one is more precise and distinct and therefore a more appropriate basis for our presented application.

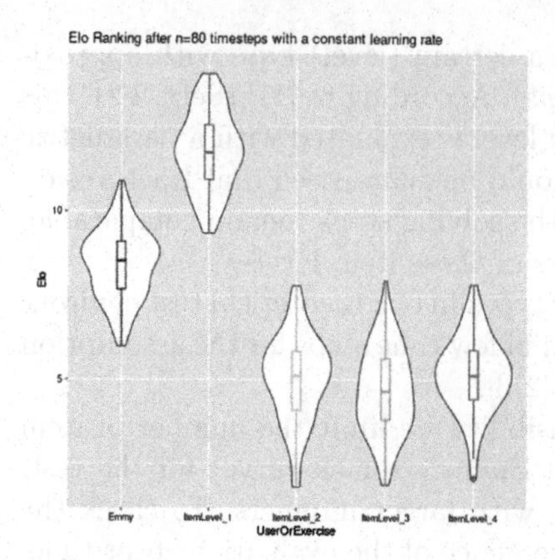

Fig. 3. Basic Elo rating: some overlap between rank-clusters.

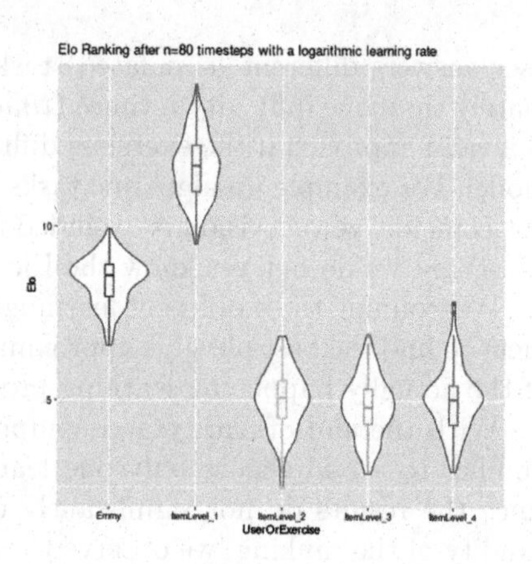

Fig. 4. Logarithmic Conv. factor: less overlap, better distinction between ranks.

As the ground truth of the limits of the ranks, we consider the expected values of the asymptotic distributions of each of the observed Elo ranks. We estimate these by taking large time values ($n = 1000$) and large running mean ranges ($l = 100$). (Note: These numbers are only used for quality assessment, not in the online use of the algorithm.)

Now, for the quality of the chosen learning-rate/running-mean pair we consider two characteristic values: i.) the standard deviations (sq. root of $\mathbb{E}[(R_A(n) - \mathbb{E}[R_A(n)])^2]$) of the empirical distributions of the simulated process for $N = 100$ simulations after $n = 80$ solved tasks (time-steps), and ii.) the standard deviation from the estimations of the exact limiting expected values (the square root of the expected value of $R_A(n) - \overline{R}_A)^2]$). In mathematical terms:

$$SD_1 := \sqrt{\frac{1}{n}\sum_{i=1}^{n}[(R_A(i) - \mathbb{E}[R_A(n)])^2]}, \text{ and } SD_2 := \sqrt{\frac{1}{n}\sum_{i=1}^{n}[(R_A(i) - \overline{R}_A)^2]}. \tag{5}$$

We note that SD_1 expresses the convergence of the rank R_A, as a sequence, while SD_2 expresses convergence of the rank *to the specific limit* $\overline{R}_A$. Both qualities are desirable for the use in the application of (person-) skill and (exercise-) level estimation. The Parameters of the formulas for $k(t)$ in Table 3 are found by optimizing (minimizing) SD_1. The results demonstrate that the pay-off of slower convergence of the running mean to the true asymptotic values (showing in larger values of SD_1) coming with the decreasing learning rate $k(t)$ is tolerable under lesser overlap of the measured distribution. This shows in particular for the item family levels, for which the common initial value (of 7) differed strongly from the asymptotic mean. The other decay types of $k(t)$ reduce their local variance SD_1 faster, but convergence to the true (expected) values (horizontal lines) does not show (SD_2 remains large). Therefore, the best approach for our use case (smartphone apps with frequently visiting and identifiable learners) emerges via

Table 3. Types of time-dependent learning rates $k(t)$ (see Table 2 for definition) and estimates of the standard deviations (SD_1 and SD_2) of Elo ranks after $n = 80$ steps: (see the violine plots in Fig. 3 and 4.). It is seen, that the logarithmic $k(n)$ succeeds over the constant $k(n)$ for both SD_1 and SD_2 for Emmy and all ItemLevels. It also succeeds over all other $k(n)$ in the case of the ItemLevels (Exercises).

Type	SD_1 (Emmy)	SD_2 (Emmy)	SD_1 (1)	SD_2 (2)	SD_1 (3)	SD_2 (3)
Const.	0.81	0.95	1.30	1.18	1.13	1.18
Arith.	0.42	0.56	0.55	1.34	0.61	1.34
Exp.	0.47	0.62	0.63	1.31	0.66	1.3
Log.	**0.67**	0.80	0.92	**1.00**	0.86	**1.00**

the logarithmic learning rate and differs from the arithmetic learning rate used in [15] in the sense of evenly balanced convergence of the running mean and the standard deviation (see Sect. 3 to the true (asymptotic) value).

5 Conclusions - Practical Use of the Ranking System

The implication for our application is insight into the number of exercises necessary to obtain reliable ratings. We are able to tolerate the reduction in the speed of the running mean's convergence in favor of observed variances of the ranks that allow for the classification of the different skill levels by using a suitable discrete scale for the available exercise groups. Again, we note the coincidence of this optimal convergence rate with the logarithmic type cooling schedule for inhomogeneous Markov chains applied to MCMC algorithms [13].

As an outlook, the knowledge about the number of exercises students have to solve to obtain a reliable ranking skill estimate lets us further develop the tools to actually improve their skills. We realize that after mapping the learner's and the exercise group's rating as good as possible onto the Elo scale, we still need to represent the gradual improvement of the problem-solving person from its educational (learning) perspective. In view of the claims made by the CLT [3,16], we are lead to ask:

Which combinations of learning rate functions $k(t)$ and averaging best suits our application of the Elo rating system under the simultaneous dynamics of changing (increasing) learning skills?

For the formidable task to also detect non-trivial learning *trends*, we aim at using assistance by machine learning methods such as deep-learning assisted non-linear trend-detection [4].

Acknowledgments. We chose the name of our fictitious learner to be 'Emmy' in honor of the great Emmy Noether who received the first habilitation in Mathematics at a German University in 1919. Emmy Noether also received private math tutoring [6].

Furthermore, we would like to add our thanks to FFG - the Forschungsförderungsgesellschaft from Austria for supporting our research financially.

References

1. Berg, A.: Statistical analysis of the Elo rating system in chess. CHANCE **33**(3), 31–38 (2020)
2. Bruner, J.S.: The Process of Education. Harvard University Press, Cambridge (2009)
3. Chandler, P., Sweller, J.: Cognitive load theory and the format of instruction. Cogn. Instr. **8**(4), 293–332 (1991)
4. Chaudhry, M.A., Kazim, E.: Artificial Intelligence in Education (AIEd): a high-level academic and industry note 2021. AI Ethics **2**, 157–165 (2021). https://doi.org/10.1007/s43681-021-00074-z
5. Chung, K.L.: Markov chains with stationary transition probabilities. Die Grundlehren der Mathematischen Wissenschaften in Einzeldarstellungen Band 104, 1st (edn.). Springer, Berlin, Heidelberg (1967). https://doi.org/10.1007/978-3-642-49686-8
6. Rowe, D.E., Koreuber, M.: Proving It Her Way: Emmy Noether, a Life in Mathematics, 1st (edn.). Springer, Cham (2020). https://doi.org/10.1007/978-3-030-62811-6
7. Desmarais, M.C., Baker, R.S.J.D.: A review of recent advances in learner and skill modeling in intelligent learning environments. User Model. User-Adap. Inter. **22**(1), 9–38 (2012). https://doi.org/10.1007/s11257-011-9106-8
8. Dobrushin, R.L.: Central limit theorem for nonstationary Markov chains I. Theor. Probab. Appl. **1**(1), 65–80 (1956)
9. Embretson, S.E., Daniel, R.C.: Understanding and quantifying cognitive complexity level in mathematical problem solving items. Psychol. Sci. **50**(3), 328 (2008)
10. Friso-van den Bos, I., van der Ven, S.H.G., Kroesbergen, E.H., van Luit, J.E.H.: Working memory and mathematics in primary school children: a meta-analysis. Educ. Res. Rev. **10**, 29–44 (2013)
11. Geerlings, H., Glas, C.A.W., Van Der Linden, W.J.: Modeling rule-based item generation. Psychometrika **76**(2), 337–359 (2011)
12. Geman, S., Geman, D.: Stochastic relaxation, Gibbs distributions, and the Bayesian restoration of images. IEEE Trans. Pattern Anal. Mach. Intell. **6**, 721–741 (1984)
13. Hajek, B.: Cooling schedules for optimal annealing. Math. Oper. Res. **13**(2), 311–329 (1988)
14. Fischer, G.H., Molenaar, I.W. (eds.) Rasch Models: Foundations, Recent Developments, and Applications, (1 edn.). Springer, New York (1995). https://doi.org/10.1007/978-1-4612-4230-7
15. Pelánek, R.: Applications of the Elo rating system in adaptive educational systems. Comput. Educ. **98**, 169–179 (2016)
16. Plass, J.L., Moreno, R., Brünken, R.: Cognitive Load Theory (2010)
17. Rittle-Johnson, B.: Developing mathematics knowledge. Child Dev. Perspect. **11**(3), 184–190 (2017)
18. Van Eck, R.: Digital game-based learning: it's not just the digital natives who are restless. EDUCAUSE Rev. **41**(2), 16 (2006)
19. Wasserman, L.: All of Statistics. STS, Springer, New York (2004). https://doi.org/10.1007/978-0-387-21736-9

20. Whyte, J., Anthony, G.: Maths anxiety: the fear factor in the mathematics classroom. New Zealand J. Teachers' Work **9**(1), 6–15 (2012)
21. Xiaolong, X., Zhang, X., Khan, M., Dou, W., Xue, S., Shui, Yu.: A balanced virtual machine scheduling method for energy-performance trade-offs in cyber-physical cloud systems. Fut. Gen. Comp. Sys. **105**, 105 (2020)

A Comparative Analysis of Anomaly Detection Methods for Predictive Maintenance in SME

Muhammad Qasim[1(✉)], Maqbool Khan[1,2(✉)], Waqar Mehmood[1,3], Florian Sobieczky[2], Mario Pichler[2], and Bernhard Moser[2]

[1] Sino-Pak Center for Artificial Intelligence, PAF-IAST, Haripur, Pakistan
{M21F0077AI002,maqbool.khan}@fecid.paf-iast.edu.pk

[2] Software Competence Center Hagenberg (SCCH), Hagenberg, Austria
{florian.sobieczky,mario.pichler,bernhard.moser}@scch.at

[3] Johannes Kepler University (JKU), Linz, Austria
waqar.mehmood@jku.at

Abstract. Predictive maintenance is a crucial strategy in smart industries and plays an important role in small and medium-sized enterprises (SMEs) to reduce the unexpected breakdown. Machine failures are due to unexpected events or anomalies in the system. Different anomaly detection methods are available in the literature for the shop floor. However, the current research lacks SME-specific results with respect to comparison between and investment in different available predictive maintenance (PdM) techniques. This applies specifically to the task of anomaly detection, which is the crucial first step in the PdM workflow. In this paper, we compared and analyzed multiple anomaly detection methods for predictive maintenance in the SME domain. The main focus of the current study is to provide an overview of different unsupervised anomaly detection algorithms which will enable researchers and developers to select appropriate algorithms for SME solutions. Different Anomaly detection algorithms are applied to a data set to compare the performance of each algorithm. Currently, the study is limited to unsupervised algorithms due to limited resources and data availability. Multiple metrics are applied to evaluate these algorithms. The experimental results show that Local Outlier Factor and One-Class SVM performed better than the rest of the algorithms.

Keywords: Predictive maintenance (PdM) · Small and medium-sized enterprises (SME) · Anomaly detection · Unsupervised algorithms · Comparative analysis · Condition monitoring · Remaining Useful Life prediction (RUL)

1 Introduction

Anomaly detection is a process of finding different data points which are away from the normal behavior of the whole data set. An outlier observation is one

Supported by SCCH , Austria.

G. Kotsis et al. (Eds.): DEXA 2022 Workshops, CCIS 1633, pp. 22–31, 2022.
https://doi.org/10.1007/978-3-031-14343-4_3

that significantly deviates from the rest of the observations. So Anomalies or outliers are those data instances that differ from other instances and are rare as compared to other instances. In the early stage, anomaly detection was used only for data cleansing purposes but with the passage of time, it got importance in many other fields like disease detection, fault detection, network intrusion detection, fraud detection, banking data, etc. The main motivation for this study is to find the Remaining Useful Life of machinery in any Manufacturing industry. This anomaly detection can be the basis for many other areas of research like network intrusion detection or fraud detection in banking data.

Predictive Maintenance in SME (Small and Medium-sized Enterprise). As we know, we are entering the era of the fourth industrial revolution with the adaptation of computers with smart and autonomous systems using data and machine learning. SMEs would get more benefits because technology is affecting their industrial goal, architecture, and Methodology. SMEs are enterprises that involve 250 workers and turnover less than 40 million. SME Maintenance plays a vital role in its financial and temporal matters. Plant downtime due to maintenance costs manufacturers therefore preventive measures, where equipment is maintained before it fails, is better than reactive maintenance in which equipment failure is fixed after a breakdown. Many SMEs now use predictive maintenance for the accurate prediction of failures before they occur. This reduces the risk of downtime and costs by replacing inefficient parts in time. It also increases the equipment lifespan through maintenance or replacement of wearing parts well in time. Anomaly Detection can play a pivotal role in fault detection and predictive maintenance in SMEs.

Anomaly Detection is gaining more importance with the passage of time. Anomaly Detection have many usage in different area of research such as in healthcare to find disease [1], in cyber-security for network intrusion detection [2], in Banking and financial sector for fraud detection [3], in manufacturing industry to find faults [4] and decay in machinery and lot more. One major use of Anomaly Detection is to find the Remaining Useful Life (RUL) [5] of machinery or assets in any industry which could be helpful in Predictive Maintenance [6]. The use of Anomaly detection in finding the Remaining Useful life of machinery helps to adopt a better approach to Predictive Maintenance which surprisingly enhances the productivity of small and medium-sized enterprises (SMEs). NASA's prognostic Centre published a data set named Commercial modular Aero Propulsion System Simulation (C-MAPSS) [7] which contains turbo fan Engine recorded data and it provides researchers enough data for every stage in machinery life with different operations and fault conditions. In this study, we have used nine different Anomaly detection algorithms on CMAPSS data set to compare the performance of each algorithm. The main motivation for detecting anomalies with different algorithms in this study is to find better algorithms for fault detection or to find Remaining Useful Life (RUL) for predictive maintenance (PdM) of machinery or asset in small and medium-sized enterprises (SMEs).

1.1 Anomaly Detection Techniques:

There are basically three main types of Anomaly Detection Techniques through machine learning [8]: Those based on i. supervised, ii. semi-supervised, iii. unsupvervised, and iv. reinforcment learning. An important aspect is the potential capability of the method to be used in online (versus offline) operational mode.

Supervised Anomaly Detection: In Supervised Anomaly Detection labeled training and test set is provided to the model. The major drawback of this approach is that it is difficult to obtain more labeled data because it is not cost-efficient and in some cases, it is not possible to have labeled data at all (see [25], Sect. 3.1).

Semi-Supervised Anomaly Detection: In Semi-Supervised Anomaly Detection training set with normal data (this means without any labeled features which distinguish anomalies) and a labeled test set are provided to the model. Again scarcity of labeled anomaly data is a major problem in this approach. [27]

Unsupervised Anomaly Detection: One data set without any labels is provided. Mostly real-world data does not provide labeled data so this approach is more applicable. In this approach, anomalies are detected on the bases of the inherent properties of the data set. Usually, distances or deviations are measured in order to distinguish what is normal and what is an anomaly (see [25], Sect. 3.2). In this study, we focus on modern unsupervised anomaly detection techniques, with relatively little prior research activity.

Probabilistic Methods: A specific sub-class of unsupervised anomaly detection methods are the methods based on the description of the signal in terms of a probabilistic model. All the so called change point detection methods from statistics (originating in sequential analysis [22,29]) are of this type. Their specific advantage is their principle capability of being applicable in an *online* setting, i.e. as opposed to processing input time series data in batches (see [25], Sect. 3.2.3). Models based on statistical comparison tests between sub-intervals of data [30], and comparison between predictions of learning models and data [23,24] have found recent interest. Bayesian models also have found online use [28], as well as models focusing on non-trivial trend detection [31], or information-theoretic approaches using (changes in) entropy for the early detection of anomalies [20], or the emergence of concept drift [21,32].

Anomaly Detection Based on Reinforcement Learning: A very recent approach to anomaly detection comes from the agent-based learning technique called reinforcement learning. In [9] a deep reinforcement learning technique is supporting a semi-supervised anomaly detection method which is claimed to outperform all previously mentioned approaches. Also, [10] suggests an anomaly detection method genuinely based on reinforcement learning, which however doesn't reach the same accuracy as the [9].

2 Literature Review

Conor McKinnon [11] presented a comparison of new anomaly detection techniques i-e OCSVM, IF, and EE for wind Turbine condition Monitoring. SCADA data set is used in this study. The author suggested that OCSVM has better performance for generic training and IF is better for specific training. Overall IF and OCSVM depicted 82% average accuracy as compared to 77% of EE.

Markus Goldstein [8] evaluated 19 different unsupervised anomaly detection algorithms on 10 different data sets. This study concluded that local anomaly detection algorithms such as LOF, COF, INFLO, and LoOP perform poorly on global anomalies. The author observed that distance measurement-based algorithms have better performance than clustering algorithms. Moreover, this report suggests that the statistical algorithm HBOS has good overall performance. This study only includes disease, handwriting, letter, and speech recognition data but no machinery-related data is used.

Shashwat Suman [12] compared 4 Unsupervised outlier detection methods: i) Local Outlier Factor ii) Connectivity Based Outlier Factor (COF) iii) Local Distance-Based Outlier Factor and iv) Influenced Outlierness on four data sets (IRIS, Spambase, Breast Cancer, Seeds). The author proposed that the Local Distance-Based Outlier Factor has better accuracy across different databases and INFLO has better speed performance as compared to other methods.

Our study is different from available studies because it focuses on facilitating predictive maintenance in small and medium-sized enterprises. Therefore this study includes the CMAPSS data set which is related to real machinery data for every stage in machinery life with different operations and fault conditions. This makes this study suitable and helpful for Predictive maintenance in small and medium-sized enterprises (SMEs) to reduce the unexpected breakdown.

3 Methodology

The general methodology for all kinds of anomaly detection is somehow similar and consists of the following basic modules or stages: i. Data Elicitation ii. Training iii. Detection. Eliciting Data includes collecting raw data from a monitored environment. In our study, we have taken the C-MAPSS data set. In the training stage, the model tries to learn normal behaviors in data through distance measurement, clustering, or any other statistical way. And finally, the detection phase model identifies those instances which are abnormal or away from the normal behavior of the data set.

3.1 Data

Commercial Modular Aero Propulsion System Simulation (C-MAPSS) [13] is published by NASA's prognostic center. In predictive maintenance, a major problem is to find enough data which encompasses an organized record of all stages of machinery life from the properly running stage to the failure stage. The CMAPSS provides researchers with enough data for every stage in the machinery life cycle with different operations and fault conditions. It is further categorized in training and test data sets. In CMAPSS turbofan engine data is recorded. Each engine goes through many wears and manufacturing variations. At the start, it operates properly but with the passage of time, it develops faults. The CMAPSS contains data from both stages. Basically, it was developed to find the number of operational cycles left in the engine which is known as Remaining Useful Life (RUL). The CMAPSS consists of 21 sensor observations along with three operational settings. Each instance in the data set depicts a single operational time cycle of the engine.

3.2 Model Description

Nine unsupervised anomaly detection algorithms were considered for this paper. Each algorithm is briefly discussed below:

One Class SVM: It is an unsupervised anomaly detection algorithm-based decision function to separate normal data from outliers. OCSVM [14] creates a boundary around normal data to distinguish outliers. OCSVM is different from normal SVM because it allows a few outlier percentages. This outlier exists between a separate hyper-plane and its origin. The remaining Data lying on the opposite side of the hyper-plane belongs to a single class; therefore this is called one Class SVM.

Isolation Forest: The algorithm of Isolation Forest [15] uses random forest decision trees and identifies outliers by taking the average number of partitions required to isolate each data point. Data points with fewer partitions are considered anomalies. Isolation Forest is based on the concept that data points which sparse have fewer partitions. It uses Ensembles of Isolation Tree.

Local Outlier Factor: The LOF [16] is also an unsupervised anomaly detection algorithm that works on the concept of how the density of a data point differs from the nearest data points. It uses the concept of local outlier first. It considers data points as an outlier if it has a remarkably lower density with respect to its local neighbors.

Cluster Based Local Outlier Factor (CBLOF): The CBLOF [8] algorithm divides the data set into many clusters on the basis of density estimation. The density of each cluster is estimated and clusters are classified into large and small clusters. Then the outlier score is calculated through the distance of each data point to the origin of its cluster or if the cluster is small then the center of the nearest large cluster is used.

Feature Bagging: The Feature Bagging [17] is an ensemble method. It is based on the Local Outlier Factor (LOF) on selected data subsets. Multiple iterations are performed and a random subset of variables is chosen each time. Finally, the cumulative sum of each iteration is calculated.

Histogram-based Outlier Detection (HBOS): The HBOS [18] is a statistical Anomaly detection Algorithm and it assumes independence of the features. A histogram for each feature and then for each instance is created. Then Density Estimation is performed using the area of bins. Speed is a major advantage of this algorithm. Multi-dimensional data fails this algorithm due to the assumption of independence in features. HBOS score for each data item is calculated by computing the inverse of the estimated density.

Angle-Based Outlier Detection (ABOD): Contrary to HBOS, Angle-Based Outlier Detection [19] is suitable for high dimensional data and its basis is to compare the angles between pairs of distance vectors to other points in order to distinguish points having similarities to other points and anomalies.

K Nearest Neighbors (KNN): Although KNN is a supervised Machine Learning Algorithm it can be used for Anomaly Detection with an unsupervised approach. Because KNN does not require learning and predetermined labeling rather it determines the cut-off value through distance measurement. If an instance is much away from the threshold value then it is considered an anomaly.

3.3 Test Description

These algorithms are applied to the CMAPSS data set and their results are evaluated to compare their performance. Before applying algorithms, data were normalized in order to eliminate redundancy, minimize data modification errors, increase security and lessen cost. The data set is split into training and testing sets to examine the performance of an algorithm. Then each algorithm was evaluated with different metrics like Accuracy, F1 Score, True Positive Rate, True Negative rate, ROC-AUC, and MCC to depict the true picture of performance comparison. Each algorithm detected the anomaly differently.

Name	MCC	F1 Score	Accuracy	TP Rate	TN Rate	PR_AUC	ROC_AUC
OCSVM	-0.006427	0.970	0.941469	0.970	0.024	0.985	0.496752
(IForest)	-0.009123	0.880	0.787879	0.880	0.111	0.932	0.495467
LOF	0.106529	0.973	0.948000	0.973	0.133	0.986	0.553265
(ABOD)	0.007595	0.048	0.920000	0.879	0.112	0.073	0.503158
(CBLOF)	0.034343	0.082	0.910000	0.879	0.112	0.105	0.516842
(FB)	-0.002287	0.043	0.912000	0.879	0.112	0.068	0.498947
(HBOS)	-0.002287	0.043	0.912000	0.879	0.112	0.068	0.498947
(KNN)	0.058231	0.095	0.924000	0.879	0.112	0.122	0.524211
Avg KNN	0.054186	0.093	0.922000	0.879	0.112	0.119	0.523158

Fig. 1. Table for performance comparison

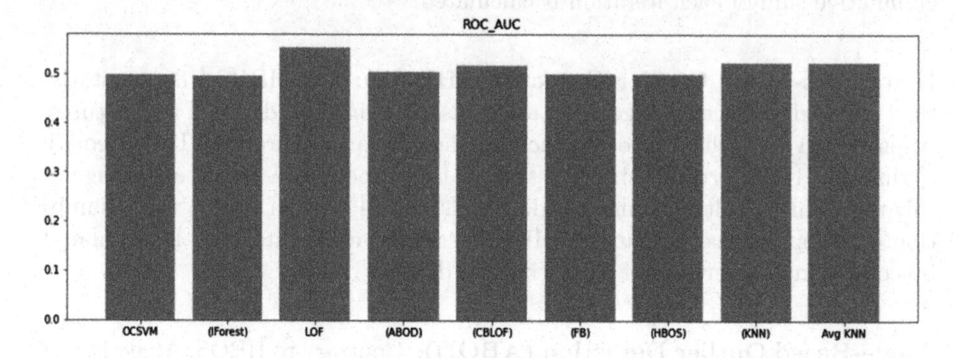

Fig. 2. ROC-AUC scores of all algorithms

4 Results

The algorithms are implemented in Python and the results are analyzed and compared for performance and accuracy.

4.1 Performance Comparison

This section of the paper highlights the comparison of performance and accuracy of nine different models with different metrics. The results presented in the table show the performance of each model according to each metrics.

The table in Fig. 1 depicts that the Local outlier factor has surpassed each model in performance with an accuracy of 94.8% and a 0.973 F1 score. This is because LOF considers the density of data points related to its neighbor and does not assume the independence of features, so this algorithm is more suitable for multidimensional data. OCSVM also performed well with 94.1% accuracy and a 0.970 F1 score (Fig. 3).

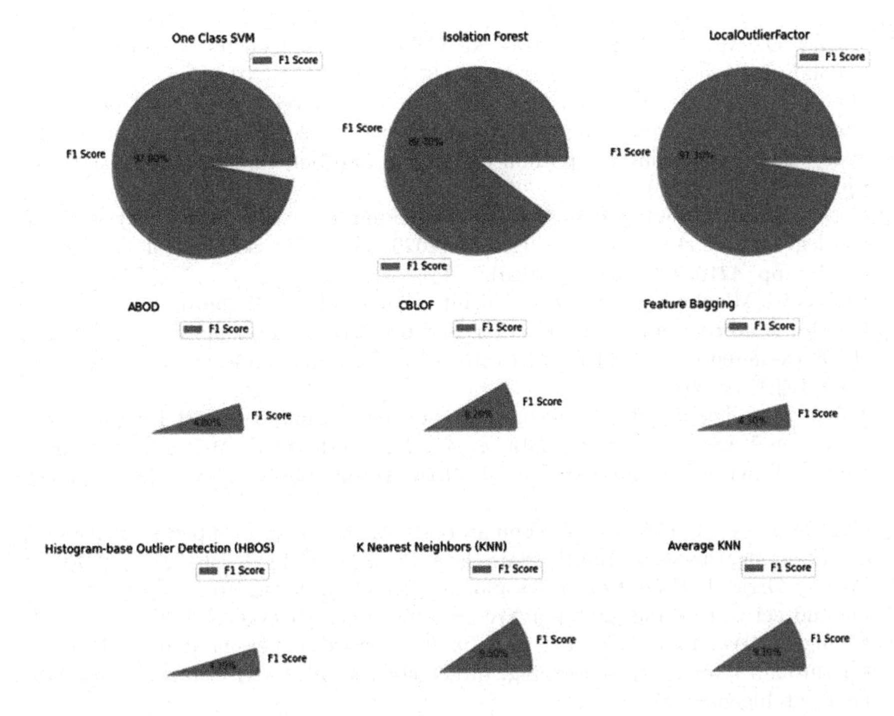

Fig. 3. F1 score comparison

5 Conclusion

This Study concludes that unsupervised anomaly detection algorithms are better for Global Outlier detection compared to others. Local Outlier Factor and Isolation Forest are the preferred choices when the time series data set is high dimensional and without output labels. We found from experiments that Local Outlier Factor, One-Class SVM, and Isolation Forest performed better than the rest of the algorithms for predictive maintenance in SMEs.

Acknowledgement. The research reported in this paper has been funded by the European Interreg Austria-Czech Republic project "PredMAIn (ATCZ279)".

References

1. Kavitha, M., Srinivas, P.V.V.S., Kalyampudi, P.L., Srinivasulu, S.: Machine learning techniques for anomaly detection in smart healthcare. In: 2021 Third International Conference on Inventive Research in Computing Applications (ICIRCA) (pp. 1350–1356). IEEE (2021)
2. Zoppi, T., Ceccarelli, A., Bondavalli, A.: Into the unknown: unsupervised machine learning algorithms for anomaly-based intrusion detection. In: 2020 50th Annual IEEE-IFIP International Conference on Dependable Systems and Networks-Supplemental Volume (DSN-S), p. 81. IEEE (2020)

3. Huang, D., Mu, D., Yang, L., Cai, X.: CoDetect: financial fraud detection with anomaly feature detection. IEEE Access **6**, 19161–19174 (2018)
4. Hong, Y., Wang, L., Kang, J., Wang, H., Gao, Z.: A novel application approach for anomaly detection and fault determination process based on machine learning. In: 2020 6th International Symposium on System and Software Reliability (ISSSR), pp. 1–5. IEEE (2020)
5. Hu, X., Li, N.: Anomaly detection and remaining useful lifetime estimation based on degradation state for bearings. In: 2020 39th Chinese Control Conference (CCC), pp. 4216–4221. IEEE (2020)
6. Farbiz, F., Miaolong, Y., Yu, Z.: A cognitive analytics based approach for machine health monitoring, anomaly detection, and predictive maintenance. In: 2020 15th IEEE Conference on Industrial Electronics and Applications (ICIEA), pp. 1104–1109. IEEE (2020)
7. Vollert, S., Theissler, A.: Challenges of machine learning-based RUL prognosis: a review on NASA's C-MAPSS data set. In: 2021 26th IEEE International Conference on Emerging Technologies and Factory Automation (ETFA), pp. 1–8. IEEE (2021)
8. Goldstein, M., Uchida, S.: A comparative evaluation of unsupervised anomaly detection algorithms for multivariate data. PLoS ONE **11**(4), e0152173 (2016)
9. Wu, T., Ortiz, J.: Rlad: time series anomaly detection through reinforcement learning and active learning (2021). arXiv preprint arXiv:2104.00543
10. Huang, C., Wu, Y., Zuo, Y., Pei, K., Min, G.: Towards experienced anomaly detector through reinforcement learning. In: Thirty-Second AAAI Conference on Artificial Intelligence (2018)
11. McKinnon, C., Carroll, J., McDonald, A., Koukoura, S., Infield, D., Soraghan, C.: Comparison of new anomaly detection technique for wind turbine condition monitoring using gearbox SCADA data. Energies **13**(19), 5152 (2020)
12. Suman, S.: Improving influenced outlierness (INFLO) outlier detection method (Doctoral dissertation) (2013)
13. Chakraborty, S., Sarkar, S., Ray, A., Phoha, S.: Symbolic identification for anomaly detection in aircraft gas turbine engines. In: Proceedings of the 2010 American Control Conference, pp. 5954–5959. IEEE (2010)
14. Budiarto, E.H., Permanasari, A.E., Fauziati, S.: Unsupervised anomaly detection using K-means, local outlier factor and one class SVM. In: 2019 5th International Conference on Science and Technology (ICST), Vol. 1, pp. 1–5. IEEE (2019)
15. Chun-Hui, X., Chen, S., Cong-Xiao, B., Xing, L.: Anomaly detection in network management system based on isolation forest. In: 2018 4th Annual International Conference on Network and Information Systems for Computers (ICNISC), pp. 56–60. IEEE (2018)
16. Su, S., et al.: ADCMO: an anomaly detection approach based on local outlier factor for continuously monitored object. In: 2019 IEEE International Conference on Parallel & Distributed Processing with Applications, Big Data & Cloud Computing, Sustainable Computing & Communications, Social Computing & Networking (ISPA/BDCloud/SocialCom/SustainCom), pp. 865–874. IEEE (2019)
17. Lazarevic, A., Kumar, V.: Feature bagging for outlier detection. In: Proceedings of the eleventh ACM SIGKDD International Conference on Knowledge discovery in data mining, pp. 157–166 (2005)
18. Kalaycı, İ., Ercan, T.: Anomaly detection in wireless sensor networks data by using histogram based outlier score method. In: 2018 2nd International Symposium on Multidisciplinary Studies and Innovative Technologies (ISMSIT), pp. 1–6. IEEE (2018)

19. Ma, Y., et al.: Real-time angle outlier detection based on robust preprocessing and feature extraction. In: 2018 37th Chinese Control Conference (CCC), pp. 4517–4524. IEEE (2018)
20. Bukovsky, I., Cejnek, M., Vrba, J., Homma, N.: Study of learning entropy for onset detection of epileptic seizures in EEG time series. In: Presented at the 2016 International Joint Conference on Neural Networks (IJCNN), Vancouver (2016)
21. Cejnek, M., Bukovsky, I.: Concept drift robust adaptive novelty detection for data streams. Neurocomputing **309**, 46–53 (2018). https://doi.org/10.1016/j.neucom.2018.04.069
22. Page, E.S.: Continuous inspection scheme. Biometrika, **41**(1/2), 100–115 (1954). hdl:10338.dmlcz/135207. JSTOR 2333009. https://doi.org/10.1093/biomet/41.1-2.100
23. Mahmoud, S., Martinez-Gil, J., Praher, P., Freudenthaler, B., Girkinger, A.: Deep learning rule for efficient changepoint detection in the presence of non-linear trends. In: Kotsis, G., et al. (eds.) DEXA 2021. CCIS, vol. 1479, pp. 184–191. Springer, Cham (2021). https://doi.org/10.1007/978-3-030-87101-7_18
24. Glock, A., Sobieczky, F., Jech, M.: Detection of anomalous events in the wear-behaviour of continuously recorded sliding friction pairs. In: Symposium Tribologie in Industrie und Forschung, pp. 30–40 (2019). ISBN 978-3-901657-62-7, 11
25. Aminikhanghahi, S., Cook, D.J.: A survey of methods for time series change point detection. Knowl. Inf. Syst. **51**(2), 339–367 (2016). https://doi.org/10.1007/s10115-016-0987-z
26. Sharma, S., Swayne, D.A., Obimbo, C.: Trend analysis and change point techniques: a survey. Energy, Ecol. Environ. **1**(3), 123–130 (2016). https://doi.org/10.1007/s40974-016-0011-1
27. Wei, L., Keogh, E.: Semi-supervised time series classification. In: 12th ACM SIGKDD International Conference on Knowledge Discovery and Data Mining-KDD 2006. ACM Press, New York, p. 748 (2006)
28. Adams, R.P., MacKay, D.J.: Bayesian online changepoint detection (2007). arXiv:0710.3742
29. Woodall, W.H., Adams, B.M.: The statistical design of CUSUM charts. Qual. Eng. **5**(4), 559–570 (1993)
30. Matteson, D.S., James, N.A.: A nonparametric approach for multiple change point analysis of multivariate data. J. Am. Stat. Assoc. **109**(505), 334–345 (2013)
31. Jan, V., Rob, H., Achim, Z., Darius, C.: Phenological change detection while accounting for abrupt and gradual trends in satellite image time series. Remote Sens. Environ. **114**(12), 2970–2980 (2010)
32. Yu, H., Liu, T., Lu, J., Zhang, G.: Automatic learning to detect concept drift (2021). arXive:2105.01419

A Comparative Study Between Rule-Based and Transformer-Based Election Prediction Approaches: 2020 US Presidential Election as a Use Case

Asif Khan[1], Huaping Zhang[1](✉), Nada Boudjellal[1](✉), Lin Dai[1], Arshad Ahmad[2,3], Jianyun Shang[1], and Philipp Haindl[4]

[1] School of Computer Science and Technology, Beijing Institute of Technology, Beijing 100081, China
{kevinzhang,nada}@bit.edu.cn

[2] Institute of Software Systems Engineering, Johannes Kepler University, 4040 Linz, Austria

[3] Department of IT and Computer Science, Pak-Austria Fachhochschule: Institute of Applied Sciences and Technology, Mang Khanpur Road, Haripur 22620, Pakistan

[4] Software Competence Center Hagenberg (SCCH), Hagenberg, Austria

Abstract. Social media platforms (SMPs) attracted people from all over the world for they allow them to discuss and share their opinions about any topic including politics. The comprehensive use of these SMPs has radically transformed new-fangled politics. Election campaigns and political discussions are increasingly held on these SMPs. Studying these discussions aids in predicting the outcomes of any political event. In this study, we analyze and predict the 2020 US Presidential Election using Twitter data. Almost 2.5 million tweets are collected and categorized into Location-considered (LC) (USA only), and Location-unconsidered (LUC) (either location not mentioned or out of USA). Two different sentiment analysis (SA) approaches are employed: dictionary-based SA, and transformers-based SA. We investigated if the deployment of deep learning techniques can improve prediction accuracy. Furthermore, we predict a vote-share for each candidate at LC and LUC levels. Afterward, the predicted results are compared with the five polls' predicted results as well as the real results of the election. The results show that dictionary-based SA outperformed all the five polls' predicted results including the transformers with MAE 0.85 at LC and LUC levels, and RMSE 0.867 and 0.858 at LC and LUC levels.

Keywords: Twitter · USA election · Sentiment analysis · Rule-based · Transformers · Deep learning · Vote-share

1 Introduction

SMPs attracted people from all over the world where they discuss and share their opinions about any topic including politics. The comprehensive use of these SMPs has radically transformed new-fangled politics. Election campaigns and political discussions are

G. Kotsis et al. (Eds.): DEXA 2022 Workshops, CCIS 1633, pp. 32–43, 2022.
https://doi.org/10.1007/978-3-031-14343-4_4

increasingly held on these SMPs. Currently, almost every political party and candidate is utilizing SMPs (i.e., Twitter) these days to connect to their constituencies and voters. As these SMPs provide a rapid and direct way to connect to them.

Numerous researchers investigated and forecasted various political events using Twitter data including election perdition as it is one of the widely used SMPs utilized by political entities [1–4]. These researchers proposed numerous methods to forecast different elections such as, sentiment analysis (SA) approaches [5–10] social network analysis (SNA) approaches [11–13], and volumetric approaches [14–16].

SA approaches are widely implemented to predict different elections [4]. Researchers utilized different SA approaches to predict elections such as, rule-based SA [10, 17–21] and supervised learning [12, 22, 23].

Recently, transformers-based SA is being in practice in several research areas such as finance [24], Italian tweets [25], stock exchange [26], and so on. To the best of our knowledge, no study has applied transformer-based SA models to predict any election. In this study, we used transformers-based SA. Furthermore, we conduct a comparative study between rule-based approaches and transformers for election prediction on Twitter data. For this purpose, we take the 2020 USA Presidential Election as a use case. For the rule-based approach, we chose VADER (Valence Aware Dictionary and sEntiment Reasoner) which is one of the widely used tools for rule-based sentiment analysis. On the other hand, we employed a sentiment analysis pipeline from the Hugging face platform for transformers-based sentiment analysis.

This research study basically addresses two questions:

1. Can tweets about a candidate predict US Presidential Election?
2. Can rule-based sentiment analysis work better than deep learning (transformers) based sentiment analysis?

In this study, we analyzed tweets related to Joe Biden and Donald Trump from July to November 2020. Tweets are categorized into two categories based on their location information; i) Categorized into location consider (LC) (USA only), and ii) Location-unconsidered (LUC) (either location not mentioned or out of USA). The tweets are further analyzed using rule-based and transformers-based SA, Next, a vote-share is predicted. Furthermore, the predicted results are compared with the five well-known polls and the actual election results.

The main contributions of this research are as follows:

- Collect the tweets—with Trump and Biden tags—in a period of four months.
- Classify the collected tweets by their polarity using two different methods: rule-based, and deep-learning-based (transformers).
- Study and predict the 2020 US Presidential Election using candidates' hashtags.
- Study and predict the 2020 US Presidential Election at LC (Location mentioned as US only) and LUC (either location not mentioned or out of USA) levels.
- Forecast the 2020 US Presidential Election and compared the results with five prominent polls as well as the actual outcomes of the election.

The rest of this paper is organized as follows. Section 2 gives an overview of the related studies. Section 3 presents the methodology, followed by results and discussion in Sect. 4. Afterward, Sect. 5 concludes this study.

2 Related Work

Since SMPs provide an easy and fast track for the politicians and political parties to connect to the voters. They run their campaigns and get direct responses from the citizens on these SMPs, which help them to understand the emotions of their voters. Likewise, they can change their campaigns according to the reactions of the voters. Twitter is one of the most widely used SMPs by political entities, which makes it a rich field for researchers to delve. Numerous researchers examined Twitter data to forecast different political events such as elections prediction. Elections are predicted mainly using three approaches: SA approach, volumetric approach, and SNA approach. SA is the most widely used approach [4].

Kristiyanti et al. [7] analyzed the election of West Java Governor candidate period 2018–2023 using Twitter data. They investigated election using supervised SA approaches (SVM, and Naïve Bayesian), discussed the comparisons between these two approaches, and claimed that NB has higher accuracy. Khan et al. [12] studied and predicted the 2018 general election using SA (KNN, NB, and SVM) and SNA approach. Jhawar et al. analyzed the Indian 2019 general election using SA (NB, and SentiWordNet) and SNA approaches [13].

The authors in [18] analyzed Indonesian Presidential Election using Twitter data. They first detect and removed the bots and then performed SA to predict the election. Jyoti et al. studied the 2016 US Presidential Election using SA. The authors labeled the tweets manually and using VADER and then they used supervised algorithms; MNB and SVM to predict the election. The authors in [27] investigated the 2020 US Presidential Election using tweets regarding Trump and Biden. They used tweets before and after the election and employed SA (NB) approach to predict the election.

All of these studies performed supervised or dictionary-based approaches and no study has analyzed tweets using transformer-based SA to predict an election. While in this research we analyzed the tweets using transformer-based SA and rule-based and we compared the results of both methods. Furthermore, we predicted a vote-share for each candidate running for the 2020 Presidential Election.

3 Methodology

This section presents the methodology of this study. Figure 1 shows the approach used for election prediction on Twitter. The major components of our model are as follows:

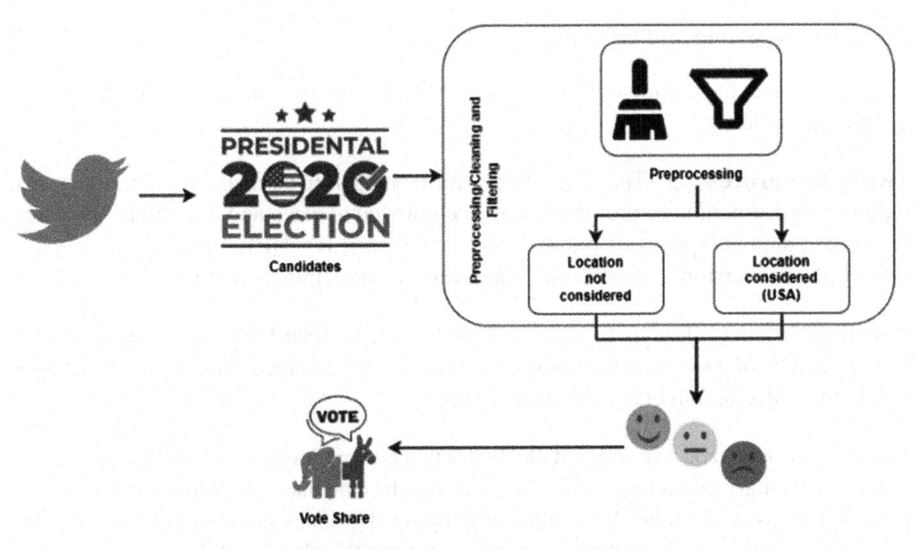

Fig. 1. Election prediction model

3.1 Data Collection

In this study, we collected tweets related to Donald Trump and Joe Biden between 28th July 2020 and 11th November 2020 to predict the 2020 US Presidential Elections. To collect tweets, we employed a Twitter Search API "Tweepy" [28]. Each collected tweet is in JSON format that contains numerous information, such as *user_id, lang, id, created_at, text, coordinates,* etc. In this study, we ignored various unnecessary attributes and focused on the ones that we need such as *user_location*, *id*, *text*, and *created_at*.

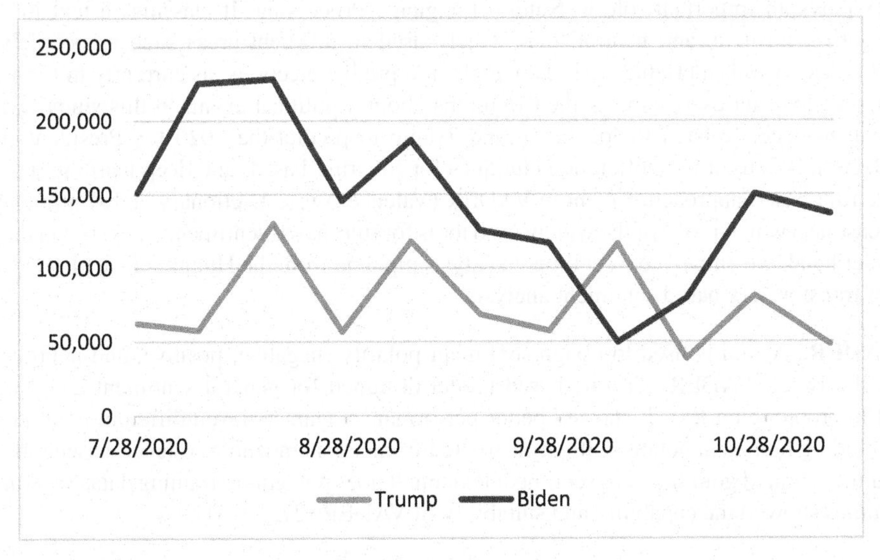

Fig. 2. Tweets distribution

3.2 Pre-processing and Filtering

The tweets are then filtered and pre-processed to enhance the performance of the prediction model.

Tweets Pre-processing. The tweets in raw form are unstructured and contain numerous redundant information, so pre-processing is employed to clean and filter them. The pre-processing step comprises of removal of hashtags (#), RT, URLs, IPs, mentions (@), stopwords, conversion to lower case, tokenization, and lemmatization.

Location. Twitter offers its users the choice to provide their locations while tweeting. We categorized the tweets into the following two categories according to the availability of location information of a user while tweeting.

Location-Considered. We selected the tweets that contain location information and used a states-dictionary containing US states to assign the state name accordingly. For example, a user has a location "VA", after employing the states-dictionary, "VA" will be matched to "Virginia" and the location information "Virginia" will be stored with the tweet, similarly OR to "Oregon", PA to "Pennsylvania", HI to "Hawaii", NV to "Nevada", and so on. Next, the tweets that contain location as "USA" or "USA state" are considered Location-considered (LC).

Location-Unconsidered. Users that do not provide location information or the location is out of the USA while tweeting is categorized as Location-unconsidered (LUC).

3.3 Sentiment Analysis

SA plays an important role in Natural Language Processing. It classifies a text into negative, positive, and neutral. It is widely used in different areas such as people's attitudes towards a product. It is also applied to predict elections, as currently millions of social media users express their opinions about a political event. In this study, we analyze tweets related to Joe Biden and Trump to predict the 2020 US Presidential Election. We used two different techniques for SA, rule-based, and deep learning. For the rule-based approach, we chose VADER (Valence Aware Dictionary and sEntiment Reasoner) which is one of the widely used tools for rule-based sentiment analysis. On the other hand, we employed a sentiment analysis pipeline from the Hugging face platform for transformers-based sentiment analysis.

VADER. A tool is used to assign sentiment polarity (negative, positive, and neutral) to the tweet. VADER is a rule-based model designed for general sentiment analysis of microblogging text. It showed better performance against eleven different methods including human annotation. It is constructed from a generalizable, valence-based, and human-curated gold standard sentiment lexicon. It does not require training data. Several studies showed the capability and suitability of VADER [27, 29–31].

This study considers positive and negative sentiments only. To handle this problem in VADER, we normalized the sentiment by distributing neutral tweets among positive and negative tweets equally. To do this, we divide neutral tweets by 2 and add them to positive and negative tweets.

Hugging Face Pipeline. The transformers pipeline used to classify the tweets into positive and negative was retrieved from the Hugging face platform. The Hugging face provides a bunch of models like BERT-based models that were deployed in many computational linguistics studies [32]. It also platform has many ready-to-use pipelines for many tasks including text classification. The sentiment analysis pipeline leverages a fine-tuned model on SST2, which is a GLUE task. It returns a "positive" or "negative" label plus a score for each tweet.

3.4 Vote-Share

The sentiment percentages are insufficient to predict an election. To improve the prediction of an election, we predict a vote-share for each candidate. Equation 1 is used to predict the vote-share [1, 33]. Furthermore, in this study, a mean absolute error (MAE) and root-mean-square error (RMSE) (See Eq. 2, and Eq. 3) are used to compare the predicted vote-share with five different polls prediction and the real US 2020 Presidential Election results.

$$\text{Vote-share Candidate (A)} = \frac{(Pos.\ A + Neg.\ B)}{(Pos.\ A + Neg.\ A + Pos.\ B + Neg.\ B)} \times 100 \tag{1}$$

$$MAE = \frac{1}{N}\sum\nolimits_{i=1}^{N} |Predicted_i - Actual_i| \tag{2}$$

$$\text{RMSE} = \sqrt{\frac{\sum_{i=1}^{N}(Predicted_i - Actual_i)^2}{N}} \tag{3}$$

4 Results and Discussion

The data was collected using the hashtags related to candidates only (See Table 1) between 28th July 2020 and 11th November 2020. Figure 2 shows the distribution of the collected tweets over the timeline. Table 1 gives an overview of the data collected in this study. 2.43 million Tweets were collected and only English-written tweets (2 million) are considered in this research study. It can be seen that 32.81% of the total English written tweets are related to Trump and 67.19% are related to Biden. The tweets were further pre-processed using NLTK [34]. Next, a sentiment analysis technique was employed to assign sentiment to each tweet. In this study, we employed two different techniques for sentiment analysis, i) the "vaderSentiment" library [35], and ii) Hugging face pipeline (transformers).

Table 1. Data collection

Candidate	Tags	No. of tweets	Total	English
Donald Trump	@realdonaldtrump	472,563	835,119	684,093
	#realdonaldtrump	111,299		
	#Trump2020	251,257		
Joe Biden	@JoeBiden	601,947	1,594,370	1,400,843
	#JoeBiden	412,303		
	#Biden2020	580,120		

4.1 Sentiment Analysis

In this step we apply two different approaches on our preprocessed data. A rule-based approach using VADER and a transformer-based approach using a sentiment analysis pipeline provided by the hugging face platform. the following subsections describe in details both methods.

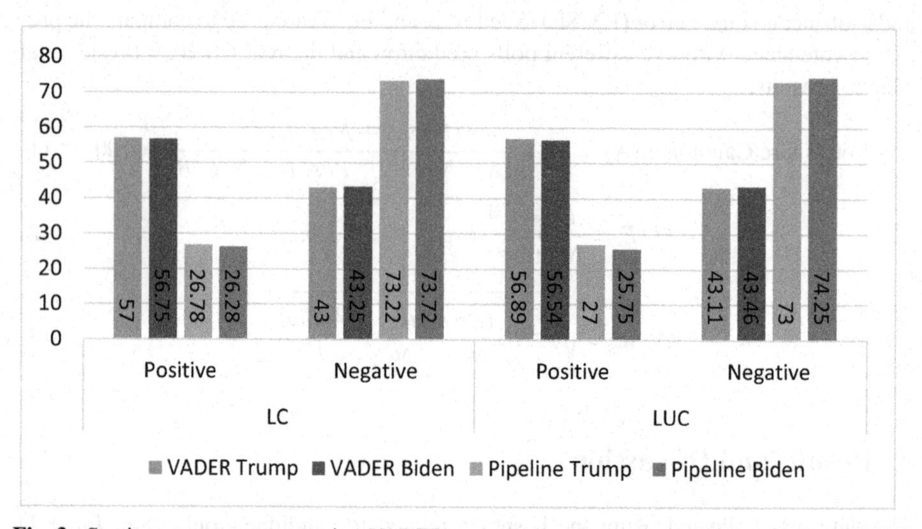

Fig. 3. Sentiments percentages using VADER and pipeline for each candidate at LC and LUC level

Sentiment Analysis Using VADER

Table 2 shows the sentiments extracted using VADER for Biden and Trump at LC and LUC. It is quite interesting that only 32% and 34% of the total tweets for Trump and Biden respectively are from USA (location defined).

VADER extracts three types of sentiments, Positive, Negative, and Neutral. As mentioned in Sect. 3.3, we consider positive and negative tweets only. To attain this, we distributed neutral sentiments among positive and negative with the same proportion as neutral does not affect positive as well as negative. We added half of the neutrals to positive tweets and half to negative tweets, the results of the VADER sentiment analysis are shown in Table 3.

Figure 3 shows the sentiment percentage for each candidate at LC and LUC levels. At the LC level, positive sentiment percentages for Biden and Trump are 56.72 and 57, and negative sentiment percentages are 43.25 and 43 respectively. At LUC the positive sentiment percentages for Biden and Trump are 56.54 and 56.89, and the negative are 43.46 and 43.11 respectively.

Table 2. VADER: Distribution of positive, negative, and neutral tweets for both candidates.

	Biden				Trump			
	Pos.	Neg.	Neu.	Total	Pos.	Neg.	Neu.	Total
LC	201,483	137,207	137,213	475,903	924,19	61,782	64,812	219,013
LUC	585,961	402,803	412,079	1,400,843	288,588	194,349	201,156	684,093

Table 3. VADER: Distribution of positive and negative tweets for both candidates (after splitting the neutral tweets)

	Biden			Trump		
	Pos.	Neg.	Total	Pos.	Neg.	Total
LC	270,089.5	205,813.5	475,903	124,825	94,188	219,013
LUC	792,000.5	608,842.5	1,400,843	389,166	294,927	684,093

Sentiment Analysis Using Pipeline. Table 4 shows the total number of tweets and the number of positive and negative tweets for Biden and Trump at LC and LUC levels. It can be seen in Fig. 3 that the negative sentiment percentage for both candidates is much higher than the positive. The negative sentiment percentages for Biden and Trump at the LC level are 73.72 and 73.22, and the positive are 26.28 and 26.78. At LUC the negative sentiment percentages are 74.25 and 73, and the positive are 25.75 and 27.

4.2 Vote-Share

The sentiment percentages (See Sect. 4.1) for both candidates do not provide an accurate prediction. As it shows that Trump is the winner since the positive sentiment percentage is higher than for Biden. To get a better prediction, we employed Eq. 1 to get vote-share for each candidate at LC and LUC. Table 5 shows the predicted vote-share (in percentage) for the candidates using both employed approaches.

The prediction of this study is compared with the five famous polls' predicted results. The prediction results are compared with the final election's results using MAE and RMSE. It is important to note that VADER on our dataset outperformed all the predictions including pipeline with RMSE at LC and LUC 0.867 and 0.858 and MAE 0.85 and 0.85 respectively. This concludes and answer our both research questions that, "yes", tweets re-garding candidates can predict US Presidential Elections, and rule-based SA works better than transformers (pipeline provided by Hugging Face).

Table 4. Transformers: Distribution of positive and negative tweets for both candidates

	Biden			Trump		
	Pos.	Neg.	Total	Pos.	Neg.	Total
LC	125,047	350,856	475,903	58,648	160,365	219,013
LUC	360,679	1,040,164	1,400,843	184,677	499,416	684,093

Table 5. Prediction results comparison

	RCP	Economist/YouGov	CNBC	Fox news	Survey USA	Our results pipeline (LC)	Our results pipeline (LUC)	Our results VADER (LC)	Our results VADER (LUC)	US Presidential Election 2020
Biden	51.2	53	52	52	52	41.90	41.25	**52.42**	**52.13**	51.40
Trump	44	43	42	44	44	58.09	58.74	**47.58**	**47.87**	46.9
RMSE	2.055	2.981	3.491	2.094	2.094	10.379	11.027	**0.867**	**0.858**	–
MAE	1.55	2.75	2.75	1.75	1.75	10.345	10.995	**0.85**	**0.85**	–

5 Conclusion

In this study, we analyzed the 2020 US Presidential Election using Twitter data. Almost 2.43 million tweets were collected and analyzed using candidates' (Trump and Biden) hashtags between 28th July 2020 and 11th November 2020. Only English-written tweets were considered. In addition, the tweets were categorized into two classes based on the location information provided by the users at the time of tweeting; Location-considered (USA only), and Location-unconsidered (either location not mentioned or out of USA). Next, two different sentiment analysis approaches were employed: dictionary-based SA, and transformers-based SA. For dictionary-based SA we employed VADER and for transformers-based SA we employed a pipeline from the Hugging Face platform. Furthermore, we predicted a vote-share for each candidate at LC and LUC levels. Afterward, our predicted results were compared with the five polls' predicted results as well as the real results of the election. The results show that dictionary-based SA (VADER) outperformed all the five polls' predicted results including the transformers with MAE 0.85 at LC and LUC levels, and RMSE 0.867 and 0.858 at LC and LUC levels.

This study observes that rule-based SA (VADER) shows better performance in predicting vote-share for a candidate than the SA pipeline provided by the Hugging Face platform.

In the future, other transformers like BERTweet needs to be utilized to analyze and compare with the current study. Social network analysis techniques need to be implemented for better performance.

References

1. Wang, L., Gan, J.Q.: Prediction of the 2017 French election based on Twitter data analysis. In: 2017 9th Computer Science and Electronic Engineering Conference, CEEC 2017 – Proceedings, pp. 89–93 (2017). https://doi.org/10.1109/CEEC.2017.8101605
2. Matalon, Y., Magdaci, O., Almozlino, A., Yamin, D.: Using sentiment analysis to predict opinion inversion in Tweets of political communication. Sci. Rep. **11**, 1–9 (2021). https://doi.org/10.1038/s41598-021-86510-w
3. Sosnkowski, A., Fung, C.J., Ramkumar, S.: An analysis of Twitter users' long term political view migration using cross-account data mining. Online Soc. Netw. Media **26**, 100177 (2021). https://doi.org/10.1016/J.OSNEM.2021.100177
4. Khan, A., Zhang, H., Boudjellal, N., et al.: Election prediction on Twitter: a systematic mapping study. Complexity **2021**, 1–27 (2021). https://doi.org/10.1155/2021/5565434
5. Beleveslis, D., Tjortjis, C., Psaradelis, D., Nikoglou, D.: A hybrid method for sentiment analysis of election related tweets. In: 2019 4th South-East Europe Design Automation, Computer Engineering, Computer Networks and Social Media Conference, SEEDA-CECNSM 2019 (2019)
6. Bose, R., Dey, R.K., Roy, S., Sarddar, D.: Analyzing political sentiment using Twitter data. In: Satapathy, S.C., Joshi, A. (eds.) Information and Communication Technology for Intelligent Systems. SIST, vol. 107, pp. 427–436. Springer, Singapore (2019). https://doi.org/10.1007/978-981-13-1747-7_41
7. Kristiyanti, D.A., Umam, A.H., Wahyudi, M., et al.: Comparison of SVM Naïve Bayes algorithm for sentiment analysis toward west Java governor candidate period 2018–2023 based on public opinion on Twitter. In: 2018 6th International Conference on Cyber and IT Service Management, CITSM 2018, pp. 1–6 (2019). https://doi.org/10.1109/CITSM.2018.8674352
8. Rezapour, R., Wang, L., Abdar, O., Diesner, J.: Identifying the Overlap between election result and candidates' ranking based on hashtag-enhanced, lexicon-based sentiment analysis. In: Proceedings of the IEEE 11th International Conference on Semantic Computing, ICSC 2017, pp. 93–96 (2017). https://doi.org/10.1109/ICSC.2017.92
9. Franco-Riquelme, J.N., Bello-Garcia, A., Ordieres-Meré, J.B.: Indicator proposal for measuring regional political support for the electoral process on Twitter: the case of Spain's 2015 and 2016 general elections. IEEE Access **7**, 62545–62560 (2019). https://doi.org/10.1109/ACCESS.2019.2917398
10. Plummer, M., Palomino, M.A., Masala, G.L.: Analysing the sentiment expressed by political audiences on Twitter: the case of the 2017 UK general election. In: Proceedings - 2017 International Conference on Computational Science and Computational Intelligence, CSCI 2017, pp. 1449–1454. IEEE (2018)
11. Castro, R., Kuffó, L., Vaca, C.: Back to #6D: predicting Venezuelan states political election results through Twitter. In: 2017 4th International Conference on eDemocracy and eGovernment, ICEDEG 2017, pp. 148–153 (2017)
12. Khan, A., Zhang, H., Shang, J., et al.: Predicting politician's supporters' network on Twitter using social network analysis and semantic analysis. Sci. Program. **2020** (2020). https://doi.org/10.1155/2020/9353120

13. Jhawar, A., Munjal, V., Ranjan, S., Karmakar, P.: Social network based sentiment and network analysis to predict elections. In: Proceedings of CONECCT 2020 - 6th IEEE International Conference on Electronics, Computing and Communication Technologies, pp. 0–5 (2020). https://doi.org/10.1109/CONECCT50063.2020.9198574
14. Sanders, E., van den Bosch, A.: A longitudinal study on Twitter-based forecasting of five Dutch national elections. In: Weber, I., et al. (eds.) SocInfo 2019. LNCS, vol. 11864, pp. 128–142. Springer, Cham (2019). https://doi.org/10.1007/978-3-030-34971-4_9
15. Coletto, M., Lucchese, C., Orlando, S., Perego, R.: Electoral predictions with Twitter: a machine-learning approach. In: CEUR Workshop Proceedings (2015)
16. Tavazoee, F., Conversano, C., Mola, F.: Recurrent random forest for the assessment of popularity in social media. Knowl. Inf. Syst. **62**, 1847–1879 (2020). https://doi.org/10.1007/s10115-019-01410-w
17. Mazumder, P., Chowdhury, N.A., Anwar-Ul-Azim Bhuiya, M., Akash, S.H., Rahman, R.M.: A fuzzy logic approach to predict the popularity of a presidential candidate. In: Sieminski, A., Kozierkiewicz, A., Nunez, M., Ha, Q.T. (eds.) Modern Approaches for Intelligent Information and Database Systems. SCI, vol. 769, pp. 63–74. Springer, Cham (2018). https://doi.org/10.1007/978-3-319-76081-0_6
18. Ibrahim, M., Abdillah, O., Wicaksono, A.F., Adriani, M.: Buzzer detection and sentiment analysis for predicting presidential election results in a Twitter nation. In: Proceedings - 15th IEEE International Conference on Data Mining Workshop, ICDMW 2015, pp. 1348–1353 (2016)
19. Kassraie, P., Modirshanechi, A., Aghajan, H.K.: Election vote share prediction using a sentiment-based fusion of Twitter data with Google trends and online polls. In: DATA 2017 – Proceedings of the 6th International Conference on Data Science, Technology and Applications, pp. 363–370 (2017). https://doi.org/10.5220/0006484303630370
20. Khan, S., Moqurrab, S.A., Sehar, R., Ayub, U.: Opinion and emotion mining for Pakistan general election 2018 on Twitter data. In: Bajwa, I.S., Kamareddine, F., Costa, A. (eds.) INTAP 2018. CCIS, vol. 932, pp. 98–109. Springer, Singapore (2019). https://doi.org/10.1007/978-981-13-6052-7_9
21. Wang, L., Gan, J.Q.: Prediction of the 2017 French election based on Twitter data analysis using term weighting. In: 2018 10th Computer Science and Electronic Engineering Conference, CEEC 2018 – Proceedings, pp. 231–235 (2019). https://doi.org/10.1109/CEEC.2018.8674188
22. Prati, R.C., Said-Hung, E.: Predicting the ideological orientation during the Spanish 24M elections in Twitter using machine learning. AI Soc. **34**, 589–598 (2019). https://doi.org/10.1007/s00146-017-0761-0
23. Oikonomou, L., Tjortjis, C.: A method for predicting the winner of the USA presidential elections using data extracted from Twitter. In: South-East Europe Design Automation, Computer Engineering, Computer Networks and Social Media Conference, SEEDA_CECNSM 2018 (2018)
24. Zhao, L., Li, L., Zheng, X., Zhang, J.: A BERT based sentiment analysis and key entity detection approach for online financial texts. In: Proceedings of the 2021 IEEE 24th International Conference on Computer Supported Cooperative Work in Design, CSCWD 2021, pp. 1233–1238 (2021)
25. Pota, M., Ventura, M., Catelli, R., Esposito, M.: An effective bert-based pipeline for Twitter sentiment analysis: a case study in Italian. Sensors (Switz.) **21**, 1–21 (2021). https://doi.org/10.3390/s21010133
26. Li, M., Chen, L., Zhao, J., Li, Q.: Sentiment analysis of Chinese stock reviews based on BERT model. Appl. Intell. **51**, 5016–5024 (2021). https://doi.org/10.1007/s10489-020-02101-8

27. Chaudhry, H.N., Javed, Y., Kulsoom, F., et al.: Sentiment analysis of before and after elections: Twitter data of U.S. Election 2020. Electron **10**, 2082 (2021). https://doi.org/10.3390/ELECTRONICS10172082
28. Tweepy. https://github.com/tweepy/tweepy. Accessed 18 Nov 2022
29. Hutto, C.J., Gilbert, E.: VADER: a parsimonious rule-based model for sentiment analysis of social media text. In: Proceedings of the 8th International Conference on Weblogs and Social Media, ICWSM 2014, pp. 216–225 (2014)
30. Ramteke, J., Shah, S., Godhia, D., Shaikh, A.: Election result prediction using Twitter sentiment analysis. In: Proceedings of the International Conference on Inventive Computation Technologies, ICICT 2016 (2016)
31. Bello, B.S., Inuwa-Dutse, I., Heckel, R.: Social media campaign strategies: analysis of the 2019 Nigerian elections. In: 2019 6th International Conference on Social Networks Analysis, Management and Security, SNAMS 2019, pp. 142–149 (2019). https://doi.org/10.1109/SNAMS.2019.8931869
32. Boudjellal, N., Zhang, H., Khan, A., et al.: ABioNER: a BERT-based model for arabic biomedical named-entity recognition. Complexity **2021**, 1–6 (2021). https://doi.org/10.1155/2021/6633213
33. Metaxas, P.T., Mustafaraj, E., Gayo-Avello, D.: How (Not) to predict elections. In: Proceedings - 2011 IEEE International Conference on Privacy, Security, Risk and Trust and IEEE International Conference on Social Computing, PASSAT/SocialCom 2011, pp. 165–171 (2011)
34. Bird, S., Klein, E., Loper, E.: Natural Language Processing with Python. O'Reilly Media (2009)
35. vaderSentiment · PyPI. https://pypi.org/project/vaderSentiment/

Detection of the 3D Ground Plane from 2D Images for Distance Measurement to the Ground

Ozan Cakiroglu[1(✉)], Volkmar Wieser[1], Werner Zellinger[1], Adriano Souza Ribeiro[2], Werner Kloihofer[2], and Florian Kromp[1]

[1] Software Competence Center Hagenberg, Softwarepark 32a, 4232 Hagenberg im Mühlkreis, Austria
ozan.cakiroglu@scch.at
[2] PKE Holding AG, Computerstraße 6, 1100 Vienna, Austria
https://scch.at
https://www.pke.at

Abstract. Obtaining 3D ground plane equations from remote sensing device data is crucial in scene-understanding tasks (e.g. camera parameters, distance of an object to the ground plane). Equations describing the orientation of the ground plane of a scene in 2D or 3D space can be reconstructed from multiple sensor output data as collected from 2D or 3D sensors such as; RGB-D cameras, T-o-F cameras or LiDAR sensors. In our work, we propose a modular and simple pipeline for 3D ground plane detection from 2D-RGB images for subsequent distance estimation of a given object to the ground plane. As the proposed algorithm can be applied on 2D-RGB images, provided by common devices such as surveillance cameras, we provide evidence that the algorithm has the potential to advance automated surveillance systems such as devices used for fall detection without the need to change existing hardware.

Keywords: 3D ground plane equation · 2D/3D plane segmentation · 2D/3D plane detection

1 Introduction

Scene-understanding can be defined as the task to find spatial and semantic information of a scene and their intrinsic interconnection. Ground plane (floor) detection, which is a sub-task of scene understanding, aims at estimating the ground plane information from the given 2D or 3D input. The ground plane information can be represented by either the classification of 2D ground plane pixels in 2D-RGB images, the classification of 3D ground plane points on 3D point cloud data or by describing the underlying 3D ground plane equation with reference to the observer sensor. Ground plane detection can be exploited in surveillance assistance systems. By estimating the distance of an object to the ground plane, it can be used as a part of the alarm system which triggers an alert

G. Kotsis et al. (Eds.): DEXA 2022 Workshops, CCIS 1633, pp. 44–54, 2022.
https://doi.org/10.1007/978-3-031-14343-4_5

if patients are not on their bed or it can act as a warning system for detecting if an elderly or impaired person fell to the floor.

2D ground plane detection from 2D-RGB images has already been tackled by multiple groups. Even though proposed methods present promising approaches showing high accuracy in the 2D pixel space, the lack of depth information and solely operating in the 2D pixel space instead of the Cartesian space does not allow to apply subsequent tasks such as person fall detection, extrinsic camera calibration or to calculate the distance of objects to the camera. In addition to methods operating on 2D image sources, methods using 3D sensor data for ground plane detection were proposed, overcoming the aforementioned limitations. Nevertheless, working on 3D sensor data implies the need to install extra hardware in most of the cases, as commonly installed surveillance cameras are usually 2D-RGB cameras. To overcome the need for new hardware, end-to-end deep learning methods utilized for 3D ground plane detection from 2D-RGB images have already been investigated. While these methods provide the 3D ground plane detection obtained from images without the requirement to install new hardware modules (e.g. 3D sensors), being end-to-end deep learning models implies that they lack explainability. Moreover, in order to adapt new scenes to their algorithms, it is required to retrain their model on such scenes.

In contrast to end-to-end trained models, our algorithm relies on a simple, modular and flexible pipeline. Depth information is estimated by a pre-trained deep learning model serving as the basis for estimating the transform of the 2D image to a 3D Cartesian space, represented as 3D point cloud data. Simple 2D-bounding points are used as input to span up vertical limits to the region of interest in the converted 3D Cartesian space. Finally, the 3D point cloud data is used to estimated the 3D ground plane equation with respect to the camera. Results are visualized on 2D-RGB images and 3D point cloud data by estimating the ground plane points from the 3D ground plane equation.

The contributions of this work are as follows:

- The algorithm relies on 2D-RGB images to overcome the need for applying 3D sensors by utilizing depth-estimation deep learning models to convert the 2D-pixel space to a 3D-Cartesian space.
- Our algorithm consists of an easy-to-interfere pipeline. While providing an end-to-end algorithm, the pipeline is modular and each module can be customized or replaced by other algorithms.
- It is straightforward to adapt the algorithm to new scenes by changing the depth estimation model (e.g. retraining the model or using another model) and without the need of retraining the whole plane detection model.

2 Related Work

2.1 Depth Estimation

Depth Estimation is a method used to estimate the distance of each pixel of a given RGB image to the sensing camera. Shah et al. estimate depth information by creating a disparity map referring to the apparent pixel difference or motion between a pair of stereo images [SA94]. However, with the recent popularity of deep learning on computer vision problems, there are new deep learning-based architectures used to estimate depth information from a single RGB image. [MGLT20] proposed a depth estimation algorithm for combining the synthetic data with the real data by estimating defocus map and all-in focus map with autoencoders. [BAW21] proposed an encoder-decoder structure: EfficientNet-based encoder and standard feature upsampling decoder. Moreover, the depth scale is discretized into histogram bins and the width and the center of the adaptive bins are evaluated with supervision. [LLD+21] utilized stereo images for estimating disparity maps by a shared feature extractor followed by a transformers with attention for depth estimation from single images. [RBK21] proposed a depth estimation from single images by splitting the input into patches, embedding each patch, feeding the patches into the transformers and fusioning of the patches.

2.2 Plane Detection

Plane Detection is defined as the task of estimating the position of a specific plane such as the ground plane in a 2D pixel-space or in a Cartesian-space for 2D sensors and 3D sensors, respectively. [CD10] estimates the 2D plane by homography while [CMP09] uses descriptors for 2D plane detection on a monocular RGB image. Both proposals use traditional image processing methods. [BKC17, KWHG20] proposed a deep learning-based solution for a 2D ground plane detection. [LYZ+17, XJS+16, VCMP17, ZEF16] propose 3D plane detection from 3D point cloud data while [HHRB11] is a 3D ground plane detection approach from a RGB-D data. Moreover, there are RGB image-based 3D plane detection methods such as [LKG+19, LYC+18] which utilize deep learning architectures for improved accuracy. Recently, Li et al. [LHL+21] proposed an algorithm to segment planar regions on 2D-RGB images of indoor scenes by utilizing self-supervised depth estimation deep learning models. The authors propose depth estimation from RGB images fed into a Manhattan normal detection and a planar region detection module, subsequently. The main focus of this approach is set to the improvement of a pre-trained depth estimation model by increasing the overall accuracy through utilizing self-supervised learning. Similar to the method of Li et al., we use a depth estimation model to create a depth map of the scene. In contrast to their approach, we then generate a 3D point cloud estimate followed by ground plane point filtering and RANSAC-based plane fitting.

3 Extraction of 3D Ground Plane Equation from RGB Images

Our approach solves the estimation of the 3D ground plane equation from a given 2D-RGB image. Then, by the given inputs and estimated depth map, the scene reconstructed in 3D. In parallel to the reconstruction of the scene, our algorithm filters the 3D point cloud by the given bounding points to enhance the RANSAC plane fitting algorithm's success [FB81].

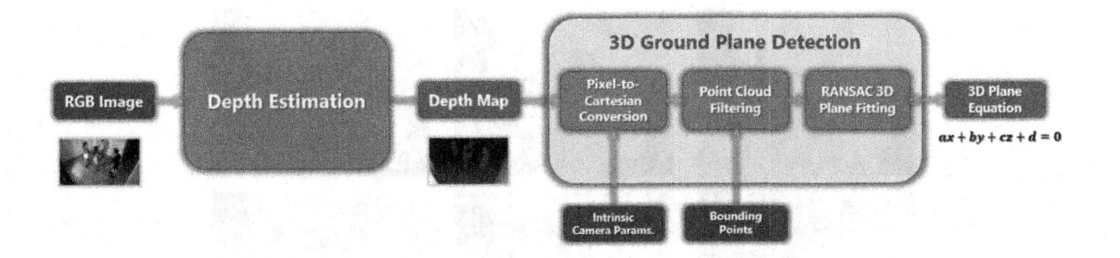

Fig. 1. 3D ground plane detection pipeline.

The detailed flow of our pipeline is demonstrated in Fig. 1 and will be explained in more detail in the next sections.

3.1 Depth Estimation

As the first step, we generate the depth map of a given 2D-RGB image with the help of a pre-trained deep learning model. Our algorithm's modularity provides the flexibility to use different deep learning models in this step. In our work, we utilized the pre-trained model of Vision Transformers for Dense Prediction [RBK21] which is a transformer-based depth estimation algorithm. The estimated depth map will be converted into 3D in the Pixel-to-Cartesian module.

3.2 3D Ground Plane Detection

Bounding Point Selection. The aim of bounding point selection is to create a 3D bounding box to filter out the dominance of noisy points for the subsequent application of a RANSAC plane fitting algorithm (e.g. non-floor points). Thus, the motivation is to force RANSAC to find a plane containing most of the selected points as ground plane but not walls for example. Bounding points vertically limit the RoI, hence, it should be selected to cover a maximum of ground plane points while including a minimum of non-ground plane points as described in Fig. 2. Thus, the ground-to-non-ground plane point ratio must be maximized. Subsequently, these two points will be converted into 3D coordinates in the Pixel-to-Cartesian Conversion module. A python-based framework was implemented allowing the user to select a number of two bounding points.

Pixel-to-Cartesian Conversion. The selected bounding points and the estimated depth map are converted into 3D Cartesian coordinates in this module, with the help of intrinsic camera parameters. This module introduces real distance metrics into our pipeline. For this purpose, we assume the intrinsic camera parameters are known. With intrinsic camera parameters $C_i = \begin{matrix} f_x & 0 & c_x \\ 0 & f_y & c_y \\ 0 & 0 & 1 \end{matrix}$ where (f_x, f_y) is focal length and (c_x, c_y) is a principle point in (x, y), a 2D point (x', y') in the 2D image with the depth value d is converted into 3D as described in the equation below.

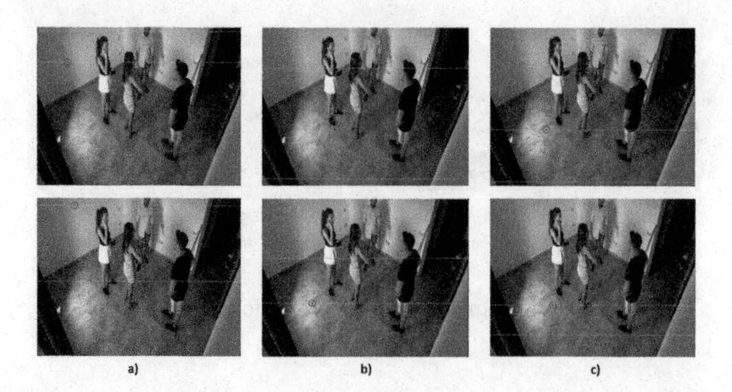

Fig. 2. Selection of bounding points. a) Invalid points: both of the bounding points should be selected on the plane. b) Moderate points: bounding points are valid yet do cover non-plane pixels in the spanned-up area. c) Valid points: bounding points are on the plane and the plane pixels are dominant in the spanned-up area.

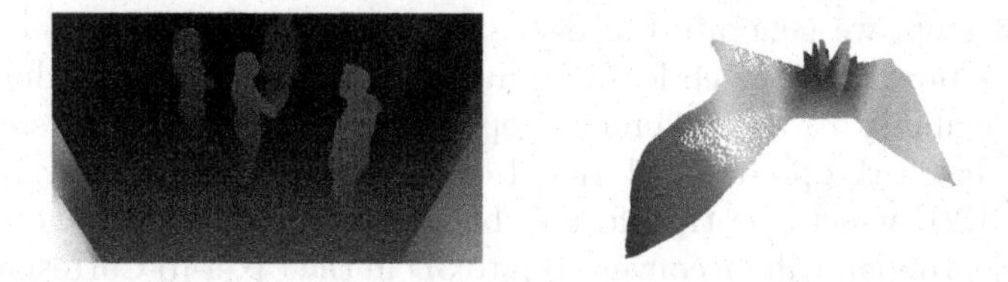

Fig. 3. Conversion of pixel coordinate to 3D Cartesian coordinate; left: estimated depth map, white-to-black color code: close-to-far depth distance from the camera; right: 3D reconstructed scene, red-to-blue color code: close-to-far distance from the camera. (Color figure online)

$x' = (x - c_x) * d/f_x$, $y' = (y - c_y) * d/f_y$, $z' = d$; where (x', y', z') represents the width, height and depth distance of the respective point to the camera (Fig. 3).

Point Cloud Filtering. The estimated 3D point cloud and the 3D bounding box are fed into the module. The 3D bounding box is used as a mask to filter the 3D point cloud by applying a binary multiplication of the point cloud and the bounding box. As the aim of the bounding box definition is to mainly include informative points (points belonging to the ground plane) while excluding non-informative points (points of other objects such as walls or furniture), the point cloud resulting from the mask operation should mainly consist of points of the ground plane. In Fig. 4, a 3D point cloud, a 3D bounding box and filtered 3D point cloud is illustrated. The filtered point cloud is subsequently fed into a RANSAC plane fitting algorithm in order to obtain the ground plane equation accurately.

RANSAC 3D Plane Fitting. RANSAC is an iterative method to find the hyperplane which contains as many points as possible while leaving the other points as outliers [FB81]. With the parameters n_p, n_i, d_t that represent the num-

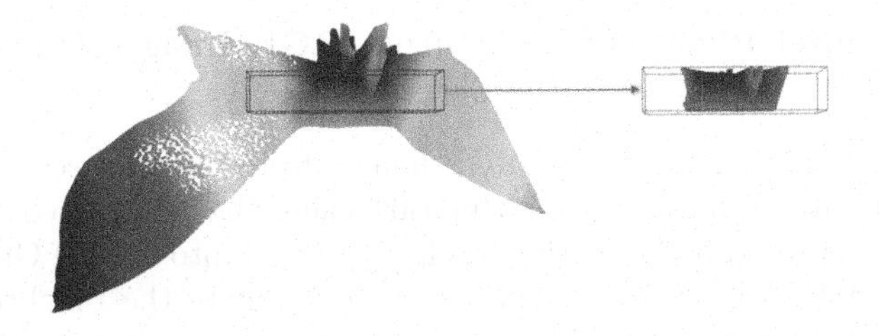

Fig. 4. Left: an example of a 3D bounding box on 3D point cloud, red-to-blue color code for close-to-far distance from the camera; right: 3D bounding box on filtered 3D point cloud. (Color figure online)

ber of points, the number of iterations and a distance threshold, the algorithm selects a random number n_p of points and generates a plane which contains those points and evaluates any point's distance to the plane. If the distance of a point is smaller than d_t, the point is assumed as an inlier. The number of inliers are kept as a score. This process iterates n_t times and the plane with the highest score (e.g. highest number of inlier points) is accepted as the best plane. RANSAC 3D plane fitting provides the equation of the plane which has the highest score as output. In Fig. 5, the single 2D-RGB input and its corresponding output is visualized.

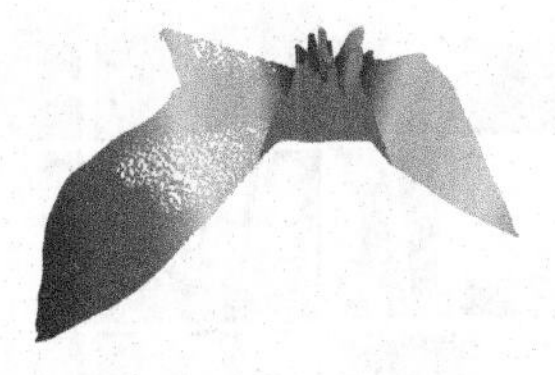

Fig. 5. Left: input image, right: visualization of the detected 3D ground plane on 3D reconstructed scene, red-to-blue color code: close-to-far distance from the camera and purple: detected 3D ground plane points. (Color figure online)

4 Experiments and Results

To evaluate our algorithm, we designed experiments based on the publicly available ScanNet Dataset [DCS+17]. The ScanNet dataset contains 1500 scenes which cover 2.5 million 3D-RGB-D images, including 3D camera poses, surface reconstructions and instance-level semantic segmentation annotations. Moreover, we exploited a transformer-based deep learning proposal for depth estimation [RBK21] with the publicly provided pre-trained model (DPT model).

4.1 3D Ground Plane Detection from RGB Images (Entire Workflow)

In this experiment, we used five 2D-RGB images from the same scene as input to our algorithm, and applied depth estimation using the pre-trained DPT model to obtain the depth map. Then, the depth map is fed into the 3D Ground Plane Detection module. We used the depth maps as input to the Pixel-to-Cartesian conversion module by using intrinsic camera parameters and then converted the 2D-RGB image and the depth map into a 3D point cloud. Then, bounding points were selected on a provided GUI. Lastly, we evaluated the 3D equation of the ground plane by taking advantage of the RANSAC 3D Plane Fitting algorithm. We used the following parameters for RANSAC 3D Plane Fitting: n_i : 1000, d_t : 0.3, n_p : 3 In order to show our qualitative results, we estimated 2D ground plane points on 2D-RGB images and 3D-ground plane points on 3D point clouds. The results are visualized in Fig. 6.

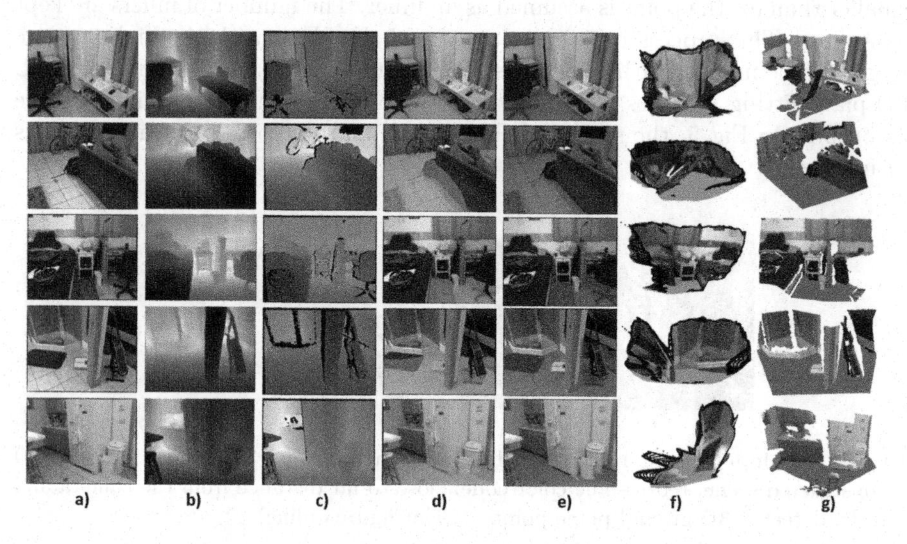

Fig. 6. Qualitative results on RGB images. a) Input RGB image. b) Predicted depth map of input image. c) Groundtruth depth map. d) Predicted 2D ground points (green) on input image. e) Groundtruth 2D ground points (red) on input image. f) Predicted 3D ground points (green) on predicted 3D point cloud, g) Groundtruth 3D ground points (red) on groundtruth 3D point cloud. (Color figure online)

It can be observed from Fig. 6d and 6f where the ground points are visualized on the corresponding 2D-RGB image and the reconstructed scene, that the ground plane detection is highly accurate. However, 3D reconstructions of the scenes, visualized in Fig. 6f are distorted and not accurate. The main reason of distorted 3D reconstructed scenes is not achieving enough accuracy on the depth estimation.

4.2 3D Ground Plane Detection from Groundtruth Depth Maps (Intermediate Workflow)

As stated in the last section, the 3D reconstruction is highly dependent on an accurate depth estimation. The architecture used for depth estimation was trained on an outdoor dataset and thus, does not deliver accurate results. To prove our method on precise depth estimation images, we used the groundtruth depth maps provided by the ScanNet dataset. In more detail, we ignored the Depth Estimation module and instead of estimating depth maps, we used groundtruth depth maps provided by the ScanNet Dataset. We used the same RGB images as in the previous experiment with their corresponding depth maps. Other parameters were kept the same as described in Sect. 4.1.

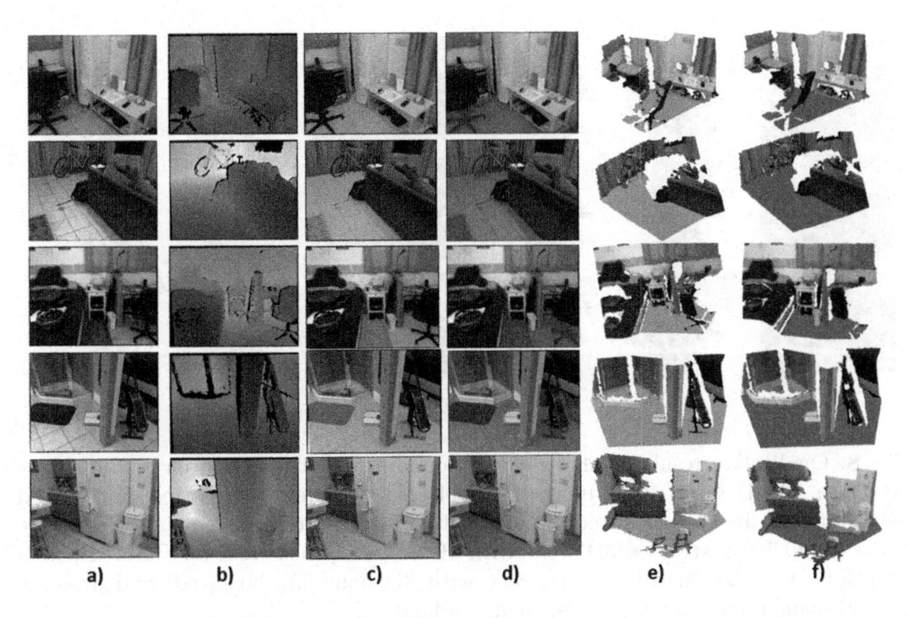

Fig. 7. Qualitative result on depth maps. a) Input RGB image. b) Groundtruth depth map. c) Predicted 2D ground points (green) on input image. d) Groundtruth 2D ground points (red) on input image. e) Predicted 3D ground points (green) on groundtruth 3D point cloud. f) Groundtruth 3D ground points (red) on groundtruth 3D point cloud. (Color figure online)

We can observe from Fig. 7e that the 3D reconstruction is highly accurate based on the provided groundtruth depth map. Thus, We can expect that our algorithm succeeds if the underlying depth estimation is accurate.

4.3 Influence of Bounding Point Selection on Ground Plane Detection

We conducted a test to examine the influence of selecting different bounding points on the ground plane detection accuracy by using one RGB image and by selecting six different bounding points in two different test scenarios (investigating the effect of selecting multiple, different bounding points on the ground plane detection as illustrated in Fig. 8a to c, and investigating the effect of including invalid points on the ground plane detection as illustrated in Fig. 8d to f).

Selection of different valid bounding points does not dramatically effect the success of ground plane detection. As long as the selected two points are in the same plane (ground), the filtered points will mostly contain the planar region without extra noise as it can be seen in Fig. 8a to c. If bounding points are invalid (the resulting filtered points contain a limited number of ground plane points while the number of non-ground plane points is high), the ground plane detection will fail (Fig. 8d-f).

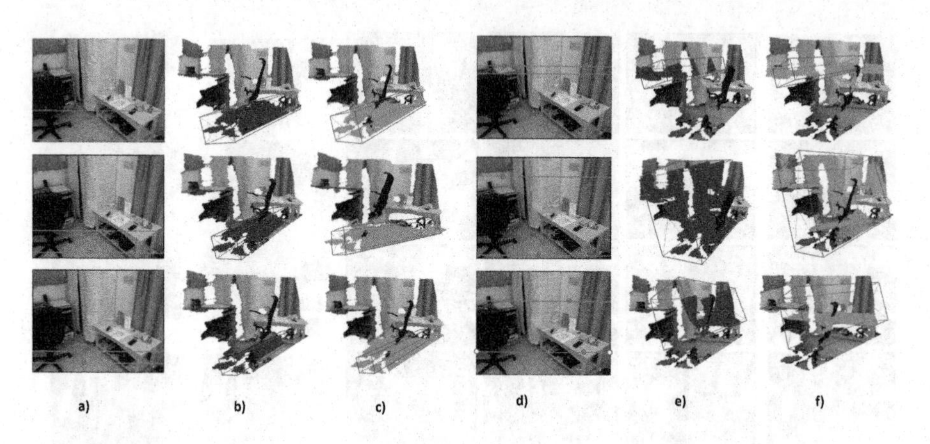

Fig. 8. Qualitative result on the effect of bounding point selection on ground plane detection. a), d) Input RGB Image with selected bounding points (red circles) and perpendicular lines to the y-axis which contain selected bounding points (light-blue lines). b), e) Reconstructed 3D scene with 3D bounding box (red) and filtered points (purple). c), f) Reconstructed 3D scene with 3D bounding box (red) and detected ground plane points (green). (Color figure online)

5 Conclusion and Outlook

In this paper, we propose a simple and flexible pipeline for estimating the 3D-ground plane equation and 2D/3D-ground point segmentation from a single 2D-RGB Image. We are exploiting 3D scene reconstruction from RGB-D images and the RANSAC 3D plane fitting algorithm. We are estimating depth maps from single 2D-RGB images and are thereby overcoming the need to use extra depth sensors. Moreover, we are estimating the 3D ground points and the 3D ground plane equation from the reconstructed 3D scene. Hence, in order to achieve a highly accurate 3D reconstruction of the scene, it is required to have an accurate depth estimation method.

This accurate model for estimating depth information is currently missing. The next steps undertaken to achieve a robust and accurate workflow are as follows: the DPT model will be retrained with the indoor ScanNet RGB-D Dataset, thereby aiming to improve the depth estimation accuracy. After that, the entire pipeline's performance will be tested and evaluated. The main focus will be put on the evaluation of the 3D ground plane equation. Thus, the ScanNet dataset will be processed and the 3D ground plane equations will be obtained from 3D ground plane points from manually labeled 3D point clouds. Moreover, we will use the cosine similarity metric to evaluate the 3D ground plane equation estimation accuracy. Lastly, we will estimate and evaluate an object's distance to the ground plane using the ScanNet Dataset to prove the applicability of the approach.

Acknowledgements. The research reported in this paper has been funded by the Federal Ministry for Climate Action, Environment, Energy, Mobility, Innovation and Technology (BMK), the Federal Ministry for Digital and Economic Affairs (BMDW), and the State of Upper Austria in the frame of SCCH, a center in the COMET - Competence Centers for Excellent Technologies Programme managed by Austrian Research Promotion Agency FFG.

References

[BAW21] Bhat, S.F., Alhashim, I., Wonka, P.: Adabins: depth estimation using adaptive bins. In: Proceedings of the IEEE/CVF Conference on Computer Vision and Pattern Recognition, pp. 4009–4018 (2021)

[BKC17] Badrinarayanan, V., Kendall, A., Cipolla, R.: Segnet: a deep convolutional encoder-decoder architecture for image segmentation. IEEE Trans. Pattern Anal. Mach. Intell. **39**(12), 2481–2495 (2017)

[CD10] Conrad, D., DeSouza, G.N.: Homography-based ground plane detection for mobile robot navigation using a modified EM algorithm. In: 2010 IEEE International Conference on Robotics and Automation, pp. 910–915. IEEE (2010)

[CMP09] Cherian, A., Morellas, V., Papanikolopoulos, N.: Accurate 3D ground plane estimation from a single image. In: 2009 IEEE International Conference on Robotics and Automation, pp. 2243–2249. IEEE (2009)

[DCS+17] Dai, A., Chang, A.X., Savva, M., Halber, M., Funkhouser, M., Nießner, M.: Scannet: richly-annotated 3D reconstructions of indoor scenes. In: Proceedings of the IEEE Conference on Computer Vision and Pattern Recognition, pp. 5828–5839 (2017)

[FB81] Fischler, M.A., Bolles, R.C.: Random sample consensus: a paradigm for model fitting with applications to image analysis and automated cartography. Commun. ACM **24**(6), 381–395 (1981)

[HHRB11] Holz, D., Holzer, S., Rusu, R.B., Behnke, S.: Real-time plane segmentation using RGB-D cameras. In: Röfer, T., Mayer, N.M., Savage, J., Saranlı, U. (eds.) RoboCup 2011. LNCS (LNAI), vol. 7416, pp. 306–317. Springer, Heidelberg (2012). https://doi.org/10.1007/978-3-642-32060-6_26

[KWHG20] Kirillov, A., Wu, Y., He, K., Girshick, R.: Pointrend: image segmentation as rendering. In: Proceedings of the IEEE/CVF Conference on Computer Vision and Pattern Recognition, pp. 9799–9808 (2020)

[LHL+21] Li, B., Huang, Y., Liu, Z., Zou, D., Yu., W.: Structdepth: leveraging the structural regularities for self-supervised indoor depth estimation. In: Proceedings of the IEEE/CVF International Conference on Computer Vision, pp. 12663–12673 (2021)

[LKG+19] Liu, C., Kim, K., Gu, J., Furukawa, Y., Kautz, J:. Planercnn: 3D plane detection and reconstruction from a single image. In: Proceedings of the IEEE/CVF Conference on Computer Vision and Pattern Recognition, pp. 4450–4459 (2019)

[LLD+21] Li, Z., et al.: Revisiting stereo depth estimation from a sequence-to-sequence perspective with transformers. In: Proceedings of the IEEE/CVF International Conference on Computer Vision, pp. 6197–6206 (2021)

[LYC+18] Liu, C., Yang, C., Ceylan, D., Yumer, E., Furukawa, Y.: Planenet: piece-wise planar reconstruction from a single rgb image. In: Proceedings of the IEEE Conference on Computer Vision and Pattern Recognition, pp. 2579–2588 (2018)

[LYZ+17] Li, L., Yang, F., Zhu, H., Li, D., Li, Y., Tang, L.: An improved RANSAC for 3D point cloud plane segmentation based on normal distribution transformation cells. Remote Sens. **9**(5), 433 (2017)

[MGLT20] Maximov, M., Galim, K., Leal-Taixé, L.: Focus on defocus: bridging the synthetic to real domain gap for depth estimation. In: Proceedings of the IEEE/CVF Conference on Computer Vision and Pattern Recognition, pp. 1071–1080 (2020)

[RBK21] Ranftl, R., Bochkovskiy, A., Koltun, V.: Vision transformers for dense prediction. In: Proceedings of the IEEE/CVF International Conference on Computer Vision, pp. 12179–12188 (2021)

[SA94] Shah, S., Aggarwal, J.K.: Depth estimation using stereo fish-eye lenses. In: Proceedings of 1st International Conference on Image Processing, vol. 2, pp. 740–744. IEEE (1994)

[VCMP17] Van Crombrugge, I., Mertens, L., Penne, R.: Fast free floor detection for range cameras. In: International Conference on Computer Vision Theory and Applications, vol. 5, pp. 509–516. SCITEPRESS (2017)

[XJS+16] Bo, X., Jiang, W., Shan, J., Zhang, J., Li, L.: Investigation on the weighted ransac approaches for building roof plane segmentation from lidar point clouds. Remote Sens. **8**(1), 5 (2016)

[ZEF16] Zeineldin, R.A., El-Fishawy, N.A.: Fast and accurate ground plane detection for the visually impaired from 3D organized point clouds. In: 2016 SAI Computing Conference (SAI), pp. 373–379. IEEE (2016)

Towards Practical Secure Privacy-Preserving Machine (Deep) Learning with Distributed Data

Mohit Kumar[1,2](✉), Bernhard Moser[1], Lukas Fischer[1], and Bernhard Freudenthaler[1]

[1] Software Competence Center Hagenberg GmbH, 4232 Hagenberg, Austria
mohit.kumar@scch.at

[2] Institute of Automation, Faculty of Computer Science and Electrical Engineering, University of Rostock, Rostock, Germany

Abstract. A methodology for practical secure privacy-preserving distributed machine (deep) learning is proposed via addressing the core issues of fully homomorphic encryption, differential privacy, and scalable fast machine learning. Considering that private data is distributed and the training data may contain directly or indirectly an information about private data, an architecture and a methodology are suggested for

1. mitigating the impracticality issue of fully homomorphic encryption (arising from large computational overhead) via very fast gate-by-gate bootstrapping and introducing a learning scheme that requires homomorphic computation of only efficient-to-evaluate functions;
2. addressing the privacy-accuracy tradeoff issue of differential privacy via optimizing the noise adding mechanism;
3. defining an information theoretic measure of privacy-leakage for the design and analysis of privacy-preserving schemes; and
4. addressing the optimal model size determination issue and computationally fast training issue of scalable and fast machine (deep) learning with an alternative approach based on variational learning.

A biomedical application example is provided to demonstrate the application potential of the proposed methodology.

Keywords: Privacy · Homomorphic encryption · Machine learning · Differential privacy · Membership-mappings

---

Supported by the Austrian Research Promotion Agency (FFG) Project PRIMAL (Privacy Preserving Machine Learning for Industrial Applications); FFG Project SMiLe (Secure Machine Learning Applications with Homomorphically Encrypted Data); FFG COMET-Modul S3AI (Security and Safety for Shared Artificial Intelligence); FFG Sub-Project PETAI (Privacy Secured Explainable and Transferable AI for Healthcare Systems); EU Horizon 2020 Project SERUMS (Securing Medical Data in Smart Patient-Centric Healthcare Systems); the BMK; the BMDW; and the Province of Upper Austria in the frame of the COMET Programme managed by FFG.

G. Kotsis et al. (Eds.): DEXA 2022 Workshops, CCIS 1633, pp. 55–66, 2022.
https://doi.org/10.1007/978-3-031-14343-4_6

1 Introduction

The emergence of cloud infrastructure not only raises the concern of protecting data in storage, but also requires an ability of performing computations on data while preserving the data privacy. There has been a recent surge in the interest on advanced privacy-preserving methods such as fully homomorphic encryption and differential privacy, however, their practical applications are still limited because of several issues. The aim of this study is to develop a methodology for practical secure privacy-preserving distributed machine (deep) learning via addressing some of the fundamental issues related to fully homomorphic encryption, differential privacy, and scalable machine (deep) learning.

1.1 The State-of-Art

Fully Homomorphic Encryption: Fully homomorphic encryption (FHE) is a solution to the privacy concerns in the cloud computing scenario. The first FHE scheme [23] is based on ideal lattices. The bootstrapping operation is performed on a ciphertext to reduce the noise contained in a ciphertext which is the computationally most expensive part of a homomorphic encryption scheme. The breakthrough of [23] was followed by several attempts to develop more practical FHE schemes. The scheme of [13] uses only elementary modulo arithmetic and is homomorphic with regard to both addition and multiplication. This scheme was improved in [12] with reduced public key size, extended in [8] to support encrypting and homomorphically processing a vector of plaintexts as a single ciphertext, and generalized to non-binary messages in [41]. Schemes based on Learning With Errors (LWE) problem [44] were also constructed. In a variant of the LWE problem, called ring learning with errors problem (RLWE) problem, the algebraic structure of the underlying hard problem reduces the key sizes and speeds up the homomorphic operations. A leveled fully homomorphic encryption scheme based on LWE or RLWE, without bootstrapping procedure, was proposed in [5]. The amount of noise contained in ciphertexts grows with homomorphic operations. For a better management of the noise growth, [5] introduced a modulus switching technique where a complete ladder of moduli is used for scaling down the ciphertext to the next modulus after each multiplication. [4] introduced a tensoring technique for LWE-based FHE that reduced ciphertext noise growth after multiplication from quadratic to linear. The scale-invariant fully homomorphic encryption scheme of [4] does no longer require the rescaling of the ciphertext. An RLWE version of the scale-invariant scheme of [4] was created in [17]. The approximate eigenvector method in which homomorphic addition and multiplication are just matrix addition and multiplication was proposed in [24]. The bootstrapping remains as the bottleneck for an efficient FHE in practice, and therefore several studies aimed at improving the bootstrapping. A much faster bootstrapping, that allows to homomorphically compute simple bit operations and bootstrap the resulting output in less than a second, was devised in [14]. Finally, the TFHE scheme was proposed in [9,10] that features a considerably more efficient bootstrapping procedure than the previous state of the art. The TFHE scheme generalizes structures and schemes over the torus (i.e., the reals modulo 1) and improves the bootstrapping dramatically. TFHE is an open-source C/C++ library [11] implementing the ring-variant of [24] together with the optimizations of [9,10,14]. TFHE library implements a very fast gate-by-gate bootstrapping and supports the homomorphic evaluation of an arbitrary boolean circuit composed of binary gates. However, the bootstrapped bit operations are still several times slower than their plaintext equivalents. Thus for an efficient secure machine learning scenario in practice, homomorphic evaluation of the function with smallest possible number of gates is one of the optimality criteria for designing learning algorithms with distributed data.

Differential Privacy: The datasets may contain sensitive information that need to be protected from *model inversion* attack [18] and from adversaries with an access to model parameters and knowledge of the training procedure. This goal has been addressed within the framework of differential privacy [1,42]. Differential Privacy [15,16] is a formalism to quantify the degree to which the privacy for each individual in the dataset is preserved while releasing the output of a data analysis algorithm. Differential privacy provides a guarantee that an adversary, by virtue of presence or absence of an individual's data in the dataset, would not be able to draw any conclusions about an individual from the released output of the analysis algorithm. This guarantee is achieved by means of a randomization of the data analysis process. In the context of machine learning, randomization is carried out via either adding random noise to the input or output of the machine learning algorithm or modifying the machine learning algorithm itself. Differential privacy preserves the privacy of the training dataset via adding random noise to ensure that an adversary can not infer any single data instance by observing model parameters or model outputs. However, the injection of noise into data would in general result in a loss of algorithm's accuracy. Therefore, design of a noise injection mechanism achieving a good trade-off between privacy and accuracy is a topic of interest [2,19–22,25,26,30,31,37]. The authors in [30] derive the probability density function of noise that minimizes the expected noise magnitude together with satisfying the sufficient conditions for $(\epsilon, \delta)-$differential privacy. This noise adding mechanism was applied for differentially private distributed deep learning in [31,37]. Differential privacy, however, doesn't always adequately limit inference about participation of a single record in the database [28]. Differential privacy requirement does not necessarily constrain the information leakage from a data set [6]. Correlation among records of a dataset would degrade the expected privacy guarantees of differential privacy mechanism [40]. These limitations of differential privacy motivate an information theoretic approach to privacy where privacy is quantified by the mutual information between sensitive information and the released data [3,6,43,45,47]. Information theoretic privacy can be optimized theoretically using a prior knowledge about data statistics. However, in practice, a prior knowledge (such as joint distributions of public and private variables) is missing and therefore a data-driven approach based on generative adversarial networks has been suggested [27]. The data-driven approach of [27] leverages generative adversarial networks to allow learning the parameters of the privatization mechanism. However, the framework of [27] is limited to only binary type of sensitive variables. A similar approach [46] applicable to arbitrary distributions (discrete, continuous, and/or multivariate) of variables employs adversarial training to perform a variational approximation of mutual information privacy. The approach of approximating mutual information via a variational lower bound was also used in [7]. The information theoretic approach of quantifying privacy-leakage in-terms of mutual information between sensitive data and released data was also considered in [33].

Scalable and Fast Machine (Deep) Learning: Deep neural networks outperform classical machine learning techniques in a wide range of applications but their training requires a large amount of data. The issues, such as determining the optimal model structure, requirement of large training dataset, and iterative time-consuming nature of numerical learning algorithms, are inherent to the neural networks based parametric deep models. The nonparametric approach on the other hand can be promising to address the issue of optimal choice of model structure. However, an analytical solution instead of iterative gradient-based numerical algorithms will be still desired for the learning of deep models. These motivations have led to the development of a nonparametric deep model [29,38,39] that is learned analytically for representing data points. The study in [29,38,39] introduces the concept of fuzzy-mapping which is about representing mappings through a fuzzy set such that the dimension of membership function increases with an increasing data size. A relevant result is that a deep autoencoder model formed via a composition of finite number of nonparametric fuzzy-mappings can be learned analytically via variational optimization technique. However, [29,38,39] didn't provide a formal mathematical framework for the conceptualization of so-called fuzzy-mapping. The study in [35] provides to fuzzy-mapping a measure-theoretic conceptualization and refers it to as *membership-mapping*. Further, an alternative idea of deep autoencoder, referred to as Bregman Divergence Based Conditionally Deep Autoencoder (that consists of layers such that each layer learns data representation at certain abstraction level through a membership-mappings based autoencoder) was introduced in [34] for data representation learning. Motivated by fuzzy theory, the notion of membership-mapping has been introduced [35] for representing data points through attribute values. Further, the membership-mapping could serve as the building block of deep models [34]. The motivation behind membership-mappings based deep learning is derived from the facts that an analytical learning solution could be derived for membership-mappings via variational optimization methodology and thus the typical issues associated to parametric deep models (such as determining the optimal model structure, smaller training dataset, and iterative time-consuming nature of numerical learning algorithms) will be automatically addressed. Although [35] provided an algorithm for the variational learning of membership-mappings via following the approach of [29,38,39], a more simple and elegant learning algorithm was provided in [32]. These developments (i.e. [29,34,35,38,39]) lead to an alternative machine (deep) learning approach addressing both the optimal model size determination issue and computationally fast training issue of scalable deep learning.

1.2 Requirements for a Practical Secure Privacy-Preserving Machine Learning

The following requirements are identified with respect to the state-of-art for a practical secure privacy-preserving machine learning:

Requirement 1: An efficient secure machine (deep) learning with fully homomorphic encryption in practice demands an approach ensuring that the homomorphic evaluation of functions would require evaluating the circuits with number of gates as small as possible.

Requirement 2: The optimal differentially private machine learning requires optimizing the privacy-accuracy tradeoff. Further, an information theoretic privacy-leakage measure, enabling the quantification of privacy-leakage in-terms of mutual information between sensitive private data and the released public data without the availability of a prior knowledge about data statistics (such as joint distributions of public and private variables), is required to be computed in practice for design and analysis of privacy-preserving machine (deep) learning algorithms.

Requirement 3: The scalable and fast machine (deep) learning demands addressing at least the following two issues: a) how to automatically determine the model size matching the complexity of the problem?, and b) how to develop computationally efficient algorithms for the training of machine (deep) models instead of relying on slow gradient-based learning algorithms?

2 Proposed Methodology

Requirements 1–3, identified for a practical secure privacy-preserving learning, can be fulfilled with an architecture as suggested in Fig. 1. Since our target are federated data sets, we make the assumptions that the complete data set on which we apply machine-learning based big data analytics, is distributed across different organizational units and that learning models are constructed and applied within each unit independently, after which these local models are combined into a global model. Therefore, the data analysis process starts, within each organizational unit, with local private data on which the learning models are to be trained and input data, which is the data of interest to be analyzed using the derived models. The privacy issue that we are addressing is concerning the protection of private/sensitive/confidential information, referred to as "local private data" in Fig. 1. The "local training data", as shown in Fig. 1, may contain information about the private data. The suggested methodology is as follows.

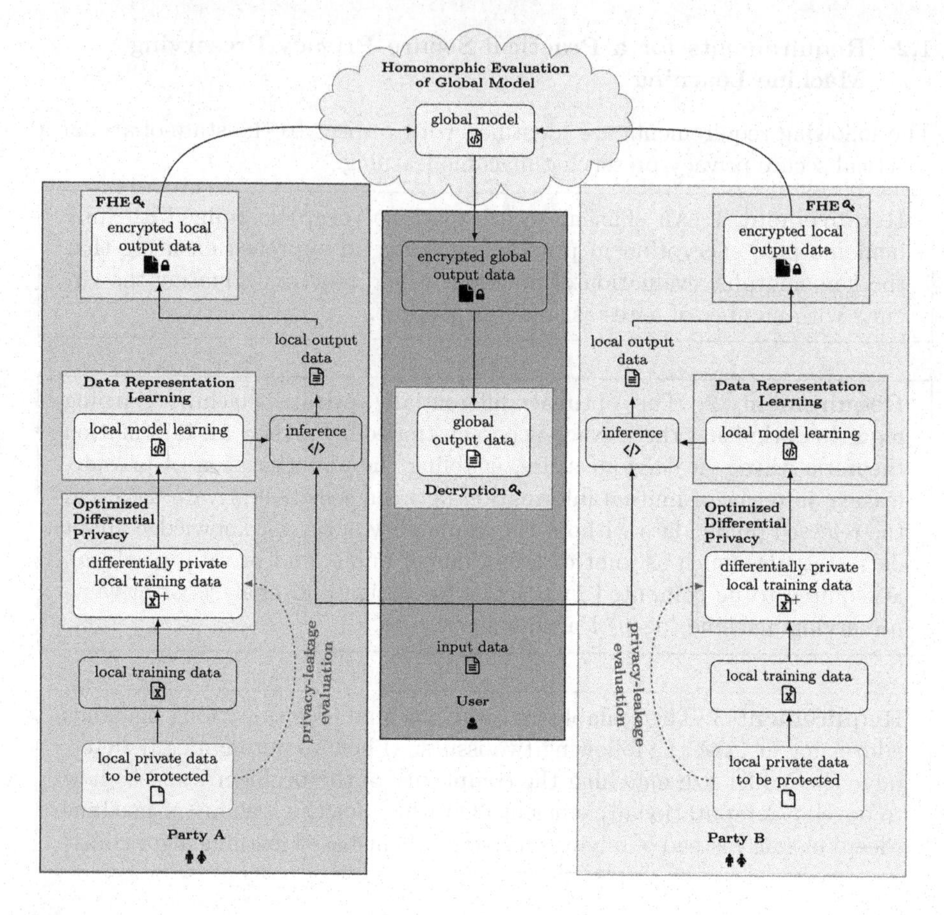

Fig. 1. An architecture for practical secure privacy-preserving distributed machine (deep) learning under the scenario that data is distributed (amongst different parties) and each party's training data (used to build models) may contain directly or indirectly an information about private data of the party.

1. From local private data, local training data is extracted, onto which we apply optimized differential privacy. The goal of this stage is to make sure that the models produced by the analytics on the local training data do not accidentally reveal sensitive private information from this data. The privacy-accuracy tradeoff issue is addressed via optimizing the noise adding mechanism [30,31]. This results in the differentially private local training data.
2. On the differentially private local training data we then perform local learning, the outcome of which is a local model for inference. It is worth noting that this model is built and trained using unencrypted data (that is, just the local training data with the appropriate level of noise added), therefore there is no computational overhead that comes from running learning algorithms on FHE encrypted data. An alternative machine (deep) learn-

ing approach [29,34,35,38,39] based on variational learning, addressing both the optimal model size determination issue and computationally fast training issue, is followed for the learning of local models using local training data. Determining the optimal model size is not only relevant for scalability but it is also related to success rates of inference attacks, as there is a correlation between overfitted and too complex models and the information leakage. Thus, finding a "minimum sized model" is also relevant for improving privacy.
3. The local model is applied to the input data, which is the data that we are actually interested in analyzing and from which we want to make inference. The result of this is the local output data. It is worth noting again that "local" in this context means "within the same organisational unit". At this point, we do not make assumptions on how the training and output data is stored within the organization (e.g. whether it is within one single storage node or across different nodes). Since the input data is processed by all of the local models for making inferences, it is assumed that confidentiality of input data is not a concern here. If the confidentiality of input data is an additional requirement, then input data could be encrypted and the local model inference function could be homomorphically evaluated, however, at the cost of additional computational cost.
4. Another step is combining local output data from different organizational units into a global output. Here the data needs to be transmitted between organizations and, therefore, encryption is needed. Each local model output is homomorphically encrypted and shared in the cloud where a global model (that combines the distributed local models) is homomorphically evaluated in an efficient manner to predict the output. The issue regarding the impracticality of homomorphic computation of global model output due to large computational overhead is mitigated via two ways:
   (a) Very fast gate-by-gate bootstrapping is implemented [11].
   (b) Combining local entities' models requires homomorphic computation of only simpler functions (such as computing minimum or maximum amongst scalars) that can be homomorphically evaluated in an efficient manner. For example, a rule-based fuzzy model can be used for combining the local models [31,37].
5. An information theoretic measure, evaluating privacy-leakage in-terms of mutual information between private data and differentially private training data, is defined [33,36] for the design and analysis of privacy-preserving schemes.

Hence, the methodology will protect the private data of each organizational unit where fully homomorphic encryption will solve the issue of "dishonest" aggregator and a suitably designed differentially private mechanism will mitigate the privacy-leakage through local models' outputs and thus through the global output data.

3 A Biomedical Application Example

The proposed methodology was implemented using MATLAB R2017b and TFHE C/C++ library [11] on a MacBook Pro machine with a 2.2 GHz Intel Core i7 processor and 16 GB of memory. As an application example, the mental stress detection problem is considered. The dataset from [39], consisting of heart rate interval measurements of different subjects, is considered for the study of individual stress detection problem. The problem is concerned with the detection of stress on an individual based on the analysis of recorded sequence of R-R intervals, $\{RR^i\}_i$. The R-R data vector at $i-$th time-index, y^i, is defined as $y^i = \left[RR^i\, RR^{i-1} \cdots RR^{i-d}\right]^T$. That is, the current interval and history of previous d intervals constitute the data vector. Assuming an average heartbeat of 72 beats per minute, d is chosen as equal to $72 \times 3 = 216$ so that R-R data vector consists of on an average 3-min long R-R intervals sequence. A dataset, say $\{y^i\}_i$, is built via 1) preprocessing the R-R interval sequence $\{RR^i\}_i$ with an impulse rejection filter for artifacts detection, and 2) excluding the R-R data vectors containing artifacts from the dataset. The dataset contains the stress-score on a scale from 0 to 100. A label of either "*no-stress*" or "*under-stress*" is assigned to each y^i based on the stress-score. Thus, we have a binary classification problem.

Private Data: Here we assume that heart rate values are private. As instantaneous heart rate is given as $HR^i = 60/RR^i$, thus an information about private data is directly contained in the R-R data vectors.

Distributed Learning Scenario: A two-party collaborative learning scenario is considered where a randomly chosen subject is considered as Party-A. While keeping Party-A fixed, the distributed learning experiments are performed independently on every other subject being considered as Party-B. For each subject, 50% of the data samples serve as training data while remaining as test data. The subjects, with data containing both the classes and at least 60 samples, were considered for experimentation. There are in total 48 such subjects.

Optimized Differentially Private Local Models: A differentially private approximation of training data samples is provided using the optimized noise adding mechanism [30,31]. Further, an information theoretic measure of the privacy-leakage is computed [33,36] corresponding to differential privacy-loss bound values as $\epsilon = 0.1$, $\epsilon = 1$, and $\epsilon = 10$. The noise added training samples are used for the variational learning of deep membership-mapping autoencoders based classifiers [32,34]. The models, that correspond to minimum privacy-leakage amongst all the models obtained corresponding to different choices of differential privacy-loss bound ϵ, serve as local classifiers for each party.

Homomorphic Evaluation of Global Classifier: The outputs of local classifiers are sent to the cloud for performing secure homomorphic computations. The global classifier uses a fuzzy rule-based model [31,37] for combining the outputs of local classifiers. The output of the global classifier is homomorphically evaluated in the cloud from the encrypted data (sent by two parties) using a boolean circuit composed of bootstrapped binary gates. The homomorphic evaluation of the global model is done with the precision of 16-bits and also 8-bits.

Table 1. A few results obtained by the proposed secure privacy-preserving distributed deep learning method in stress detection experiments on a dataset consisting of heart rate interval measurements.

Precision of homomorphic computations	Computational time for homomorphic evaluation of global model	Accuracy obtained by proposed method	Accuracy obtained by non-private baseline model without collaborative learning
8-bits	3.1571 ± 0.0282 s	0.8325 ± 0.1054	0.7407 ± 0.1014
16-bits	4.8610 ± 0.0639 s	0.9591 ± 0.0302	0.7407 ± 0.1014

The proposed method is evaluated in-terms of test data classification accuracy and computational time required for secure homomorphic computations in the cloud for computing the encrypted global output for a given input. For a reference, the non-private local classifier of Party-A serves as the baseline model to evaluate the gain in accuracy as a result of collaborative learning. Table 1 reports the mean and standard deviation values obtained in distributed learning experiments on all of 48 subjects. It is observed that the averaged classification accuracy increased from 0.7407 to 0.9591 as a result of the collaboration between two parties and the average time required for secure homomorphic evaluation of the global model is 4.8610 s seconds on a MacBook Pro machine with a 2.2 GHz Intel Core i7 processor and 16 GB of memory, which may be acceptable for the practical applications. Thus, the proposed methodology is promising for this practical application.

4 Concluding Remarks

This study has reviewed the state-of-art to identify the various requirements of practical secure privacy-preserving distributed machine (deep) learning. The identified requirements have been fulfilled via several optimizations including computational time optimization, privacy-accuracy tradeoff optimization, information theoretic optimization of privacy-leakage, and optimization of machine (deep) models for data representation learning. The proposed methodology has been validated on a practical biomedical application related to individual stress detection.

References

1. Abadi, M., et al.: Deep learning with differential privacy. In: Proceedings of the 2016 ACM SIGSAC Conference on Computer and Communications Security, pp. 308–318. Association for Computing Machinery, New York (2016)
2. Balle, B., Wang, Y.: Improving the gaussian mechanism for differential privacy: analytical calibration and optimal denoising. CoRR abs/1805.06530 (2018)
3. Basciftci, Y.O., Wang, Y., Ishwar, P.: On privacy-utility tradeoffs for constrained data release mechanisms. In: 2016 Information Theory and Applications Workshop (ITA), pp. 1–6 (2016)
4. Brakerski, Z.: Fully homomorphic encryption without modulus switching from classical GapSVP. In: Safavi-Naini, R., Canetti, R. (eds.) CRYPTO 2012. LNCS, vol. 7417, pp. 868–886. Springer, Heidelberg (2012). https://doi.org/10.1007/978-3-642-32009-5_50
5. Brakerski, Z., Gentry, C., Vaikuntanathan, V.: (leveled) fully homomorphic encryption without bootstrapping. In: Proceedings of the 3rd Innovations in Theoretical Computer Science Conference, ITCS 2012, pp. 309–325. Association for Computing Machinery, New York (2012)
6. Calmon, F.D.P., Fawaz, N.: Privacy against statistical inference. In: Proceedings of the 50th Annual Allerton Conference on Communication, Control, and Computing, Allerton 2012 (2012)
7. Chen, X., Duan, Y., Houthooft, R., Schulman, J., Sutskever, I., Abbeel, P.: Infogan: interpretable representation learning by information maximizing generative adversarial nets. In: Lee, D.D., Sugiyama, M., Luxburg, U.V., Guyon, I., Garnett, R. (eds.) Advances in Neural Information Processing Systems, vol. 29, pp. 2172–2180. Curran Associates, Inc. (2016)
8. Cheon, J.H., Coron, J.-S., Kim, J., Lee, M.S., Lepoint, T., Tibouchi, M., Yun, A.: Batch fully homomorphic encryption over the integers. In: Johansson, T., Nguyen, P.Q. (eds.) EUROCRYPT 2013. LNCS, vol. 7881, pp. 315–335. Springer, Heidelberg (2013). https://doi.org/10.1007/978-3-642-38348-9_20
9. Chillotti, I., Gama, N., Georgieva, M., Izabachène, M.: Faster fully homomorphic encryption: bootstrapping in less than 0.1 seconds. In: Cheon, J.H., Takagi, T. (eds.) ASIACRYPT 2016. LNCS, vol. 10031, pp. 3–33. Springer, Heidelberg (2016). https://doi.org/10.1007/978-3-662-53887-6_1
10. Chillotti, I., Gama, N., Georgieva, M., Izabachène, M.: Faster packed homomorphic operations and efficient circuit bootstrapping for TFHE. In: Takagi, T., Peyrin, T. (eds.) ASIACRYPT 2017. LNCS, vol. 10624, pp. 377–408. Springer, Cham (2017). https://doi.org/10.1007/978-3-319-70694-8_14
11. Chillotti, I., Gama, N., Georgieva, M., Izabachène, M.: TFHE: fast fully homomorphic encryption library (2016). https://tfhe.github.io/tfhe/
12. Coron, J.-S., Mandal, A., Naccache, D., Tibouchi, M.: Fully homomorphic encryption over the integers with shorter public keys. In: Rogaway, P. (ed.) CRYPTO 2011. LNCS, vol. 6841, pp. 487–504. Springer, Heidelberg (2011). https://doi.org/10.1007/978-3-642-22792-9_28
13. van Dijk, M., Gentry, C., Halevi, S., Vaikuntanathan, V.: Fully homomorphic encryption over the integers. In: Gilbert, H. (ed.) EUROCRYPT 2010. LNCS, vol. 6110, pp. 24–43. Springer, Heidelberg (2010). https://doi.org/10.1007/978-3-642-13190-5_2
14. Ducas, L., Micciancio, D.: FHEW: bootstrapping homomorphic encryption in less than a second. In: Oswald, E., Fischlin, M. (eds.) EUROCRYPT 2015. LNCS, vol. 9056, pp. 617–640. Springer, Heidelberg (2015). https://doi.org/10.1007/978-3-662-46800-5_24

15. Dwork, C., Kenthapadi, K., McSherry, F., Mironov, I., Naor, M.: Our data, ourselves: privacy via distributed noise generation. In: Vaudenay, S. (ed.) EUROCRYPT 2006. LNCS, vol. 4004, pp. 486–503. Springer, Heidelberg (2006). https://doi.org/10.1007/11761679_29
16. Dwork, C., Roth, A.: The algorithmic foundations of differential privacy. Found. Trends Theor. Comput. Sci. **9**(3–4), 211–407 (2014)
17. Fan, J., Vercauteren, F.: Somewhat practical fully homomorphic encryption. IACR Cryptol. ePrint Arch. 2012, 144 (2012). http://eprint.iacr.org/2012/144
18. Fredrikson, M., Jha, S., Ristenpart, T.: Model inversion attacks that exploit confidence information and basic countermeasures. In: Proceedings of the 22Nd ACM SIGSAC Conference on Computer and Communications Security, CCS 2015, pp. 1322–1333. ACM, New York (2015)
19. Geng, Q., Kairouz, P., Oh, S., Viswanath, P.: The staircase mechanism in differential privacy. IEEE J. Sel. Topics Signal Process. **9**(7), 1176–1184 (2015)
20. Geng, Q., Viswanath, P.: The optimal noise-adding mechanism in differential privacy. IEEE Trans. Inf. Theory **62**(2), 925–951 (2016)
21. Geng, Q., Viswanath, P.: Optimal noise adding mechanisms for approximate differential privacy. IEEE Trans. Inf.Theory **62**(2), 952–969 (2016)
22. Geng, Q., Ding, W., Guo, R., Kumar, S.: Optimal noise-adding mechanism in additive differential privacy. CoRR abs/1809.10224 (2018)
23. Gentry, C.: Fully homomorphic encryption using ideal lattices. In: STOC 2009, pp. 169–178. Association for Computing Machinery, New York (2009)
24. Gentry, C., Sahai, A., Waters, B.: Homomorphic encryption from learning with errors: conceptually-simpler, asymptotically-faster, attribute-based. In: Canetti, R., Garay, J.A. (eds.) CRYPTO 2013. LNCS, vol. 8042, pp. 75–92. Springer, Heidelberg (2013). https://doi.org/10.1007/978-3-642-40041-4_5
25. Ghosh, A., Roughgarden, T., Sundararajan, M.: Universally utility-maximizing privacy mechanisms. SIAM J. Comput. **41**(6), 1673–1693 (2012)
26. Gupte, M., Sundararajan, M.: Universally optimal privacy mechanisms for minimax agents. In: Proceedings of the Twenty-ninth ACM SIGMOD-SIGACT-SIGART Symposium on Principles of Database Systems, PODS 2010, pp. 135–146. ACM, New York (2010)
27. Huang, C., Kairouz, P., Chen, X., Sankar, L., Rajagopal, R.: Context-aware generative adversarial privacy. Entropy **19**(12), 656 (2017)
28. Kifer, D., Machanavajjhala, A.: No free lunch in data privacy. In: Proceedings of the 2011 ACM SIGMOD International Conference on Management of Data, SIGMOD 2011, pp. 193–204. Association for Computing Machinery, New York (2011)
29. Kumar, M., Freudenthaler, B.: Fuzzy membership functional analysis for nonparametric deep models of image features. IEEE Trans. Fuzzy Syst. **28**(12), 3345–3359 (2020)
30. Kumar, M., Rossbory, M., Moser, B.A., Freudenthaler, B.: Deriving an optimal noise adding mechanism for privacy-preserving machine learning. In: Anderst-Kotsis, G., et al. (eds.) DEXA 2019. CCIS, vol. 1062, pp. 108–118. Springer, Cham (2019). https://doi.org/10.1007/978-3-030-27684-3_15
31. Kumar, M., Rossbory, M., Moser, B.A., Freudenthaler, B.: An optimal $(\epsilon, \delta)-$differentially private learning of distributed deep fuzzy models. Inf. Sci. **546**, 87–120 (2021)
32. Kumar, M.: Differentially private transferrable deep learning with membership-mappings. CoRR abs/2105.04615 (2021). https://arxiv.org/abs/2105.04615v6

33. Kumar, M., Brunner, D., Moser, B.A., Freudenthaler, B.: Variational optimization of informational privacy. In: Kotsis, G., et al. (eds.) DEXA 2020. CCIS, vol. 1285, pp. 32–47. Springer, Cham (2020). https://doi.org/10.1007/978-3-030-59028-4_4
34. Kumar, M., Moser, B., Fischer, L., Freudenthaler, B.: Membership-mappings for data representation learning: a bregman divergence based conditionally deep autoencoder. In: Kotsis, G., et al. (eds.) DEXA 2021. CCIS, vol. 1479, pp. 138–147. Springer, Cham (2021). https://doi.org/10.1007/978-3-030-87101-7_14
35. Kumar, M., Moser, B., Fischer, L., Freudenthaler, B.: Membership-mappings for data representation learning: measure theoretic conceptualization. In: Kotsis, G., et al. (eds.) DEXA 2021. CCIS, vol. 1479, pp. 127–137. Springer, Cham (2021). https://doi.org/10.1007/978-3-030-87101-7_13
36. Kumar, M., Moser, B.A., Fischer, L., Freudenthaler, B.: Information theoretic evaluation of privacy-leakage, interpretability, and transferability for trustworthy AI. CoRR abs/2106.06046 (2021). https://arxiv.org/abs/2106.06046v5
37. Kumar, M., Rossbory, M., Moser, B.A., Freudenthaler, B.: Differentially private learning of distributed deep models. In: Adjunct Publication of the 28th ACM Conference on User Modeling, Adaptation and Personalization, UMAP 2020 Adjunct, pp. 193–200. Association for Computing Machinery, New York (2020)
38. Kumar, M., Singh, S., Freudenthaler, B.: Gaussian fuzzy theoretic analysis for variational learning of nested compositions. Int. J. Approx. Reas. **131**, 1–29 (2021)
39. Kumar, M., Zhang, W., Weippert, M., Freudenthaler, B.: An explainable fuzzy theoretic nonparametric deep model for stress assessment using heartbeat intervals analysis. IEEE Trans. Fuzzy Syst. **29**(12), 3873–3886 (2021)
40. Liu, C., Chakraborty, S., Mittal, P.: Dependence makes you vulnberable: Differential privacy under dependent tuples. In: 23rd Annual Network and Distributed System Security Symposium, NDSS 2016, San Diego, California, USA, 21–24 February 2016. The Internet Society (2016)
41. Nuida, K., Kurosawa, K.: (Batch) fully homomorphic encryption over integers for non-binary message spaces. In: Oswald, E., Fischlin, M. (eds.) EUROCRYPT 2015. LNCS, vol. 9056, pp. 537–555. Springer, Heidelberg (2015). https://doi.org/10.1007/978-3-662-46800-5_21
42. Phan, N., Wang, Y., Wu, X., Dou, D.: Differential privacy preservation for deep auto-encoders: An application of human behavior prediction. In: Proceedings of the Thirtieth AAAI Conference on Artificial Intelligence, AAAI 2016, pp. 1309–1316. AAAI Press (2016)
43. Rebollo-Monedero, D., Forné, J., Domingo-Ferrer, J.: From t-closeness-like privacy to postrandomization via information theory. IEEE Trans. Knowl. Data Eng. **22**(11), 1623–1636 (2010)
44. Regev, O.: On lattices, learning with errors, random linear codes, and cryptography. In: Proceedings of the Thirty-Seventh Annual ACM Symposium on Theory of Computing, STOC 2005, pp. 84–93. Association for Computing Machinery, New York (2005)
45. Sankar, L., Rajagopalan, S.R., Poor, H.V.: Utility-privacy tradeoffs in databases: an information-theoretic approach. IEEE Trans. Inf. Forensics Secur. **8**(6), 838–852 (2013)
46. Tripathy, A., Wang, Y., Ishwar, P.: Privacy-preserving adversarial networks. In: 2019 57th Annual Allerton Conference on Communication, Control, and Computing (Allerton), pp. 495–505 (2019)
47. Wang, Y., Basciftci, Y.O., Ishwar, P.: Privacy-utility tradeoffs under constrained data release mechanisms. CoRR abs/1710.09295 (2017). http://arxiv.org/abs/1710.09295

Applied Research, Technology Transfer and Knowledge Exchange in Software and Data Science

Collaborative Aspects of Solving Rail-Track Multi-sensor Data Fusion

Florian Kromp[1(✉)], Fabian Hinterberger[2], Datta Konanur[2], and Volkmar Wieser[1]

[1] Software Competence Center Hagenberg GmbH, Hagenberg, Austria
florian.kromp@scch.at
[2] TMC, Linz, Austria

Abstract. Multi-sensor data fusion depicts a challenge if the data collected from different remote sensors present with diverging properties such as point density, noise and outliers. To overcome a time-consuming, manual sensor-registration procedure in rail-track data, the TransMVS COMET project was initiated in a joint collaboration between the company Track Machines Connected and the research institute Software Competence Center Hagenberg. One of the project aims was to develop a semi-automated and robust data fusion workflow allowing to combine multi-sensor data and to extract the underlying matrix transformation solving the multi-sensor registration problem. In addition, the buildup and transfer of knowledge with respect to 3D point cloud data analysis and registration was desired. Within a highly interactive approach, a semi-automated workflow fulfilling all requirements could be developed, relying on a close collaboration between the partners. The knowledge gained within the project was transferred in multiple partner meetings, leading to a knowledge leap in 3D point cloud data analysis and registration for both parties.

Keywords: Remote sensing · Point cloud analysis · Multi-sensor data fusion · Interactive collaboration · Knowledge buildup and transfer

1 Introduction

Fusion of 3D point cloud data collected from different remote sensing devices depicts a common use-case in multiple domains such as autonomous driving [1], creation of digital twins [2], mobile robotics [3] or smart manufacturing [4]. Remote sensing devices can be laser scanners such as LiDAR sensors, depth cameras or stereo vision systems, generating digital representations of natural scenes. These sensors differ in multiple aspects including but not limited to the datatype output, accuracy, range and resolution. Common challenges such as diverging point density, resolution, orientation and scale have to be solved while integrating data from different sensors [5]. Although 3D sensor registration and data fusion is a widely studied field of research [8,9], the registration of data from sensors with different characteristics, also called cross-source registration, still depicts a challenge [10].

G. Kotsis et al. (Eds.): DEXA 2022 Workshops, CCIS 1633, pp. 69–78, 2022.
https://doi.org/10.1007/978-3-031-14343-4_7

The company Track Machines Connected (TMC) uses multiple laserscanner sensors in assistance systems for railroad construction. In more detail, sensors including a LiDAR sensor collecting global rail data and linescanner sensors collecting locally limited but high-resolved data are mounted on a tamping machine. Tamping machines are rail-mounted machines developed to pack the track ballast under railway tracks. The tamping machine is then used to collect multi-sensor data along a distinct track section. The collected datasets serve as the basis to investigate data fusion. Thereby, each of the sensors generates 3D point cloud data representing different but partially overlapping parts of the rail-track section scanned. The underlying aim to use multiple remote sensors is to enhance the resolution of distinct regions while keeping the global context.

2 Problem Statement

The nature of the point cloud data originating from LiDAR and linescanner sensors used by TMC is diverse. The linescanner sensor data contain dense points representing rail-track parts with consistent shape of a locally limited, high-resolved scanning region. In contrast, the LiDAR sensor data collects data of the entire rail-track scene and consists of points with varying density, mainly dependent on the distance of the points to the scanner. Moreover, the shapes of scanned objects are less consistent than that of the linescanner sensor data. Landmark points, defined as points that can be distinctly identified in each sensor data and that are commonly used as reference points for registration, miss in point cloud rail-track data. In order to fuse point cloud data of these sensors, a cross-source registration problem has to be solved. Thus, challenges of cross-source registration such as unequal noise and outliers, difference in point density and partial overlap between the data had to be overcome.

The FFG COMET project TransMVS was initiated as a collaborative project between TMC and the Software Competence Center Hagenberg (SCCH). One of multiple aims of the project was to solve the multi-sensor data fusion problem in an automated fashion, allowing a subsequent extraction of the transformation matrix to be used for sensor calibration. At project start, the setting was as follows:

- Registration of LiDAR and linescanner sensor data was performed by relying on manual measurement of distances and angles between the scanners.
- Only a few datasets collected from the different sensors were available for use within the project.
- No registration groundtruth was available.
- Knowledge on point cloud analysis including deep learning-based object detection and automated 3D data fusion was limited.

The aim of the project was to solve the multi-sensor fusion problem using a semi-automated workflow. Thereby, the requirements were defined as follows:

- The workflow has to be robust. Thus, all sensors need to be integrated.
- The workflow has to be semi-automated with fast execution time. Thus, tasks such as registration or object segmentation as the basis for subsequent registration must be automated including a human-in-the-loop to guarantee completeness.

- The integrated data has to be accurate, with a maximal deviation of each registered sensor data from a groundtruth registration of below 2 cm in each direction.
- The project is intended to be highly interactive and collaborative.
- Within the project, knowledge in point cloud analysis and data fusion shall be gained and transferred between the partners.

3 Collaborative Problem Solving Approach

To develop a cross-source registration of multi-sensor point cloud data fulfilling the aforementioned requirements, TMC and SCCH executed the project in a collaborative setting, aiming at using resources best complimenting each other.

3.1 Roles and Competences

The TransMVS project team was assembled as follows: on TMC side, experts in using the tamping machine and the manual sensor registration procedure, machine learning, software development and a project manager were involved (see Fig. 1 for a more detailed description of the levels of expertise assigned to the present TMC roles). On SCCH side, the project team consisted of a project manager and data scientists with expertise in software development, machine learning and experience in point cloud analysis (see Fig. 2 for a more detailed description of the levels of expertise assigned to the present SCCH roles). Experience in 2D image registration was available, but experience in 3D point cloud data registration had to be gained within the project.

TMC			
Short	Role	Competences	Expected tasks
TMC-1	Product owner	Project management	Project management, project controlling
TMC-2	Software developer	Manual sensor calibration L3, Software development L3	Data recording, Assessment of workflow towards integration into TMC software
TMC-3	Software developer & data scientist	Software development L3, Machine learning L2, Computer vision L1	Point cloud registration, object detection and segmentation
TMC-4	Lead AI developer	Machine learning L3, Software development L3, Point cloud data analysis L2	Assessment of workflows and architectures with respect to integration into TMC software

Levels of expertise: L1 ... basic knowledge, L2 ... advanced, L3 ... expert

Fig. 1. Roles and compentences in the TransMVS TMC project team.

SCCH			
Short	Role	Competences	Expected tasks
SCCH-1	Senior research project manager data science	Project management, Machine learning L3, Computer vision L3	Project management, Project controlling
SCCH-2	Researcher and senior data scientist	Point cloud data analysis L2, Machine learning L3, Computer vision L3	Point cloud registration, object detection and segmentation, supervision SCCH-3
SCCH-3	Researcher and data scientist	Machine learning L2, Point cloud data analysis L1	Point cloud object detection and segmentation
SCCH-4	Researcher and data scientist	Machine learning L1, Software development L3	Graphical user interface for quality assessment and adaption of automated object detection results

Levels of expertise: L1 ... basic knowledge, L2 ... advanced, L3 ... expert

Fig. 2. Roles and compentences in the TransMVS SCCH project team.

3.2 Collaborative Development of a Calibration Object

To solve the problem of cross-source registration without landmarks available, the project team aimed at designing a calibration object, placed within the visible area of each sensor and used for registration between each linescanner sensor data and the LiDAR sensor data. The workflow carried out to define and produce the final calibration object is visualized in Fig. 3.

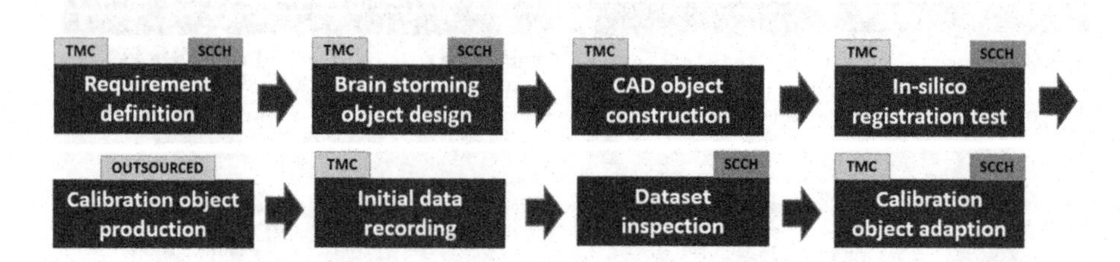

Fig. 3. Workflow to design, produce and adapt the final calibration object used within the TransMVS project. The contribution of the partners to each of the steps is highlighted in the colored boxes.

Before designing a calibration object used for multi-source sensor data fusion, a meeting was organized to define the necessary requirements. To successfully register point cloud data of different sensors, the requirements were defined as follows:

- One calibration object has to be visible for each of the linescanner sensors.

- The shape of the calibration object needs to be consistent irrespective of the viewpoint of the sensor and thus, the angle to the object.
- The calibration object needs to present a shape that is suitable for registration. In more detail, the orientation of the calibration objects has to be unambigous.

Based on these requirements, a brainstorming meeting was set up to work out the design of the calibration object. Within the meeting, a calibration object design was proposed: an object consisting of three cylinders of different height and 3 spheres on top of the cylinders. The idea of the present design was that the spheres are visible irrespective of the view angle, while their orientation can be estimated based on the height difference of the three spheres. Upon CAD object construction, a joint in-silico registration test was carried out simulating registration by cropping parts of the calibration object and applying distortion, scaling and translation to test the registration between a calibration object and a modified calibration object. As the registration tests have proven to be robust, the calibration object was produced and initial data was recorded. Upon inspecting the datasets, diffuse reflection was noted in settings with direct light insolation. Thus, the calibration object was adapted by using a black lacquer spray to obtain the final calibration object.

3.3 Feedback-driven, Iterative Data Recording

As one of the requirements of the TransMVS project was to create a semi-automated workflow, and a calibration object was used to enable cross-source registration, the calibration object had to be detected in an automated fashion. Deep learning architectures for 3D point cloud processing or 2D representations of 3D point cloud data are considered as state-of-the-art for object detection or segmentation. To enable the training of such architectures to detect the calibration object in previously unseen data, architectures have to be trained using an annotated dataset. The dataset must contain point cloud raw data or 2D-representations of point cloud data along with annotation masks, reflecting the points representing the calibration objects. As commonly known, deep learning architectures need large datasets for training to generalize to unseen data while inference [6]. As annotated data representing rail track data has not been publicly available, it had to be collected and annotated. As the optimal placement of the calibration objects with respect to object detection depicts a crucial step towards a successful registration, data was recorded in a highly interactive fashion. see Fig. 4.

To this end, a live-recording session between TMC and SCCH was organized. TMC booked a timeslot on the tamping machine to enable the test of multiple placements. At the same time, a live remote meeting with SCCH was initiated, allowing to update each partner of the current status. One round of placement was carried out as follows: 1. TMC placed the calibration objects on the rails. 2. Data from all sensors was collected. 3. The data was transferred to SCCH and TMC waited for feedback. 4. SCCH carried out segmentation and registration on

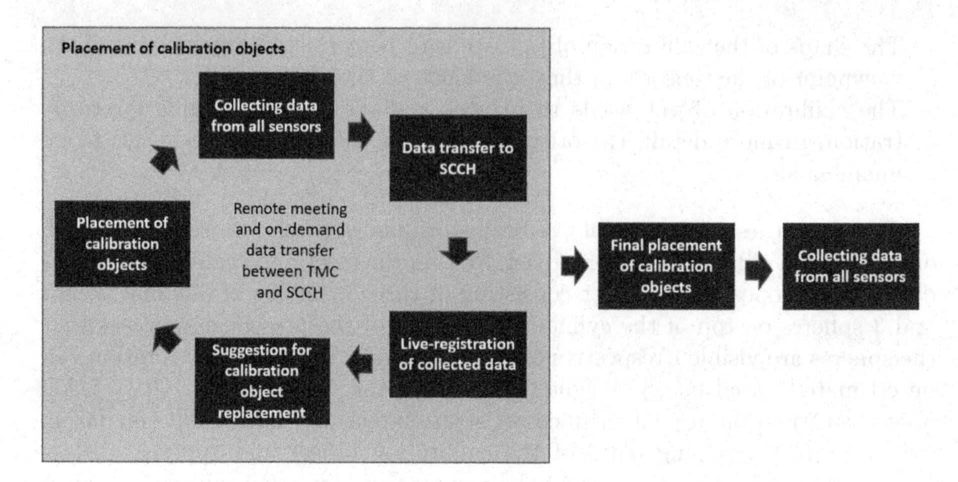

Fig. 4. Workflow to find the optimal placement of calibration objects.

the received datasets. 5. Based on the results and observations, SCCH suggested to modify the placement of calibration objects. These circles were carried out until a perfect placement of calibration objects had been achieved. Using this placement setting, multiple datasets of different parts of the track were collected and annotated, now serving to train deep learning architectures for point cloud calibration object detection.

3.4 Project Results

The collaborative execution of the project resulted in a workflow to fuse mutli-sensor point cloud data, allowing to extract the underlying transformation matrix for each of the linescanner sensors used to fuse the data. The workflow consists of a robust semi-automated, interactive pipeline including a human-in-the-loop, relying on automated calibration object detection and segmentation, manual quality assessment and adaption and registration. The automated detection was achieved by applying 2D projections of the 3D point cloud data in multiple views, and using a 2D state-of-the-art object detection model (YOLO-P6 [7]). Manual quality assessment and adaption of automated object detection results was solved using a simple graphical user interface. Robust registration of multi-sensor data was ensured relying on a same-source registration approach utilizing a manually calibrated reference dataset.

4 Knowledge Buildup and Transfer

One of the TransMVS project aims was the buildup and transfer of knowledge within the project, with respect to point cloud analysis and registration. The knowledge gained within the project is visualized in Fig. 5. Based thereon and based on knowledge already existing at project start, an interactive knowledge transfer was achieved in multiple partner meetings (Fig. 6). The main focus of the multi-sensor calibration was the development of a robust sensor data fusion pipeline. At project start, basic knowledge about 3D point cloud registration was gained at TMC and transferred to SCCH (from TMC-2 to SCCH-2). Within the project, knowledge about 3D point cloud registration methods was developed and applied to design the calibration workflow. Based on multiple meetings jointly executing and discussing the code developments, knowledge about advanced registration methods was transferred from SCCH to TMC (from SCCH-2 to TMC-2). With respect to object detection and registration, expertise in 2D image object detection and segmentation and machine learning existed (SSCH-2 and SCCH-3). At project end, comprehensive knowledge about 3D point cloud object detection and segmentation was gathered and transferred to TMC in regular status meetings (from SCCH-2, SCCH-3 to TMC-3, TMC-4). A crucial aspect depicted the transformation of the 3D point cloud annotation labels to 2D bounding boxes for the training of a 2D object detection model, based on heavy data augmentation of the 3D labeled data. This method was developed in an interactive fashion, including the transfer of knowledge on 2D computer vision methods from SCCH to TMC (from SCCH-2 to TMC-3).

Knowledge building within TransMVS			
Role-short	**Topic**	**Expertise project start**	**Expertise project end**
TMC-3	Point cloud registration	Beginner level.	Registration of cross-source objects using popular algorithms as RANSAC or ICP.
SCCH-2	Point cloud registration	Familiar with 2D image registration, no point cloud registration experience.	Same-source and multi-source point cloud registration using Open3D functions or advanced registration methods (GMM-Tree, Bayesian Coherent Point Drift).
SCCH-2	Point cloud object segmentation	Familiar with deep learning-based 2D image object detection.	Application of point-based deep learning architectures and architectures operating on 2D-representations of 3D point cloud data.
SCCH-3	Point cloud object detection	Familiar with 2D image object detection.	Application of PointNet on point cloud data for object detection.
SCCH-4	Graphical user interface	Familiar with software development including GUI design.	Application of SimpleGUI for effective python prototyping.

Fig. 5. Development of knowledge in registration and analysis of 3D point cloud data.

5 Insights Gained Within the Project

Within the runtime of the TransMVS project, insights in how to solve semi-automated multi-sensor registration approaches was gained.

Knowledge transfer within TransMVS		
Topic	Expertise existing or developed within TransMVS	Knowledge transfer
3D point cloud data registration	Basic registration between simulated and cropped CAD objects	TMC-3 → SCCH-2
3D point cloud data registration	Advanced cross-source and same-source registration of distorted virtual objects and real objects.	SCCH-2 → TMC-3
Point cloud object detection and segmentation	Object detection and segmentation using point-based deep learning architectures and architectures operating on 2D projections of 3D point cloud data.	SCCH-2, SCCH-3 → TMC-3, TMC-4
Computer vision	Conversion of 3D point cloud masks into 2D bounding boxes and classification into different classes according to position.	SCCH-2 → TMC-3

Fig. 6. Transfer of knowledge in registration and analysis of 3D point cloud data and computer vision algorithms.

5.1 Avoiding Cross-Source Registration by Using a Reference Dataset

Multi-sensor data fusion depicts a cross-source registration challenge. Due to the utilization of remote sensors generating 3D point cloud data with diverging properties, the application of cross-source registration methods is error prone (varying point density, diverging object shapes, irregular noise and outliers). We developed a registration strategy avoiding cross-source registration by generating a manually calibrated reference dataset. Registration within the TransMVS final workflow is based on calibration objects, placed on the rails, scanned by the sensors and segmented using a semi-automated object detection and segmentation workflow. By using a manually calibrated reference point cloud dataset that includes calibration objects collected from similar sensors, a registration strategy based on same-source registration could be developed. Thus, segmented calibration objects collected from a particular sensor type are always registered to data of the same source (same-source registration). In combination with a graphical user interface used to correct automated calibration object detection results, a fast and robust workflow could be proposed.

5.2 Generating Datasets Serving for Point Cloud Segmentation Model Training

A crucial aspect towards a successful fusion of multi-sensor point cloud data depicted the accurate segmentation of calibration objects. At an early phase of the project, multi-sensor data was collected including the calibration objects placed at arbitrary positions. The best performing placement of calibration objects with respect to a subsequent segmentation and registration was yet unknown. In addition, the objects' location was changed from the left rail to the right rail in multiple cases. Moreover, the linescanner sensors come along with specific features such as noise suppression, which was partially activated or deactivated. As only a small number of datasets could be collected in total (40 datasets) due to a limited availability of the tamping machines, all records were annotated and used for deep learning object detection or segmentation training. The result was a non-uniform, noisy and highly heterogeneous dataset, consisting of data from different sensors with the aformentioned properties, leading to a highly unstable training performance and low detection accuracies on a validation set. This was even amplified by a low object-to-background ratio of calibration object points to non-calibration object points. By excluding datasets showing high divergence to most of the datasets and by relying on 2D projections in multiple views and a sliding-window approach, thereby reducing the overall amount of data points while increasing the object-to-background ratio, these challenges could be overcome.

5.3 Collaborative Problem Solving Approach

The project was carried out in a highly interactive manner, including multiple hands-on sessions targeting data recording and algorithm development. In addition, regular meetings were carried out to transfer knowledge gained between the partners. An important aspect that contributed to the success of the project was the interactive data recording session, allowing to perform a real-time evaluation of the data collected and thus, to incrementally adjust the experiment settings towards an optimal placement of the calibration objects with respect to subsequent object detection and data fusion. Hence, this interactive data recording session helped to solve the multi-sensor data fusion challenge in a short period of time by generating high quality data for the given task, and will be considered in future projects.

Acknowledements. The research reported in this paper has been funded by the Federal Ministry for Climate Action, Environment, Energy, Mobility, Innovation and Technology (BMK), the Federal Ministry for Digital and Economic Affairs (BMDW), and the State of Upper Austria in the frame of SCCH, a center in the COMET - Competence Centers for Excellent Technologies Programme managed by Austrian Research Promotion Agency FFG.

References

1. Wang, Z.: Multi-sensor fusion in automated driving: a survey. IEEE Access **8**, 2847–2868 (2020)
2. W. Hu. Digital twin: a state-of-the-art review of its enabling technologies, applications and challenges. J. Intell. Manuf. Spec. Equip. **2**(1) (2021)
3. Pomerleau, F., et al.: A review of point cloud registration algorithms for mobile robotics. Found. Trends Rob. **4**(1), 1–104 (2015)

4. Tsanousa, A., et al.: A review of multisensor data fusion solutions in smart manufacturing: systems and trends. Sensors **22**, 1734 (2022)
5. Huang, X., et al.: A comprehensive survey on point cloud registration. arXiv (2021)
6. Fischer, L.: AI system engineering-key challenges and lessons learned. Mach. Learn. Knowl. Extr. **3**, 56–83 (2021)
7. Wang, C., et al.: You only learn one representation: unified network for multiple tasks. arXiv (2021)
8. Cheng, L., et al.: Registration of laser scanning point clouds: a review. MDPI Sensors **18**, 1641 (2018)
9. Cui, Y., et al.: Deep learning for image and point cloud fusion in autonomous driving: a review. arXiv (2020)
10. Bracci, F., et al.: Challenges in fusion of heterogeneous point clouds. ISPRS Int. Arch. Photogram. Remote Sens. Spat. Inf. Sci. **XLII-2**, 155–162 (2018)

From Data to Decisions - Developing Data Analytics Use-Cases in Process Industry

Johannes Himmelbauer(✉), Michael Mayr, and Sabrina Luftensteiner

Software Competence Center Hagenberg, Hagenberg, Austria
{johannes.himmelbauer,michael.mayr}@scch.at
http://www.scch.at

Abstract. Nowadays, large amounts of data are generated in the manufacturing industry. In order to make these data usable for data-driven analysis tasks such as smart data discovery, a suitable system needs to be developed in a multi-stage process - starting with data acquisition and storage, data processing and analysis, suitable definition of use cases and project goals, and finally utilization and integration of the analysis results to the productive system. Experience from different industrial projects shows that close interaction between all these sub-tasks over the whole process and intensive and steady knowledge transfer between domain experts and data experts are essential for successful implementation. This paper proposes a stakeholder-aware methodology for developing data-driven analytics use-cases by combining an optimal project-development strategy with a generic data analytics infrastructure. The focus lies on including all stakeholders in every part of the use-case development. Using the example of a concrete industry project where we work towards a system for monitoring process stability of the whole machinery at the customer side, we show best practice guidance and lessons learned for this kind of digitalization process in industry.

Keywords: Industrial data analytics · Digitalization process · Human and AI

1 Introduction

The Industry 4.0 revolution enables a massive source of complex data in industrial manufacturing [8]. Fast-changing market requirements, as well as the need for rapid decision making, pose significant challenges for such companies [12]. An intelligent analysis of all available data sources is a necessity for companies to stay scientifically, technologically, and commercially competitive. Especially in industrial manufacturing, intelligent data analytics combined with specific use case definitions are often the basis for crucial business decisions to gain an advantage over the competition. According to *Gartner*, a research institute that

The research reported in this paper has been funded by BMK, BMDW, and the State of Upper Austria in the frame of the COMET Programme managed by FFG.

G. Kotsis et al. (Eds.): DEXA 2022 Workshops, CCIS 1633, pp. 79–89, 2022.
https://doi.org/10.1007/978-3-031-14343-4_8

focuses on market research and analysis of developments in the IT sector estimates that 80% of analytical insights will not deliver business outcomes [1]. At the conference *Transform 2019* of *Venture Beat* experts even estimated that 87% of data science projects never make it into production [2]. According to [5] et al., key reasons that influence the probability of failure for data-driven analytical projects are most frequently attributed to *data-related* challenges like data- *quality*, *access*, *understanding* and *preparation*. Furthermore, *process-related* challenges like *budget/time*, *cultural resistance* and *unstructured project execution*, as well as, *people-related* challenges including the lack of *technical expertise* pose significant impacts on the non-success of data-driven projects. The research also highlights a conceptual distance between business strategies and the implementation of analytics solutions in more than half of their respondents. This conceptual shift shows the necessity of addressing data, user and business-related aspects during the conceptualization of the data analytics use-case and has proven to be challenging [10], According to a study conducted by *Forbes Insights* and *EY*, clear communication between different stakeholders and an enterprise-wide digitization strategy play a major role in the successful realization of effective and value-generating data analytics use-cases for the industrial manufacturing domain [11]. In this paper, we investigate and document the steps needed for defining, developing and deploying data analytics use-cases, describe common pitfalls and provide concrete real-world examples in the industrial manufacturing domain. The paper is structured as follows: Sect. 2 describes the industrial manufacturing company *starlim* and the project vision developed for our collaboration. Sect. 3 covers the digitalization workflow from data to decisions. We describe the overall methodology to get from project definition (Sect. 3.1) and data provision (Sect. 3.2) to data analysis and the integration as well as exploitation of all project results (Sect. 3.3) utilizing the knowledge of different stakeholders. Sect. 4 concludes with a brief discussion of the work provided and future work.

2 Use Case Description - Project Vision

In this paper, we will repeatedly illustrate our statements with concrete examples from our several years of cooperation with our project partner *starlim*, which is an Austrian company and the world market leader in the field of liquid silicone injection molding. 14 billions of high-quality silicone parts are produced annually for the industry, life sciences and mobility sectors, with several hundred machines in operation every day. In elastomer injection molding, the material is propelled by a screw and injected into the molds (= *cavities*) that correspond to individual products. During one injection of the material (= *shot*) multiple cavities are filled. Only in the mold is there a sudden rise in temperature to about 200 °C, which is used for vulcanization (conversion of the liquid silicone raw material into the silicone rubber), followed by removing the products from the cavities. At our partner company, this process runs on several hundreds of machines producing various products, where each of the machines produces thousands up to hundreds of thousands of product pieces per day. With each shot, a lot of data (i.e. several hundred data points) is collected at each individual machine, referring to the physical state of the process, e.g., temperature and pressure, the current configuration of the machine as well as to various properties of the products, e.g., volumes. Parts of this data are on the one hand directly used for machine operation and monitoring by the operator through display on the machine and on the other hand are also available to machine experts for historical analyses (i.e. by visual data inspection and statistical evaluations).

A few years ago, the company decided to intensify the digitalization process in the direction of automated and intelligent data use in order to better exploit the potential inherent in the existing large volumes of available data. It was clear to both the IT experts and the machine experts of *starlim* from the very beginning that there are a number of challenges to tackle in this process and that standard IT solutions alone cannot achieve the goal. The main reasons for this included the following: Injection molding represents a very complex physical-chemical process and there exists a wide variety of situations that can lead to a bad, or at least unstable, condition of a machine as well as its production process. Considering that there are usually more than one hundred machines simultaneously in operation that produce several dozens of different types of products allows to understand how difficult the work towards a generic solution for an all-encompassing machine monitoring has to be. Finally, a one-to-one allocation of the production process and produced quality is infeasible.

An exemplary business usage of the available data would be to create a system that monitors and evaluates the production quality of all machines in use online, simultaneously, and in an all-encompassing manner.Given the circumstances described above, working in this direction at our corporate partner seems extremely challenging, if not impossible. The risk of failure has to be considered very high for such complex and challenging project goals. Therefore, instead of heading directly for the produced product quality, we want to focus in our project on monitoring the process stability. The following vision goal was defined for our collaboration:

Project Vision: *A system is to be created that is capable of assessing the stability of the processes of the machine park and the associated product quality and providing trained machine operators with information on which machines processes need to be optimized.*

This goal should not be understood as a global objective of the project that must be fully implemented in a predefined project time frame, but rather as a vision that is to be approached step by step in a process embedded in the project work.

3 Workflow for a Digitalization Process

Studies like [1] and [2] as well as literature in data science [5,6] claim that data-driven analytical projects in the industrial manufacturing sector still tend to have a high risk of failure. The main reasons given in this context [5] coincide with the typical challenges that also we as data scientists at *SCCH* face in our daily work in the industrial environment. Based on our experience, we are convinced that successful interaction between the business experts and the data experts is indispensable for overcoming all these challenges. In this chapter, we want to point out ways, but also pitfalls, which can favor or hinder a successful implementation of this in practice. The results presented here are based on the experience gained from a wide variety of industrial projects and the strategies for successful project implementation developed thereby over many years. Furthermore, for a complete implementation of an industrial data analysis project, from data acquisition to the integration of project results into the productive system, a number of objectives usually need to be successfully implemented. In view of this fact, it should be mentioned that independent, sequential processing of these sub-tasks is not advisable here either, but that a consistent, interlinked project strategy can represent an important, even decisive criterion for project success. Figure 1 shows a very generic, ideal, all-encompassing workflow that addresses the above conditions. On the vertical line we see the time sequence of the individual work steps and their dependencies and interactions with each other. The horizontal level represents the interaction between the project members during the individual sub-tasks. For each work step, suitable people from both the team of data experts and the team of the industry partner work actively on the respective task in a core team (marked red in the graphic). In addition, it should also be defined which colleagues are to be available to this core team in an advisory capacity and/or as stakeholders (marked in black). Since the respective experts will be mainly involved in each project phase, the core team will usually change during the course of the project.

Before, in the next subsections, we touch on insights gained from our project experience for the individual work steps during a digitalization process, the following, in our opinion, important note should be given: It is beneficial if all project participants are aware that, in addition to achieving explicit project goals, there are also implicit results throughout the work process that can additionally be very valuable for the company. The following two prominent examples may be mentioned here:

- **Continuous improvement of data quality:** Nowadays, large amounts of data are generated in industry throughout the entire production process. During data recording, a variety of problems can occur with regard to data quality, such as incorrect or missing measurements, measurements with inappropriate resolution (both in terms of measurement unit and time), incorrect allocations or time stamps, etc. These problems often only become apparent when corresponding data are actually used in analysis evaluations, and only after they have been realized can work be done on solving such problems. As a consequence, you end up with data of significantly improved quality, which is not only beneficial for the project in question, but also improves any further use of the data in future.

- **Enrichment of process knowledge and data (analysis) understanding:** As we will explain below, the best possible knowledge exchange between domain experts and data experts is essential for the successful implementation of a data analysis project. The use of this knowledge built up during this process - whether in explicit form e.g. through documentation, or in implicit form (i.e., further development of company employees) - is again not limited to the project itself. Experiences on past projects show, for example, that domain experts on the company side learn about ways to get certain questions about the process (state) answered through appropriate data visualizations.

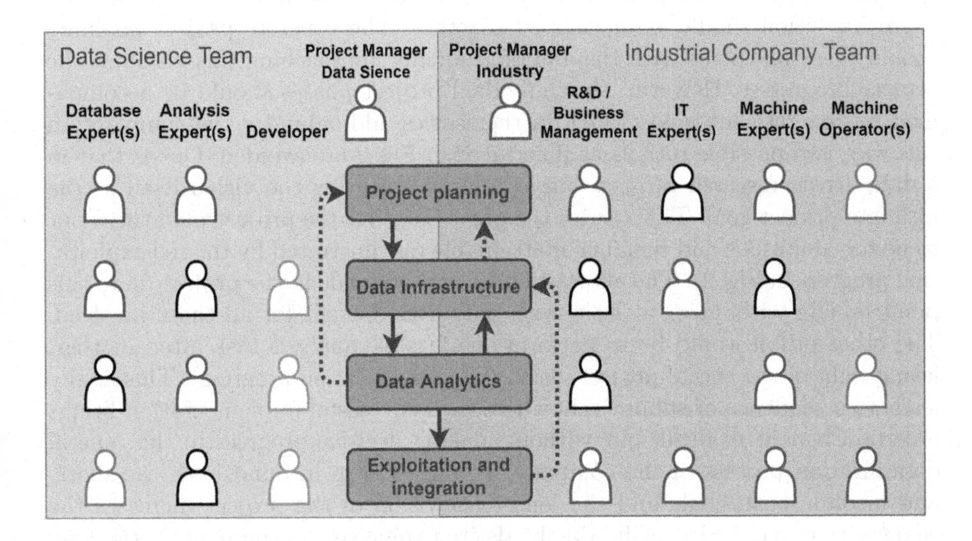

Fig. 1. Project workflow

3.1 Project Outline - Definition of Project Vision, Project Goals and Use Cases

As everywhere else, it is also essential in data-driven industrial projects to carry out good project planning at the beginning of the project. If you come to the conclusion that the defined project goals are reachable and that the path to achieving them seems relatively clear, then it makes sense to create a project plan that is as precise and time-defined as possible and then go for its implementation. In this field, however, it is usually the case that both the definition of concrete project goals and the estimation of their achievability prove to be very challenging and complex, if not impossible. According to our observations, the reason for this is typically a combination of the following problem characteristics: Firstly, *There's no picnic*; i.e. data analysis can only answer questions posed by domain experts if the relevant information needed to answer the question is also present in the data, whether explicitly or implicitly. Secondly, there are *mutual dependencies between the project steps*; they can not be seen as independent, sequential tasks. Finally, there are potentially many different person roles involved in a data-driven analytics project. For the overall success of the project, it is important that the inputs and interests of all these stakeholders are taken into account. This is even more challenging as the group of people is expected to be very heterogeneous and a common understanding needs to be built. If for the above reasons one finds oneself in the situation of not being able to predict concrete final results and if it cannot be assumed that proved standard solutions can lead to the goal, all persons involved must be aware that ultimately this is an applied research project. That is, the planning of the project and its outcome must be designed openly. In this case, we propose the following two-track approach for project development: At the beginning of the project, a project vision should be developed together. The concrete project planning should then take place with the aim of gradually approaching this vision as the project progresses. However, the individual project phases should be accompanied by concrete sub-goals which in themselves add value for the company. In this way, two possible pitfalls as illustrated in Fig. 2 are avoided: One is that in a data-driven research project, one is tempted of using the vision itself as the ultimate project goal. This carries the great risk that the project plan turns out to be too ambitious and remains unattainable (as illustrated by the red exploitation progress in Fig. 2). The efforts that have been made in the project up to this point remain without or at least with only little benefit for all those involved. The other pitfall would be to perform one feasible analysis task after another, losing sight of the actual project vision (grey exploitation progress). This finally leads to a sequence of subprojects where each successful implementation brings a certain benefit in itself, but without making decisive progress in the general digitalization process of the company. Trying to keep in mind both the short- and medium-term goals and the long-term vision of the project increases the chances that the project will add the desired value to the company in the long run (green exploitation progress).

Within our collaboration with *starlim* we tried to go after the presented strategy in the following way: After developing the project vision (see Sect. 2), we worked intensively on the concept and prototyping of a suitable data infrastructure with the participation of both the IT department and the machine experts on the company side and database experts as well as data analysts on the part of *SCCH* (see Sect. 3.2). In parallel, work on the first analysis use cases was started at an early stage. On the one hand, the focus was on building a better understanding of the production process and the data generated in the process. On the other hand, this enabled the data analysts to provide input to the database experts on future requirements for the emerging data infrastructure. In the course of these analyses, for example, a visualization tool was created that can also be used by the machine experts for data viewing or as support for labeling process states. For the next project phase, the domain experts at *starlim* chose the automatic detection of a known, sporadically occurring problem in the injection molding process as a use case to run through from definition to result integration into the productive system for the first time (see Sect. 3.3).

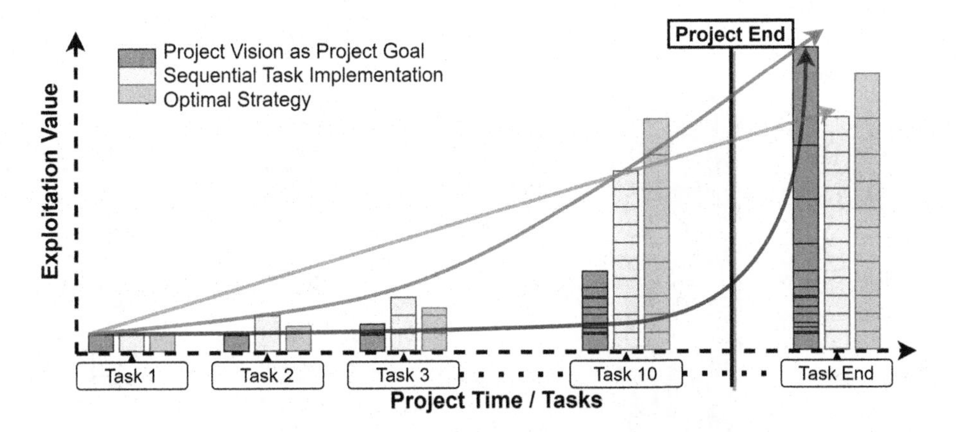

Fig. 2. Three projects strategies and their corresponding exploitation value over time.

3.2 Data Analytics Infrastructure - Building Up, Data Management and Preparation

A well-defined data analytics infrastructure is crucial in driving data analytics use-cases toward production-ready systems. Figure 3 shows a high-level view of the technology stack and the information flow between components, use-cases and stakeholders. We design this infrastructure so that it theoretically is capable of deploying the high-level project vision (see Sect. 2) as well as lower-level subgoals (e.g. see Sect. 3.3) in the context of industrial process monitoring. We emphasize the importance of including the expertise of various stakeholders in the development of the infrastructure and in the definition, interpretation, and tuning of the data analytics use-cases.

In the concrete use-case example of *starlim* (see Sect. 2), we are dealing with high volumes of data due to the sheer amount of produced products in combination with a lot of different sensors governing the manufacturing process. In addition, data produces fast, and it is essential, especially for real-time monitoring use-cases, that data routes throughout the different technological layers with minimal delay. Unfortunately, industrial partners often overlook the importance of high-quality data, i.e. data veracity, making pre-processing necessary to conduct complex and interpretable data analytics. In the example of *starlim*, we identified several data-quality related issues, where the most prominent issue manifested itself in the form of semantic shifts in the gathered data (see [4] for more details on how we tackled this problem). Concerning data variety, we solely deal with structured data in the *starlim* use-case in the form of process-state and product-state data.

To timely route data to the different components of the stack, we utilize *Apache Kafka*, an open-source event streaming platform. The distributed nature of *Kafka* coupled with high throughput, low latency and high scalability are key reasons why we chose *Kafka* as our message broker. In addition, *Kafka* features a technology called *Kafka Connect* for sourcing and sinking data streams to and

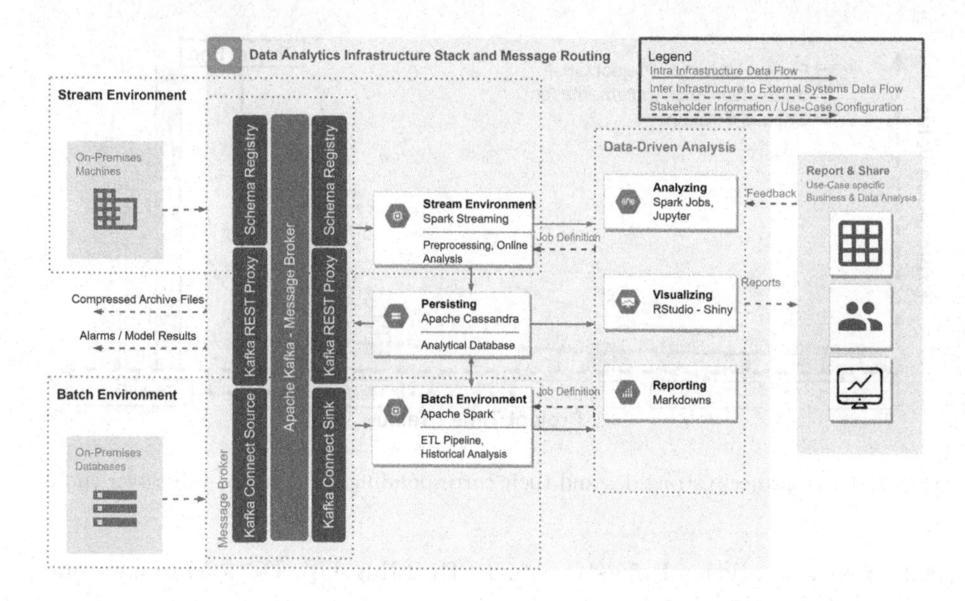

Fig. 3. Concept of data analytics infrastructure for *Starlim*

from (external) databases, abstracting low-level database code with user-friendly connector configurations. *Kafka* routes data using a queue-like structure called *topics* to different components of the infrastructure. One may enforce individual data schemes for such *topics* utilizing the *Schema Registry*. In our infrastructure stack, we settle with *Apache Avro* for schema definition and serialization due to prior experience in other projects and its efficient binary serialization capabilities. Querying the serialized data is impractical for analytical purposes. We split data persistence into hot and cold storage to increase latency and analytical interpretability while decreasing storage costs. In this paper, we refer to hot storage as storage that holds deserialized data that needs to be accessed often and without much latency by different stakeholders for analytical purposes, e.g. data of the ongoing year, whereas cold storage acts as an efficient, i.e. serialized, data-store for historic or less demanded data. We have chosen *Apache Cassandra*, a distributed storage system for high volumes of structured data, for the persistence layer of this infrastructure stack, as *Cassandra* is highly scalable, has no single point of failure and handles high write throughput without sacrificing read efficiency [7]. Analyzing the high volumes of data requires a distributed computational layer that is capable of conducting heavy computational analysis with high volumes of data, e.g. offline model training on historical data, as well as near-real-time analysis tasks like online predictions and outlier detection. Different use-cases may require different processing tools; however, in process monitoring for industrial manufacturing, a reasonable assumption is that tools support near-real-time stream and batch analysis with a good integration of machine learning frameworks. In the context of near real-time process monitoring, *Apache Spark* is a reasonable state-of-the-art choices with scale-able streaming and batching capabilities, high throughput and low latency [9].

We deploy the data analysis infrastructure on five *CentOS* nodes, where two of them act as management nodes running *Apache Kafka*, *Kafka Connect* and *Schema Registry*. The other three are used for distributed analysis, i.e. *Apache Spark* and persistence, i.e. *Apache Cassandra*. To provide failure-tolerant data persistence, we replicate data on all three *Apache Cassandra* instances in order to limit the amount of downtime and data loss risks. We retrieve newly generated and binary serialized process- and product-state data using *Kafka-Connect*. Data is de-serialized and pre-processed in distributed *Spark Streaming* applications. Compressed serialized Avro data exports to disk acting as a backup and additionally persists as cold data in *Cassandra* tables. Use-case specific analysis tasks are divided upon the *Spark* batching and streaming environment and are configured by analysis experts as well as continuously adapted using domain knowledge of machine operators and machine experts. Analysis results are published for business management and other stakeholders in the form of, e.g. Markdowns or Shiny applications, as well as in real-time using consumable *Kafka* topics.

3.3 Data Analysis - Development of Data-Driven Approaches and Exploitation of Analysis Results

When all relevant data is available, after a prior data review, a selection process (e.g. in the form of joint workshops) usually has to take place first, in which a number of problems relevant to the company are to be presented by the domain experts. These possible use cases should then be jointly reviewed in terms of effort, feasibility and potential for exploitation, and a selection of these can be included in the project plan by order of priority. All participants should be aware that this step can be decisive for success in the further course of the project.

The use case presented here is about a known, sporadically occurring problem in the injection molding process where due to different possible factors one or more cavities within one shot are filled with improper amount of material (= *underfilling / overfilling*). Such cases are recognizable in the data as multidimensional outliers, i.e., extreme or atypical observations in one or more recorded variables. The ideal goal of our application was to implement an approach for the automatic real time detection of such cases, simultaneously from the stream of all different production machines. However, considering the high heterogeneity of the data to be monitored (in terms of different machines, machine states and products produced) and several characteristics regarding the outliers to be detected (described in [3]), this undertaking had to be considered as very challenging and risky. In fact, it was very soon clear to the data analysts that standard outlier detection methods will not lead to success. Therefore, we decided to approach the key goal step by step whereby it was already worked out in advance with the domain experts what benefits can already be derived from the work when a respective step is achieved. As a first step, we worked on a batch algorithm that would enable production periods per machine to be evaluated historically with regard to the defect pattern. We were very confident that we could achieve sufficient performance (i.e., detect a large number of problem periods without triggering too many false alarms) so that in production the results could be used to automatically highlight production periods from the previous day that needed careful quality inspection. In the end, we not only achieved very good results with the batch algorithm, but the experience gained during development helped us come up with a promising idea of how we could also successfully implement an online algorithm. Finally, we really managed to come up with a streaming solution that gives very promising results over the whole range of machines and products. The integration of the algorithm into the productive system is ongoing work that will be mainly done by a data expert of *starlim*. In this context, it should be mentioned that from the start, an explicit project goal was defined as building up internal data expertise at the company side by means of best possible knowledge transfer during project work, so that in the future more and more data-driven tasks can be carried out directly in the company.

4 Discussion and Conclusion

In this work, we have described the main challenges that generally need to be overcome to implement industrial data analytics projects successfully. Based on our experience in practice, we have come up with a stakeholder-aware project development strategy that supports to meet these challenges successfully. We are aware that our recommendations cannot or should not always be implemented one-to-one. On the one hand, not all prerequisites may be in place, e.g. not all stakeholders may be available when needed. On the other hand, each project ultimately has its own specific requirements, which should also be met flexibly. However, we think it is an important support for project planning and implementation to have a general method or strategy and best practices for typical project challenges available. This can serve as a helpful basis for making strategically important decisions in each specific project. We therefore recommend that the points raised in this paper be actively addressed with the industry partner at the start of a data-driven project. In relation to our use case, both our corporate partner *starlim* and we see awareness of this from the beginning as a key factor for our fruitful collaboration.

We see any scientific effort that helps support the successful implementation of industrial data analytics projects as a promising research direction. Be it towards further standardization of the workflows involved or the automated use of existing domain knowledge.

References

1. Our top data and analytics predicts for 2019, January 2019. https://blogs.gartner.com/andrew_white/2019/01/03/our-top-data-and-analytics-predicts-for-2019/
2. Why do 87% of data science projects never make it into production?, July 2019. https://venturebeat.com/2019/07/19/why-do-87-of-data-science-projects-never-make-it-into-production/

3. Bechny, M., Himmelbauer, J.: Unsupervised approach for online outlier detection in industrial process data. Procedia. Comput. Sci. **200**, 257–266 (2022). https://doi.org/10.1016/j.procs.2022.01.224
4. Ehrlinger, L., Lettner, C., Himmelbauer, J.: Tackling Semantic Shift in Industrial Streaming Data Over Time (2020)
5. Ermakova, T., Blume, J., Fabian, B., Fomenko, E., Berlin, M., Hauswirth, M.: Beyond the Hype: Why Do Data-Driven Projects Fail?, January 2021. https://doi.org/10.24251/HICSS.2021.619
6. Hoch, T., et al.: Teaming.ai: Enabling human-ai teaming intelligence in manufacturing. In: I-ESA Workshop AI beyond efficiency: Interoperability towards Industry 5.0 (2022)
7. Lakshman, A., Malik, P.: Cassandra: a decentralized structured storage system. ACM SIGOPS Oper. Syst. Rev. **44**(2), 35–40 (2010). https://doi.org/10.1145/1773912.1773922, https://doi.org/10.1145/1773912.1773922
8. Lasi, H., Fettke, P., Kemper, H.G., Feld, T., Hoffmann, M.: Industry 4.0. Bus. Inform. Syst. Eng. 6(4), 239–242 (2014). https://doi.org/10.1007/s12599-014-0334-4, https://doi.org/10.1007/s12599-014-0334-4
9. Luftensteiner, S., Mayr, M., Chasparis, G.C., Pichler, M.: AVUBDI: a versatile usable big data infrastructure and its monitoring approaches for process industry. Front. Chem. Eng. **3** (2021). https://www.frontiersin.org/article/10.3389/fceng.2021.665545
10. Nalchigar, S., Yu, E.: Business-driven data analytics: a conceptual modeling framework. Data Knowl. Eng. **117**, 359–372 (2018). https://doi.org/10.1016/j.datak.2018.04.006, https://www.sciencedirect.com/science/article/pii/S0169023X18301691
11. Rogers, B., Maguire, E., Nishi, A., Joach, A., Wandycz Moreno, K.: Forbes insights: data & advanced analytics: high stakes, high rewards. https://www.forbes.com/forbesinsights/ey_data_analytics_2017/
12. Zeng, Y., Yin, Y.: Virtual and physical systems intra-referenced modelling for smart factory. Procedia CIRP **63**, 378–383 (Jan 2017). https://doi.org/10.1016/j.procir.2017.03.105, http://www.sciencedirect.com/science/article/pii/S2212827117302512

Challenges in Mass Flow Estimation on Conveyor Belts in the Mining Industry: A Case Study

Bernhard Heinzl[1(✉)], Christian Hinterreiter[2], Michael Roßbory[1], and Christian Hinterdorfer[2]

[1] Software Competence Center Hagenberg GmbH, Hagenberg, Austria
{bernhard.heinzl,michael.rossbory}@scch.at
[2] RUBBLE MASTER HMH GmbH, Linz, Austria
{christian.hinterreiter,christian.hinterdorfer}@rubblemaster.com

Abstract. This paper presents a case study in indirect mass flow estimation of bulk material on conveyor belts, based on measuring the electric net energy demand of the drive motor. The aim is to replace traditional expensive measurement hardware, which results in benefits such as lowering overall costs as well as the possibility of working under harsh environmental conditions, such as dust, vibrations, weather, humidity, or temperature fluctuations. The data-driven model uses a dynamic estimation of the idle power in order to take into account time-varying influences. The case study has been developed in close collaboration between industry and scientific partners. Experiences gained from a first field prototype were used and incorporated to create an improved prototype setup, including a modular software infrastructure for automatically capturing all relevant measurement data. We discuss some of the challenges in development, like data quality, as well as our experiences in academia-industry collaboration. The presented case study showcases the importance to bring research into real-world applications for generating technology innovations.

Keywords: Conveyor belt · Dynamic mass flow estimation · Data-driven model · Academia-industry collaboration

1 Introduction

In the context of industrial machinery, a belt scale is a sensor that measures the mass flow of material that is being transported on a conveyor belt. The mass flow is measured by weighing the current belt load and measuring the belt speed at the same time. However, this kind of specialized industry-grade sensor hardware is typically quite expensive when required to achieve high accuracy under harsh conditions present e.g. in the mining industry [16].

Especially when dealing with harsh operating conditions, dust, dirt, vibrations, fluctuating temperatures or humidity, it makes sense to try to avoid such equipment, which would also reduce the overall cost [9]. For this reason, we

G. Kotsis et al. (Eds.): DEXA 2022 Workshops, CCIS 1633, pp. 90–99, 2022.
https://doi.org/10.1007/978-3-031-14343-4_9

devised a case study where we try to indirectly estimate the mass flow on a conveyor belt in an online setting, by measuring the electrical power consumption of the belt drive motor (using inexpensive measurement equipment). In theory, this electrical power should correlate with the weight on the belt at any given time. The goal is to develop "virtual" belt scale that is able to replace conventional expensive built-in belt scale systems. Possible application of this solution include e.g. mineral processing, to monitor stone crushers by detecting how much material has already been crushed during a given working period.

The case study presents an example of transferring research results into application and has been developed in close collaboration between scientific and industrial partners. Scientific methods are tested in field tests and real-world data is collected to evaluate their performance and accuracy, supported by the domain and process knowledge from the industry. The end result will be a production-grade software solution.

2 Related Work

Nowadays, belt scales are one of the most common mass flow equipment used in material processing. Some of the existing literature using other kinds of sensors include machine vision [1,5,10,11], laser profilometers [2,13], load cell sensors [15], tachometers [12], or radiation-based sensors [3]. Estimating mass flow indirectly from energy demand can be a promising method to avoid expensive sensor equipment and make mass flow measurements available at lower cost. Several other works investigate approaches to use power measurements to detect mass flow [7,8,14] or load profiles [4] of conveyor belts. They typically suffer from calibration issues in real-world settings, especially if the estimations are based on static models. This is why we aim to investigate alternative estimation models that take into account time-varying environmental influences.

3 Case Study

3.1 Initial System Design

The first system design which was used consisted of a industrial car PC for logging and processing all machine data, a belt scale as reference for conveyed materials in kg/h and a laser-based volumetric measuring system for providing material data concerning its material flow in m^3/h. For the estimation of conveyed material via power measurement, a three-phase power measurement device was installed.

The first tests using this prototype have shown that the data collected via modbus TCP and CAN bus caused severe difficulties for evaluation and estimation of transported material through analyzing the power draw of the belt motor. Time shifts of supposedly synced data like material/volume flow and power draw, inaccurate detection of belt velocities by the belt scale and uncertain information states whether material was on the conveyor or not made development and research difficult. In addition, the system was running in tests without human supervision and therefore without test protocols. Severe preparation of the collected data was needed to develop a system which was able to predict the amount of transported material within a certain accuracy. The difficulties in the system of the first iteration have led to needed improvements of a updated system design described in the next paragraph.

3.2 New Design

To collect more accurate and reliable data in field tests, the following parts of original system have been modified and substituted. The data collection and processing software is now running on an industrial edge AI computer and the laser-based volumetric measurements system was exchanged with a stereoscopic volumetric measurement system, also capable of providing image data of the belt. The belt scale was substituted with a more accurate one, resulting in better mass flows detection and accuracy in belt speed measurements. The three-phase electric power measurement device was also upgraded to a higher precision class. In addition to the exchanged components, the physical position of the stereoscopic volumetric measurement camera was adjusted to deliver data correctly synced to the power draw measurements.

The described field setup of the measurements on the conveyor is illustrated in Fig. 1.

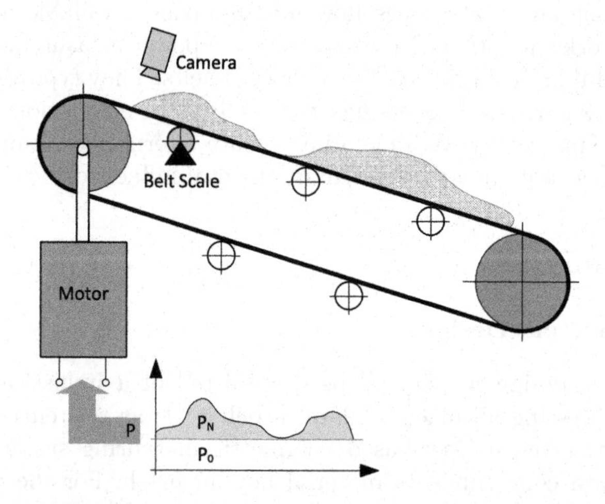

Fig. 1. Conveyor belt with electric drive motor. Belt scale and camera system provide reference measurements.

In addition to these hardware improvements, also the software for overall data collection and handling has been upgraded. A new and improved modular software platform has been developed and implemented that provides a Big Data infrastructure to record all relevant measurement data in the field, including control commands and operator input, which can then be stored on edge as well as transmitted remotely to the cloud.

3.3 Data Aquisition

With the help of the new hardware and software design, new heavily supervised field test have been carried out. Tests with different material loads on the conveyor have been executed strictly following prior defined test cases. The installed belt scale has been calibrated prior to all tests with the help of a bridge scale at the mineral processing plant for assuring ground truth. All necessary data, including 15 images per second of the belt have been collected.

The field tests were carried out under supervision of engineers in the field and researchers remotely following the test via VPN connection, live video streams and live visualisation of all necessary data. Intense collaboration and communication between the edge and cloud team was necessary to guarantee perfect data acquisition in the field. Due to the improvements of the hard- and software system and the detailed planned an executed field tests all prior existing difficulties of the first design and tests have been addressed.

4 Mass Flow Estimation Model and Results

The method for estimating the mass flow on an inclined conveyor belt builds on the physically motivated idea that the transported mass M during a time interval Δ is proportional to the net energy input E_N during that time:

$$E_N(\Delta) = k \cdot M(\Delta), \tag{1}$$

where the coefficient

$$k = k(\rho, \vartheta, \ldots) \tag{2}$$

depends on the friction ρ, temperature ϑ and possibly other external factors. Consequently, by measuring the power consumption P_{total} of the belt drive motor and subtracting the idle power P_0, one can estimate the transported mass $M(\Delta)$ according to

$$M(\Delta) = \frac{1}{k} E_N(\Delta) = \frac{1}{k} \int_{t_0}^{t_0+\Delta} P_{\text{total}}(t) - P_0(t)\, dt. \tag{3}$$

The idle power P_0, which is defined as the electric power demand for moving the empty conveyor belt without material, is affected by external influences such as weather conditions (temperature, humidity, rain, etc.), not all of which can be measured precisely, especially since these external factors are changing over time. Consequently, P_0 cannot be assumed constant, but has to be estimated dynamically over time. This is achieved by, in a first step, identifying idle phases during operation, i.e. time intervals in which the belt is empty and where the measured total active power is equal to the idle power, and then, in a second step, using a regression model to interpolate between these idle phases.

As part of the first proof of concept, a reference measurement system is available, consisting of a camera sensor and a belt scale, that help to identify idle points during operation. In later production use, however, the reference measurement system will not be available and instead the idle point will be defined either via an automated controller or manually by a user-triggered re-calibration.

Using a piece-wise robust regression, combined with extrapolation in the outer areas, the idle power is then estimated based on the identified idle points during operation with material on the belt. Due to recurring idle phases and the resulting support points for regression, this estimate is automatically adapted on an ongoing basis.

Figure 2 illustrates this idle power estimation on an exemplary operation period. Idle points are indicated in red. The plot shows that the idle power is not constant over time, even with repeated idling, which is why a sophisticated and robust estimation of the idle power is crucial. Every now and then, longer idle phases (>1 min) occur, which provide robust indications for the idle power. But also shorter phases are visible in between, e.g. due to discontinuity of the material feed flow, that can provide additional support points.

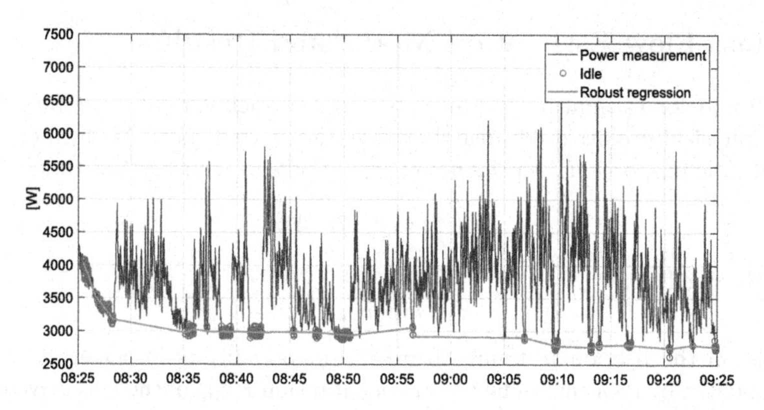

Fig. 2. Power measurement during an example working period with idle points and estimated idle power using robust piece-wise regression.

Another challenge, besides estimating P_0 is how to model the k factor. Comparison of different approaches, including constant and time-dependent models, have shown that a non-linear state-dependent model

$$k = k(P_0) = \alpha_0 + \alpha_1 P_0 + \alpha_2 P_0^2 \tag{4}$$

dependent on P_0 delivers the most promising results. In [6], various models were investigated in this regard and evaluated for their suitability. The model in Eq. 4 is motivated by the assumption that external factors like temperature etc. (see Eq. 2) affect the total power consumption in the same manner as the idle power, and that therefore these influences are also indirectly included with

P_0. Given measurement data for E_{total} and the mass M_{meas} (from the reference measurement system), the model parameters α_i can easily be estimated using linear regression over E_N:

$$E_N = E_{\text{total}} - E_0 \sim (\alpha_0 + \alpha_1 P_0 + \alpha_2 P_0^2) \cdot M_{\text{meas}}. \tag{5}$$

More details on the estimation model and evaluation results are presented in [6].

5 Discussion

5.1 Challenges

Based on the experiences of the case study, several key challenges were identified, which will be briefly discussed below.

Data Quality. A typical problem with data-based models is the need for sufficient high-quality data and measurement scope for model training as well as validation. The data quality also poses great challenges to the measurement equipment, especially when data is to be measured in field operation under harsh conditions like in the mining industry. Judging from the experience gained, data quality was not only the most important challenge in our case study, but also the one having the most significant impact on a successful project outcome, which is why it is of particular importance to assess together with the industry partner how data quality can be maximized in a feasible manner, and together come up with a suitable and realistic test plan for the field studies.

For this reason, various measures were considered to improve data quality, including cleansing, synchronization and redundancy. Based on the experiences gained from the first prototype (see Sect. 3), with data acquisition that included the most relevant features, in the improved design attention was paid not only to expand the scope of data acquisition, taking into account all potentially relevant measurement points (see Big Data infrastructure), but also to improve the accuracy and resolution. This applies e.g. to the belt scale, which now not only provides higher resolution mass flow measurement, but also includes an improved measurement of belt speed, which is crucial for being able to accurately synchronize different time series, as described below.

To further improve the data quality, the measured time series data were cleaned, e.g. outliers were filtered out. This is done on the one hand by explicit restrictions to plausible value ranges (e.g. mass flow to $0 \ldots 400\,\text{t/h}$), on the other hand by automatic detection of faulty measurements or data transmission. Among other things, consistency checks can be made with correlated time series. This can counteract a distortion of the subsequent P_0 estimation due to faulty measuring points.

The installed reference measuring system is also designed redundantly (belt scale and camera measure mass and volume flow, respectively) in order to be able to detect and correct faulty measurements. This is an important lesson learnt from the first prototype, where doubts arose about the accuracy and reliability of the measurements of the belt scale used, what prevented to be able to make reliable statements regarding idle periods, which finally also distorted the P_0 estimation. Therefore, it was decided to perform additional determination of the volume flow.

The redundant measurement of the mass and volume flow allows to carry out consistency checks, which on the one hand led to an increase in data quality for later analyses, and, on the other hand, could show the weaknesses of the conventional systems. For example, it was possible to identify individual peaks in the mass flow that were not due to an actually increased band load, but could be concluded as measurement errors (since these peaks are not reflected in either the power or the volume measurement).

Instead of a file-based data exchange, all time series data from the test series are now collected in an InfluxDB database and are available for replay, further analysis and evaluation. This fusion also makes it much easier to clean up data, to annotate it for documentation purposes and to continue the development of new data-driven models.

The resulting database was not only a necessary prerequisite for a data-driven development of the virtual belt scale, but also represents a significantly improved basis for any further use of the data (e.g. data understanding, visualizations, data evaluations for customers, or further data analyses such as wear detection).

Synchronisation. Especially the synchronization of the measured time series data proved to be difficult in the first prototype iteration. The different features have different time stamps and sampling rates, which made an alignment and synchronization necessary, including interpolation and re-sampling.

There is also a time offset between the mass flow measurements of the belt scale and the camera due to the offset positioning (see Fig. 1). This time offset is of course not constant, but dependent on the belt speed. Consequently, the accuracy of the synchronization also depends on the measurement accuracy of the belt speed.

Calibration and Re-calibration. In the first series of tests it has been shown that, over an extended period of time (e.g. several weeks), there are different external influences that affect the stability of the mass estimation, but not all of them can be accurately recorded or modeled. For example, change of position and location when the machine is moved, different inclination, ambient temperature and humidity influence mechanical stress of the belt (in a non-linear manner) and thus the friction losses.

Partial countermeasures can be taken by appropriate modeling of the features, e.g. by selecting the k factor as a function of P_0 to capture influences that also have an effect on P_0.

Furthermore, this problem is mitigated by periodic re-calibrations. For example, recurring short idle phases can be inserted throughout the working period, e.g. 30 s every 4 h, which then act as supporting points for the P_0 estimation. This is especially necessary at the beginning of a new work period or after changeover/realignment of the machine.

The necessity for re-calibration can be seen as a disadvantage of the existing virtual belt scale approach. However, the experience of the industry partner has shown that even cost-intensive conventional belt scales in operation over a longer period of time usually do not get by with a one-time calibration but also have to be re-calibrated periodically (which often has to be done manually) in order to maintain sufficient accuracy.

5.2 Academia-Industry Collaboration

A big success factor for fruitful collaboration between industry and academia, besides having a clearly defined common goal, is to have a frequent exchange of knowledge and experience in close coordination. In periodic weekly or bi-weekly meetings, industrial and academic partners compare not only the progress of the project in a continuous manner, but also give updates on any changes in objectives. For example, after the first series of tests on the prototype, it became apparent that the volume and quality of the measurement data were not sufficient, whereupon joint discussions were held to work out which measures could be implemented with reasonable effort and if these envisioned measures would be sufficient so that the data-bases models could meet their performance targets.

It is difficult to say in advance which methods for predicting the mass flow are going work well and deliver satisfying accuracy. This is why it is important on both sides to be willing to iterate on goals and progress rather than expecting a one-time linear technology transfer from academia into application. Instead of focusing on a final solution, it is, like everywhere in software engineering, more important to put in place an operational workflow as well as infrastructure to iteratively extend and improve the solution. The individual iteration steps are extensively tested to detect problems and learn from them in an effort to mitigate similar problems in the future.

During the development, the industrial partners contribute their domain expertise from the mining industry, the production processes and equipment, and the academic partners support with their experience in data science methods. This creates a knowledge transfer that benefits both sides, allowing to combine data-driven methods from academia with physics-based approaches to achieve better results. Results from research can be transferred into practice and can be used in a real application and bring innovation and added value to the industry.

6 Conclusions

In this work, we have presented a method for online estimation of the mass flow on a conveyor belt in mineral processing applications, which has been developed in collaboration between industry and academia. By leveraging the skillsets of both parties, methodical results from research are brought to real-world applications.

One of the key challenges in the development of a virtual belt scale is that it needs to work robustly in the presence environmental influences and disturbance factors, especially when trying to estimate the idle power P_0. Observations have shown that idle power can change significantly, sometimes rapidly, during operation and cannot be assumed to be even approximately constant [6]. The more P_0 support points available, the better the estimate. In practice however, during operation one would like to get along with as few P_0 support points as possible.

Over the course of different working periods (e.g. on different days with different weather conditions) an accurate estimation of the mass flow is difficult. This is why we want to improve robustness and accuracy of the model in the future, by continuing close collaboration between scientific and industrial partners to collect more field data over an extended period of time in order to perform a more comprehensive statistical analysis on additional model features (e.g. temperature). It is to be expected that a greater quantity and quality of collected data could provide us with an improvement on accuracy.

Acknowledgements. The research presented in this paper has been funded by the Austrian Federal Ministry for Climate Action, Environment, Energy, Mobility, Innovation and Technology (BMK), the Federal Ministry for Digital and Economic Affairs (BMDW), and the Province of Upper Austria in the frame of the COMET Competence Centers for Excellent Technologies Programme and the COMET Module S3AI managed by the Austrian Research Promotion Agency FFG.

References

1. Al-Thyabat, S., Miles, N., Koh, T.: Estimation of the size distribution of particles moving on a conveyor belt. Miner. Eng. **20**(1), 72–83 (2007)
2. Deng, F., Liu, J., Deng, J., Fung, K.S., Lam, E.Y.: A three-dimensional imaging system for surface profilometry of moving objects. In: 2013 IEEE International Conference on Imaging Systems and Techniques (IST), pp. 343–347. IEEE (2013)
3. Djokorayono, R., Arfittariah, Priantama, D.B., Biyanto, T.R.: Design of belt conveyor weight scale using gamma radiation technique. In: AIP Conference Proceedings, vol. 2088, p. 020048. AIP Publishing LLC (2019)
4. Gebler, O.F., Hicks, B., Yon, J., Barker, M.: Characterising conveyor belt system usage from drive motor power consumption and rotational speed: a feasibility study. In: European Conference of the Prognostics and Health Management Society (PHM Society)
5. Guyot, O., Monredon, T., LaRosa, D., Broussaud, A.: Visiorock, an integrated vision technology for advanced control of comminution circuits. Miner. Eng. **17**(11–12), 1227–1235 (2004)

6. Heinzl, B., Martinez-Gil, J., Himmelbauer, J., Roßbory, M., Hinterdorfer, C., Hinterreiter, C.: Indirect mass flow estimation based on power measurements of conveyor belts in mineral processing applications. In: Proceedings of the 19th International Conference on Industrial Informatics (INDIN), pp. 1–6 (2021). https://doi.org/10.1109/INDIN45523.2021.9557482
7. Hulthén, E.: Real-Time Optimization of Cone Crushers. Chalmers Tekniska Hogskola, Sweden (2010)
8. Itävuo, P., Hulthén, E., Yahyaei, M., Vilkko, M.: Mass balance control of crushing circuits, vol. 135, pp. 37–47. https://doi.org/10.1016/j.mineng.2019.02.033
9. Jeinsch, T., Sader, M., Ding, S., Engel, P., Jahn, W., Niemz, R.: A model-based information system for simulation and monitoring of belt conveyor systems. IFAC Proc. Volumes **33**(26), 637–642 (2000)
10. Kaartinen, J.: Machine Vision in Measurement and Control of Mineral Concentration Process. http://lib.tkk.fi/Diss/2009/isbn9789512299553/
11. Mkwelo, S.: A machine vision-based approach to measuring the size distribution of rocks on a conveyor belt. Ph.D. thesis, University of Cape Town (2004)
12. Rehman, T., Tahir, W., Lim, W.: Kalman filtering for precise mass flow estimation on a conveyor belt weigh system. In: Zhang, D., Wei, B. (eds.) Mechatronics and Robotics Engineering for Advanced and Intelligent Manufacturing. LNME, pp. 329–338. Springer, Cham (2017). https://doi.org/10.1007/978-3-319-33581-0_25
13. Taskinen, A., Vilkko, M., Itävuo, P., Jaatinen, A.: Fast size distribution estimation of crushed rock aggregate using laserprofilometry. IFAC Proc. Volumes **44**(1), 12132–12137 (2011)
14. Väyrynen, T., Itävuo, P., Vilkko, M., Jaatinen, A., Peltonen, M.: Mass-Flow Estimation in Mineral-Processing Applicationsm vol. 46(16), pp. 271–276. https://doi.org/10.3182/20130825-4-US-2038.00023,www.sciencedirect.com/science/article/pii/S1474667016313210
15. Yamazaki, T., Sakurai, Y., Ohnishi, H., Kobayashi, M., Kurosu, S.: Continuous mass measurement in checkweighers and conveyor belt scales. In: Proceedings of the 41st SICE Annual Conference. SICE 2002, vol. 1, pp. 470–474. IEEE (2002)
16. Zhao, L.: Typical failure analysis and processing of belt conveyor. Proc. Eng. **26**, 942–946 (2011)

A Table Extraction Solution for Financial Spreading

Duc-Tuyen Ta([✉]), Siwar Jendoubi, and Aurélien Baelde

Upskills R&D, 16 Rue Marc Sangnier, 94600 Choisy-le-Roi, France
{ductuyen.ta,siwar.jendoubi,aurelien.baelde}@upskills.ai
http://www.upskills.com

Abstract. Financial spreading is a necessary exercise for financial institutions to break up the analysis of financial data in making decisions like investment advisories, credit appraisals, and more. It refers to the collection of data from financial statements, where their extraction capabilities are largely manual. In today's fast-paced banking environment, inefficient manual data extraction is a major obstacle, as it is time-consuming and error-prone. In this paper, we, therefore, address the problem of automatically extracting data for Financial Spreading. More specifically, we propose a solution to extract financial tables including Balance Sheet, Income Statement and Cash Flow Statement from financial reports in Portable Document Format (PDF). First, we propose a new extraction diagram to detect and extract financial tables from documents like annual reports; second, we build a system to extract the table using machine learning and post-processing algorithms; and third, we propose an evaluation method for assessing the performance of the extraction system.

Keywords: Table extraction · Deep learning · Financial spreading

1 Introduction

Financial Spreading refers to the collection of data from financial statements for various analyses of banks and financial institutions. This data is essential in making decisions like investment advisories, credit appraisals, etc. Presently, it is increasingly becoming a significant challenge since their extraction capabilities are largely manual. Nowadays, the inefficient data extraction is a major impediment, as it is time-consuming and error-prone. Then, there is a dire need to address this issue. In general, financial data is often presented in tabular form and is appreciated by financial workers. However, extracting tabular information manually is often a tedious and time-consuming process. For these reasons, financial institutions are trying to find an automated solution to extract financial tables for their needs.

In the past two decades, a few methods and tools have been devised for table extraction from documents. However, these solutions contains some shortages. First, such solutions cannot automatically detect the page with relevant financial

G. Kotsis et al. (Eds.): DEXA 2022 Workshops, CCIS 1633, pp. 100–111, 2022.
https://doi.org/10.1007/978-3-031-14343-4_10

table. They need a manual input from the user. Second, these methods detect tables page-by-page, and cannot process tables that are spread on multiple pages. Finally, the table structure is not well restructured.

In this paper, a solution to solve the problem of automatic table extraction for financial spreading is introduced. More specifically, financial tables such as balance sheets, income statements and cash flow statements are automatically identified, extracted and corrected from PDF-formatted financial reports. The contributions of this work are the following:

1. A complete solution for financial table extraction is presented. It can automatically locate the financial table inside the document, extract, process and reconstruct the table.
2. A new similarity metric, *ExactMatchSim*, and an evaluation method are introduced to evaluate the extracted financial tables with the ground truth.
3. Experiments on real-world data are implemented to confirm the performance of the proposed system.

In the following, table extraction for financial spreading related work is reviewed in Sect. 2. Session 3 provides detailed information on the table extraction method from a document and the complete pipeline for financial table extraction. Section 4 introduces the evaluation measure and method. Then it details the considered datasets and the system performance results. Finally, Sect. 5 concludes and summarizes our work.

2 Related Work

This section, first, familiarizes the reader with an overview of table extraction technologies, the financial tables extraction related work, and the available extraction evaluation methods.

2.1 Table Extraction Technologies

Table extraction methods generally focus on the processing of: i) text-based only PDF document and ii) the general document with both text and image. For the text-based document, table extraction methods are generally based on the layout of the text in the PDF page (e.g., text boxes coordinates and/or vertical/horizontal lines structure). Several tools are proposed, such as Camelot [2] and Tabula [13]. In these tools, two common extraction methods are Stream (which uses the text boxes coordinates to reconstruct the table) and Lattice (which is based on image processing to detect the vertical/horizontal lines and reconstruct the table).

In the general case of table extraction, Deep-learning based architectures are considered [3,4,15]. The proposed solutions include accurate detection of the tabular region within an image, and subsequently detecting and extracting information from the rows and columns of the detected table. These solutions include CDecNet [1], TableNet [8], CascadeTabNet [9] and DeepDeSRT [11].

However, these solutions need powerful hardware resources and are not optimized for the case of long and complex tables like financial tables. In [7], the authors proposed the FinTabNet dataset for financial table recognition and a method to extract the content of these tables.

In general, financial documents contains the following characteristics: (i) financial tables are not clearly marked - there is no information showing where the financial tables are located in the document; (ii) there is no common template for financial tables; (iii), tables can be spread over many pages; and (iv) financial documents can be large, ranging from tens to hundreds of pages. It means, to solve the problem of automatically extracting financial statements from a document, a table extraction solution needs to be able to automatically locate the considered table in the document. Unfortunately, the above solutions work only with page numbers as input, which means that the location of the table has to be provided manually by the user. The same issue is encountered with cloud solutions like Amazon Textract[1] or Azure Form Recognizer[2].

2.2 Quality Evaluation Methods

The problem of table extraction quality evaluation is not well studied in the literature. In fact, there are few works that addressed this problem and proposed an adaptation of precision, recall and F1-score to tables [5,6]. To compute these measures, the table structure is transformed to "cell adjacency relations". It defines the relation between each content cell with its nearest neighbor in horizontal and vertical directions, no adjacency relations are generated between blank cells or a blank cell and a content cell [5]. Cells adjacency relations of the ground truth tables are then compared to cells adjacency relations of the predicted tables to compute the recall, precision and F1 score. However, these measures have two main drawbacks: 1) Only immediate adjacency relations between content cells are considered, thus, this representation cannot detect errors caused by empty cells and misalignment of cells. 2) These measures consider the exact match of the content, which is not always relevant for the evaluation [15].

To avoid these problems, another measure was introduced: Tree-Edit-Distance-based Similarity (TEDS) [15]. This measure uses HTML tags to represent the table, and consider comparing such representation of the ground truth and the predicted table. However, this representation could lead to a wrong structure if a tag is missing or if there is some difference in row/col span, this wrong structure will consider the predicted table as different from the ground truth.

In general, available table extraction evaluation measures need to transform the table to an intermediate structure. This intermediate structure could contain information like cells relations, cells boundaries, etc. This structure is complicated to gather, and existent table extraction solution do not produce it. This transformation complicates the use of state-of-the-art measures to evaluate a given table extraction solution.

1 https://aws.amazon.com/textract/.

2 https://azure.microsoft.com/services/form-recognizer/.

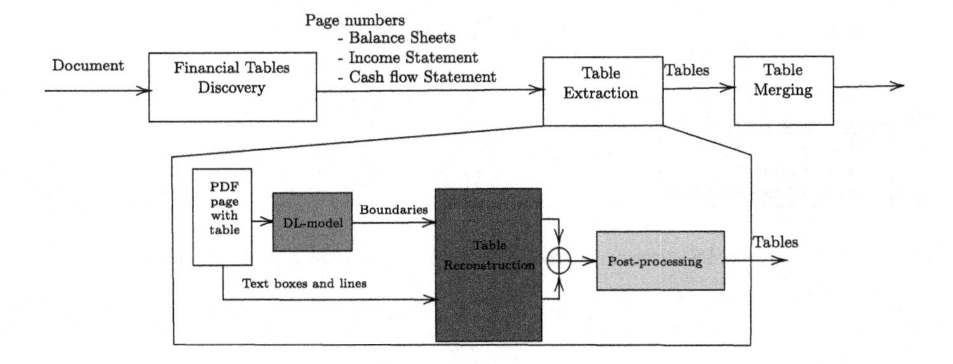

Fig. 1. The overall pipeline of table extraction system from financial report.

3 Table Extraction for Financial Spreading

This section explains the pipeline to extract financial tables for financial spreading process. Then, it presents the page identification and the extraction of financial tables from the document.

3.1 Extraction Pipeline for Financial Spreading

We propose a financial tables' extraction pipeline defined in Fig. 1. First, the financial table is identified using the financial table discovery process. Results contain table types (e.g., balance sheet, income statement, etc.) and corresponding pages. These results are then the input to the table extraction process, where the table is extracted from each document's page. Finally, extracted tables are merged to create the full financial table.

3.2 How to Identify Financial Tables

Figure 2 presents the method to identify financial tables from the document. For the financial spreading process, three types of financial tables are considered: Balance Sheet, Income Statement and Cash Flow Statement. To identify these financial tables, the following process were considered: first, matching patterns are defined for each table type, these patterns are useful to define regular expressions filters that find these patterns in the document. For example, a page with a balance sheet often contains "*consolidated balance sheets*" or "*consolidated financial statements*". These regular expression filters are then used to get the set of potential pages[3]. Finally, potential pages are filtered by the consecutive

[3] The table's title and its text are collected, and similar patterns are then merged to create matching patterns. In the real system, many regular expression patterns are designed. One matching pattern is used in the filter to detect a potential page with financial tables. Others are then used to filter these pages into the potential balance sheet, potential income statement, and potential cash flow statement through the respective regular expression filters.

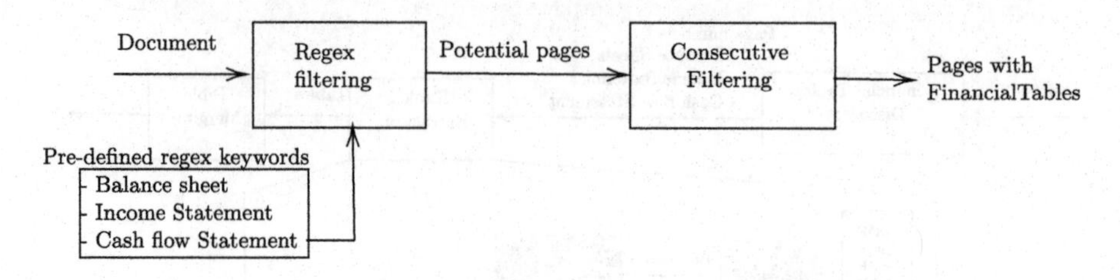

Fig. 2. Financial tables discovery process from the financial report. Regex means regular expression.

filtering to obtain the final result that thanks for their consecutive presentation[4] in the document.

3.3 How to Extract a Table from a Document

When the table's page is determined, we then study the question of how to extract the table's content. Such a question contains two main tasks:

- First, table detection: This task aims to identify the page region that contains a table.
- Second, table structure reconstruction: This task aims to recover the table into its components (e.g., headers, columns and rows structures, correct allocation of data units, etc.) as close to the original table as possible.

Current work detects and extracts the table by using the coordinates of text boxes and lines in the page, such as Camelot [2] or Tabula [13]. Other works [4,11,12,14] directly use deep learning (DL) models to detect the table and corresponding cells inside, thereby reconstructing the table structure.

In this paper, we propose a hybrid table extraction method, highlighted in the diagram in Fig. 1. First, a DL model is used to detect the table's boundaries. Next, from the detected boundaries, we combine the coordinates of the text boxes and lines inside the detected region to identify and recreate the table structure. We use the DL model to detect table boundaries by first converting the document to an image and then detecting the table's region. This result is normalized according to the PDF document page size, creating the corresponding boundary of the table. Yolo v3 [10] was chosen due to the trade-off between model size, computation time and accuracy. The model is then trained on the FinTabNet [14] dataset since such a dataset is specially designed for the task of extracting tables in financial statements. Compared to other models like CascadeTabNet [9], the trained Yolo model is much smaller (30 Mb *v.s.* 1 Gb). As a result, the inference time is smaller and no special hardware like GPU is needed to run the model. In addition, the trained model is better able to detect the boundaries of large tables, such as financial statements, than the CascadeTabNet model.

[4] The financial tables are presented consecutively, however, in no fixed order.

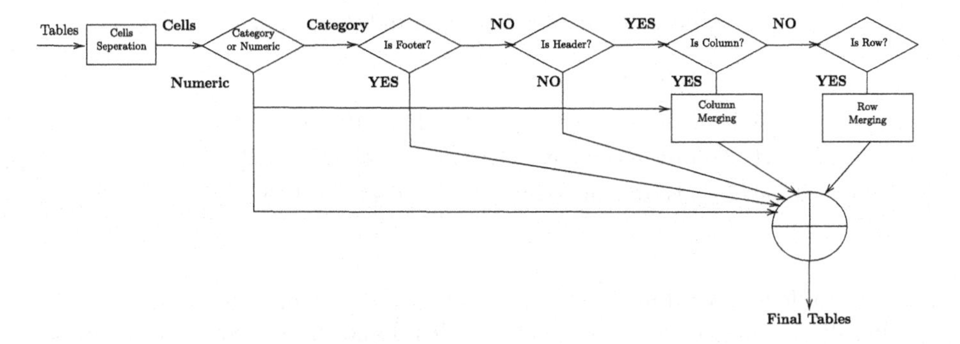

Fig. 3. The post-processing algorithm.

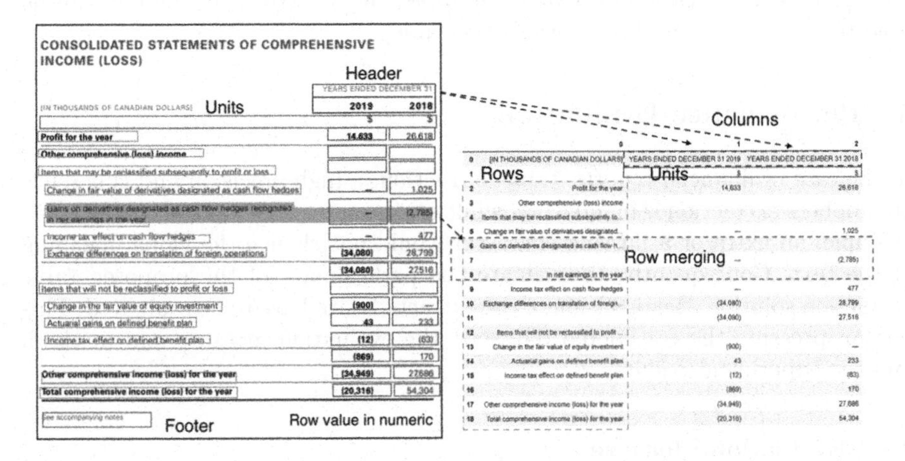

Fig. 4. An example of the post-processing process for the financial table. The left side is the original table structure. The right side is the extracted table with the post-processing steps that needs to be applied to correct the table.

Next, the predicted boundaries are used to reconstruct the table. These values are normalized to the corresponding value on the pdf document by comparing the image size and page size. Using the Lattice and Stream extractors from [2], we first compare the extracted results from two extractors with and without table's boundary. If the results are the same, then we have the table. Otherwise, we compute the overlap between these tables and combine them together.

Finally, the post-processing is used to clean the extracted table. Indeed, the extracted table may contain some extra text like the table name or the table's footer, the header with nested structure, row that is separated into multiple lines, etc. Therefore, a post-processing step is required to repair and clean the table. Step-by-step post-processing process is showed in Fig. 3. The idea behind is that, by identifying the type of each cell in the table, we can clean extra content, and separate fused cells with different content or combine cells belonging to the same row, to reconstruct the table. An example of the post-processing for a

Table 1. Table transformation to lists example

Table			Lists of values
1	23	50	List of rows: [[1, 23, 50], [2, 24, 10], [4, 30, 20]]
2	24	10	List of columns: [[1, 2, 4], [23, 24, 30], [50, 10, 20]]
4	30	20	List of numbers: [1, 23, 50, 2, 24, 10, 4, 30, 20]

financial table is showed in Fig. 4. Note that such a post-processing algorithm is specifically designed for financial tables, which do not have a common template but represent the same type of information. Special table elements, such as time, unit, header, the location of text in the table, the type of each cell, etc. can be used to post-process and reconstruct the table.

4 Performance Evaluation

In the case of financial reports' table extraction, a high precision and coherence on numbers extraction is mandatory, while errors on text are more acceptable, for example, an extra or a missing character would not change the global meaning of the text. Consequently, to evaluate the performance of the proposed table extraction solution, the precision of the extracted data have to be evaluated. This section presents the considered datasets, the evaluation measure and method, and the experiment results.

4.1 Evaluation Measure

This section presents the similarity measure that is used to evaluate the extracted table. This measure considers the table as lists of values and compares the matching between these lists. For example, a given table can be seen as a list of rows and evaluated through comparing rows. Or, lists could be lists of numbers or texts to study the performance over each data type separately. Table 1 presents an example of table transformation to list.

The exact matching similarity measure is a similarity measure that is simple and fast to compute. It allows comparing the content of two lists of values, and could be used to compare the content of two tables.

The exact matching similarity between two lists of values is defined as the proportion of common unique values from the total number of unique values, as follows:

$$ExactMatchSim(P1, P2) = \frac{2 * unique(P1 \cap P2)}{unique(P1) + unique(P2)} \tag{1}$$

This measure is useful to compare table rows, columns, list of numbers and list of texts. It considers the exact matching inside the given two lists and returns its proportion compared to all unique values in the two lists. When this evaluation

Table 2. Computation example of the *ExactMatchSim* measure between ground truth and extracted rows.

Ground truth	"North America", 2000, 30, 10
Extracted	"North America", 2000, 30, 10, 10
ExactMatchSim	$8/8 = 1$

measure is equal to 100%, it means that the two compared lists contain the same unique values.

Table 2 presents a computation example of the *ExactMatchSim* measure. It returns 100% because the two rows contain the same unique value. Besides, the extra value is redundant and has no effect with this measure.

4.2 Evaluation Method

ExactMatchSim similarity measure is used to evaluate the similarity between two tables through four different evaluation approaches:

1. *Tables similarity according to rows:* each row in the ground truth table is matched with its most similar row in the target table. Then, the mean rows' similarity is considered as the global similarity between the two tables.
2. *Tables similarity according to columns:* each column in the ground truth table is matched with its most similar column in the target table. The mean columns' similarity is considered as the global similarity between them.
3. *Tables similarity according to numbers:* for each table, the numbers are extracted on a separate list. Next, number lists are compared using the similarity measure.
4. *Tables similarity according to texts:* for each table, the texts are extracted on a separate list. Next, texts lists are compared using the similarity measure.

4.3 Datasets

Current available public datasets are not designed for the task of extracting financial tables. The closest dataset to our work, FinTabnet [14], is designed for table detection and recognition but not for table discovery. Additionally, the use of matching tokens between the PDF and HTML versions of each table in FintabNet does not represent the table structure correctly in some cases, especially with complex tables. To evaluate the proposed financial table extraction, we therefore manually annotate a dataset of 100 different companies' annual reports in different languages containing 291 full tables. The data contains pages with the Balance Sheet, Income Statement, and Cash Flow Statement tables as well. This dataset is called Dataset-A. Detailed types and quantity distribution of tables are shown in Table 3.

Table 3. Table distribution of our benchmark dataset.

Table types	Number of tables	Single-page table	Multiple pages table
Balance sheet	98	55	43
Income statement	93	41	52
Cash flow statement	100	57	43
Total	**291**	**153**	**138**

Table 4. Table page discovery evaluation results on dataset-A.

Table	# of table Pages	Total documents	Detection accuracy
Balance sheet	1	**55**	**95%**
	2	**27**	**90%**
	3	**16**	**88%**
Income statement	1	**41**	**95%**
	2	**37**	**87%**
	3	**15**	**85%**
Cash flow statement	1	**57**	**95%**
	2	**35**	**87%**
	3	**18**	**85%**

For evaluating the extraction content, twelve companies' annual reports in English[5] were selected from dataset-A. From these documents, three types of tables were considered: Balance sheet (14 table pages were found in these reports), Income statement (21 table pages) and Cash flow statement (15 table pages). Next, 50 tables were extracted from these reports and manually annotated to get the table content in Excel format for each considered table. This dataset is called dataset-B. To evaluate the proposed system performances on this dataset, the table extraction system was used to extract the 50 annotated tables. In the experiments, the system failed to detect and extract one table and extracted 49 from the 50.

4.4 Results and Discussion

First, we evaluate the table page discovery algorithm. According to Table 4, we find that this algorithm works well when tables span over 1 or 2 pages. For the case of a 3-page table, the accuracy is reduced to 85%. We guess this happens since some pages only contain a word like "(*continued*)" without a header. The table extraction process is then applied by using the proposed table reconstruction algorithm.

[5] Due to the lack of human resources, only English reports were selected to be able to accurately annotate the table contents.

Table 5. Table extraction results on dataset-A.

	Page with 1 table	Page with 2 tables	Page with 3 tables
Number of tables	253	74	3
Number of detected tables	**244**	**72**	**3**
Percent of detected tables	**96%**	**97%**	**100%**

Table 6. PDF files evaluation results on dataset-B, the mean similarity over all extracted tables were considered.

	ExactMatchSim (%)
Rows	92.1
Columns	94
Numbers	95.7
Texts	91.5

Next, we consider the extract capability of the table extraction algorithm. Table 5 shows the number of extracted tables for each document's page in different cases. According to this experiment, the system detected more than 96% of tables. The detection error mainly occurs when the financial table contains two subsections with gaps, then the system detects only part of the table instead of the whole table.

In the next experiments, extracted tables were compared to their corresponding annotated tables (ground truth). For simplicity of annotation, we only evaluate tables spanning over only one page, but not merged tables from multiple pages. Overall experimented tables of dataset-B, the system missed one table and extracted 49 tables from 50. The 49 tables were evaluated, and mean similarities over all files are shown in Table 6. According to Table 6, the measured exact matching similarity scores are above 92% on rows and columns, which means that the extracted tables content is close to the ground truth. Results on Numbers are higher than 95% which confirm that the content is well extracted. Besides, results for Texts are equals to 91.5%. According to this experiment, data extraction from PDF files shows good quality results for rows and columns similarity, extracted numbers and texts similarity. The performance is not 100% because some tables were not fully detected, and their extraction is missing some data.

5 Conclusions and Perspectives

In this paper, we propose a full process to automatically extract financial statements from PDF files in the context of financial spreading. Meanwhile, we propose a new similarity measure to evaluate the performance of the system. We also discuss its performance and some possible difficulties by extracting tables from financial statements.

For future work, we will focus on the case of financial table extraction from scanned PDF files, and we will focus also on the evaluation and work on a measure that penalize the extra and missing data in the extraction results.

References

1. Agarwal, M., Mondal, A., Jawahar, C.V.: Cdec-net: composite deformable cascade network for table detection in document images. In: 2020 25th International Conference on Pattern Recognition (ICPR), pp. 9491–9498. IEEE (2021)
2. Camelot: PDF Table Extraction for Humans. http://camelot-py.readthedocs.io/
3. Cesarini, F., Marinai, S., Sarti, L., Soda, G.: Trainable table location in document images. In: Object Recognition Supported by User Interaction For Service Robots, vol. 3, pp. 236–240. IEEE (2002)
4. Gilani, A., Qasim, S.R., Malik, I., Shafait, F.: Table detection using deep learning. In: 2017 14th IAPR International Conference on Document Analysis and Recognition (ICDAR), vol. 1, pp. 771–776. IEEE (2017)
5. Göbel, M., Hassan, T., Oro, E., Orsi, G.: A methodology for evaluating algorithms for table understanding in pdf documents. In: Proceedings of the 2012 ACM symposium on Document Engineering, pp. 45–48 (2012)
6. Li, M., Cui, L., Huang, S., Wei, F., Zhou, M., Li, Z.: Tablebank: table benchmark for image-based table detection and recognition. In: Proceedings of the 12th Language Resources and Evaluation Conference, pp. 1918–1925 (2020)
7. Li, Y., Huang, Z., Yan, J., Zhou, Y., Ye, F., Liu, X.: GFTE: graph-based financial table extraction. In: Del Bimbo, A., et al. (eds.) ICPR 2021. LNCS, vol. 12662, pp. 644–658. Springer, Cham (2021). https://doi.org/10.1007/978-3-030-68790-8_50
8. Paliwal, S.S., Vishwanath, D., Rahul, R., Sharma, M., Vig, L.: Tablenet: deep learning model for end-to-end table detection and tabular data extraction from scanned document images. In: 2019 International Conference on Document Analysis and Recognition (ICDAR), pp. 128–133. IEEE (2019)
9. Prasad, D., Gadpal, A., Kapadni, K., Visave, M., Sultanpure, K.: CascadeTabNet: an approach for end to end table detection and structure recognition from image-based documents. In: Proceedings of the IEEE/CVF Conference on Computer Vision and Pattern Recognition Workshops, pp. 572–573 (2020)
10. Redmon, J., Divvala, S., Girshick, R., Farhadi, A.: You only look once: Unified, real-time object detection. In: Proceedings of the IEEE Conference on Computer Vision and Pattern Recognition, pp. 779–788 (2016)
11. Schreiber, S., Agne, S., Wolf, I., Dengel, A., Ahmed, S.: Deepdesrt: Deep learning for detection and structure recognition of tables in document images. In: 2017 14th IAPR International Conference on Document Analysis and Recognition (ICDAR), vol. 1, pp. 1162–1167. IEEE (2017)
12. e Silva, A.C., Jorge, A.M., Torgo, L.: Design of an end-to-end method to extract information from tables. Int. J. Doc. Anal. Recogn. (IJDAR) **8**(2–3), 144–171 (2006)
13. Tabula-py. https://tabula-py.readthedocs.io/en/latest/

14. Zheng, X., Burdick, D., Popa, L., Zhong, X., Wang, N.X.R.: Global table extractor (GTE): A framework for joint table identification and cell structure recognition using visual context. In: Proceedings of the IEEE/CVF Winter Conference on Applications of Computer Vision, pp. 697–706 (2021)
15. Zhong, X., ShafieiBavani, E., Jimeno Yepes, A.: Image-based table recognition: data, model, and evaluation. In: Vedaldi, A., Bischof, H., Brox, T., Frahm, J.-M. (eds.) ECCV 2020. LNCS, vol. 12366, pp. 564–580. Springer, Cham (2020). https://doi.org/10.1007/978-3-030-58589-1_34

Synthetic Data in Automatic Number Plate Recognition

David Brunner[1](✉) and Fabian Schmid[2]

[1] Software Competence Center Hagenberg, Softwarepark 32a, 4232 Hagenberg im Mühlkreis, Austria
david.brunner@scch.at

[2] Kapsch TrafficCom, Am Europlatz 2, 1120 Vienna, Austria

Abstract. Machine learning has proven to be an enormous asset in industrial settings time and time again. While these methods are responsible for some of the most impressive technical advancements in recent years, machine learning and in particular deep learning, still heavily rely on big datasets, containing all the necessary information to learn a particular task. However, the procurement of useful data often imposes costly adjustments in production in case of internal collection, or has copyright implications in case of external collection. In some cases, the collection fails due to insufficient data quality, or simply availability. Moreover, privacy can be an ethical as well as a legal concern. A promising approach that deals with all of these challenges is to artificially generate data. Unlike real-world data, purely synthetic data does not prompt privacy considerations, allows for better quality control, and in many cases the number of synthetic datapoints is theoretically unlimited. In this work, we explore the utility of synthetic data in industrial settings by outlining several use-cases in the field of Automatic Number Plate Recognition. In all cases synthetic data has the potential of improving the results of the respective deep learning algorithms, substantially reducing the time and effort of data acquisition and preprocessing, and eliminating privacy concerns in a field as sensitive as Automatic Number Plate Recognition.

Keywords: Synthetic data · Automatic number plate recognition · Deep learning

1 Introduction

Research in deep learning, for the most part, is benchmark driven and new results are reported on long established datasets [2,8,10]. However, these datasets are not always an accurate or complete representation of the real world, making it difficult to transfer promising results in academia to industrial settings. In these cases, companies are forced to curate their own datasets, which can be extraordinarily challenging for a multitude of reasons. For one, the real-world class distributions might not mirror those in an idealized setting, creating class-imbalances and skewing the results in the direction of the more abundant classes.

G. Kotsis et al. (Eds.): DEXA 2022 Workshops, CCIS 1633, pp. 112–118, 2022.
https://doi.org/10.1007/978-3-031-14343-4_11

Furthermore, even if available, by virtue of its nature the data might come with serious implications on the privacy of the companies customers. Lastly, in contrast to carefully curated benchmark datasets, real-world data also often is messy and requires additional processing to remediate imperfections.

Kapsch TrafficCom AG (KTC) offers an example for an industrial player, affected by all these challenges. Their portfolio includes intelligent traffic solutions, of which Automatic Number Plate Recognition (ANPR) is an important constituent. ANPR operates on images of vehicles carrying number plates, which are inherently privacy sensitive. Their collection also requires specialized hardware and they often show qualitative flaws nonetheless. A means to battle all of these issues that currently receives much attention is synthetic data [13]. Instead of laboriously collecting samples in the real world, artificial representations of the data in demand are generated. In many cases, the process of generation offers much more control over the quality and distribution of the resulting data. In a classification scenario the number of samples per class can be determined before generation. Since the samples are completely fabricated, no issues regarding the privacy of individuals can occur. To satisfy quality stipulations, botched samples can easily be discarded and new samples created in their stead. The options for generating synthetic data are diverse. In composition [19] existing real-world samples are mixed, resulting in new, artificial samples. The much studied Generative Adversarial Net (GAN) [4] uses the power of deep learning to create completely synthetic samples, following the distribution of the real data. An especially elaborate technique of producing prior unseen samples, is by way of simulation [1,6,14,15,17].

In this paper, we present intermediate results of KTC's ongoing collaboration with the Software Competence Center Hagenberg (SCCH), aiming at exploiting the benefits of synthetic data for KTC's efforts in the field of ANPR. We delineate use-cases in which the procurement of real-world data is problematic for the described reasons, and discuss how synthetic data poses a feasible remedy. Our goal is to emphasize the potential and versatility of synthetic data in real-world settings. We end by providing details on the collaboration between SCCH and KTC.

2 Methodology

2.1 Automatic Number Plate Recognition

ANPR is an important task in traffic analysis. It opens the door to a wide variety of applications, for example automatic tolling on motorways, law enforcement purposes (e.g. the search for stolen vehicles) or the management of private car parks. ANPR is typically split into multiple stages. Frequently, the crucial character recognition stage is preceded by a plate localisation and character segmentation stage. Optionally, the pipeline can also include an initial vehicle detection step, see [9] for an example of such a full pipeline. While advancements in machine learning and computer vision have made it trivial to implement a rudimentary form of ANPR with decent performance (e.g. [16]), the challenge lies in the fact that in a real-world setting ANPR becomes viable only when very high standards of recognition accuracy are met. The collection of massive amounts of data, as required to meet these high standards, is afflicted by many of the discussed problems.

2.2 Use Cases

Font Generation. In terms of data availability, rare characters are a known issue in ANPR. In the real world, not all characters are equally likely to be featured on a number plate, in fact, some characters are even specific to certain countries. Training on an imbalanced dataset makes the ANPR system error prone, whenever rare characters are involved. A simple solution is the generation of synthetic characters, balancing the dataset. This can be accomplished by training a conditional GAN (cGAN) [12] on the rare characters, which can then produce convincing new samples. Figure 1 shows synthesized outputs for rare character classes.

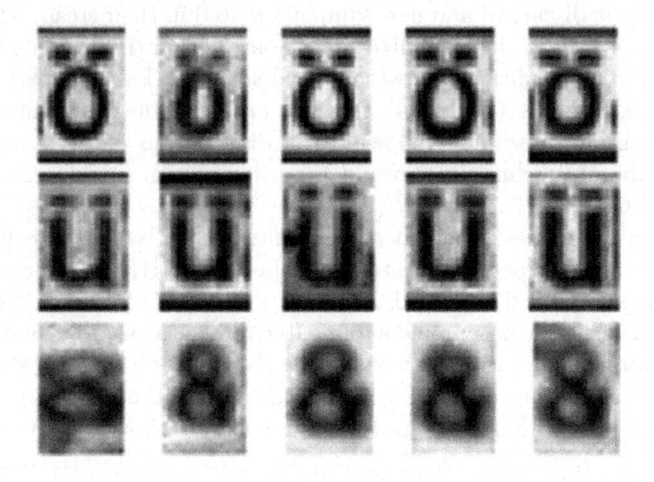

Fig. 1. Examples for the generation of rare characters.

Number Plate Generation. Unlike the final character recognition stage, which is trained on character snippets, all prior stages operate on whole number plates. Number plates are highly sensitive in terms of privacy, which means both acquisition and storage of images containing them have legal implications. Conversely, synthetic number plates are free of such considerations, since they do not correspond to individuals in the real world. Much of the prior work uses number plate templates, created with a vision library like OpenCV and filled with realistic combinations of characters. These are either used directly [11], or transformed via image-to-image translation to increase realism [5,20]. While the former lack said realism, the diversity of the latter is also constrained by the templates used. We on the other hand, present preliminary results of synthesizing number plates purely from noise, similar to the font generation described above. For this purpose, we employ StyleGAN2 [7], a powerful state-of-the-art GAN architecture. The difficulty of this approach lies in the reduced level of supervision. Since no shape priors are given, the network has to learn the shapes of characters in training. Figure 2 demonstrates StyleGAN2's capability of doing so. While getting rid of the necessity of templates, one drawback of this approach is that like with a real world dataset, subsequent annotation is necessary.

Fig. 2. An arbitrary selection of synthetic number plates generated with StyleGAN2. For the unlikely case of collisions with real plates we anonymize the plates.

Number Plate Cleaning. Aside from data availability and privacy, another issue ANPR is confronted with is data quality. In the real world number plates can be damaged, worn out, covered in dirt or simply occluded. At times these imperfections can be fairly subtle and still lead to a drastic decline in recognition accuracy. Inspired by [18], we set out to "clean" faulty samples with the help of CycleGAN [21]. For each category of fault, we build a faulty and a clean dataset, and train an instance of CycleGAN to transform one into the other. GANs are known to struggle with geometrical transformations [3, 21], whereas they excel in transferring image styles. This is reflected in our experiments, where we find that removing specs of dirt or a film of dust, distorting character shapes and character to plate contrast respectively, can be accomplished by our CycleGAN models, while morphological transformations like straightening bent plates poses a challenge. We find, however, that the removal of minor occlusions and subsequent reconstruction of the occluded part of the plate works surprisingly well. Figure 3 shows examples for the discussed transformations. We acknowledge that, although the cleaned samples are the end-product of a generative process, it is somewhat misleading to refer to them as synthetic data, especially since they so closely resemble their unprocessed counterparts.

Fig. 3. Results of the plate cleaning method. Top row: Faulty samples, bottom row: their cleaned counterparts. Note how in the leftmost example, the cGAN did not only fix the contrast but also removed the specs above the Y and P character.

3 Outlook

The goal of this paper is to showcase the versatility of synthetic data by illustrating concrete examples for their application in real-world settings. Their preliminary nature does not admit of a detailed quantitative analysis at the moment, however, qualitatively, they are promising. Future work will include detailed analyses of the described methods.

4 Collaboration and Knowledge Transfer

In this section, we outline the collaborative process by which the described methods were achieved. KTC has amounted enormous quantities of data related to ANPR. However, its privacy sensitive nature does not allow excessive external sharing, which means that effective research is only possible through close collaboration between KTC and SCCH. SCCH develops algorithms and conducts experiments on a representative and GDPR compliant data subset, which are then evaluated on large-scale datasets by KTC.

A more detailed illustration of the process is shown in Fig. 4, and described in the following. After the initial problem statement by KTC, literature review by SCCH, and subsequent joint formulation of a work package, KTC proceeds by creating a suitable dataset, which is shared via a cloud storage infrastructure. As a next step, SCCH analyzes and preprocesses the data. For the experiments - mostly in the realm of deep learning - a high-performance-computing cluster is utilized. Depending on the task at hand, KTC may run experiments in parallel. The implemented algorithms and their results are documented in an instance of Atlassian Confluence[1] and the produced code uploaded to a Git-repository,

[1] https://www.atlassian.com/software/confluence.

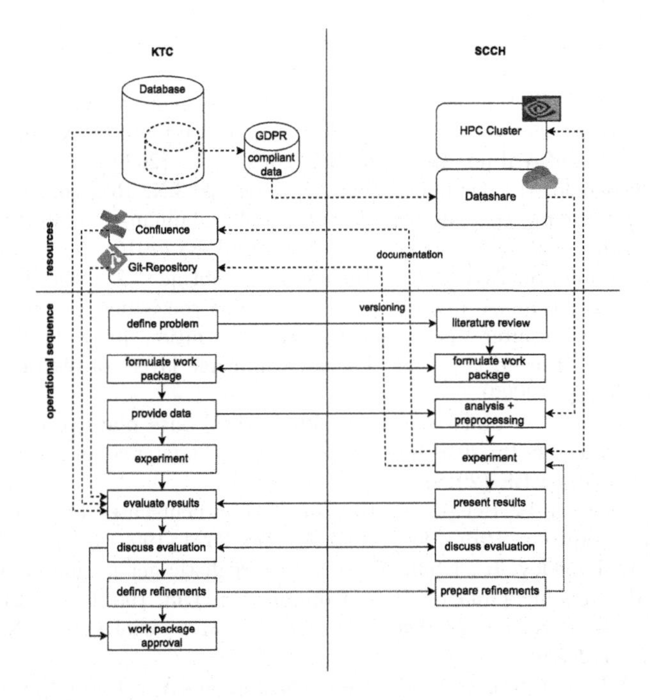

Fig. 4. Collaboration workflow. Solid lines show the operational sequence, dashed lines signify a flow of resources and dotted lines mark optional events.

both hosted by KTC. Furthermore, the results are discussed in a meeting of the teams of both institutions. Discussion, together with the produced code and documentation, represents the most important means for knowledge transfer. Experience has shown, that discussing results once a week is most productive, leaving enough time to conduct meaningful experiments, while constraining the experiment's scope so that in-depth discussions about the subject matter are viable. To verify the generalizability of the developed algorithms, KTC repeats the experiments on their large-scale datasets. If the results are satisfying the work package is finalized, else possible refinements of the algorithms are discussed and further experiments conducted.

References

1. Cabon, Y., Murray, N., Humenberger, M.: Virtual kitti 2. arXiv preprint arXiv:2001.10773 (2020)
2. Deng, J., Dong, W., Socher, R., Li, L.J., Li, K., Fei-Fei, L.: Imagenet: a large-scale hierarchical image database. In: 2009 IEEE Conference On Computer Vision and Pattern Recognition, pp. 248–255. IEEE (2009)
3. Gokaslan, A., Ramanujan, V., Ritchie, D., Kim, K.I., Tompkin, J.: Improving shape deformation in unsupervised image-to-image translation. In: Ferrari, V., Hebert, M., Sminchisescu, C., Weiss, Y. (eds.) ECCV 2018. LNCS, vol. 11216, pp. 662–678. Springer, Cham (2018). https://doi.org/10.1007/978-3-030-01258-8_40

4. Goodfellow, I., et al.: Generative adversarial nets. In: Advances in neural information processing systems, vol. 27 (2014)
5. Han, B.G., Lee, J.T., Lim, K.T., Choi, D.H.: License plate image generation using generative adversarial networks for end-to-end license plate character recognition from a small set of real images. Appl. Sci. **10**(8), 2780 (2020)
6. Johnson-Roberson, M., Barto, C., Mehta, R., Sridhar, S.N., Rosaen, K., Vasudevan, R.: Driving in the matrix: can virtual worlds replace human-generated annotations for real world tasks?. arXiv preprint arXiv:1610.01983 (2016)
7. Karras, T., Laine, S., Aittala, M., Hellsten, J., Lehtinen, J., Aila, T.: Analyzing and improving the image quality of stylegan. In: Proceedings of the IEEE/CVF Conference on Computer Vision and Pattern Recognition, pp. 8110–8119 (2020)
8. Krizhevsky, A., Hinton, G., et al.: Learning multiple layers of features from tiny images. Tech. rep. (2009)
9. Laroca, R., et al.: A robust real-time automatic license plate recognition based on the yolo detector. In: 2018 International Joint Conference on Neural Networks (ijcnn), pp. 1–10. IEEE (2018)
10. LeCun, Y., Bottou, L., Bengio, Y., Haffner, P.: Gradient-based learning applied to document recognition. Proc. IEEE **86**(11), 2278–2324 (1998)
11. Maltsev, A., Lebedev, R., Khanzhina, N.: On realistic generation of new format license plate on vehicle images. Proc. Comput. Sci. **193**, 190–199 (2021)
12. Mirza, M., Osindero, S.: Conditional generative adversarial nets. arXiv preprint arXiv:1411.1784 (2014)
13. Nikolenko, S.I.: Synthetic Data for Deep Learning. SOIA, vol. 174. Springer, Cham (2021). https://doi.org/10.1007/978-3-030-75178-4
14. Richter, S.R., Hayder, Z., Koltun, V.: Playing for benchmarks. In: Proceedings of the IEEE International Conference on Computer Vision, pp. 2213–2222 (2017)
15. Ros, G., Sellart, L., Materzynska, J., Vazquez, D., Lopez, A.M.: The synthia dataset: a large collection of synthetic images for semantic segmentation of urban scenes. In: Proceedings of the IEEE Conference on Computer Vision and Pattern Recognition, pp. 3234–3243 (2016)
16. Rosebrock, A.: Opencv: Automatic license/number plate recognition (anpr) with python (2015). https://pyimagesearch.com/2020/09/21/opencv-automatic-license-number-plate-recognition-anpr-with-python
17. Saleh, F.S., Aliakbarian, M.S., Salzmann, M., Petersson, L., Alvarez, J.M.: Effective use of synthetic data for urban scene semantic segmentation. In: Ferrari, V., Hebert, M., Sminchisescu, C., Weiss, Y. (eds.) ECCV 2018. LNCS, vol. 11206, pp. 86–103. Springer, Cham (2018). https://doi.org/10.1007/978-3-030-01216-8_6
18. Sharma, M., Verma, A., Vig, L.: Learning to clean: a GAN perspective. In: Carneiro, G., You, S. (eds.) ACCV 2018. LNCS, vol. 11367, pp. 174–185. Springer, Cham (2019). https://doi.org/10.1007/978-3-030-21074-8_14
19. Tripathi, S., Chandra, S., Agrawal, A., Tyagi, A., Rehg, J.M., Chari, V.: Learning to generate synthetic data via compositing. In: Proceedings of the IEEE/CVF Conference on Computer Vision and Pattern Recognition, pp. 461–470 (2019)
20. Wang, X., Man, Z., You, M., Shen, C.: Adversarial generation of training examples: applications to moving vehicle license plate recognition. arXiv preprint arXiv:1707.03124 (2017)
21. Zhu, J.Y., Park, T., Isola, P., Efros, A.A.: Unpaired image-to-image translation using cycle-consistent adversarial networks. In: Proceedings of the IEEE International Conference on Computer Vision, pp. 2223–2232 (2017)

An Untold Tale of Scientific Collaboration: SCCH and AC^{2}T

Somayeh Kargaran[1], Anna-Christina Glock[1(✉)], Bernhard Freudenthaler[1(✉)], Manuel Freudenberger[2(✉)], and Martin Jech[2(✉)]

[1] Software Competence Center Hagenberg, GmbH (SCCH), Hagenberg, Austria
anna-christina.glock@scch.at
[2] AC2T Research GmbH (AC2T), Wiener Neustadt, Austria

Abstract. In the last two decades, the application of artificial intelligence (AI) in different fields has increased significantly. As an interdisciplinary field, AI methods are improving toward efficiency and applicability due to a vast number of high-quality research and adopting themselves with multiple use cases. Therefore, the industry is investing in AI-based companies to overcome their use cases efficiently. This paper discusses the collaboration between two research-based companies with different expertise, one in the fields of Data and Software Science and the other in tribology. We show how both companies benefit from such collaboration to tackle the problem at hand.

Keywords: Company · Research-based centers · COMET program · Research collaboration

1 Introduction

When talking about the critical factors for the success of the companies, innovation and product development are at the head of the list [8]. For the companies serving global markets, it is essential to collaborate with other companies and benefit from their expertise to offer attractive products to their customers. The collaborations can be created between companies and customers (use cases), other companies with different expertise, competitors, and universities to help with the research [6,11,24,26]. The partnering success depends on several factors, e.g., compatible ideas and goals, fair collaboration from all sides, commitment, motivations, trust, and realistic expectations [7,9]. Another critical factor is the management of the shared projects between several companies. Hiring inexperienced managers can lead the project to a failure [10]. In [14,17], the authors proposed research regarding the percentage of the successful and failed projects based on the rule of the management.

Relationships between the companies always bring new ideas, expertise, technologies, productivity, and access to novel and/or larger markets [5,21]. In [20], the authors describe a study based on the literature findings on common business collaboration concepts, analyses of their characteristics, and an evaluation of implications from a company's perspective.

G. Kotsis et al. (Eds.): DEXA 2022 Workshops, CCIS 1633, pp. 119–128, 2022.
https://doi.org/10.1007/978-3-031-14343-4_12

In this paper, we describe the collaboration experience between two research-oriented companies **Software Competence Center Hagenberg (SCCH)** and **Austrian Competence Center for Tribology (AC^2T)**.

In Sect. 2, we briefly explain the history of the companies. In Sect. 3, the projects and achievements are described. Moreover, we discuss the next steps of the collaboration on the involved projects. In Sect. 4, we consider the quality of the collaboration between the two companies, and we discuss the critical points, the lessons learned and the advantages of such collaborations. Section 5 is dedicated to the conclusion.

2 Involved Partners

This section briefly explains the activities and expertise of the two COMET centers, **Software Competence Center Hagenberg (SCCH)** and **Austrian Competence Centre for Tribology (AC^2T). COMET Competence Centers for Excellent Technologies** is a central funding program of the Austrian technology politics that creates competence centers in Austria for different fields of science. The strategy of the COMET program is to join academic scientists and industrial researchers to work on research projects that are closer to the industry than academic works. About 50% of the company's budget comes from the funding, and the other 45% must be obtained by companies. Also, 5% of the budget is in-kind contributions from the scientific partners. The COMET funding is divided into three classes, K1, K2, and K, based on the size, budget, funding duration, and international cooperation. For instance, K2 centers receive the funding for eight years, K1 centers for seven years, and K centers are funded for three to five years by the COMET program [2].

2.1 SCCH

SCCH is funded by the Austrian COMET K1 program and is the only COMET research-based center focusing on data and software science. Due to its highly-qualified staff, SCCH is one of the leading research centers in AI and software science in upper Austria. The most important research topics in SCCH are big data analysis, data management, computer vision, software science, deep/machine/transfer learning, and privacy and security. More than one hundred employees are currently working at SCCH in several positions responsible for research projects and publications, company use cases, software development, administration, and management tasks [3].

2.2 AC^2T

AC^2T is funded by the Austrian COMET K2 program. The company is the central node for national and international research activities in tribology and preserving as the "European Center of Tribology". The company has around one hundred and forty employees. AC^2T has substantial experience in theory and experiments, such as tribological testing, surface, and material characterisation, lubricant analysis, and lubrication, fluid mechanics and heat transfer, wear measurement and wear protection technologies, and modeling and simulation on various size scales. Thus, AC^{2}T can solve truly interdisciplinary challenges and bring together application-oriented research and industrial development. This provides necessary unique insights that simultaneously consider industrial requirements and scientific procedures [1].

2.3 History of the Collaboration

Due to the significant increase in the volume of the data, improvements in the software and hardware of the computers, and advancement of the recent AI methods, application of these algorithms in different fields of science and technology has increased significantly during the last decade [13,16,18,19]. Most companies are interested in using AI algorithms in their application due to the possibility of automation, and efficiency to extract relevant features from datasets to recognize real-life patterns.

SCCH, with the expertise in data and software science, and AC^{2}T , with the expertise in tribology, started a collaboration to adopt each other's expertise to apply advanced AI algorithms in the tribology applications. Figure 1 is a visualisation of this cooperation. Later on, the collaboration between the two companies led to proposing original and fresh topics for research, especially master's and Ph.D. projects. The two companies have done multiple research projects by funding students/employees in the last few years. For instance, the master thesis with the title, "offline change point detection in RIC (Radioactive Isotope Concentration) /wear progress" , was defended by **Anna-Christina Glock** in 2020 [12]. Moreover, SCCH and AC^{2}T have/had a comprehensive collaboration in using AI methods to analyse the wear progress in various tribology datasets.

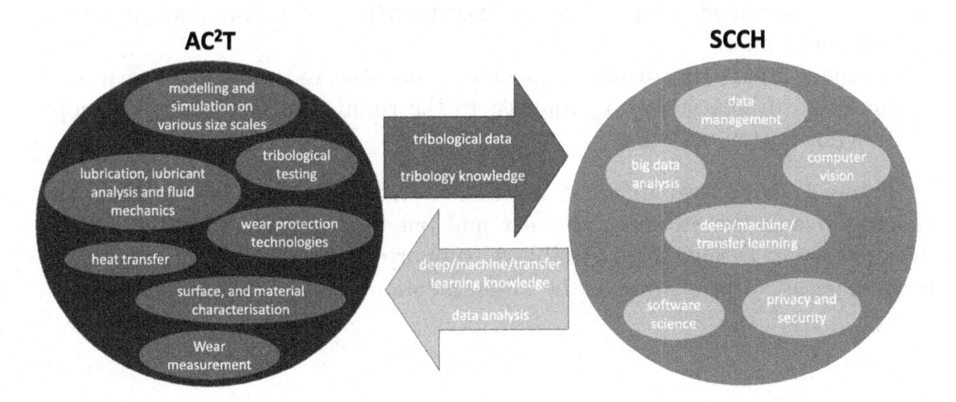

Fig. 1. A visualisation of the cooperation between AC^{2}T and SCCH showing their respective expertise and the knowledge transferred occurring in the collaboration.

3 Discussion on the Projects

Currently, AC^2T and SCCH have funded two Ph.D. students to work on two different research projects since 2020. The first one started in July 2020 with the "online change point detection in RIC/wear progress" title. The second one is titled "feature detection via transfer learning". This project began in November 2020.

3.1 Proj N.1: Online Change Point Detection in RIC/Wear Progress

This project is a continuation of the master thesis and aims to develop a solution to detect change points in continuous wear data in an online setup.

Radio isotope concentration (RIC) is a method patented by AC^2T to continuously measure wear in tribological experiments with two objects sliding on each other and fluid between them. The fluid runs through the experiment in a circuit to a gamma-ray detector. The data produced in this experiment shows how the wear progressed over time. Wear in such experiments can be demonstrated as in Fig. 2, depending on the experimental design (such as loading conditions, lubricant, and material) and the measurement set-up (such as applied isotopes), the resulting curve and noise vary. A change in the slope indicates a change in the wear behaviour. The following three different wear behaviours are considered in this project:

- running-in: this kind of wear behaviour occurs at the beginning of the experiment. It is characterised by a slope with an exponentially decreasing wear rate.
- steady-state wear: A steady-state regime can be observed after the running-in phase. A steady-state wear may be characterised by a steady-state wear rate, which is dominant compared to the exponentially decreasing wear rate of the running-in phase.
- divergent wear: the divergent wear can be observed when the wear rate increases (dramatically). In contrast to the running-in behaviour, the slope does not flatten over time.

Figure 2 shows two different noticeable change points, one between the running-in and the steady-state wear and one between the steady-state and divergent wear. Developing a possible AI-supported method to find those points, as an experiment is running, is the initial mentioned goal of this cooperation.

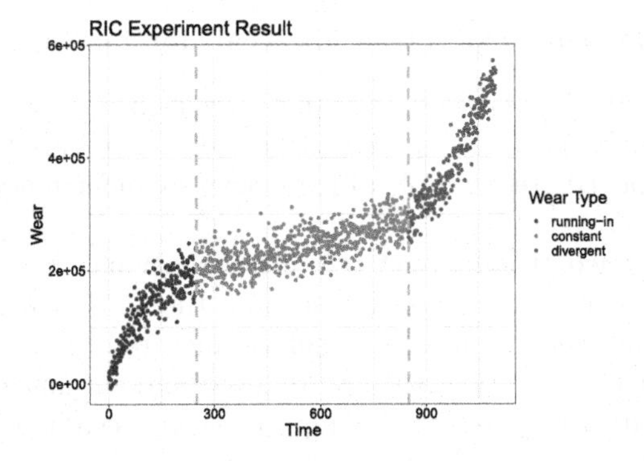

Fig. 2. A plot that shows what results of a RIC Experiment might look like. The colour of the data points signifies which kind of wear is recorded. The change points are marked with two orange dashed lines. (Color figure online)

Discussion on the Data

Data Acquirement - The RIC data provided for this project by AC^{2}T was acquired from multiple different tribological experiments. Each of these experiments produces a univariate non-stationary time series. The run time of the experiments differs as the setup, material and goal are different for each experiment. Furthermore, some wear behaviour is not observable in the data from some experiments. For example, not all experiments were run long enough that divergent wear is present in the data. Therefore, it is important for the tribology experts from AC^{2}T to label the data to indicate which part corresponds with which wear behaviour. After that step AC^{2}T provided 107 different time series for this Ph.D. project.

The Quality of the Data - The quality of the provided RIC data is sufficient to be used in our AI algorithms, especially when pre-knowledge about the reasonable wear trends is taken into account.

Results and Further Progress

Current result - During this project, we evaluated different change point detection methods on our dataset, e.g., Cusum [22], bfast [27], and a Bayesian-based method [4]. Currently, we are working on publishing the results of this analysis.

Ongoing Work - After finding the best change point detection method for our problem, the next step is finding the best parameter set for general usage of this method, especially for online detection of divergent wear behaviour to anticipate failure. Furthermore, the found change points must be classified to determine which type of change occurred (running-in to steady-state or steady-state to divergent).

3.2 Proj N.2: Feature Detection via Transfer Learning

The multidimensional data (e.g., wear profile, surface image, chemical spectrum) were recorded during the triboexperiments in AC^{2}T. In this project, data are related to a V-Grooved steel wheel rolling over a cylinder made from a harder steel [25].

The aforementioned "time-series" of multidimensional data is hard to analyse due to the large volume of the dataset and the small incremental changes between subsequent steps of measurement or pictures. Hence, AI algorithms should be applied to analyse the wear progress fast and automatically. Therefore, we analysed the data using advanced computer vision techniques, e.g., semantic segmentation in the first step. Moreover, the project's primary goal is to detect the wear tracks on the cylinder using both images and wear profiles efficiently and automatically. Thus, in the next step, a transfer learning algorithm based on deep learning should be designed to detect the wear tracks for all datasets with the same nature, using a small dataset to adapt the deep neural network to the new dataset. On the other hand, since the privacy of the data is essential for the AC^{2}T customers, we will secure the network using the privacy aspect that is one of the expertise of SCCH.

Discussion on the Data

Data Acquirement - AC^{2}T is using a reciprocating tribometer to carry out the various effects of using different pair materials in a cylinder-wheel contact. The wheel rolls with a defined force 80,000 times unidirectional over the fixed cylinder. During the experiment, two wear tracks appear on both sides of the cylinder which grow in each cycle, and the process is automatically documented via images per cycle, see Fig. 3 (left). Moreover, the wear profiles are recorded using a line laser scanner every five cycles. For this Ph.D. project, AC^{2}T provided four data sets with various cylinder-wheel materials and angles between cylinder and wheel.

The Quality of the Data - The performance of deep/machine learning models highly depends on the quality of the data. Hence, in the first step, we had to perform proper preprocessing, such as alignment of different pictures as well as correction of artifacts, e.g., due to stray light, etc., to increase the quality of the data and prepare the data for our deep neural network.

Results and Further Progress

Current Result - As usual in this field, we spend a significant amount of time increasing the data quality and labeling them using an in-house labeling tool. Moreover, as the size of the dataset is massive and hand-labeling all the samples are extremely time-consuming, to perform the semantic segmentation, we used a common pre-trained network and transfer learning to obtain our AI model. The training process does not require a massively large dataset using the techniques mentioned above. We employed U-Net [23] for the segmentation task to detect the wear tracks using the manually labeled data. It is worth mentioning

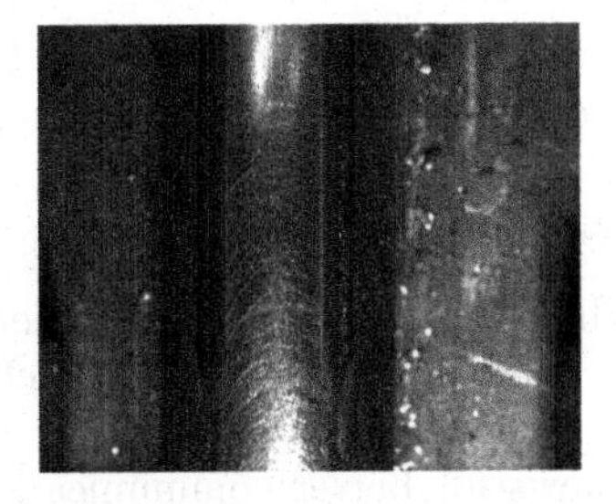
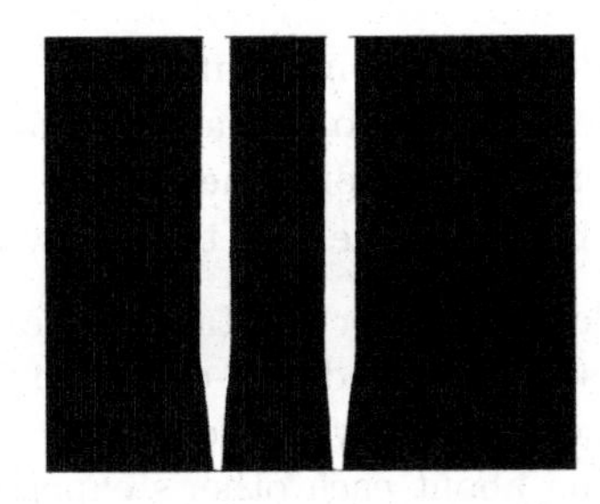

Fig. 3. A cylinder-wheel recorded image (left), and its corresponding wear track detection using U-net (right).

that nnU-Net [15] would also be appropriate to full fill the segmentation task. A small set of raw images and their masks (obtained manually) were used as the input of the U-net. Moreover, the output of the U-net gives us the masks for each raw image. Figure 3, demonstrates an image and its corresponding mask.

We used the masks to compute the width of the wear tracks in each image. Using the width of the wear tracks, we could analyse the wear progress. Moreover, the wear progress has been analysed using the wear profiles separately, and we could obtain the correlation between the growing of the wear tracks in the images and the obtained wear progress in the wear profiles. Analysing the data was essential since one of the project's goals was to identify the best material pairing for the cylinder-wheel contact.

Ongoing Work - Currently, we are designing a transfer learning algorithm to be able to learn from the trained networks of the experiments of the same nature and reduce the effort in the data acquirement. Also, computing the depth of the wear tracks is one of the goals in this Ph.D. project.

4 Quality of the Collaboration

As we mentioned in Sect. 2.3, the idea of this collaboration was to use AI methods to overcome challenges in tribology. AC^{2}T could provide large datasets and the necessary pre-knowledge related to each problem. Furthermore, understanding the tribology concept is vital for selecting the suitable and tailor-made method for the problem and analysing their results. Moreover, the tribology experts can help with the quality improvement of the acquired datasets. On the other hand, data scientists can also help with data acquirement and analysing the data using AI algorithms. The involved companies have different expertise, and at the beginning, they might have limited knowledge about the expertise of the other partner. For example, each research field's specific vocabularies and technical terms need to be learned and understood mutually. The problems mentioned above can be improved by organizing regular meetings and discussions. For instance, we set up visits regarding the data acquirement in AC^{2}T to better understand where the data comes from and the challenges of the data recording. However, due to Covid-19, these visits could only take place one year after the start of the project. Besides the online meetings, the plan is to meet every half year to show the progress of the projects and exchange knowledge now that the pandemic has less influence on personal meetings.

From these experiences, the following lessons can be extracted:

- Knowledge transfer is vital in choosing the right solution for a problem.
- Data scientists and the data providing experts need to work together to increase the quality of the data.
- Learning about each other's expertise allows for better communication and limits the number of misunderstandings.
- Regular meetings are an important tool to increase the knowledge transfer between the partners. Regardless if these meetings take place in person or online.

5 Conclusion

This paper discussed the importance of the companies' cooperation and overcoming to merge their expertise to obtain better solutions for real-world problems. More specifically, we described the collaboration between two COMET centers AC^2T and SCCH, and how they use their expertise in several research projects. The idea of such collaboration was using AI methods to analyse the tribology data. Moreover, we discussed the advantages, the difficulties, and the solutions to overcome the possible problems. Based on the experiences that we gained through working with other partners, especially AC^2T , we conclude that nowadays, companies can not only rely on their own competence, since the real-world applications require different expertise that can be difficult to achieve in a single center. Therefore, they need to engage with other companies to overcome their challenges quickly and efficiently. The success of the shared projects depends on several factors like compatible ideas and goals, fair collaboration from all partners, commitment, motivation, trust, and realistic expectations. Moreover, the management of the project plays an essential role in the success or the failure of the project. On the other hand, the companies' cooperation adds more innovation into the involved projects and leads the companies to more and more significant markets. However, not all the projects may be solved by the presented collaboration, and we are looking forward to strengthening our research base with further partners.

Acknowledgement. This work is mainly supported by the "Austrian COMET-Programme" (Project InTribology, no. 872176). Furthermore, a part of the research has been funded by the Federal Ministry for Climate Action, Environment, Energy, Mobility, Innovation and Technology (BMK), the Federal Ministry for Digital and Economic Affairs (BMDW), and the Province of Upper Austria in the frame of the COMET-Competence Centers for Excellent Technologies Programme and the COMET Module S3AI managed by Austrian Research Promotion Agency FFG. We would like to thank Dr. Florian Sobieczky, Dr. Bernhard Moser, and Dr. Volkmar Wieser from SCCH and Dr. Josef Prost, Dr. Markus Verga, and Dr. Georg Vorlaufer from AC^2T for supporting the cooperation.

References

1. AC2T research GMBH (AC^{2}T). https://www.ac2t.at/
2. COMET program. https://www.comet.ucar.edu/index.php
3. Software competence center hagenberg (SCCH). https://www.scch.at/
4. Adams, R.P., MacKay, D.J.C.: Bayesian Online Changepoint Detection (2007). arXiv:0710.3742 [stat]
5. Barnes, C., Blake, H., Pinder, D.: Creating and delivering your value proposition: Managing customer experience for profit. Kogan Page Publishers (2009)
6. Belderbos, R., Carree, M., Lokshin, B.: Cooperative r&d and firm performance. Res. Policy **33**(10), 1477–1492 (2004)
7. Brouthers, K.D., Brouthers, L.E., Wilkinson, T.J.: Strategic alliances: choose your partners. Long Range Plan. **28**(3), 2–25 (1995)
8. Chesbrough, H.W.: Open innovation: the new imperative for creating and profiting from technology. Harvard Business Press (2003)
9. Distanont, A., Haapasalo, H., Kamolvej, T., Meeampol, S., et al.: Interaction patterns in collaborative product development (CPD). Int. J. Synergy Res. **1**(2), 21–44 (2013)
10. Eveleens, J.L., Verhoef, C.: The rise and fall of the chaos report figures. IEEE Softw. **27**(1), 30 (2010)
11. Flipse, S.M., van der Sanden, M.C.A., Osseweijer, P.: Setting up spaces for collaboration in industry between researchers from the natural and social sciences. Sci. Eng. Ethics **20**(1), 7–22 (2013). https://doi.org/10.1007/s11948-013-9434-7
12. Glock, A.C.: Offline change point detection in RIC (Radioactive Isotope Concentration) /wear progress. Master's thesis, University of Applied Science Upper Austria - Hagenberg (2020)
13. Goodfellow, I., Bengio, Y., Courville, A.: Deep learning. MIT press (2016)
14. Guide, A.: Project management body of knowledge (pmbok® guide). In: Project Management Institute, vol. 11, pp. 7–8 (2001)
15. Isensee, F., et al.: nnu-net: self-adapting framework for u-net-based medical image segmentation (2018)
16. Jordan, M.I., Mitchell, T.M.: Machine learning: trends, perspectives, and prospects. Science **349**(6245), 255–260 (2015)
17. Jørgensen, M., Moløkken-Østvold, K.: How large are software cost overruns? a review of the 1994 chaos report. Inf. Softw. Technol. **48**(4), 297–301 (2006)
18. LeCun, Y., Bengio, Y., Hinton, G.: Deep learning. Nature **521**(7553), 436–444 (2015)
19. LeCun, Y., et al.: Backpropagation applied to handwritten zip code recognition. Neural Comput. **1**(4), 541–551 (1989)
20. Majava, J., Isoherranen, V., Kess, P.: Business collaboration concepts and implications for companies. Int. J. Synergy Res. **2**, 23–40 (2013)
21. Meade, L., Liles, D., Sarkis, J.: Justifying strategic alliances and partnering: a prerequisite for virtual enterprising. Omega **25**(1), 29–42 (1997)
22. Page, E.S.: Continuous Inspection Schemes. Biometrika **41**, 100 (1954)
23. Ronneberger, O., Fischer, P., Brox, T.: U-net: convolutional networks for biomedical image segmentation (2015)
24. Swink, M.: Building collaborative innovation capability. Res. Technol. Manag. **49**(2), 37–47 (2006)

25. Trausmuth, A., Lebersorger, T., Badisch, E., Scheriau, S., Brantner, H.: Influence of heat treatment and surface condition on early-damaging of rail materials. In: Proceedings of the Second International Conference on Railway Technology: Research, Development and Maintenance, vol. 188 (2014)
26. Un, C.A., Cuervo-Cazurra, A., Asakawa, K.: R&d collaborations and product innovation. J. Prod. Innov. Manag. **27**(5), 673–689 (2010)
27. Verbesselt, J., Zeileis, A., Herold, M.: Near real-time disturbance detection using satellite image time series. Remote Sens. Environ. **123**, 98–108 (2012)

On the Creation and Maintenance of a Documentation Generator in an Applied Research Context

Bernhard Dorninger[1], Michael Moser[1(✉)], Josef Pichler[2], Michael Rappl[3], and Jakob Sautter[1]

[1] Software Competence Center Hagenberg GmbH, Hagenberg, Austria
{bernhard.dorninger,michael.moser,jakob.sautter}@scch.at
[2] University of Applied Sciences, Upper Austria, Hagenberg, Austria
josef.pichler@fh-hagenberg.at
[3] Österreichische Gesundheitskasse, Linz, Austria
michael.rappl@oegk.at
http://www.scch.at, https://www.fh-ooe.at, https://www.gesundheitskasse.at

Abstract. Reverse engineering-based documentation generation extracts facts from software artefacts to generate suitable representations in another level of abstraction. Although the tool perspective in documentation generation has been studied before by many others, these studies mostly report on constructive aspects from case studies, e.g. how tools are built and evaluated. However, we believe a long-term perspective is important to cover issues that arise after initial deployment of a tool.

In this paper, we present challenges and observations made during prototyping, development and maintenance of a documentation generator in an applied research project. Insights are drawn from different project phases over a period of 4-years and cover topics related to tool implementation as well as topics related to knowledge transfer in an applied research project. A key observation is that the maintenance of the system to be documented often triggers maintenance effort on the documentation generator.

Keywords: Reverse engineering · Documentation generation · Prototyping · Software maintenance · Research project

1 Introduction

Software documentation is an important means to support maintenance and evolution of software. Documentation generators automate production of documentation and ease keeping that documentation up-to-date. Documentation generators based on reverse engineering [2,8,9] have been used since decades for understanding and documenting legacy software. API documentation generators [7] have become popular with *JavaDoc* [3], which is extracting information from Java source code comments annotated with specific tags. Contemporary

G. Kotsis et al. (Eds.): DEXA 2022 Workshops, CCIS 1633, pp. 129–140, 2022.
https://doi.org/10.1007/978-3-031-14343-4_13

tools for the production of technical software documentation are available for many programming languages and the use of such tools is omnipresent in software development. Recent research explores new ways to extract human-readable documentation from source code using machine learning and natural language processing (NLP) techniques. For instance, McBurney et al. [5] leverages NLP to generate natural language descriptions of source code.

These approaches and tools mainly focus on providing technological documentation for the analyzed software, e.g. describing API signatures. In other words, such documentation generators mainly address implementation, design and architectural aspects. At the *SCCH*, we have developed documentation generators specifically for software systems in domains such as banking, insurance, and engineering. We have focused on providing software documentation emphasizing the subject-specific or "business" aspect, with the targeted audience being domain specialists, e.g., business analysts, process managers, electrical engineers, with little or no IT skills.

In the following, we report on experiences made and lessons learned during the development of a documentation generator for a public health insurance company. The development has been triggered by the migration of a legacy system to a newer technology. As the development proceeded in the context of a research project, we will elaborate on the challenges, obstacles, but also synergies we experienced between the priorities of research goals and customer requirements.

This paper is structured as follows. In Sect. 2 we outline the industrial and research context of the presented work. Section 3 summarizes activities in prototyping, development and maintenance of a documentation generator. In Sect. 4 we present observations and challenges from tool development and project specifics resulting from its character as a scientific research project, most notably the application and transfer of the knowledge gained. Finally, Sect. 5 concludes our work.

2 Context

The scope of this study is the development of a documentation generator (*ReDoc*) for a public health insurance company (*ÖGK*) in the context of the applied research project *REEDS*.

2.1 Research Context

The development of the presented documentation generator was conducted within an applied research project funded by the COMET funding program. The research project *Reverse Engineering Based Software Evolution and Documentation* (*REEDS*) exploits reverse engineering-based approaches for the automated generation of software documentation. *REEDS* builds upon methods for static and dynamic source code analysis with the goals to recover knowledge from software systems (1), represent that knowledge by domain- and stakeholder-specific documentation models and visualizations (2), efficiently generate documentation (3) that may be generated on-demand (4), conforms to standards and legal regulations (5) or facilitates the analysis of system evolution (6).

As the nature of *REEDS* is applied research, it involves a number of partner companies from industry contributing to the funding and providing requirements as well as real-world use cases. Solutions provided to the project partners are specific to their industrial or business context and usually contribute to the research goals described above. As for the documentation generator being the subject of this paper, already the process of conception and development of the tool has contributed to the definition of stakeholder-specific documentation models and the efficient creation of documentation. *ReDoc* itself is still in productive usage at *ÖGK*.

2.2 Industry Context

The documentation generator targets a large-scale legacy software system (*lgkk*), with roughly 4M SLOC, providing business services to calculate insurance claims for more than a million clients of *ÖGK*. Modules of the software systems can be classified as either batch programs or online modules. Online modules provide access to business services called from client applications. Batch programs on the contrary run in a predefined sequence and are executed on demand or scheduled periodically. Batch programs of *lgkk* were originally implemented in COBOL and needed to be migrated to Java in recent years. To support the migration of significant parts of *lgkk* from COBOL to Java, up-to-date documentation was requested. As manual update and re-documentation of the software system seemed not economically feasible, a project for automatic re-documentation was initiated. Requirements for *ReDoc* and the generated documentation can be summarized as follows:

1. Fully automated documentation generation, no manual re-documentation of source level.
2. Moving abstraction level of the generated documentation from source code towards business scenarios.
3. Support for different stakeholders, i.e. developers, project managers, business analysts.
4. Support for the COBOL and Java implementations of the software system.
5. Applicability of the documentation generator to similar software systems at *ÖGK*.

2.3 Documentation Generator - *ReDoc*

Figure 1 shows a screenshot of the HTML documentation generated by *ReDoc*. *ReDoc* lists batch programs, online services, messages exchanged between services, and text files and database tables accessed from batch programs and service. For each of these, detailed information is generated, which may be reached via links.

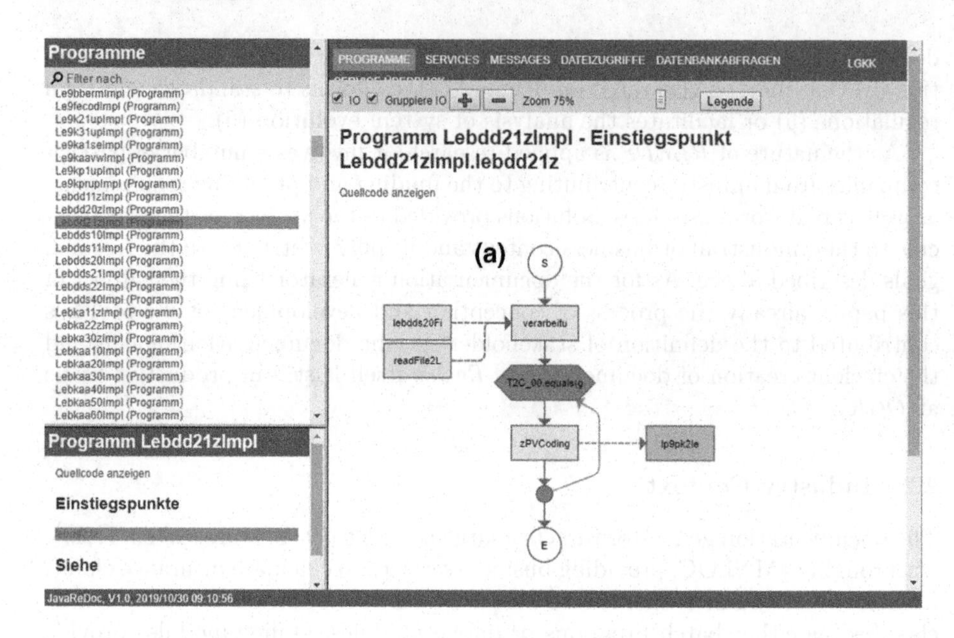

Fig. 1. Data- and control-flow documentation generated by *ReDoc*

The focus is on the functional behavior of batch programs and online services. This is described by flow diagrams, which show the control-flow starting from a defined entry-point. The control-flow is extended with nodes depicting read or write operations to database tables, files, or messages. These I/O operations are aligned in separate swim lanes on the left (input operation) and right (output operation) of the program control-flow. Documentation of messages and file accesses show usage information and provide description on the precise data formats used. Access to database tables is documented by SQL statements recovered from program sources and XML definitions. Generally, all entities of the documentation are linked with corresponding source code elements.

3 Project Phases

This section provides an overview of the phases in the development process of *ReDoc* within the applied research project REEDS.

Figure 2 shows the time-line for the first four years plus the major milestones and outcome of the project phases during that time. The phases include: (1) Exploratory prototyping of a documentation generator, split into two main iterations, (2) development of a documentation generator, and (3) maintenance of the created generator.

At the start of the project, stakeholders at *ÖGK* have expressed a profound need for a thorough documentation of *lgkk*, however the required content, structure and format of that documentation have been unclear. Similarly unclear

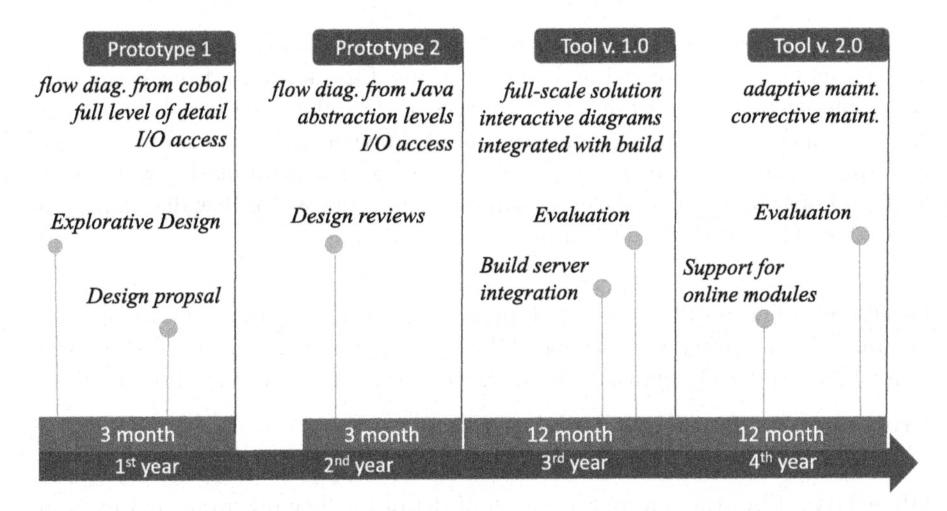

Fig. 2. Major milestones of documentation generations

has been the extent to which documentation could be generated automatically. Therefore, the project started off with a feasibility study on documentation generation for the COBOL and Java version of *lgkk*. During this project phase two versions of an exploratory tool prototype targeting COBOL and Java source code have been developed. Technical details on this prototype and challenges in multi-language re-documentation have been documented in [1]. Approval of concepts and tools developed during prototyping phase has triggered a development project which resulted in a high-quality documentation generator for *lgkk*.

3.1 Prototyping

Prototyping phase can be split in two main iterations. While the first prototyping phase answered which documentation content is needed, the second phase dealt with finding the proper level of detail.

Prototype 1 - The Proper Documentation Content

Objective: Identify documentation content generating high value for all stakeholders, which can be generated automatically from source code.

Work Progress: To identify documentation content, we started with generating documentation by utilizing prefabricated software components. Reused software components are part of the software platform *eknows* [6], which facilitates rapid development of documentation generators and reverse engineering tools. Main purpose of this step was to show capabilities of automated documentation generation. First drafts of the documentation were generated from a selected set of COBOL programs. Analyzing the first results revealed the importance of I/O operations. Therefore, early versions of the documentation already listed files and database tables accessed by I/O operations. First feedback from stakeholders emphasized the importance to comprehend data flow between data sources and data sinks (i.e. files and database tables). Further discussion led to the idea to display control flow for batch programs and focus control blocks, which perform I/O operations. Therefore, an initial design proposal for flow diagrams was created which guided the development of the tool prototype.

Outcome: The result of this first phase was documentation containing flow diagrams for a limited set of seven COBOL programs. Flow diagrams displayed control flow and I/O operations in separate swim lanes at full level of detail.

Prototype 2 - The Right Level of Detail

Objective: Find the appropriate level of detail for flow diagrams and expand documentation generation to Java sources.

Work Progress: To handle complexity of flowchart diagrams, the concept of abstraction levels was introduced. We reduced the number of nodes by folding nodes of lower abstraction levels into nodes of higher levels. Due to a generic language independent representation of source code, additional effort to process and document Java source code was low. Obviously, analysis to detect file and database operations had to be adapted for Java-specific patterns.

Outcome: The outcome of this phase is represented by documentation for the Java versions of the batch programs selected in the first prototyping phase. Due to limited resources for prototype development, we generated images for each abstraction level and linked the different levels from documentation. Results were discussed with business analysts and developers at *ÖGK*. Feedback on the introduction of abstraction levels was generally positive and the benefits of the presented approach were recognized.

3.2 Development

Positive feedback on the tool prototypes led to a substantially increased funding of the project, which allowed to start further development of *ReDoc*. The following goals were pursued in the development phase.

Objectives: First, develop a documentation generator that can be applied on the entire Java implementation of *lgkk*. Second, overcome shortcomings of static flowchart diagrams and third, integrate documentation generation with the continuous build environment at *ÖGK*.

Work Progress: Initial work packages focused on the completion of analysis components in order to correctly process all Java sources. As during prototyping only selected examples of batch programs were discussed, the expansion on the complete system required substantial effort. Effort was rooted in inefficiencies of the provided analysis and implementation patterns not contained within example files. As agreed in prototyping, presentation of flowchart diagrams needed to be changed from plain images to interactive JavaScript-based visualizations. After nine months, *ReDoc* was integrated with the continuous build environment and the first complete version of *lgkk* documentation was evaluated by selected software developers and business analysts. The results of the evaluation created a product backlog which was subsequently worked off during a stabilization phase.

Outcome: By the end of the first year of development, a feature-complete version of the documentation generator was released. The version was integrated with a continuous build chain and was capable of continuously generating up-to-date documentation of all batch programs contained in the Java implementation of *lgkk*.

3.3 Maintenance

Maintenance phase started a year after development was launched. Maintenance work for *ReDoc* was integrated in the overall quality assurance process of *lgkk*. It was decided to continue cooperation with *ÖGK* within *REEDS* to evaluate tool development from various perspectives, e.g. technical completeness, stability, user acceptance.

Objectives: Integrate feedback and bug reports from users, keep documentation generator compatible with modifications in the source code base of *lgkk*, roll out *ReDoc* to additional software modules.

Work Progress: During the first year of maintenance 73 issues were reported. Reported issues fall into all four maintenance categories, i.e. adaptive (32), corrective (28), perfective (8) and preventive (3) maintenance tasks. Adaptive tasks resulted either from evolution of *lgkk* or from the application of the documentation generator to additional software modules of *lgkk*. Therefore, analysis had to be updated with new features, e.g. documentation of database access via Java persistence API (JPA) or extending documentation scope from batch programs to online modules. Corrective tasks resulted from missing content in documentation due to incomplete or incorrect analysis and output generation. Fine tuning of the documentation generator, especially the generation of flow diagrams, was the cause for perfective maintenance tasks. Preventive tasks were clearly the exception and can be categorized as refactoring or removal of code duplicates.

Outcome: The results of the maintenance phase were updated versions of *ReDoc* producing improved documentation for *lgkk*. From a research perspective, evaluation of the maintenance phase led to valuable insights and lessons learned, also documented in this report.

4 Observations and Insights

In the following, we present observations and insights gained during different project phases. The statements address challenges from a tool development perspective as well from conducting development within an applied research project.

4.1 Prototyping

Determine the Proper Content of Documentation - Stakeholders often have only a vague idea of the content of a system's documentation, let alone the possibilities of automated generation. Existing documentation of a system - if available at all - may help to identify valuable knowledge that should be extracted from source code. However, as such documentation usually was written manually, the content is always defined by the constraint that the effort must be low enough so that it can be written (and maintained) for the entire software system. This kind of constraint is, by definition, obsolete for an automated approach and may restrict ideas. Results from previous research projects are also often used to discuss possible content of documentation. In our experience, however, the key question is how to transfer problem context and applied solution from one project to the another. Software documentation of a different project that targets a specific domain and/or stakeholder is difficult to transfer to another project.

Automatically Generate Documentation from the Beginning - Showing capabilities of automated documentation generation is important, as users do not know what is possible. However, it is still a challenge to provide out-of-box parsing/analysis/generation tool-chain. With the result of prototype 1 we could show that valuable knowledge can be automatically extracted from the source code. Although the resulting documentation contained too many details, developers familiar with the set of selected COBOL programs were able to understand the generated diagrams.

Difficult to Provide Fast Results and Usable Presentation At the Same Time - As discussed above, it is important to quickly show what is possible, in order to stimulate discussion and new ideas. The usage of prefabricated reverse engineering components for extraction and visualization is inevitable to provide quick results. Still, output of generic tools will not necessarily meet expectations of all stakeholders. Therefore, it is important to explain ideas and potential behind visualizations and postpone high-quality implementation to later project phases.

Research Perspective: Valuable Insights from Prototyping Phase - The creation of a prototype is a proven approach to transfer results from research to industry. The software engineering research and especially the reveres engineering community are no exception. Prototypes are a good means to showcase research approaches and demonstrate applicability in a specific industrial context. For example, the prototypes described in this paper have demonstrated the feasibility of supporting COBOL and Java through a single documentation generator, which was questioned in advance by the project partner. However, close collaboration between academia and industry is required during prototyping phase to elicit requirements, explain and clarify known limitations, and obtain feedback from industry.

4.2 Development

Low Effort for Requirement Analysis - Due to two prototyping phases no dedicated requirement analysis phase was done at the start of the development phase. The project team assumed that this was not necessary. In retrospective, this assumption can be confirmed. The set of requirements available at the start of the development phase was stable during the whole development phase - not least because of the preceding prototyping phase and its gained insights.

Effort for Full-Scale Solution - Effort to expand the scope from selected example files to the entire code base was significant. Although, selected examples were representative in terms of complexity and style of programming, adding additional source code meant additional development effort. This was due to the fact new implementation patterns being identified, e.g. the usage of new Java API, and therefore visualizations had to be adapted to accommodate a larger code base. Running cross-checks on different software modules could be a solution to estimate this additional effort. We learned, that effort for rolling out a full-scale solution is significant and must be planned.

Applicability for Entire Software - After requirements have been gathered and validated (e.g. by prototyping), tool features are unrolled for the entire software to be documented. When choosing an interactive approach, as in our case, every feature added (e.g. database write access) must be available for all batch programs immediately. This is necessary in order to bring the entire development team on board and enables stimulation of further ideas. In this phase of the project - unlike during prototyping - stakeholders no longer advocate for compromises in terms of content, visualization, and interaction.

Research Perspective: Valuable Insights from Development Phase - Applied research must create real-world solutions that provide practical benefits to industry and business. In the case of reverse engineering tools and especially documentation generators, software prototypes provide valuable concepts, but only the development phase reveals the shortcomings of the proposed solutions. For example, standard flowcharts may work well in prototypes, but are useless for large source files with thousands of nodes. Applied research must be challenged to overcome these shortcomings and incorporate insights for future research.

4.3 Maintenance

Evolution of System Under Documentation - As any other software system, documentation generators require maintenance. Considering laws of software evolution as described in [4], this is not surprising. However, it is interesting to see, that evolution in the software to be documented often triggers evolution of the generator itself. The more specific a documentation generator is, with respect to the targeted software, the more prone to change it will be. For instance, restructuring of *lgkk* led to the introduction of new types of online services, which were not recognized by *ReDoc*. Alignment and integration of the maintenance process of *ReDoc* with the software quality management of *lgkk* can be a means to control this process.

Feedback Loop from Users - Getting feedback from users is a challenge, especially for documentation generators. Demands and expectations of users towards a tool, which provides assistance and comprehension support, might significantly differ from expectations on software which users require for their daily work. This can lead to the situation that missing features or errors do not necessarily bubble up to the surface immediately. Poor quality of documentation may just lead to documentation that is not used anymore. Therefore, we requested user feedback not only during development but as well during maintenance of *ReDoc*.

Regression Testing - Minor changes to analysis components might impact documentation generation in unforeseen ways. Testing regression of documentation after maintenance activities involves substantial effort. As automated regression testing of documentation is difficult, manual inspection is the dominant testing method for *ReDoc*. To decide whether documentation is correct and complete, availability of program sources is inevitable. Versioning documentation alongside with source code is a means to detect regressions in documentation.

Research Perspective: Valuable Insights from Maintenance Phase - Developing stable solutions being deployable in a productive environment is a key factor in *applied* research projects. Whether this holds true for maintaining the delivered solutions, may be debated. In our opinion it depends on the field of research and the concrete research goals. The goals of *REEDS* include the consideration of software maintenance, as it is clearly a part of a software system's lifecycle and thus part of its evolution. In addition, the targeted research community has a strong interest in topics related to software maintenance. Being involved with the maintenance process of *ReDoc* has granted us access to facts and figures potentially serving as valuable input for scientific studies on the topic of software maintenance. Obviously, if insights from maintenance phase had not contributed to research goals, we would have transferred maintenance tasks to a commercial, non-scientific maintenance project, as we have done in the past in other cooperations.

5 Conclusion

In this paper, we report challenges encountered and observations made during development and maintenance of a documentation generator intended to assist the migration of a large scale software system. As the tasks have been carried out in the context of an applied research project, we have outlined two main contributions: Firstly, we have illustrated the process-related and technological aspects encountered in the making of a documentation tool. Secondly, we have reflected on the results and insights from the perspective of the research goals. While reports on reverse engineering tools mainly cover design aspects during tool development, we have selected a long-term (four-year) period of observation which as well includes aspects of tool maintenance. An interesting observation here has been the development of the system to be documented (*lgkk*) having triggered the development of the documentation generator (*ReDoc*). In our case, a long-term commitment has been beneficial to successfully complete the transfer of results and knowledge gained from research. The foundation for this has been formed by a close, stable and long-term partnership with the involved partner company (*ÖGK*). In addition, another finding is that valuable insights for fulfilling research goals can be gained, when initial ideas from the prototyping phase are tested on real requirements in later development phases.

Acknowledgements. The research reported in this paper has been supported by the Austrian ministries BMVIT and BMDW, and the Province of Upper Austria in terms of the COMET - Competence Centers for Excellent Technologies Programme managed by FFG.

References

1. Dorninger, B., Moser, M., Pichler, J.: Multi-language re-documentation to support a COBOL to java migration project. In: 2017 IEEE 24th International Conference on Software Analysis, Evolution and Reengineering (SANER), pp. 536–540 (2017). https://doi.org/10.1109/SANER.2017.7884669
2. Kienle, H.M., Müller, H.A.: The tools perspective on software reverse engineering: requirements, construction, and evaluation. In: Advances in Computers, vol. 79, pp. 189–290. Elsevier (2010)
3. Kramer, D.: API documentation from source code comments: a case study of Javadoc. In: Proceedings of the 17th Annual International Conference on Computer Documentation, pp. 147–153 (1999)
4. Lehman, M.M.: Programs, life cycles, and laws of software evolution. Proc. IEEE **68**(9), 1060–1076 (1980)

5. McBurney, P.W., McMillan, C.: Automatic source code summarization of context for Java methods. IEEE Trans. Softw. Eng. **42**(2), 103–119 (2015)
6. Moser, M., Pichler, J.: eknows: platform for multi-language reverse engineering and documentation generation. In: 2021 IEEE International Conference on Software Maintenance and Evolution (ICSME), pp. 559–568 (2021). https://doi.org/10.1109/ICSME52107.2021.00057
7. Nybom, K., Ashraf, A., Porres, I.: A systematic mapping study on API documentation generation approaches. In: 2018 44th Euromicro Conference on Software Engineering and Advanced Applications (SEAA), pp. 462–469. IEEE (2018)
8. Storey, M.A.D., Wong, K., Müller, H.A.: Rigi: a visualization environment for reverse engineering. In: Proceedings of the 19th International Conference on Software Engineering, pp. 606–607 (1997)
9. Van Deursen, A., Kuipers, T.: Building documentation generators. In: Proceedings IEEE International Conference on Software Maintenance-1999 (ICSM 1999). 'Software Maintenance for Business Change' (Cat. No. 99CB36360), pp. 40–49. IEEE (1999)

Towards the Digitalization of Additive Manufacturing

Carlos González-Val, Christian Eike Precker, and Santiago Muíños-Landín(✉)

Smart Systems and Smart Manufacturing, Artificial Intelligence and Data Analytics Laboratory, AIMEN Technology Centre, PI. Cataboi, 36418 Pontevedra, Spain
{carlos.gonzalez,santiago.muinos}@aimen.es
https://www.aimen.es

Abstract. Additive manufacturing (AM) is a trending technology that is being adopted by many companies around the globe. The high level of product customization that this technology can provide, added to its link with key green targets such as the reduction of emissions or materials waste, makes AM a very attractive vehicle towards the transition to more adaptive and sustainable manufacturing. However, such a level of customization and this fast acceptance, raise new needs and challenges on how to monitor and digitalize the AM product life cycles and processes, which are essential features of a flexible factory that address adaptive and first-time-right manufacturing through the exploitation of knowledge gathered with the deep analysis of large amounts of data. How to organize and transfer such amounts of information becomes particularly complex in AM given not just its volume but also its level of heterogeneity. This work proposes a common methodology matching with specific data formats to solve the integration of all the information from AM processes in industrial digital frameworks. The scenario proposed in this work deals with the AM of metallic parts as a specially complex process due to the thermal properties of metals and the difficulties of predicting defects within their manipulation, making metal AM particularly challenging for stability and repeatability reasons but at the same time, a hot topic within AM research in general due to the large impact of such customized production in sectors like aeronautical, automotive, or medical. Also, in this work, we present a dataset developed following the proposed methodology that constitutes the first public available one of multi-process Metal AM components.

Keywords: Additive manufacturing · Digitalization · Dataset

1 Introduction

The transition from a mass production model to a mass customization approach has brought a new landscape for manufacturing frameworks, triggering the evolution from the classic production lines into real adaptive systems. These

G. Kotsis et al. (Eds.): DEXA 2022 Workshops, CCIS 1633, pp. 141–154, 2022.
https://doi.org/10.1007/978-3-031-14343-4_14

systems are composed of large collections of different elements along the factory, interacting through information flows composed of different types of data. Such interactions make possible the reconfiguration of production lines or processes responding to different customer demands, design improvements, or the detection of product failures related to manufacturing parameters. Considering that such adaptability of manufacturing systems is possible thanks to the efficient exchange of large amounts of heterogeneous data, it is fair to say that the digitalization of manufacturing processes is responsible for the emergence of fully automated and connected factories, and consequently, the existence of smart systems and smart manufacturing [1].

Additive manufacturing (AM) is a manufacturing process capable of producing complex parts through the layer-by-layer deposition of a material, which increases the design freedom of the desired part when compared to conventional methods. AM is driven by computational methods, which is a key point for its development due to the ease of computing, data manipulation, and product prototyping in the development phase [2–4]. Some of the advantages of this process are, e.g., making parts on-demand, reducing storage needs, less waste of materials, and fewer pollutant emissions. For all these reasons AM is one of the best-known actors in the current digitalization of factories and the transition to smart manufacturing.

Nowadays, the presence of AM in the industry is a reality. One can have an intuition about how wide the spectrum of AM applications can be just by taking a look at the different length scales and materials used in the manufacturing of parts by additive technologies. Many examples of the application of AM can be found in sectors such as aeronautics, automotive, medical, biotechnology, or consumer products [5,6]. However, the high level of customization that AM brings, results also in a large amount of data whose heterogeneity in terms of software, operations or process, makes complex the extraction knowledge. The knowledge that can be extracted from such a vast amount of data coming from different branches of the whole AM process like simulation, quality, path planning, or process, can be exploited to improve the manufacturing itself through the use of advanced data analytics or even machine learning methods within different phases of the manufacturing process from design to quality control [7]. But the definition of a common ontology that relates all the elements involved in such a complex process in a digital layer, and a methodology to exploit a common data architecture based on it is still missing.

This work presents a common ontology based on product, process, and resources relations (PPR) for the optimization of information flows within the AM of metallic parts. The work develops such ontology in a digital framework and integrates all the elements in a data architecture. Relying on the exposed approach, an open dataset is presented as the first available in an open format of multiprocess AM of metallic parts. The outcome of this work can be exploited for the optimization and interoperability of AM processes in the industry in general.

2 Problem Definition

Conventional manufacturing (CM) processes to produce metallic parts such as subtractive manufacturing, electrical discharge machining (EDM), or laser cutting, share a set of common steps in their workflow as design, path planning, simulation, manufacturing, quality inspection, and product validation. However, in contrast with the lack of freedom to fabricate free-form shapes [8] through CM methods, AM represents a revolutionary process that builds objects by adding material one layer at a time until the desired part is complete. For instance, one can characterize AM processes in terms of the feed stock, namely powder bed, powder feed, and wire feed systems, possessing energy sources like electron beam, laser, or arc [9] with all their characteristic parameters in terms of materials and also thermodynamics of the system to be considered. Which means that it introduces a remarkable level of complexity in all the data associated with all those steps within the workflow. The integration of all this information result crucial to ensure traceability and repeatability.

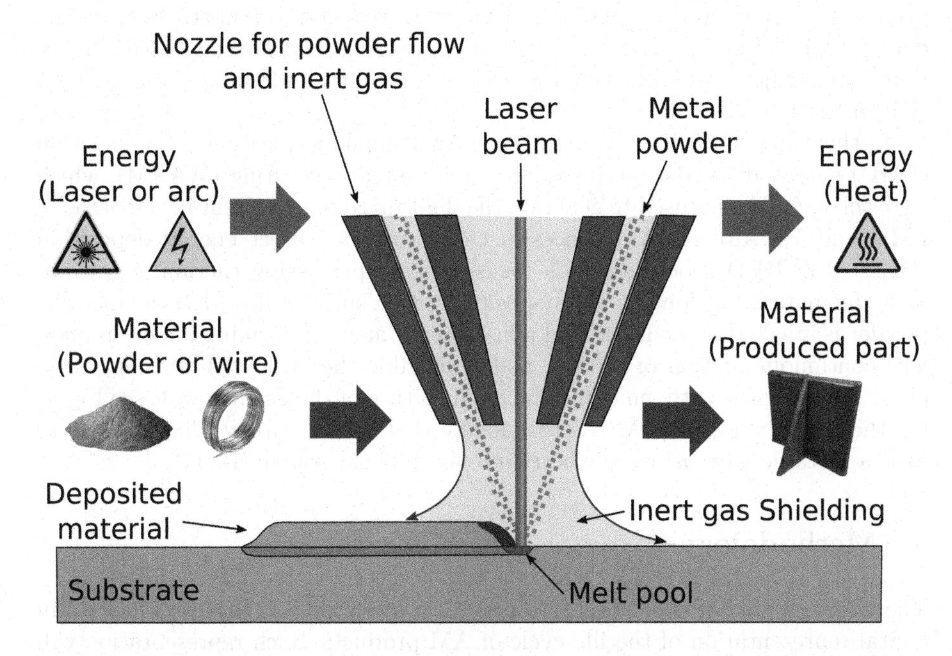

Fig. 1. The central part of the figure shows a typical DED machine nozzle, which is mounted on a multi-axis arm. It deposits powder or wire material onto the specified surface and into the path of an energy beam, where it melts and solidifies. Data collection of DED processes relies on the kind of energy employed (laser or arc), the feedstock (material and shape), and sensors to acquire heat information (infrared) and produced part shape (tomographic data).

In particular, the complexity of AM of metallic parts has led to intense developments in very different areas of knowledge. One evident point of remarkable research is related to the clear need for advanced analytical and computational methods to address the study of temperature flows along the pieces within the manufacturing process that results critical. This thermal response and the thermal leaks within the AM have been identified as one of the main reasons for the emergence of defects such as pores or cracks [10], but also instabilities within the process that lead to issues to achieve the repeatability of processes. This first step leads to other areas of research such as the quality control methods for the analysis of the manufactured pieces, the path planning that optimizes the manufacturing of the piece, or the correlation through disparity analysis of the final piece and its design. All these areas have associated different models and data that need to be integrated into a common Masterfile that follows a well-defined data architecture [11,12]. This way, such vast and heterogeneous information involves for instance material's data containing its type, form, and feed lot; part geometry's data containing feature variability, and CAD tools; processing data, holding the parameters that control the build process; post-processing for the machining of surface roughness, and heating to reduce residual stresses; testing data containing physical and non-destructive testing, and physics-based simulations for certification; end-of-life data for recyclability, and maintenance, can be efficiently exchanged and exploited.

In this work, we will refer to only two AM techniques, laser metal deposition (LMD) wire and powder, and wire arc additive manufacturing (WAAM), which were the technologies used to generate the digitalized data present in the dataset. LMD and WAAM are AM processes categorized as direct energy deposition (DED) [13]. DED processes use a focused beam possessing thermal energy to melt the material as long as it is deposited onto a surface. In LMD, the metallic powder is injected at a surface and a laser beam melts it, forming a molten pool. The continuous motion of a robot makes possible the layer-by-layer deposition along the designed path, enabling the manufacture of the desired part, see Fig. 1. On the other hand, WAAM is a technology that uses an arc welding process to melt a metallic wire using an electric arc as the heat source [14,15].

3 Methodology

The essential goal of this work is to present a methodology that provides a full digital representation of the life-cycle of AM products. Such representation will hold all the data from design to the manufacturing and use, in order to create a holistic information model that can be used to, on the one hand, ensure process repeatability, and on the other improve the manufacturing process itself by means of further methods relying on advance analytics or even artificial intelligence that are our of the scope of this work.

Our proposed methodology is based on two main pillars: on one side, the modeling and traceability of all the digital assets produced during the design, engineering, and production of AM components, and on the other side the generation of a common digital representation of the physical operations in distinct AM processes.

3.1 Digitalization of the AM Data Chain

One of the direct consequences of the use of AM is the flexibility introduced in the production lines and processes. Where the engineering phases of traditional manufacturing are more streamlined, the different options in AM results in multiple ways of producing the same pieces. Also, the freedom of design associated with AM results in new ways to optimize the manufacturing of pieces. These characteristics result in a much more iterative and unstructured engineering process, where, the results of the fabrication can introduce new changes in the process parameters or the piece's physical design. To be able of representing the AM products' life-cycle it is needed to capture these feedback loops and iterative operations in a consensual format.

To achieve this goal, we propose the use of an information model derived from the commonly used in CM Product, Process, and Resource (PPR) Modeling, adapted for the digital assets produced during the different phases of AM. In traditional PPR, as shown in Fig. 2, we have three basic structures: products, that represent the physical items in a manufacturing domain; resources, that represent the assets in a manufacturing plant; and processes, that represents an action performed by a resource on a product. Using these three blocks, it is possible to model most of the manufacturing processes.

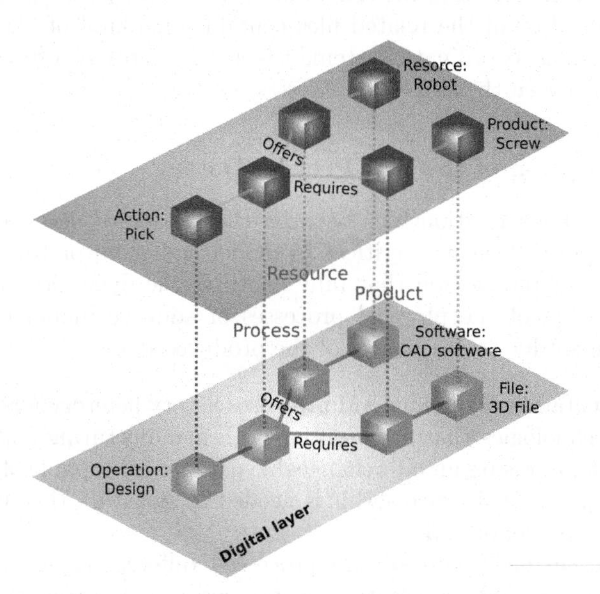

Fig. 2. Representation of the translation of the PPR model in the upper layer in gray to a digital domain in the blue lower layer. (Color figure online)

However in AM, many of these operations are performed in a digital domain, so PPR Modeling needs to be adapted to this type of cycles. In the modified PPR for AM, the files created along the engineering phases are considered products, that are generated and consumed by the processes, or the digital operations associated with the files (e.g., Design). Finally, these operations are related to the Resources, that correspond to the software tools used for that digital operation.

Using this information model, we can represent the flow of data and information in the engineering and production phases while keeping track of the feedback loops where, for example, the results of the simulation force the re-design of the piece, or a defect in production causes the re-evaluation of the production phase. The aggregation of this information can help to identify patterns and optimize the operations needed in the pre-production of an AM piece.

To implement this information model in an interoperable and flexible format, we have used AutomationML [18], a known XML schema for industrial systems. While AutomationML is specialized in the modeling of cyber-physical systems, it also offers support for PPR models, which makes it suitable for this application. Using this format we are able to represent the proposed model as interrelated collections of Operations, Files and Software that represent the flow of information in those phases (see Fig. 3). Therefore, all the information related to an AM component can be summarized in a single AutomationML file, also known as Masterfile, where the relationship between the different files and a link to the files themselves are kept in a way that facilitates the location and traceability of the designs.

Since the Masterfile only contains a link to the digital files with the data from the stages, the resultant file is light and can be easily used for the exchange of information. Most of the related files contain some kind of 3D information, based on the same coordinate system, which facilitates to establish casualty relationships between the stages and files.

3.2 Digitalization of the AM Process Data

In order to establish relationships between the produced pieces and the other stages of the production it is critical to model and monitor the physical AM process with a representation that fully captures the dynamics of the system. This digitalization of the physical processes of additive manufacturing raises several challenges due to the nature of the produced data.

- The representation of distinct AM processes is very heterogeneous, due to the different technologies that are used for both manufacturing and monitorization. In order to create an interchangeable data model capable of representing multiple types of AM processes, it is needed to structure this data following the abstract model of AM.
- Due to the flexible nature of AM processes, different types of sensors and systems can be used to monitor the processes. This raises the need for a flexible data structure capable of aggregating multi-modal information from different sources or with different sample rates.

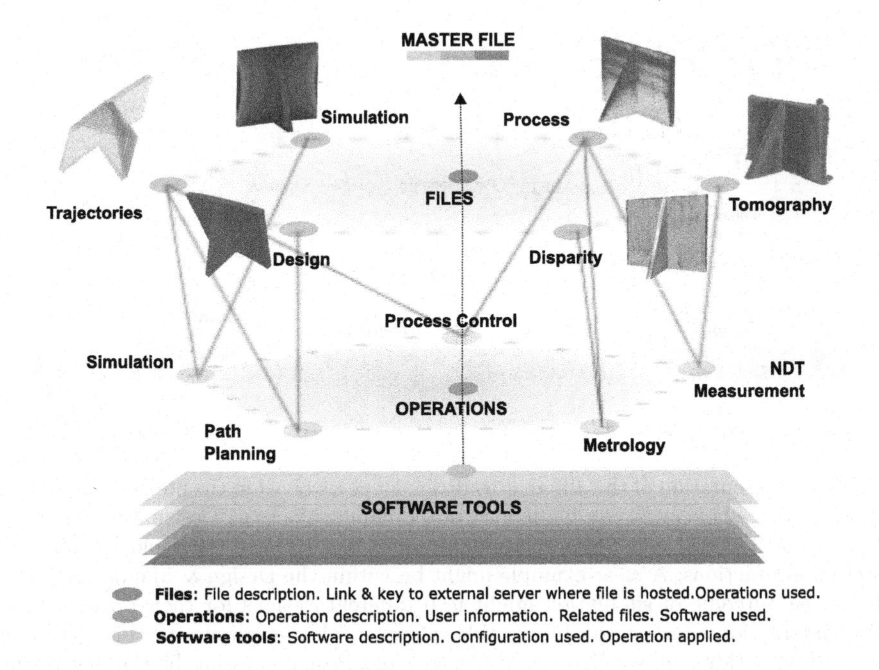

Fig. 3. Representation of the proposed main structure of the Masterfile to ensure the complete traceability of an AM process. Three main blocks of information describe all the processes. All the operations (blue layer) have a correspondence with a given software (orange layers) that drives those operations and at the same time are translated into files (red layer) where operations turn into digital information. These files are self-descriptive. The operations hold information about their relations with different types of files and their location and software tools information provides details on parameters configuration among the corresponding software and at the same time relations with different operations. (Color figure online)

- It is needed to establish a correlation of the monitorization data with specific regions of the produced piece, in order to link this production information with other stages of the piece's life-cycle like dimensional or quality control.

To address these challenges, we propose the use of a hierarchical data structure to model the AM process as a sequence of synchronized measurements over the duration of the process. To that end, the proposed format uses a dynamic array to represent the temporal axis associated with the sequential nature of AM processes. This array stores common attributes of AM processes such as position, power input, or speed, which are then related by a link to other tables that contain measurements of the process structured in their own types and formats (e.g. thermography imaging). As shown in Fig. 5, the end result is a main structured array of "snapshots" of the process associated with a set of 3D coordinates with references to other tables of sensor data that can be recorded synchronously or asynchronously to the process (Fig. 4).

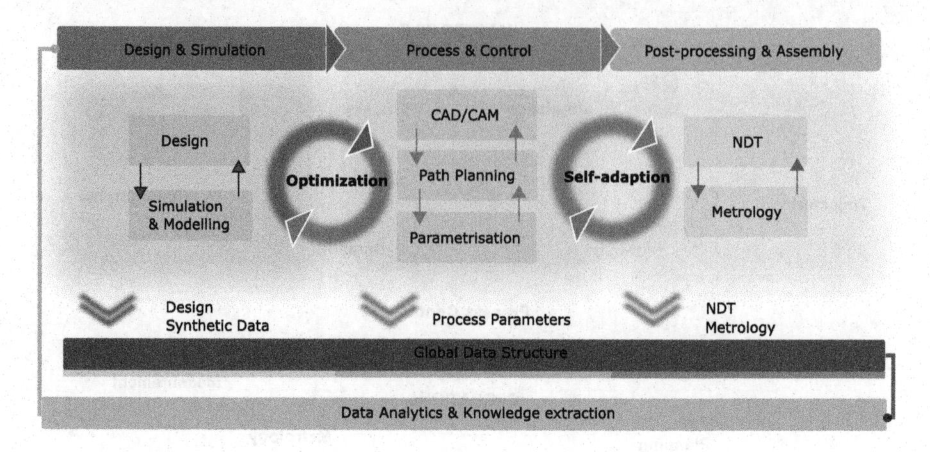

Fig. 4. Representation of the different feedback loops involved in the life-cycle of an AM product. The whole cycle can be divided into different blocks as Design & Simulation, Process and Control and Post-processing & assemble. Each of these blocks can hold internal interactions. A clear example might be within the Design & Simulation block where the outcome of simulation might feed different designs for their optimization. Also, interactions among the blocks take place. A clear example, in this case, can be the interactions among Process & Control and Post-processing blocks, where the outcome of NDT testing might feed process control to correct parameters related to the emergence of a given defect. All these interactions take place within a digital layer that relies on a common data structure (black box). Such a Global data structure has all the elements to perform advanced statistics or machine learning methods to optimize the whole AM process.

The use of this ontology to model the data from AM processes offers several technical advantages while being compatible with most databases and data collection systems. The use of a structured format facilitates the storage, compression, and access of the tables, which can reach considerable size, while keeping a flexible and scalable structure. Also, its abstract design makes it able to represent any type of AM process regardless of the machines and sensors being used for the modeling.

The representation of an AM process as a series of multi-modal "snapshots" provides also more insights about the status of the process, allowing the representation of one or multiple modes of information as a temporal series or a 3D structure. These methods of representing the process information create a straightforward way to find correlations between the modes of the information, like for example, thermodynamics of the process and the manufactured geometries.

This structure was implemented using the Hierarchical Data Format (HDF) standard, which supports the hierarchical relationships needed by the model and allows us to keep the "one file per piece" mentality present in the holistic data model. Another big advantage of this standard is that is widely supported by scientific communities and libraries.

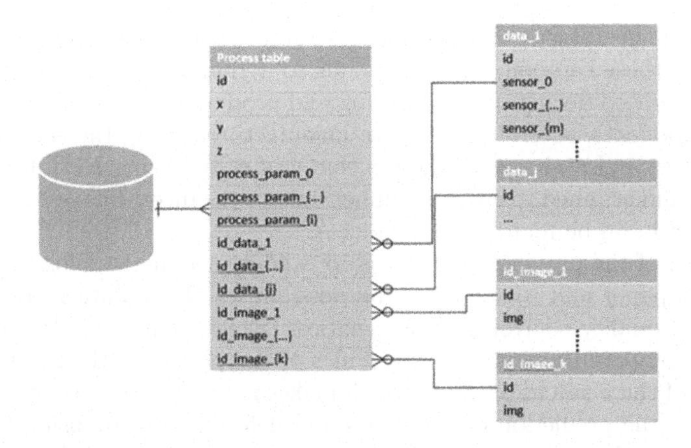

Fig. 5. Hierarchical structure of the AM process data. The overall AM recording contains a group of processes, that are tables that contain a series of 3D points referenced to the piece origin of coordinates. Besides, the process table contains relevant process parameters that are measured directly (such as speed) and references to sub-tables indexes that contain measurements from sensors and cameras. This decoupling from the process and the sensors, introduces a lot of flexibility in the pattern, since the sensor tables can be recorded at a different rate and removes any limitation in the number of sensors to monitor the process. Following the sequence of points in the main table it is possible to completely reproduce the process and all the related data, which enables the use of this structure for the analysis of the manufacturing afterwards.

4 Dataset

Following the methodologies described in Sect. 2, we have created the first dataset [16] of AM components that contains not only information regarding the building of the piece, but also information from the previous (design and engineering) and posterior (Inspection and validation) phases. Each component of the dataset contains information regarding the design, path planning, manufacturing, inspection, and validation of the produced pieces.

This dataset also provides a baseline to compare the effects of process parameters on the overall process dynamics and the quality of the resulted pieces. Furthermore, the dataset comprises pieces from three different AM processes recorded in the same format, which allows a direct comparison between both the produced pieces and the manufacturing data. The process data is stored in the HDF format specified in Sect. 3, and a software tool is provided to interact with the process data [19].

As shown in Table 1, the dataset is structured as a series of components. Each one of these components corresponds to a global data chain and contains the files generated during the engineering and production of the pieces. For each of the components, several pieces were manufactured using the same or different processes, and each one of these pieces was tested using NDT tomography inspection and scanned to check its dimensional tolerances. Images of the pieces of the dataset can be found in Appendix A.

The goal of the presented dataset is to validate the methodology exposed in the present paper and to showcase the possibilities of the data model for production and quality control in AM scenarios. Using the process data model we can easily compare the state and dynamics of the process with the quality and tolerances of the resulted product, which makes it possible to develop automatic methods for the prediction of anomalies and defects. Also, by using the global information model, it is also possible to trace back these anomalies and defects to the engineering stages, which makes possible the implementation of methodologies and checks to preventively assert the good quality of the components.

Table 1. Specimens of pieces found in the dataset

Name	Number of pieces	Process type
T Coupons	3	LMD - Powder
Jet Engine	2	LMD - Powder
CC Coupons - A	13	LMD - Powder
CC Coupons - I	3	LMD - Wire
CC Coupons - M	8	WAAM
CC Coupons - W	4	WAAM

5 Conclusion

This work has presented a methodology to optimize the flow of information within AM of metallic parts. An assessment of a PPR ontology has been done to find its equivalent in a digital domain where our methodology has been implemented. In addition, as a validation of this methodology, a dataset has been presented representing the very first one made open for AM of metal components based on multiple processes. Predictive analysis for the emergence of defects within metallic manufactured parts, design optimization to overcome systematic process issues or trajectory optimization could be some of the immediate ways of exploitation of the presented data structure. Also, the optimization in terms of data storage and transferring, that this methodology offers might provide significant progress in terms of traceability that would trigger relevant breakthrough towards repeatability of the AM process. While the methodology presented was conceived and might be directly useful for the AM in metalworking, the data structure would be valid for the manufacturing of parts using any other material.

Particularly interesting might be how this type of structure opens the door to materials development that, based on generative approaches [17], could address material discovery based on desired properties or fitness with AM processes. Also as a future step, physical models of the system might be added to the master file to promote interactions with process and simulation data in order to promote the exploitation of Physics Informed Systems for the optimization of the manufacturing process.

Acknowledgments. This research has received funding from the European Union's Horizon 2020 research and innovation programme under the project INTEGRADDE with Grant Agreement 820776. The authors want to thank the comments and fruitful discussions with all the members of the Artificial Intelligence and Data Analytics Lab (AIDA-Lab) of the Smart Systems and Smart Manufacturing (S3M) department and the people from the Advanced Manufacturing Processes department of the AIMEN Technology Centre.

Appendix A

Pieces on Metal Additive Manufacturing Open Repository

In this appendix, Fig. 6 shows some pictures of the physical specimens manufactured for the dataset are displayed to facilitate the visual understanding of the manufactured components.

- The T-Coupon is a simple geometry that combines a curved wall with a rib manufactured with stainless steel as a first demonstrator of the capabilities of the presented methodology. Three unique specimens with different process parameters were manufactured.
- The Jet Engine is a complex geometry that combines a cylindrical body with smaller subcomponents around it manufactured with stainless steel to showcase the use of the methodology with large real-world components. Two unique specimens were manufactured in different sizes.
- The CC-Coupon is a simple geometry that combines two curved walls with three flat walls manufactured with stainless steel to demonstrate the capabilities of the presented methodology and formats to represent different processes and machines. These coupons were manufactured in four unique machines and locations, making a total of 28 different coupons created with completely different processes.

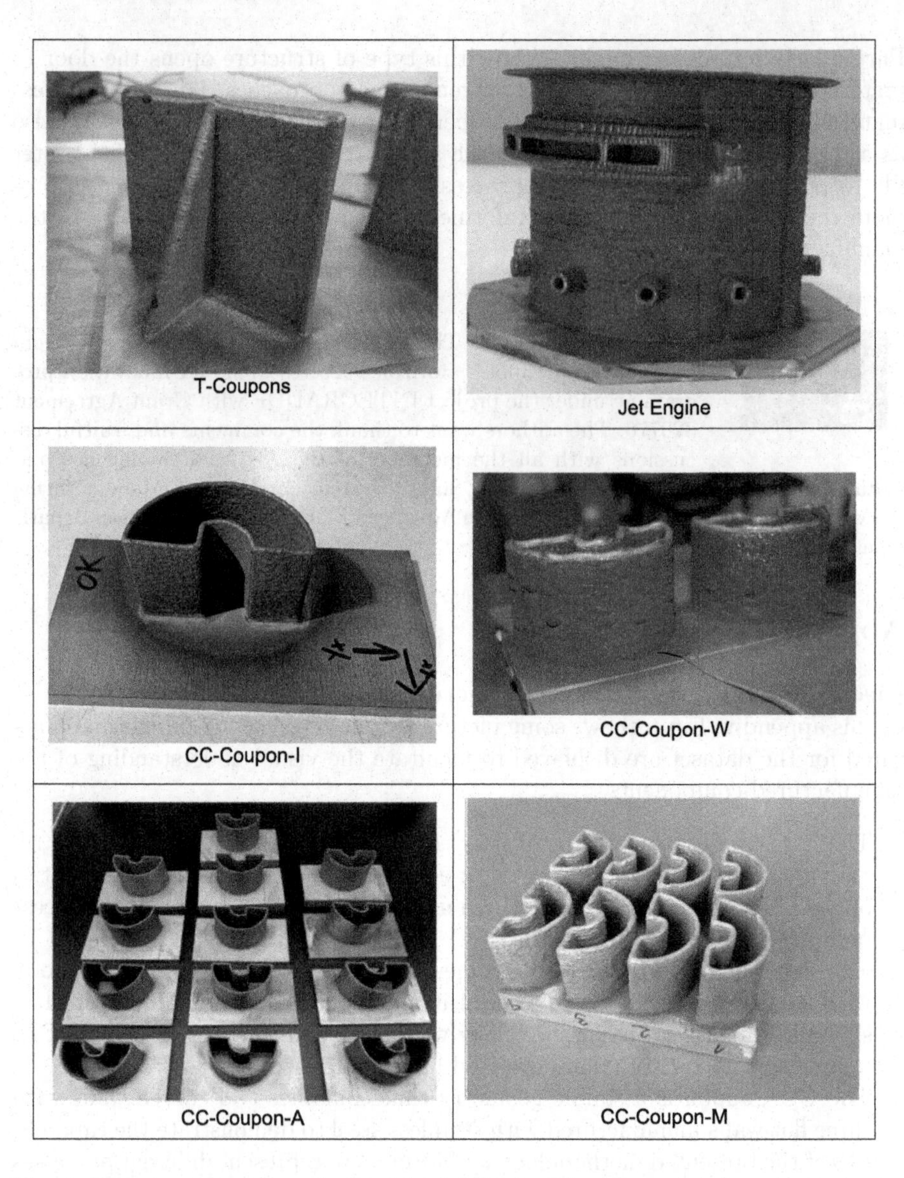

Fig. 6. Pictures of the pieces in the Metal Additive Manufacturing Open Repository

References

1. Phuyal, S., Bista, D., Bista, R.: Challenges, opportunities and future directions of smart manufacturing: a state of art review. Sustain. Futures **2**, 100023 (2020). https://doi.org/10.1016/j.sftr.2020.100023. ISSN 2666-1888
2. Vafadar, A., et al.: Advances in metal additive manufacturing: a review of common processes, industrial applications, and current challenges. Appl. Sci. **11**, 1213 (2021). https://doi.org/10.3390/app11031213

3. Blakey-Milner, B., et al.: Metal additive manufacturing in aerospace: a review. Mater. Des. **209**, 110008 (2021). https://doi.org/10.1016/j.matdes.2021.110008. ISSN 0264-1275
4. Wiberg, A., Persson, J., Ölvander, J.: Design for additive manufacturing - a review of available design methods and software. Rapid Prototyp. J. **25**(6), 1080–1094 (2019). https://doi.org/10.1108/RPJ-10-2018-0262
5. Butt, J.: Exploring the interrelationship between additive manufacturing and Industry 4.0. Designs **4**, 13 (2020). https://doi.org/10.3390/designs4020013. ISSN 2411-9660
6. González-Val, C., Muíños-Landín, S.: Generative design for social manufacturing. In: Proceedings of the Workshop on Applied Deep Generative Networks, Zenodo, vol. 2692 (2020). https://doi.org/10.5281/zenodo.4597558
7. Mies, D.D., Marsden, W., Warde, S.: Overview of additive manufacturing informatics: "a digital thread". Integr. Mater. Manuf. Innov. **5**, 114–142 (2016). https://doi.org/10.1186/s40192-016-0050-7
8. Ahn, D.-G.: Direct metal additive manufacturing processes and their sustainable applications for green technology: a review. Int. J. Precis. Eng. Manuf. Green Technol. **3**(4), 381–395 (2016). https://doi.org/10.1007/s40684-016-0048-9
9. Ngo, T.D., et al.: Additive manufacturing (3D printing): a review of materials, methods, applications and challenges. Compos. B Eng. **143**, 172–196 (2018). https://doi.org/10.1016/j.compositesb.2018.02.012
10. Taheri, H., et al.: Powder-based additive manufacturing - a review of types of defects, generation mechanisms, detection, property evaluation and metrology. Int. J. Addit. Subtract. Mater. Manuf. **1**(2), 172–209 (2017). https://doi.org/10.1504/ijasmm.2017.088204. ISSN 2057-4975
11. Habib, M.A., Khoda, B.: Hierarchical scanning data structure for additive manufacturing. Procedia Manuf. **10**, 1043–1053 (2017). https://doi.org/10.1016/j.promfg.2017.07.095. ISSN 2351-9789
12. Bonnard, R., et al.: Hierarchical object-oriented model (HOOM) for additive manufacturing digital thread. J. Manuf. Syst. **50**, 36–52 (2019). https://doi.org/10.1016/j.jmsy.2018.11.003. ISSN 0278-6125
13. Tepylo, N., Huang, X., Patnaik, P.C.: Laser-based additive manufacturing technologies for aerospace applications. Adv. Eng. Mater. **21**, 1900617 (2019). https://doi.org/10.1002/adem.201900617
14. Garcia-Colomo, A., et al.: A comparison framework to support the selection of the best additive manufacturing process for specific aerospace applications. Int. J. Rapid Manuf. **9**, 194–211 (2020). https://doi.org/10.1504/IJRAPIDM.2020.107736. ISSN 1757-8817
15. Jin, W., et al.: Wire arc additive manufacturing of stainless steels: a review. Appl. Sci. **10**, 1563 (2020). https://doi.org/10.3390/app10051563. ISSN 2076-3417
16. Gonzalez-Val, C., Baltasar, L., Marcos, D.: Metal additive manufacturing open repository (2.0) [Data set]. Zenodo (2020). https://doi.org/10.5281/zenodo.5031586
17. Precker, C.E., Gregores-Coto, A., Muíños-Landín, S.: Generative adversarial networks for generation of synthetic high entropy alloys. Zenodo (2021). https://doi.org/10.5281/zenodo.5582109

18. Drath, R., Luder, A., Peschke, J., Hundt, L.: AutomationML - the glue for seamless automation engineering. In: IEEE International Conference on Emerging Technologies and Factory Automation, vol. 2008, pp. 616–623 (2008). https://doi.org/10.1109/ETFA.2008.4638461
19. González Val, C., Molanes, R.F., Señarís, B.L.: Visualizer (3.0.1). Zenodo (2021). https://doi.org/10.5281/zenodo.5469946

Twenty Years of Successful Translational Research: A Case Study of Three COMET Centers

Katja Bühler[1(✉)], Cornelia Travniceck[1], Veronika Nowak[2], Edgar Weippl[2,3], Lukas Fischer[4], Rudolf Ramler[4], and Robert Wille[4]

[1] VRVis Zentrum für Virtual Reality und Visualisierung Forschungs-GmbH, Donau-City-Straße 11, 1220 Wien, Austria
{buehler,travniceck}@vrvis.at

[2] SBA Research gGmbH, Floragasse 7, 1040 Vienna, Austria
{vnowak,eweippl}@sba-research.org

[3] University of Vienna, Kolingasse 14-16, 1090 Vienna, Austria
edgar.weippl@univie.ac.at

[4] Software Competence Center Hagenberg GmbH (SCCH), Softwarepark 32a, 4232 Hagenberg, Austria
{lukas.fischer,rudolf.ramler,robert.wille}@scch.at

Abstract. The term 'translational research' is traditionally applied to medicine referring to a transfer of scientific results into practice, namely from 'bench to bedside'. Only in recent years there have been attempts to define *Translational (research in) Computer Science* (TCS), aside from applied or basic research and mere commercialisation. In practice, however, funding programs for academia-industry collaboration in several European countries – like the Austrian COMET Program and its predecessors which include opportunities for Computer Science institutions – already provided a unique framework for TCS for over two decades. Although the COMET Program was initially set up as a means of temporary funding, a majority of the partaking institutions have managed to stay in the program over several funding periods – turning it in a de facto long-term funding framework that provides a successful structure for academia-industry collaboration. How to (i) identify the key factors to the success of individual Competence Centers and (ii) maintain fruitful relationships with industry and other academic partners are the main aims of this paper.

Keywords: Computer science · Translational research · Technology transfer · Research centers · Academia-industry collaboration

1 Motivation

The transfer of technology knowledge and innovative results from universities and research institutions to industry was identified by the European Union as one out of ten key areas for action in 2007 [1]. With that intention, Austria established its first competence center program already in 1998 to increase scientific

G. Kotsis et al. (Eds.): DEXA 2022 Workshops, CCIS 1633, pp. 155–166, 2022.
https://doi.org/10.1007/978-3-031-14343-4_15

and industrial competitiveness by fostering knowledge and innovation transfer from science to industry [2]. The follow-up program COMET (Competence Centers for Excellent Technologies) supports 25 competence centers in five sectors: digitization, information and communication technologies (ICT), energy & environment, life sciences, mobility, and materials & production.

The implementation of technology transfer in the sense of the COMET goals is particularly demanding for computer science as a cross-sectional technology characterized by highly interdisciplinary projects that require a great deal of flexibility, mutual understanding and education from all participants. Only recently, Computer Science has begun to adopt the translational research model [3] originally established by the medical domain to provide a formalized framework for the entire translation process from theory to practical use. Generally, literature highlighting implementation- and management strategies of such interdisciplinary IT projects at the border between science and industry from a research centers' perspective is still sparse. We aim at filling this gap by sharing and discussing insights, best practices and challenges summarizing over 20 years of experience of three COMET research centers in Austria performing successful research in Visual Computing,[1] Cybersecurity,[2] and Software and Data Science.[3]

2 The COMET Program

Since 1998, the COMET predecessor programs *Kplus* and *K_ind/K_net* had built up central research competencies in cooperation between science and industry in 45 centers and networks across Austria; the 'best-of' of the two programs has continued under the name COMET since 2005/2006.

COMET centers are funded by the Republic of Austria – specifically by the Federal Ministry for Climate Action, Environment, Energy, Mobility, Innovation and Technology (BMK) and the Federal Ministry for Digital and Economic Affairs (BMDW), the participating federal states, and by industry as well as scientific partners. Together, a medium- to long-term research program is defined, with the goal to strengthen the cooperation between industry and science and to promote the development of joint research competences and their scientific and economic utilization. COMET centers are embedded in the national and federal technology strategies and in scientific, economic and political ecosystems as shown in Fig. 1. Furthermore, extending beyond challenges and problems formulated from purely market-based strategies, issues arising from digitization, climate change, aspects of sustainability as well as ethics and health, which are increasingly becoming the focus of society, are of ever greater importance to the research questions addressed in the competence centers.

According to the COMET Programme Evaluation Report [4] published in June 2021,[4] around EUR 50 million from federal funds and around EUR 25

[1] https://www.vrvis.at.
[2] https://www.sba-research.org.
[3] https://www.scch.at.
[4] https://www.bmk.gv.at/themen/innovation/publikationen/evaluierungen/comet_evaluierung.html.

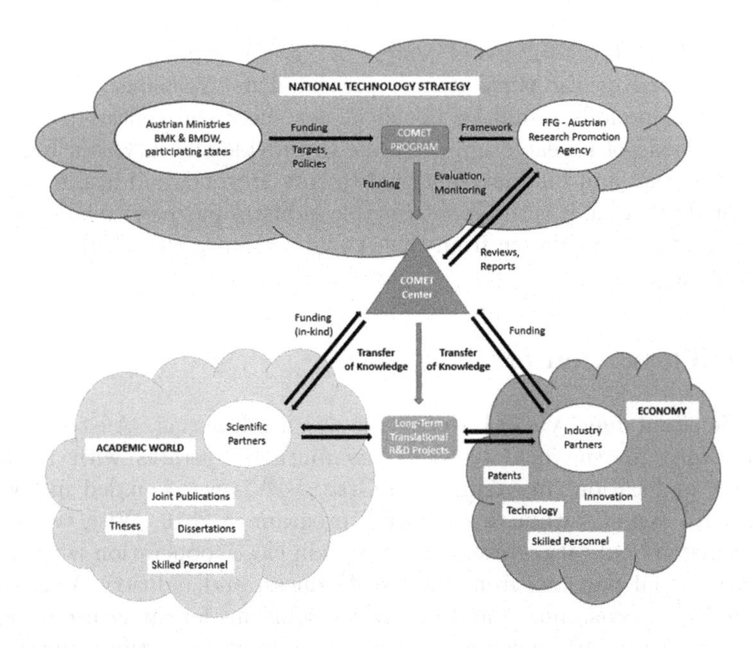

Fig. 1. Illustrating the network of relationships of COMET centers.

million from state funds flow into COMET funding. Furthermore, the report underpins the COMET funding program's goal of establishing long-term research collaborations by illustrating the durability of established centres already funded by the initial K programs. This applies to 64% of the active COMET centres (16 out of 25). The focus of the COMET centres on applied research and the associated knowledge transfer is also underpinned by the report: 38% of project resources are spent on basic research, 50% on industrial research and 12% on experimental development. The competence centres currently employ 1,742 full-time equivalents (2,389 researchers), which in turn underlines the contribution to retaining and promoting excellent researchers in Austria.

The numbers of publications, associated theses, patent application, and the generation of revenue outside COMET are strong indicators for the success of the program. Every year, there are now around 2,200 publications connected to COMET centers, about 700 of them in reviewed journals and conference proceedings. About 7 % of PhD theses on relevant topics under progress at Austrian Universities are already associated with COMET and in the review period from 2015–2019 centers reported over 50 new patent applications. In addition, centers triple each Euro they get in COMET funding in revenue. Feedback by a majority of industry partners indicated enhancements of existing products as well as product, process, and service innovations through COMET [4]. The much needed certainty for translational style research and a sustainable academic-industry partnership can be achieved via back-to-back applications within the COMET program granting currently funding for research centers in phases of eight years. This allows successful centers to enter a de-facto long-term funding (as demonstrated by the three centers presented in this paper). Nevertheless, longer funding periods would be beneficial for successful centers to avoid the massive overhead of repeated applications, to provide stability as a R&D partner and to attract and bind international talents. Beyond, and in the context of translational research, it is the goal to enable industry partners to independently further the advantages drawn from their collaboration with COMET centers in the long term.

3 The Three Centers

VRVis Zentrum fuer Virtual Reality und Visualisierung, Austria's leading research center in the field of visual computing, operates with around 70 researchers at locations in Vienna and Graz. VRVis was founded in 2000 and became a K1 center within the COMET program in 2010. With two decades of experience in application-oriented research in close cooperation with national and international partners from science, business, and industry, VRVis' main goal is to develop customized and innovative solutions for direct use in practice [5], using the newest technologies in visual data analytics, artificial intelligence, extended reality (XR), image processing, simulation, and digital twins.

SBA Research, located in Vienna, was originally founded in 2006 and currently employs about 130 people. This COMET center covers both the scientific- and industry aspects of information security. In addition to the COMET flagship program *SBA-K1*, SBA continuously conducts between 20 and 30 national and international research projects. The topics range from IoT- and Industry 4.0- to Internet- and software security. Questions like privacy protection or the long-term effects of digitization for our society - always with a special focus on cyber-security - are a central concern. SBA's network encompasses national and international institutions from academia and industry alike. A special emphasis is on the creation of bridges between theoretical, scientific high-quality research and immediately usable results for company partners. In the field of professional services, the focus is on security analyses, both in the organizational and technical areas. In this regard, SBA has been working with ministries, public authorities, large companies, and SMEs for many years.

SCCH (Software Competence Center Hagenberg), is a non-university research center founded in 1999 by the Johannes Kepler University Linz. Since then, the center has continuously been growing and currently employs over 100 people. Since 2008, SCCH is supported as a K1 center within the COMET program. SCCH drives innovation in the creation and application of software by integrating fundamental research with the solution of complex application problems. SCCH acts as an interface between international research and regional industry and business. The center conducts research along the full AI system engineering life cycle focusing on linking Software and Data Science [6] in the areas of intelligent software testing, software evolution and documentation, user-centred software engineering, big/stream data processing, smart data discovery, fault detection and identification, and predictive analytics and optimization.

4 Factors for Successful Translational Research Within the COMET Framework

Close collaboration with industry partners is one of the core tenets of the COMET funding program. Companies joining a COMET center not only contribute money, but also the time and expertise of their employees as well as information about their business, R&D efforts, and research needs. Therefore, mutual trust between the COMET center and the company partner is crucial, and a lot of time and effort is allocated to building these relationships. In the following, we want to outline how we establish partnerships, mutual understanding, trust, and knowledge and technology exchange with long-term impact. We report on our experiences and established practices, how we manage sector-specific goals and tackle the challenges arising from fast-changing business environments and new digital technologies. As competence centers, we also maintain relationships with other scientific institutions, which is a further topic we will place a focus on. To conclude, we will pay particular attention to the topic of knowledge transfer between the scientific world and industry.

4.1 Initiating Successful Industry Collaborations

The first steps towards a successful long-term relationship are made when the collaboration is initiated. In all of the three centers, similar patterns have emerged for supporting industry partners in finding their way to joint R&D activities from individual and often very different starting points.

Individual Contact Points for Companies. What does it take for a company to join the COMET program? Written agreements regarding, e.g., IPRs and rights to publish, are typical legal cornerstones. However, ultimately it is the strong personal relationship that staff of COMET centers establish with their industry counterparts which drives those partnerships. While it may occur that a company approaches a COMET center directly, the initial contact usually unfolds via personal networking at industry events, attending low-threshold formats like Meetups, or using the services of dedicated platforms (e.g., Austria Start-ups) or semi-public bodies like the Vienna Business Agency; the latter provides 'match-making' in both directions, supporting companies with research needs as well as research centers in search for a industry partner suitable for a given scientific endeavor. Given that the industry partners range from local SMEs and Start-ups to large multi-national corporations, COMET centers and their administrative and scientific employees are well-versed in adjusting to a variety of company cultures, complying with very diverse procedural requirements, and meeting research needs in all shapes and sizes.

A Joint Path to R&D Activities. If the first contact sparked mutual interest - and assuming that the company in question has not yet worked with a research institution - the following unfolds: First, acquainting the company with the world of research and the related funding landscape. This often takes place via small-scale funding schemes which facilitate projects running three to twelve months (a path to structured R&D activities, e.g. funded by FFG) and are independent from the COMET program. They can usually be applied for with comparably little effort and funding decisions are swift. Here, COMET centers share their extensive know-how regarding grant applications, since especially start-ups and SMEs often do not have first-hand experience with such matters. On occasion, especially bigger companies might even commission research at a COMET center, with the same effect of building a first bridge between these two parties from science and industry.

Common Language and Mutual Understanding. Secondly, embarking on the process of clearly defining the research and/or innovation need, establishing common ground and a shared vocabulary, so that all people involved eventually speak the same language. This may take place through in-depth workshops between research groups and industry experts during which problems are defined, available state-of-the-art solutions are discussed and evaluated, and initial R&D roadmaps are agreed upon. In other cases, a center's management and/or specialized personnel manage this initial phase and see to it that researchers, practitioners, and managers come to the same understanding of a given research endeavor. In the mid-term, these efforts lead to small- to medium-sized projects which give all parties involved a sense of how cooperation works out and if a more long-term collaboration - as in the COMET program - is of interest. Lastly, it should be mentioned that even working on a grant application together, regardless of if the project gets funded or not, is a very effective way to build a rapport between science and industry. In more than one instances, 'failed' project proposals have led to mutually beneficial partnerships in the long run.

4.2 Maintaining Long Term Relationships with Industry Partners

Given its runtime - eight years per call, and successful centers usually get granted back-to-back funding - the COMET program offers an excellent precondition for long-term partnerships between science and industry. The COMET centers featured in this paper have been up and running for more than 20 (VRVis, SCCH) resp. 15 (SBA) years each, and some of their company partners have been with them from the very beginning. How is this long-term relationship accomplished, especially in the fast-paced world of ICT?

Flexibly in Adapting to Cooperation Requirements. First and foremost, investing time and effort in building company-center relationships and, thus, firmly establishing mutual trust is the very foundation of successful long-term partnerships. Taking the time to align diverging mental models, to foster a joint understanding of and respect for the needs, abilities, goals, and constraints of all involved parties and to be responsive and accommodating should issues arise has proven crucial in maintaining a center's relationship with their industrial partners. Secondly, although it might take time for a company to be willing to become a COMET partner, once this decision is made, the further procedures are straightforward. A company can join a center, and joint work on R&D challenges can start the following day. Nevertheless, it is also quite unbureaucratic to scale a company's involvement up or down throughout the project, or to even release an industry partner from its obligations. Avoiding a lock-in situation is essential for industry partners, since they have to compete in a dynamic market environment where strategic plans are subject to change when new business opportunities arise.

Stability of the Research Program. Being able to handle the discrepancy of how research and industry operate differently in terms of available funds, time and personnel resources is a central asset of COMET centers. This is on the one hand thanks to the program's structure which favors multi-firm projects to exploit synergies and promote cross-company knowledge transfer, as well as to ensure the continuity of a COMET project, even if an industry partner should leave. On the other hand, industrial partners are provided with a stable group of contacts at COMET centers which employ their staff via permanent contracts (as opposed to universities which usually have substantial fluctuation at the MSc and PhD level). Personal contacts are essential in maintaining long-term working relationships between organizations. Good personal relationships and a sense of how the other party works is also very important when it comes to other possible fields of tension, most notably with regards to IPR strategies vs. a researcher's need to publish.

Impact by Translational Research. COMET centers and their partners contribute substantially to getting research results from the *laboratory* to the *locale*: while the centers further develop along the edge of the state of the art, companies do apply the research outcome in their field of business. Thanks to the close collaboration, feedback from the application of research results in practice is immediate, and lessons can be learned quickly. Through incremental research strategies, known scientific methods can be translated into industry practices, and experiences from these practices are fed back to the researchers. While industry does depend on cutting-edge research by scientific institutions, those institutions are in turn dependent on insights into the issues and needs of companies 'in the wild' - a COMET center working in the field of ICT must be able to reorient itself constantly to stay at the forefront of research and meet the needs of its industry partners. However, this to-and-from does not happen via joint projects and knowledge exchange alone. COMET centers are a unique place where people are trained on the interface of science and industry, thus making career changes between those two domains easier. With personnel, one may postulate that also knowledge and expertise translate from the laboratory to the locale and back.

4.3 Teaming Up with Academia

In general, COMET centers build upon stable networks of national and international scientific partners who are directly involved in the COMET program. These networks form the basis for transferring knowledge from basic research and ensure through close collaboration the firm embedding of the COMET center's research agenda in a national and international context. The composition of academic partners is shaped on one hand by the need for a strong local network of academic partners supporting directly the research agenda of the COMET center, and on the other hand by the need for complementary expertise and exchange on an international level.

Complementary Scientific Contributions from Academic Partners. As specified by the COMET program, *key researchers* are selected based on the core topics of and recruited from these institutions. These are recognised scientists who, with their expertise as advisors, scientifically accompany research projects and the centre management in strategic decisions. They also often take over the academic supervision of students and doctoral candidates working on research projects at the center and, thus, contribute to a direct knowledge transfer between the university and the research centre.

The core network of academic partners is complemented by an extended network of associated academic partners with whom cooperation takes place either within the framework of non-COMET projects or within the framework of the strategic work on a more informal level, e.g., in the form of joint publications, co-supervision of doctoral students, and exchange of scientists.

Conversely, the scientific partners also benefit from insights into practical problems in the context of industry. COMET centers add a complementary application-specific perspective to the academic view on a research topic: the challenges of transferring theoretical results and proof-of-concepts from a laboratory setting into the typically much more complex environment of the real-world application.

Application-Relevant Knowledge and Talent Exchange. Effective collaboration between academia and a research center is frequently implemented in form of PhD- and master students involved in project work with industry and supervised by research partners. Four-year funding periods, including strategic projects, allow for PhD work at our academic partners with some distance to the 'daily business' of industry projects. This enables deep-dives into specific research topics, while supported with practical research questions and real use cases for evaluation.

Direct talent transfer from universities and scientific partners to the COMET centers ensures inflow of fresh academic knowledge and is quite common due to attractive working conditions and unlimited working contracts. The transfer from personnel back to the universities, still rather infrequent, is supported by all centers in order to enable all kind of research careers and to further strengthen the relationships to academia.

All three COMET centers are also active research institutions by themselves, showing a strong presence of the centers' key personnel in the respective research communities in terms of scientific contributions and engagement in scientific events like the organization of conferences and workshops at national and international level, being editors of special issues in journals, and participation in reviewing, scientific committees and associations.

4.4 Conducting Knowledge and Technology Transfer

The transfer of knowledge originating from basic research to the application domain, which is in the focus for industry, is a core task as well as a core competence of COMET centers. In the course of the many years of experience of the three centers described above, several best practices have been established to master this multifaceted challenge.

Direct Collaboration in Joint Projects. One of the most important and frequent opportunities for knowledge transfer results from direct collaboration in joint projects – whether within the COMET program, other funded research projects, or directly commissioned projects. Knowledge transfer happens through regular exchange at project meetings, working together on technical and scientific tasks, regular content-related discussions and retrospectives, as well as jointly prepared reports and scientific publications. Direct collaboration in the context of ongoing project work provides the opportunity for exchanging experience and best practices regarding processes, methods and tools, scientific approaches and technologies. Employees of company partners participate in the project to learn about new practices and technologies, e.g., from the subject area of ML/AI, Visual Analytics, or security.

Frequency and intensity of exchange activities vary and need to be customized according to research topic, project and funding type, company culture and size. In multi-firm projects, knowledge transfer also happens through COMET center staff being involved in the collaboration with different company partners over time and promoting spill-over effects. In such a cross-company settings, several companies may have the same, similar or complementary research needs, but at (slightly) different times, and synchronization is not always possible in these cases. COMET project staff therefore 'buffers' relevant knowledge and transfers it between companies when needed. Furthermore, in addition to collaboration in the course of project work, dedicated knowledge transfer workshops are organized not only for individual companies but in the form of joint events in which all companies of a multi-firm project take part.

Transfer of Personnel. Knowledge transfer is also supported by personnel exchange during the project duration, i.e., a COMET employee may stay with the partner company for a limited period of time or vice versa. Student interns working in collaborative projects are considered especially helpful in translating scientific knowledge [7]. In general, access to excellent personnel is seen as a major advantage by corporate partners. Regardless of the personnel transfer, whether from the research center to the company or vice versa, this strengthens mutual relationships and even paves the way for new projects and partnerships.

Transfer Events Beyond Project Collaboration. In addition to joint projects with company partners, special funding programs such as *Innovation Camps* of the Austrian Research Promotion Agency (FFG) allow to translate COMET research results into workshops and trainings for a broader audience, beyond the partners in established collaborations. Several other instruments for knowledge transfer include cluster initiatives, joint hackathons, community building events such as meetups or tech talks, special interest groups and networks that enable knowledge exchange with industry outside of dedicated projects.

In a long term perspective, being able to convey scientific research to diverse target audiences and stakeholder groups, ranging from youth to laypeople, other researchers, practitioners, management and many more are key success factors for knowledge transfer to industry and society in general.

5 Conclusion

VRVis, SBA Research, and SCCH are three instances of research centers operating in the frame of the Austrian COMET program. All three centers have a long and successful history of knowledge and technology transfer in collaborating with industry. In this paper we showed that *translational research*, a concept adapted from medicine to Computer Science, has in fact been practiced by Austrian COMET centers for the last two decades. The following cornerstones facilitate such successful translation of research results *from laboratory to locale* and into industrial applications:

- The first step towards a successful long-term industry collaboration is to offer **individual contact points for companies** starting into R&D activities with different levels of experience, background, and culture.
- Furthermore, defining the **joint path to R&D activities** together with prospective company partners helps them to get acquainted with the world of research and the related funding landscape through joint small-scale projects or commissioned research.
- In the course of such endeavors, strong working relationships are built and **a common language and mutual understanding** is established as basis for long-term collaborations with industry partners.

- Maintaining **flexibility in adapting to cooperation requirements**, also in the course of ongoing collaborations, is a key strength of COMET centers, which allows to accommodate the inevitable changes companies are facing in their usually highly dynamic business environments.
- At the same time, the COMET centers' **stability regarding their research program** over at least four years guarantees the necessary continuity to stay focused on a clear research vision and agenda.
- **Impact by translational research** is accomplished through the centers' further developing along the edge of the state of the art, while the companies are applying the research results in their domain. Close collaboration allows for immediate feedback from applying research results in practice.
- **Complementary scientific contributions from academic partners** support the high scientific standards and output of the centers' research projects. Active scientific partner networks also ensure intense knowledge transfer between universities and research centers.
- **Application-relevant knowledge and talent exchange** offer on the one hand a complementary application-oriented perspective about academic research; on the other hand, direct talent transfer from universities benefits the COMET centers, while transfer from personnel back to the universities is supported as well.
- Knowledge and technology transfer to industry is typically achieved via **direct collaboration in joint projects** at regular project meetings as well as through active and direct technical and scientific collaboration. Transfer activities are tailored to the respective research topic, project type, as well as company size and culture.
- Additional **knowledge transfer via exchanging personnel** usually takes place during a given project. Access to excellent personnel is seen as a particular advantage by industry partners, and such exchange strengthens the mutual relationships between the centers and their partners.
- **Transfer events beyond project collaboration** serve the dissemination of project results and novel solutions to a larger audience. Instruments of knowledge transfer include, but are not limited to the participation in open science initiatives, community networks, and public events.

Acknowledgements. The research reported in this paper has been funded by BMK, BMDW, and the State of Upper Austria in the frame of SCCH (865891), part of the COMET Programme managed by FFG. VRVis is funded by BMK, BMDW, Styria, SFG, Tyrol and Vienna Business Agency in the scope of COMET - Competence Centers for Excellent Technologies (879730) which is managed by FFG. SBA Research is a COMET Centre within the COMET - Competence Centers for Excellent Technologies Programme and funded by BMK, BMDW, and the City of Vienna. The COMET Programme is managed by FFG.

References

1. European Commission: Improving knowledge transfer between research institutions and industry across Europe. EUR 22836 EN (2007)
2. Biegelbauer, P.: Learning from abroad: the Austrian competence Centre Programme Kplus. Sci. Public Policy **34**(9), 606–618 (2007)
3. Abramson, D., Parashar, M.: Translational research in computer science. Computer **52**(9), 16–23 (2019)
4. Warta, K., et al.: Evaluierung des COMET Programms, Endbericht (2021)
5. Shneiderman, B.: The New ABCs of Research, pp. 196–197. Oxford University Press (2016)
6. Ramler, R., Ziebermayr, T.: Software Competence Center Hagenberg: Wissens- und Technologietransfer in Software und Data Science. Gesellschaft für Informatik, Software Engineering (2020)
7. de Wit-de Vries, E., Dolfsma, W.A., van der Windt, H.J., et al.: Knowledge transfer in university-industry research partnerships: a review. J. Technol. Transf. **44**, 1236–1255 (2019)

Data Integration, Management, and Quality: From Basic Research to Industrial Application

Lisa Ehrlinger[1,2](✉), Christian Lettner[1](✉), Werner Fragner[5], Günter Gsellmann[7], Susanne Nestelberger[7], Franz Rauchenzauner[5], Stefan Schützeneder[3], Martin Tiefengrabner[4], and Jürgen Zeindl[6]

[1] Software Competence Center Hagenberg GmbH, Hagenberg, Austria
{lisa.ehrlinger,christian.lettner}@scch.at
[2] Johannes Kepler University Linz, Linz, Austria
[3] Borealis Polyolefine GmbH, Schwechat, Austria
{guenter.gsellmann,susanne.nestelberger,
stefan.schuetzeneder}@borealisgroup.com
[4] PIERER Innovation GmbH, Wels, Austria
martin.tiefengrabner@pierer-innovation.com
[5] STIWA Automation GmbH, Attnang-Puchheim, Austria
{werner.fragner,franz.rauchenzauner}@stiwa.com
[6] voestalpine Stahl GmbH, Linz, Austria
[7] Borealis Polyolefine GmbH, Linz, Austria
juergen.zeindl@voestalpine.com

Abstract. Data integration, data management, and data quality assurance are essential tasks in any data science project. However, these tasks are often not treated with the same priority as core data analytics tasks, such as the training of statistical models. One reason is that data analytics generate directly reportable results and data management is only the precondition without clear notion about its corporate value. Yet, the success of both aspects is strongly connected and in practice many data science projects fail since too little emphasis is put on the integration, management, and quality assurance of the data to be analyzed.

In this paper, we motivate the importance of data integration, data management, and data quality by means of four industrial use cases that highlight key challenges in industrial applied-research projects. Based on the use cases, we present our approach on how to successfully conduct such projects: how to start the project by asking the right questions, and how to apply and develop appropriate tools that solve the aforementioned challenges. To this end, we summarize our lessons learned and open research challenges to facilitate further research in this area.

Keywords: Data management · Data quality · Data integration · Metadata management · Applied research

L. Ehrlinger and C. Lettner—Equal contribution.

G. Kotsis et al. (Eds.): DEXA 2022 Workshops, CCIS 1633, pp. 167–178, 2022.
https://doi.org/10.1007/978-3-031-14343-4_16

1 Introduction

Data (and consequently its management) is crucial for decision making in enterprises and organizations, such as predictive maintenance in smart factories [16]. While there is a gaining interest in the analysis of data with machine learning (ML) or deep learning (DL) models, the core tasks of integrating, managing, and improving data quality (DQ) is often neglected [1].

In the course of the multi-firm and applied research project Sebista,[1] we recognized and solved common data challenges over the last four years. Figure 1 illustrates the typical data flow and tasks in such a project, where real-world data from a production process (depicted in the box to the left) is (T1) stored in multiple heterogeneous data sources [1]. This data is then usually (T2) integrated into a (central or distributed) storage. The data requires (T3) preprocessing and DQ assurance (cf. [1,14,35]) to be (T5) analyzed by data scientists. (T4) Metadata management is a parallel activity, which is required for all other tasks: technical metadata can be stored for finding data in distributed data sources, DQ measurements need to be stored over time, and information about ML models (e.g., features, training parameters) need to be stored to ensure an end-to-end ML process. In such applied research projects, common assumptions often made by basic research do not hold. For instance, data is usually not independent and identically distributed (cf. iid assumption) [5], it might be sparse for certain distributions, redundant [35], or too much for in-memory processing [8].

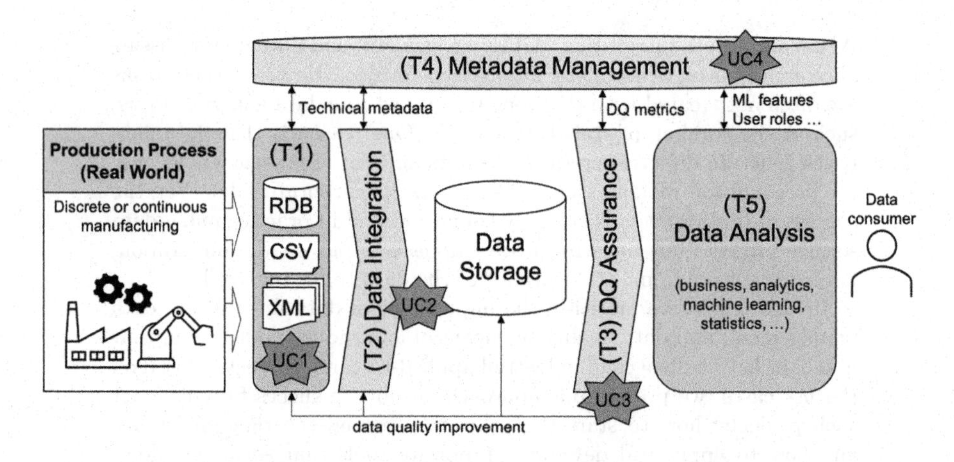

Fig. 1. Data Integration, Management, and Quality Project Environment in Practice

In this paper, we review four industrial uses cases (UC1–4) in Sect. 2. Each use case is provided by one of our company partners and can be assigned to one topic within the data project overview in Fig. 1: (UC1) on graph databases,

[1] https://www.scch.at/data-science/projekte/detail/comet-projekt-sebista-2019-2022.

(UC2) on data integration and storage, (UC3) on DQ measurement and improvement, and (UC4) on metadata management. The use cases are depicted as green stars in Fig. 1. In Sect. 3, we propose methods and tools that solve these challenges, specifically we contributed to the three colored data tasks (T2) data integration, (T3) data quality assurance, and (T4) metadata management. We summarize our lessons learned in Sect. 4.

2 Industrial Use Cases

In this section, we review four uses from our industrial partners: STIWA group in Subsect. 2.1, Borealis group in Subsect. 2.2, voestalpine Stahl GmbH in Subsect. 2.3, and PIERER Innovation GmbH in Subsect. 2.4.

2.1 UC1 – Data Integration for Discrete Manufacturing

The STIWA group [27,28] develops and provides high performance automation solutions for manufacturing. In a typical setting, single or multiple workpieces simultaneously pass through various discrete production steps until they are finished. The parts built into the workpieces are not provided individually, but in batches containing a large number of unsorted parts. This makes it difficult to trace which parts are installed in which workpieces. As workpieces can be considered as parts of other workpieces, this leads to complex bill of material hierarchies. To overcome this problem, we are currently developing a solution that aims to improve the traceability of parts installed in workpieces. The challenges and characteristics of the data we faced during our research can be summarized as follows:

- The use case requires a focus on the relationships between installed parts and workpieces.
- As workpieces are allowed to pass different production steps, this leads to heterogeneous datasets that are considered to be very sparse.
- Parts provided in batches (unsorted, large number) introduce uncertainty into the production process, which must be represented accordingly in the data model.

2.2 UC2 – Data Integration for Continuous Manufacturing

The Borealis group is a leading provider of innovative solutions in the fields of polyolefins, basic chemicals, and plant nutrients. The aim of our collaboration was the analysis of used raw materials and process parameters with statistical methods to identify critical process parameters, which have a significant impact on various key performance indicators (e.g., product quality). To allow this analysis, raw material must be traceable throughout the entire production process. Due to the continuous nature of the production process, raw material that is fed into the process, is blended. Data recorded during the various process steps is available in the form of time-series data and must be integrated for the analyses. The challenges and characteristics of the data we faced during our research can be summarized as follows:

- Recording of a time-series value may be triggered in regular intervals, on value change, or as a mix of both. To perform data analysis easily, time-series data must be sampled uniformly.
- When working with time-series data, specific emphasis needs to be put on an efficient processing due to the large amounts of data.
- For smart data discovery use cases [4], data segmentation represents a core method that must be supported efficiently.

2.3 UC3 – Missing and Outlying Data in Steel Production

In the steel mill of the voestalpine Stahl GmbH, raw steel is casted into slabs. Those slabs are about 12 m long and 200 mm thick. The slabs are heated and rolled in the hot rolling mill to receive steel coils with the desired thickness for further processing. After the hot rolling, the steel is cooled down with water and wound into coils. To monitor the steel processing in the hot rolling mill, sensors measure the temperature and many other information about the water cooling system [5,8], resulting in a high dimensional measurement dataset.

The applied statistics department at voestalpine GmbH recognized issues with data analytics (e.g., bias or errors in the results) on historical data with a large proportion of missing or outlying values. Thus the aim of our research collaboration was first an investigation of patterns in missing data (MD), and second, an evaluation of approaches on how to impute these missing values. Third, we surveyed approaches on how to detect outliers and invalid data in such high-dimensional industrial data sets. The data sets used for the evaluation of our methods contain sensor data from the hot rolling mill. We faced the following challenges and characteristics of the data [5,8]:

- The provided data sets can be attributed as *high dimensional* since our evaluation data set, which represents only an excerpt of the entire hot rolling mill measurement DB, contains 193,700 records and 557 variables. Such large amounts of data highlight the need for automated approaches and novel ways to display the results.
- The data is automatically collected *sensor data*, which is why real-time solutions to handle MD are required.
- Since we mostly deal with *numerical data*, approaches for survey MD that include suggestions how to handle textual or discrete data are not relevant.

Most existing literature on MD stems from the social sciences (e.g., sociology, economics) and therefore has a focus on survey data [29], i.e., dealing with single data sets with a focus on textual data and specific MD patterns like the monotone pattern [24]. The monotone MD pattern is observed when participants leave a study that is conducted over time. Such data characteristics have no or very little significance in the industrial domain, which was not only confirmed by our case study but also by prior work from the chemical industry [19].

2.4 UC4 – Metadata Management for Manufacturing

PIERER Innovation GmbH (PInno) as part of the PIERER Mobility AG Group is the central contact point for digital transformation and the application of innovative methods within the group, especially for KTM AG. The company aims at developing new products, services, and new business models around the digital ecosystem of vehicles. In the future, start-ups, experts, and partners should be enabled to utilize data and metadata from PInno. Collaboration without detailed knowledge about the data landscape should be enabled.

As an exemplary use case, external partners (e.g., start ups, technology experts) should be enabled to analyze the cause of motorcycle component defects that lead to warranty claims. In other words, it should be possible to execute high-level queries like "which motorcycle has which warranty claims?". Currently, deep knowledge of the data structure (in terms of databases and business processes) is necessary to answer such queries. In a first step, the motorcycle models equipped with the affected components must be retrieved from the bill of materials. Using this list of motorcycle models, the relevant warranty claims can then be extracted from KTM's dealer management system. To allow the analysis of the respective defect in detail, diagnostic data recorded at the workshops (as part of warranty processing) must be extracted from the motorcycle diagnostic system and made available to the data scientists and analysts.

3 Knowledge Transfer: Methods and Tools to Solve Data Challenges

In this section, we first highlight the need to start a data project with the right questions and subsequently discuss methods and tools for the three subtasks.

3.1 Project Initialization: Ask the Right Questions

Initially, it is important to understand the goal of the data science project, to assess the current data situation and to identify possible data challenges that may emerge during the project. Thus, it is crucial to ask the right questions at project startup. The following list covers questions that we compiled over the years to assess the readiness of an organization to start a data science project:

- Which types of data are available?
- Is historical data available?
- How can the data be linked together?
- How structured is the data (image, spreadsheets, databases, ...)?
- Assessment of data quality for each data source?
- What documentation (if any) is available for each data source?
- Which concrete questions (use cases) should be answered?
- Which models already exist (model-based vs. data-driven)?
- Which model would provide the greatest benefit?

The answers to these questions give an overview on the availability of data, how the data is connected, and which type of analysis is possible to achieve the objectives of the project.

3.2 Methods and Tools to Address Data Integration

Different methods and tool were used to address the discrete and continuous manufacturing data integration use cases. For discrete manufacturing, we used a graph database to model the complex relationship between parts and workpieces. The graph model allows to model a highly hierarchical structure with an uneven number of levels (ragged hierarchy) [33]. The query language which is optimized for graph models allows fast traversals of the graph in an intuitive way [33].

For continuous manufacturing, we used a dedicated time-series database, since it comes with built-in features like data sampling or support for big data [3]. To analyze time-series data in a meaningful way, the data needs to put into context. Therefore, we used a separate database to provide this context and connected it to the time-series database. We used Python for data prepossessing. For orchestration of the Python scripts we used Apache Airflow.

For all data management and integration projects, we claim that a systematic approach is required. In accordance with software development, where DevOps is the de-facto standard to operate code, DataOps is an emerging approach advocated by practitioners to tackle data management challenges for data science projects. DataOps is a method to automatically manage the entire data life cycle from data identification, cleaning, integration to analysis and reporting [26].

3.3 Methods and Tools to Address Data Quality

During the time of this project, we developed three different DQ tools for measuring data quality due to different requirements by customers.

- *DaQL* (Data Quality Library) [9], which allows to continuously measure DQ in ML applications with user-defined rules.
- *QuaIIe* (Quality Assessment for Integrated Information Environments, pronounced ['kvalə]) [12,13], which implements 12 DQ metrics for 8 different DQ dimensions at the data-level and the schema-level.
- *DQ-MeeRKat* (Automated Data Quality Measurement using a Reference-Data-Profile-Annotated Knowledge Graph) [7], which implements a new method to data-profiling-based DQ measurement.

In addition to DQ tools, we also developed a method for missing data imputation in the industrial domain [8], a method on how to detect specific patterns in missing data [5], as well as a method to overcome semantic shift in industrial streaming data over time [10]. All three methods address specific DQ issues (completeness and semantic shift respectively).

Wang and Strong [36] define *data quality* as "data that are fit for use by data consumer" along the line of the original definition for the more general term *quality* ("fitness for use") by Juran and Godfrey [20] in 1998. This definition highlights the importance of the consumers viewpoint and the context in which the data is used. In our research [15], we perceived two opposing approaches to tackle the topic: on the one hand, it is possible to strive for the highest possible automation [14] (e.g., using DQ-MeeRKat [7]), or deliberately choose a user-centric approach that follows the human-in-the-loop principle (e.g., using DaQL [25]). In [25], we used so-called "entity models", which allow a user to define DQ rules directly on business objects without any knowledge about the database infrastructure.

3.4 Methods and Tools to Address Metadata Management

In literature [11,17,22], the term data catalog and consequently the scope of data catalog tools is not clearly defined (cf. [23]). When data catalogs were first mentioned in the early 2000s [17], the focus was on the *cataloging* functionality, whereas a broader view has been established in the meantime. According to the systematic literature review by Ehrlinger et al. [11], a data catalog consists of four conceptual components: metadata management, business context, data responsibility roles, and the FAIR principles.

Metadata management describes the collection, creation and maintenance of metadata [31]. Metadata can be divided into three types: descriptive metadata (e.g., title, description), administrative metadata (e.g., file format, encoding), and structural metadata (i.e., how resources relate to each other) [11,32]. In this paper, we refer to all metadata that already exists in a company and only needs to be collected automatically by a data catalog as "technical metadata". The counterpart of technical metadata is the *business context*, which describes the high-level view of business users on the data and is independent of specific data sources. Business context needs to be established manually by defining company-wide agreed-on *business terms* (cf. [11,23] for details). In summary, technical metadata reflects the current state of data sources within a company and business context enriches technical metadata with domain knowledge [11].

The third component, data responsibility roles, describes the establishment of roles (such as data steward) and the assignment of these roles to people and data assets [11,23]. While data responsibility roles are often discussed in the context of data catalogs, they are clearly part of the broader topic data governance [11].

The FAIR principles stand for Findability, Accessibility, Interopability, and Re-use of data [37]. FAIRness can be evaluated by verifying the fulfillment of the FAIR principles published by [37] and summarized at the GO FAIR webpage.[2]

There are several commercial tools for metadata management, which cover these components in different detail, for example, Informatica Enterprise Data Catalog,[3] Collibra's Data Catalog,[4] or the DataHub Metadata Platform.[5] A detailed overview on the most popular vendors is provided in [18].

4 Lessons Learned

Based on our experience with the use cases and their challenges, we identified a list of five learned lessons, which are discussed in the following subsections.

[2] https://www.go-fair.org/fair-principles.

[3] https://www.informatica.com/de/products/data-catalog.

[4] https://www.collibra.com/us/en/platform/data-catalog.

[5] https://datahubproject.io.

4.1 Awareness for Different Types of Data Assets

All kinds of data, such as metadata, master data, transaction data, time-series data, text, image, etc., have specific characteristics [6,38]. For almost all types of data exist systems that are specifically tailored to them and provide data-type specific features. To get the most benefit from these dedicated systems, we claim that they should be reused and only glue code should be newly developed that connects these systems. When selecting a platform, it must be considered as early as possible if batch processing, event based processing, or both is required [30]. This lesson was learned from all four use cases discussed in this paper and relates to all tasks in Fig. 1, it is therefore considered fundamental for any data project.

4.2 Distinction Between Data Quality, Product Quality, and Outliers

When customers list their "data quality issues", we realized that it is not often clear what to understand by "data quality". The concepts of data quality, product quality, and outliers are often confused. Consequently, it is important to distinguish these terms clearly to use the right methods for their measurement and improvement. In our perception, the following distinction holds:

- *Outliers* are data points that deviate from the norm [2].
- *Data quality problems* are present if the data does not properly represent its real-world representation [36].
- *Product quality problems* are present, when there are issues in the production process and the real-world products are faulty. These errors might or might not be represented in the data.

The distinction between data quality, product quality, and outliers was primarily investigated in the steel production use case UC3 and relates to (T3) in terms of DQ assurance and (T5) with respect to product quality or maintenance.

4.3 Determination of the Correct Level of Metadata Management

To select a data catalog that meets a customer's requirements, it is initially necessary to specify the relative importance of every single component (cf. Subsect. 3.4) for the current project. Since many state-of-the-art data catalogs do not implement this wide range of functionalities outlined in [11], we developed a generic ontology layer which maps the semantics of data in form of a (business) ontology to the technical metadata provided by an existing data catalog.[6] The importance for the correct level of metadata management was primarily investigated in UC4 and relates to task 4 in Fig. 1.

[6] http://dqm.faw.jku.at/ontologies/GOLDCASE/index-en.html.

4.4 Consideration of Process-Related Uncertainty

As described in Sect. 2, parts to be assembled are often provided in large and unsorted batches, or raw material is blended in continuous production processes. Therefore, a clear mapping from input to output is often difficult. Similarly, data quality interpreted as "fitness for use" [36] also introduces context dependency. Here it is important to accept that there is no single version of the truth.

To obtain a representation that is as close to reality as possible, the process-related uncertainties should be modeled explicitly in the data model. When mapping input data to output data, or when calculating DQ metrics for raw data, the data integration should be designed as performant as possible. Nevertheless, from our experience, designing data integration traceable and providing measures for process-related uncertainties pays off.

The consideration of process-related uncertainty was observed and researched in UC1 on discrete manufacturing and UC2 on continuous manufacturing.

4.5 Use a Feature Store for Feature Engineering

A feature store[7] is used to store, manage, and share features for all phases of a machine learning project, i.e., model training, testing, and maintaining [21]. Based on latency requirements, it can be distinguished between online and offline feature stores. Feature stores help to retrieve features efficiently and in a consistent way, which is the foundation for traceable and repeatable ML. Establishing a feature store is considered a data management task, since knowledge about data modeling is required.

The importance for storing ML features was investigated in UC2 on continuous manufacturing and relates to tasks (T4) on metadata management and (T5) on data analysis in Fig. 1.

5 Conclusion and Future Work

In this paper, we observe current challenges on the topics of data integration, data management, and data quality within applied research projects. By means of four case studies, we present solutions on how to address these challenges and discuss the following lessons learned:

1. Awareness for different types of data assets
2. Distinction between data quality, product quality, and outliers
3. Determination of the correct level of metadata management
4. Consideration of process-related uncertainty
5. Employment of a feature store for feature engineering

In the future, we will investigate the impact of metadata management in more detail. We believe that metadata is the basis for all kinds of data governance

[7] https://www.featurestore.org.

tasks, such as, data quality assurance or access security [34] and in the end, also feature stores are only a specific form of metadata. In addition, we plan to further study the measurement of process-related uncertainty and its integration to explainable ML tools.

Acknowledgements. The research reported in this paper has been funded by BMK, BMDW, and the State of Upper Austria in the frame of the COMET Programme managed by FFG.

References

1. Abadi, D., et al.: The Seattle report on database research. ACM SIGMOD Record **48**(4), 44–53 (2019)
2. Aggarwal, C.C.: Outlier Analysis. Springer, Cham (2017). https://doi.org/10.1007/978-3-319-47578-3
3. Bader, A., Kopp, O., Falkenthal, M.: Survey and comparison of open source time series databases. Datenbanksysteme für Business, Technologie und Web (2017)
4. Bechnỳ, M., Himmelbauer, J.: Unsupervised approach for online outlier detection in industrial process data. Procedia Computer Science **200**, 257–266 (2022)
5. Bechny, M., Sobieczky, F., Zeindl, J., Ehrlinger, L.: Missing data patterns: from theory to an application in the steel industry, pp. 214–219. ACM, New York (2021)
6. Cleven, A., Wortmann, F.: Uncovering four strategies to approach master data management. In: 43rd HICCS, pp. 1–10. IEEE (2010)
7. Ehrlinger., L., Gindlhumer., A., Huber., L., Wöß, W.: DQ-MeeRKat: automating data quality monitoring with a reference-data-profile-annotated knowledge graph. In: Proceedings of the 10th International Conference on Data Science, Technology and Applications - DATA, pp. 215–222. SciTePress (2021)
8. Ehrlinger, L., Grubinger, T., Varga, B., Pichler, M., Natschläger, T., Zeindl, J.: Treating missing data in industrial data analytics. In: Thirteenth International Conference on Digital Information Management (ICDIM 2018), pp. 148–155. IEEE, Berlin, September 2018
9. Ehrlinger, L., Haunschmid, V., Palazzini, D., Lettner, C.: A daql to monitor the quality of machine data. In: DEXA 2019. Lecture Notes in Computer Science, vol. 11706, pp. 227–237. Springer, Cham, Switzerland (2019)
10. Ehrlinger, L., Lettner, C., Himmelbauer, J.: Tackling semantic shift in industrial streaming data over time. In: DBKDA 2020, pp. 36–39. IARIA, Portugal (2020)
11. Ehrlinger, Lisa, Schrott, Johannes, Melichar, Martin, Kirchmayr, Nicolas, Wöß, Wolfram: Data catalogs: a systematic literature review and guidelines to implementation. In: Kotsis, G., et al. (eds.) DEXA 2021. CCIS, vol. 1479, pp. 148–158. Springer, Cham (2021). https://doi.org/10.1007/978-3-030-87101-7_15
12. Ehrlinger, L., Werth, B., Wöß, W.: Automated continuous data quality measurement with QuaIIe. International Journal on Advances in Software **11**(3 & 4), 400–417 (2018)
13. Ehrlinger, L., Werth, B., Wöß, W.: QuaIIe: a data quality assessment tool for integrated information systems. In: DBKDA 2018, pp. 21–31. IARIA, France (2018)
14. Ehrlinger, L., Wöß, W.: Automated data quality monitoring. In: 22nd MIT International Conference on Information Quality, AR, USA, pp. 15.1–15.9 (2017)
15. Ehrlinger, L., Wöß, W.: A survey of data quality measurement and monitoring tools. Front. Big Data, 28 (2022)

16. Fischer, L., Ehrlinger, L., Geist, V., Ramler, R., Sobieczky, F., Zellinger, W., Brunner, D., Kumar, M., Moser, B.: AI System Engineering-Key Challenges and Lessons Learned. Machine Learning and Knowledge Extraction **3**(1), 56–83 (2021)
17. Franklin, M., Halevy, A., Maier, D.: From databases to dataspaces: A new abstraction for information management. SIGMOD Rec. **34**(4), 27–33 (2005)
18. Guido De Simoni, M.B., et al.: Gartner magic quadrant for metadata management solutions. Technical report. Gartner, Inc., November 2020
19. Imtiaz, S.A., Shah, S.L.: Treatment of missing values in process data analysis. The Canadian Journal of Chemical Engineering **86**(5), 838–858 (2008)
20. Joseph, J., Godfrey, A.B.: Juran's Quality Handbook (1998)
21. Kakantousis, T., Kouzoupis, A., Buso, F., Berthou, G., Dowling, J., Haridi, S.: Horizontally scalable ml pipelines with a feature store. In: Proceedings of 2nd SysML Conference, Palo Alto, USA (2019)
22. Korte, T., Fadler, M., Spiekermann, M., Legner, C., Otto, B.: Data Catalogs - Integrated Platforms for Matching Data Supply and Demand. Reference Model and Market Analysis (Version 1.0). Fraunhofer Verlag, Stuttgart (2019)
23. Labadie, C., Legner, C., Eurich, M., Fadler, M.: Fair enough? Enhancing the usage of enterprise data with data catalogs. In: 2020 IEEE 22nd Conference on Business Informatics (CBI), vol. 1, pp. 201–210, June 2020
24. de Leeuw, E.D., Hox, J., Huisman, M.: Prevention and treatment of item nonresponse. Journal of Official Statistics **19**(2), 153–176 (2003)
25. Lettner, C., Stumptner, R., Fragner, W., Rauchenzauner, F., Ehrlinger, L.: Daql 20: Measure data quality based on entity models. Procedia Computer Science 180, 772–777 (2021)
26. Mainali, K., Ehrlinger, L., Himmelbauer, J., Matskin, M.: Discovering DataOps: a comprehensive review of definitions, use cases, and tools. In: Data Analytics 2021, pp. 61–69. IARIA, Spain (2021)
27. Martinez-Gil, J., Stumpner, R., Lettner, C., Pichler, M., Fragner, W.: Design and implementation of a graph-based solution for tracking manufacturing products. In: European Conference on Advances in Databases and Information Systems. pp. 417–423. Springer (2019)
28. Martinez-Gil, J., Stumptner, R., Lettner, C., Pichler, M., Mahmoud, S., Praher, P., Freudenthaler, B.: General model for tracking manufacturing products using graph databases. In: Data-Driven Process Discovery and Analysis, pp. 86–100. Springer (2018)
29. Messner, S.F.: Exploring the consequences of erratic data reporting for cross-national research on homicide. J. Quant. Criminol. **8**(2), 155–173 (1992)
30. Pfandzelter, T., Bermbach, D.: IoT data processing in the fog: functions, streams, or batch processing? In: International Conference on Fog Computing (ICFC), pp. 201–206. IEEE (2019)
31. Quimbert, E., Jeffery, K., Martens, C., Martin, P., Zhao, Z.: Data Cataloguing, pp. 140–161. Springer International Publishing, Cham (2020)
32. Riley, J.: Understanding metadata: what is metadata, and what is it for? National Information Standards Organization (NISO) (2017). https://groups.niso.org/apps/group_public/download.php/17446/Understanding%20Met%E2%80%A6
33. Robinson, I., Webber, J., Eifrem, E.: Graph Databases. O'Reilly Media, Inc., Sebastopol (2015)
34. Talburt, J.: Data speaks for itself: data littering (2022). https://tdan.com/data-speaks-for-itself-data-littering/29122

35. Talburt, J.R., Sarkhi, A.K., Claassens, L., Pullen, D., Wang, R.: An iterative, self-assessing entity resolution system: first steps toward a data washing machine. Int. J. Adv. Comput. Sci. Appl. **11**(12), 680–689 (2020)
36. Wang, R.Y., Strong, D.M.: Beyond accuracy: What data quality means to data consumers. J. Manag. Inf. Syst. **12**(4), 5–33 (1996)
37. Wilkinson, M., Dumontier, M., Aalbersberg, I., et al.: The FAIR guiding principles for scientific data management and stewardship. Scientific Data **3**(1), 1–9 (2016)
38. Zhag, A.: Data types from a machine learning perspective with examples (2018). https://towardsdatascience.com/data-types-from-a-machine-learning-perspective-with-examples-111ac679e8bc

Building a YouTube Channel for Science Communication

Frank Elberzhager(✉), Patrick Mennig, and Phil Stüpfert

Fraunhofer Platz 1, 67663 Kaiserslautern, Germany
{frank.elberzhager,patrick.mennig,
phil.stuepfert}@iese.fraunhofer.de

Abstract. Sharing results of research projects with the public and transferring solutions into practice is part of research projects. Typical ways are scientific publications, presentations at industry events, or project websites. However, in the last one and a half years, we have created videos and shared video material on our YouTube channel as a new way of disseminating results in our project setting. We have observed that many research projects do not follow this path, do this only with a very limited number of videos, or provide rather poor quality of the videos. In this article, we want to share our experience and our first steps, together with open issues for new and more videos we are planning to produce. We would also like to encourage discussions in the research community on whether this is a worthwhile direction to pursue further in the future.

Keywords: Knowledge transfer · Videos · Climate-neutral smart city district

1 Introduction

In the EnStadt:Pfaff research project, which started in 2017, we aim at developing a climate-neutral smart city district. Pfaff was a former big company for sewing machines in the city of Kaiserslautern where thousands of people worked over decades before the company had to declare insolvency. This means that Pfaff is no longer a company in our city, but the name (and the district where the new climate-neutral smart city district is currently being developed) is known to many people. Many of them have personal relationships with the company and with this area, and are thus interested in the progress of the district and the results of the project.

In the project, various topics, such as energy, mobility, home, or community, are addressed by the eight project partners. We are responsible, in particular, for digital solutions, i.e., we face the question of how the digital transformation can support climate-friendly behavior. Furthermore, our interest is in the future citizens of this urban district. As climate change is an increasingly relevant topic that affects almost everybody from common citizens to companies, it is a huge motivation for us to share the results we have developed in the project. To summarize, besides the general interest of many citizens in climate topics, a lot of people in our city are also genuinely interested in what happens

G. Kotsis et al. (Eds.): DEXA 2022 Workshops, CCIS 1633, pp. 179–188, 2022.
https://doi.org/10.1007/978-3-031-14343-4_17

in the district due to their fond memories of Pfaff. Our goal is to include these people, and one important aspect is to share information.

In this paper, we will first provide an overview of related work and the background of our project (Sect. 2). Sections 3 and 4 present what we have done so far with respect to sharing project information and results. We faced several challenges and want to show the initial lessons we learned when we implemented a YouTube channel for sharing project results. Section 5 concludes our paper.

2 Background and Related Work

2.1 Knowledge Transfer and Traditional Ways of Sharing Knowledge in Our Research Project

Sharing project results from research projects is a common task for researchers. One of the main ways to do this is at scientific venues. Research communities have a certain knowledge background and can give valuable feedback to project results. However, research is not done as an end in itself, but ultimately to solve certain problems for different stakeholders. Brennan states that "scientific information is a key ingredient needed to tackle global challenges like climate change, but to do this it must be communicated in ways that are accessible to diverse groups, and that go beyond traditional methods", such as peer-reviewed publications [1]. The climate change example he mentions is a topic that is not only relevant to a particular company or a small number of people, but to society in general and thus affects almost everyone. This is an indication that there may be various topics that are interesting for society and that are worthwhile communicating to people without a scientific background.

As already mentioned in the introduction, our research project is also about climate topics in conjunction with the digital transformation. In addition, many citizens of the city of Kaiserslautern have a personal relationship with the district where we can implement new digital solutions. Therefore, we decided to include citizens as much as possible right from the beginning. This means that scientific results have to be presented in a way that citizens without a scientific background can follow our presentations. We started, for example, with the following traditional ways (for more details, we refer to Elberzhager et al. [2]):

- In the city of Kaiserslautern, there is a public event called "Night of Science", which takes place every two years. In an early project phase, we shared general project ideas and talked to citizens of different age, gender, and social background. We listened to personal stories about the citizens' relationship with the Pfaff company and the area of the factory, and gained insights into their expectations and concerns. Moreover, we answered their questions about the area and the project.
- Several times a year, we posted blog articles on our institute's website to share project results.
- We performed several hackathons. We wanted to enable citizens to contribute their ideas to the project and at the same time learn about their needs and wishes. Therefore, we invited everyone interested in the project to participate in a 24-h event. In contrast to a typical hackathon, no programming skills or any other specific skills

were required. The focus was on design tasks to which everyone could contribute. We gained interesting ideas for digital services, and shared project ideas with more than 30 people during the event. As the first PFAFF HACK in 2018 was a success, we decided to organize another hackathon a year later. Now in 2022, we are currently organizing the fifth edition. By now, the focus has shifted from a design-oriented event to an implementation-oriented one. The last two hackathons were also streamed, and videos were produced and shared after the event. Many different citizens and students participated.
- In 2021, we performed an expert symposium where all project partners shared the main results achieved in the project so far. This event was held at city hall and simultaneously streamed via YouTube so we reached many people.
- Of course, traditional ways of sharing knowledge and results with society were also implemented, such as a project website or articles in printed newspapers.

The above-mentioned ways of communication with citizens were mainly events with direct interaction with the participants at a specific venue. This kind of events only allows a limited number of participants. In other words, many citizens are excluded from the information and the opportunity to contribute. We learned from peers and business contacts that many people have not heard anything yet about the project, its results, and the contribution to the city of Kaiserslautern, even though they are interested in it. Therefore, and due to the restrictions for conducting in-place events as a result of the Covid-19-pandemic, we rethought our way of informing citizens, local organizations, and companies as well as other municipalities.

2.2 Videos and YouTube as a Way of Doing Science Communication

Every minute, 694,000 h of video content are streamed via YouTube [3]. As of November 2021, YouTube was the second most visited website worldwide with 13.34 billion total visits [4]. According to Ceci, 91.5% of the adult population in Germany can be reached by advertisers via YouTube [5].

YouTube as a platform for video content allows users to share various types of videos with other users. These include entertainment videos covering topics such as gaming, comedy, and short movies; tutorial videos for beauty, woodworking, and the use of specific technologies; review videos covering many different hardware and software products; as well as educational videos, e.g., for math problems. Also, popular science web videos (science videos) focus on communicating scientific content to a broad audience [6]. Certain video channels focus on publishing sets of science videos. Himma-Kadakas et al. refer to people running such channels as "YouTubers" if they regularly upload videos, are popular due to their activity on the social media platform, and have a regular group of followers [7]. This does not distinguish them from YouTubers who are running a channel focused on other topics not related to science and indicates that the platform is a worthwhile outlet for science communication.

Several studies have analyzed such science-focused video channels on YouTube. According to Morcillo et al. [6], over 70% of the science video channels use deliberate and complex montage, i.e., producing videos with three or more continuous video recordings (takes) and assembling them in a purposeful order to build a narrative. Beautemps and Bresges conducted a survey (n = 3,881) with YouTube users, 61% of whom stated that they are not pursuing or have completed an education in science, but over 80% indicated that they are interested or very interested in science [8]. This shows that YouTube as a platform can be used to reach people outside the scientific community and is therefore a worthwhile tool for science communication to the general public.

There are also analyses on what factors influence the success of a science video channel on YouTube. According to the study of Beautemps and Bresges [8], six factors are important for successful science videos:

- the video structure (mentioning the topic at the beginning and coming to a conclusion at the end, mentioning facts, including experts, summarizing experts' statements);
- the reliability of the video (referring to sources);
- the video quality (audio and video quality);
- community integration (answering comments, doing votings in the comment section);
- the presenter, as YouTube channels are strongly linked to the people who present them;
- and the video topic (90% of respondents indicated that they watch videos about a topic that they are interested in; i.e., entertainment seems to play a major role when choosing videos to watch).

Using the example of science communication about whale watching, Finkler and León present the framework "SciCommercial" for science videos based on lessons learned from the marketing domain. According to "SciCommercial", this is how videos should present ideas: simple; in an unexpected way; concretely to make them understandable; credibly so viewers believe them; emotionally; connected with science; and with storytelling to convey the overall message [9].

Brennan shares lessons learned from running a science video channel, explicitly referring to the videos as "do-it-yourself" videos [1]. He emphasizes that YouTube videos gain views depending on the topic (i.e., how many YouTube users search for that particular topic) but also get more views over time. This stresses the importance of YouTube as a tool for science communication, as – unlike one-off public events – many interested people can be reached over a longer period of time.

The great potential of knowledge transfer via YouTube videos has also been recognized by scientists at UCL. For this reason, they offer a lecture for physics students that teaches the participants how to present scientific content in videos in such a way that it can reach a broad audience. They teach that you cannot simply record a scientific talk and publish it – you have to do a lot more in terms of preparation and execution [10].

The research we have presented shows that science videos on YouTube are a topic of ongoing research. General guidelines and frameworks exist that one can follow to create science videos. So far, we have not found any published research covering the use of science videos in the context of our concrete research project, i.e., how to inform the general public and local citizens about ongoing research in a multi-disciplinary project. In the following, we will present our approach, which we adapted from the work presented in this section.

3 First Steps Regarding Videos and Streaming

3.1 Objective

We aim at communicating intermediate scientific results to society. Our goal is to inform interested viewers about the project itself and about its results, potentially raising further interest in science in general. Our main goal is to reach a large number of people remotely. Therefore, we considered current ways used by other researchers and other professionals to share content. Offline tools such as flyers were no option to us since they do not fit the media used by younger citizens, add to pollution, and are therefore not environmentally friendly. Online means of communication such as text on websites requires time and motivation to read. In the end, our choice were videos and the platform YouTube. Videos are one of the most frequently used ways for people to obtain information.

3.2 Idea

For this purpose, we created the project's YouTube channel, which is called "Quartierswerkstatt"[1] (district workshop), as this is the same name as a planned, but still unfinished, neighborhood center. The focus is on interaction with and learning from interested citizens. Hence, we developed the video communication concept with the idea of the neighborhood center as a related place of gathering in mind. With the videos, we also want to invite viewers to interact with scientists involved in the research project.

3.3 Implementation

We have been building up the video studio over the course of one and a half years, in parallel to the other activities in the research project. It is an incremental process, where we started with an initial vision of the studio, then further detailed and shaped it once we learned more about video production along the way. None of our project members had any previous professional experience with video production. Initially, we used online learning resources (e.g., YouTube videos about video production) to learn about the topic in general, identify relevant sub-topics, and find new resources to learn more specifics.

4 Experiences Regarding Videos and Streaming

4.1 Lessons Learned on Hardware

One major area of learning included the main hardware required for quality video production. Today, even smartphones offer photo and video capabilities, which, at first glance, match dedicated photo and video hardware such as camcorders, video cameras, digital single-lens reflex (DSLR) cameras, or digital single-lens mirrorless (DSLM) cameras. When we looked at the details, however, we found that dedicated video hardware offers better image quality compared to smartphones, but also more versatility. The impression

[1] https://www.youtube.com/channel/UCK8LjvcvaCHBF-voo0qcqqA.

of an image depends on the whole system used to capture it, i.e., the lens (quality, aperture, focal length) as well as the camera's sensor size (in general, the larger the sensor, the better the image quality). Some dedicated video hardware additionally allows remote control. Therefore, we decided to buy two dedicated video cameras[2] instead of using other cameras.

To achieve a good-looking scene, illumination plays a major role as well. We use studio lights with soft boxes as the scene's main light source and combine them with colored lights for background illumination. For interviews and presentations, we use a background system that uses a stretched piece of fabric to provide a uniformly colored background (black and green), as can be seen in Fig. 1.

Fig. 1. Studio with two separate recording positions and control room area. (Color figure online)

Besides the video image, audio quality is also relevant for video production. Dedicated microphones (not camera in-built) are very much required to achieve a pleasant sound. One can either use camera-mounted microphones, overhead microphones attached to booms and pointing at the speaker, or microphones attached to the speaker. The most dominant factor in audio quality is the proximity of the microphone to the speaker. So-called lavalier microphones are attached to the speaker's clothes and therefore provide good audio quality with relative ease, as nobody is needed to operate the boom. Their disadvantage is that they are visible to the viewer if not carefully hidden, an aspect we deemed neglectable. We use a wireless microphone system[3] that transmits the audio signal either directly to the camera or to a video mixer[4] that we use for live production and camera control.

[2] Blackmagic Pocket Cinema Camera 4K, cf. https://www.blackmagicdesign.com/products/blackmagicpocketcinemacamera.

[3] Røde Wireless Go II, cf. https://rode.com/de/microphones/wireless/wirelessgoii.

[4] Blackmagic Atem Mini Pro Iso https://www.blackmagicdesign.com/de/products/atemmini.

The video mixer allows us to remotely control the cameras (e.g., adjust aperture, focus, brightness) and input their video and audio signal. We can then combine this with other HDMI-based sources such as computers showing presentation slides, or other cameras if more angles are required (e.g., action cameras for wide-angle effects or spare DSLR cameras). The video mixer allows producing videos with cuts and even picture-in-picture effects. Such final videos can be recorded to an external solid-state drive as well as separately to all video and audio streams. Additionally, it allows for direct live streaming to several platforms, such as YouTube.

4.2 Lessons Learned on Style and Content

Another relevant challenge for us is what the videos should look like in terms of style and content. We intentionally decided not to limit the way strictly at the beginning in order to be able to explore different possibilities. We started by interviewing different project members about their topics and concrete project results, because they are the experts in the project and we wanted to put them in the center of attention. The interviews were also a comfortable way for them to get uses to being video-recorded, as they could talk freely about their fields of expertise. We conducted open interviews, just using leading questions to stimulate the interviewees to talk about their topic of expertise in an open and passionate manner. Moreover, we also recorded some videos addressed to researchers and other professionals in the field. The videos reflect a typical conference presentation (in English) or are from the expert symposium. Most of the videos are in German, as our primary initial target audience are the local citizens and those of other municipalities in Germany.

Interview topics were the development process of the game MiniLautern, new mobility concepts for the urban district, or hackathon results. The interviewer was responsible for preparing the topic and the interview questions. A second person operated the cameras and checked the audio levels; i.e., took care of the technical aspects of the video recording.

Afterwards, all the video material was viewed and cut into a video that aims at telling a story; i.e., the person cutting the videos decided how to use the recorded material to assemble a video that people can follow and learn from. The story itself was built out of parts of the interviewees' answers, not necessarily using the whole answer to a question or following the original order. We experienced that some videos were easier to cut than others, depending on the fit of the interviewees' answers for building a certain storyline. Hence, we learned that a more scripted approach to the interviews, where we already establish a baseline story that we "fill" with the interview questions, would be preferable. Following the guidelines shown in Sect. 2, we decided to make the videos rather short, roughly about 5–15 min. One concrete lesson was that we recorded much more material than we could use, often due to various quality issues, such as bad noise, bad light, questions that led to unusable answers, or questions being answered in a way that would be too complicated for our primary audience.

4.3 Analysis of the Current Video Content

Since we started the Quartierswerkstatt YouTube channel in October 2020, we have published 22 videos so far, of which four are interview videos. We did not follow a regular schedule but published new videos whenever we finished producing one. The channel and videos were only advertised on our research institute's homepage through blog posts. At the time of writing this work, the channel has 11 subscribers and all videos together have been viewed 657 times, with the interview videos accounting for 271 views. This adds up to 21.6 h of playback time. On average, our videos have been viewed for 1:58 min or 23.9% of their length.

Right now, with our video studio setup, we are able to start a video recording within 30–60 min. For our 5–15-min interview videos, we record around 30–60 min of footage. Reviewing, cutting, color grading, audio correction, and feedback loops take about 1–2 workdays of effort, depending on how easily the video's storyline can be built from the recorded footage. All videos are produced by scientists regularly working on the research project and who have no prior professional knowledge about video production; hence these efforts include the time required to learn working with the software and hardware. We expect the effort required for such videos to eventually settle at one day.

The videos' topics are selected based on the availability of information that we identified as relevant to the general public, e.g., novel research results, new developments regarding the construction progress, and political developments.

Each video is concerned with preferably one narrow aspect of the research project with regards to the future climate neutral smart city district. As we are an active research partner in the project, we generally have a good understanding of the different topics. Based on that, we produce the videos following a process shown in Fig. 2. Our general goal is to regularly release a new video at least once per month, preferring shorter intervals of two to three weeks if possible. Due to the nature of the research project and due to our team consisting of researchers concerned with usually more than one project at a time, we schedule video recordings en bloc and distribute the remainder of the production process over several weeks.

Assuming the average effort required (0.75 h of preparation, 0.75 h of interview, 12 h of post-production), all four interview videos together amount to 54 h of effort spent. Per view, this accounts for approx. 0.2 h (i.e., 12 min) of effort. The average duration of the interview videos is 10 min. Hence, when we analyze only the interview videos we have published, they currently produce a negative net return on investment. But this neglects several aspects: The effort required to produce videos includes training and will decrease with more experience with the software and hardware tools while the quality will increase. The videos will remain available, so more views will be accumulated. Some of the videos had to be recorded anyway for remote conferences, so they can also be used for the project's YouTube channel. This dual use is a cost-efficient way of producing more content for our YouTube channel. To date, we have focused on one research project and mainly interview-style videos, but the video studio has also been used for live streaming of online events with a professional look.

Due to the small number of subscribers and views on our videos yet, we do not have access to advanced features of YouTube analytics, as these data are only available once a certain threshold of views is reached. Therefore, we cannot rely on YouTube's analytics functionality for analyzing the impact or deducing potential measures for increasing the outreach of our videos. Nevertheless, we plan to conduct marketing efforts such as installing billboards advertising our YouTube channel at the outside of the city district's construction site and cooperation with local newspapers, e.g., articles covering our efforts to further increase the impact of our YouTube channel. Additional research is required to understand how research projects can reach the general public as an audience, ensuring efficient science communication.

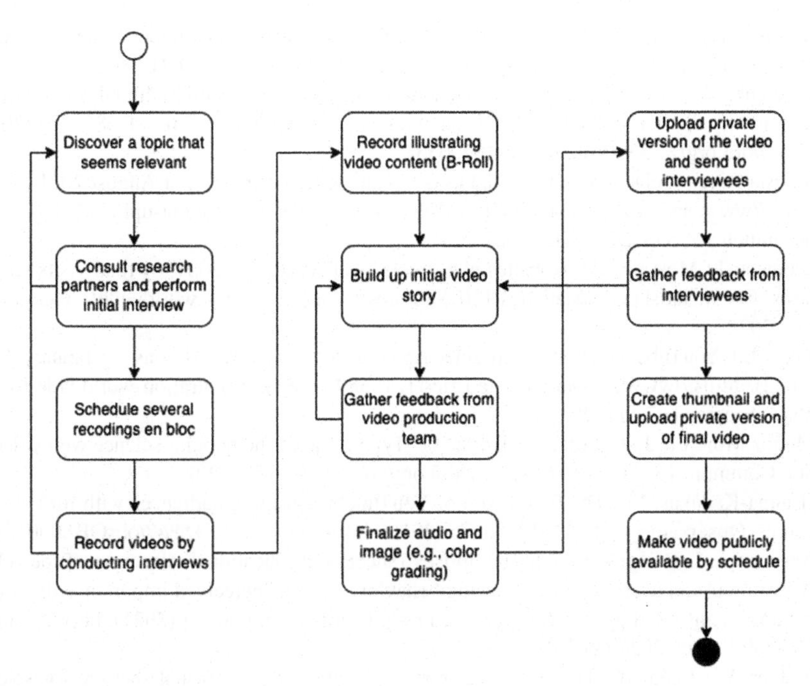

Fig. 2. Video production process as used for the "Quartierswerkstatt" YouTube channel.

5 Summary and Conclusion

With the "Quartierswerkstatt" YouTube channel, we have added another tool to the portfolio of communication activities we are performing in the context of a long-running research project. Without prior experience, we have created a dedicated video studio and produced 22 videos covering the results of research activities. Our goal is to communicate the results and context of scientific work to the citizens of Kaiserslautern. In the future, we plan to add more types of videos and investigate further how other science-related YouTube channels do science communication, even though their focus might not be on a single research project. Overall, we see a lot of potential for regular science communication as part of publicly funded research projects.

Acknowledgment. The research described in this paper was performed in the EnStadt:Pfaff project (grant no. 03SBE112D and 03SBE112G) of the German Federal Ministry for Economic Affairs and Climate Action (BMWK) and the Federal Ministry of Education and Research (BMBF). We thank Sonnhild Namingha for proofreading.

References

1. Brennan, E.: Why should scientists be on YouTube? It's all about bamboo, oil and ice cream. Front. Commun. **6**, 586297 (2021). https://doi.org/10.3389/fcomm.2021.586297
2. Elberzhager, F., Mennig, P., Polst, S., Scherr, S., Stüpfert, P.: Towards a digital ecosystem for a smart city district: procedure, results, and lessons learned. Smart Cities **4**, 686–716 (2021). https://doi.org/10.3390/smartcities4020035
3. Statista Research Department. Media usage in an internet minute as of August 2021 (2021). https://www.statista.com/statistics/195140/new-user-generated-content-uploaded-by-users-per-minute/. Accessed 27 Mar 2022
4. Clement, J.: Most popular websites worldwide as of November 2021, by total visits (2022). https://www.statista.com/statistics/1201880/most-visited-websites-worldwide/. Accessed 27 Mar 2022
5. Ceci, L.: YouTube penetration in selected countries and territories as of January 2022 (2022). https://www.statista.com/statistics/1219589/youtube-penetration-worldwide-by-country/. Accessed 27 Mar 2022
6. Muñoz Morcillo, J., Czurda, K., Trotha, C.: Typologies of the popular science web video. J. Sci. Commun. **15**, A02 (2016). https://doi.org/10.22323/2.15040202
7. Himma-Kadakas, M.: The food chain of YouTubers: engaging audiences with formats and genres. Observatorio (OBS*) **12**, 54–75 (2018). https://doi.org/10.15847/obsOBS0001385
8. Beautemps, J., Bresges, A.: What comprises a successful educational science YouTube video? A five-thousand user survey on viewing behaviors and self-perceived importance of various variables controlled by content creators. Front. Commun. **5**, 600595 (2021). https://doi.org/10.3389/fcomm.2020.600595
9. Finkler, W., León, B.: The power of storytelling and video: a visual rhetoric for science communication. J. Sci. Commun. **18**(05), 1–23 (2019). https://doi.org/10.22323/2.18050202
10. Coates, R.L., Kuhai, A., Turlej, L.Z.J., Rivlin, T., McKemmish, L.K.: Phys FilmMakers: teaching science students how to make YouTube-style videos. Eur. J. Phys. **39**(1), 015706 (2017). https://doi.org/10.1088/1361-6404/aa93bc

Introduction of Visual Regression Testing in Collaboration Between Industry and Academia

Thomas Wetzlmaier[1](✉), Claus Klammer[1], and Hermann Haslauer[2]

[1] Software Competence Center Hagenberg GmbH, Softwarepark 32, 4232 Hagenberg, Austria
{thomas.wetzlmaier,claus.klammer}@scch.at

[2] Palfinger Europe GmbH, Franz-Wolfram-Schererstraße 24, 5020 Salzburg, Austria
h.haslauer@palfinger.com

Abstract. Cyber-physical systems (CPSs) connect the computer world with the physical world. Increasing scale, complexity, and heterogeneity makes it more and more challenging to test them to guarantee reliability, security, and safety. In this paper we report about a successful industry-academia collaboration to improve an existing hardware-in-the-loop (HIL) test system with visual regression testing capabilities. For this purpose, we tried to follow an existing model for technology transfer from academia to industry and show its applicability in the context of CPSs. In addition, we provide a list of special collaboration challenges that should be considered in this context for successful industry-academia collaboration. And we discuss some key success factors of our collaboration, emphasizing that the CPS's *system and expertise knowledge* are of great importance.

Keywords: Industry-academia collaboration · Cyber-physical systems · Hardware-in-the-loop testing · Technology transfer

1 Introduction

For some years now, the wave of digitalization has also been sweeping through companies from more traditional domains like in the field of mechanical engineering. For companies in these domains, software has not been a top priority in the past because it represents only a small percentage of the total cost of the product. Nevertheless, in the last two decades the requirements and the complexity of the software have increased enormously, even in the domain of mechanical engineering. Furthermore, advances in industry 4.0, internet of things (IoT) and massive investments in digitalization in general, led to a transformation of products to support new business models and improved customer support, which increases the complexity of these systems [8]. There is an obvious relationship between the complexity of a system and its maintenance costs [7]. Although this problem is well known for a long time [6], still the complexity of the systems grows

G. Kotsis et al. (Eds.): DEXA 2022 Workshops, CCIS 1633, pp. 189–198, 2022.
https://doi.org/10.1007/978-3-031-14343-4_18

faster than the ability to control it. To keep pace with this fast-growing software complexity and to ensure specified and predefined quality criteria, requires more and more sophisticated testing methods. Test automation is one important key to address this increasing testing burden, but unfortunately automation is much more difficult to implement when dealing with physical goods and products [1]. In addition, existing test research approaches are also often not directly applicable to large and complex cyber-physical systems with a legacy of technology that evolved over decades [3].

Hence, external test experts and the domain-experts of the CPS are needed to successfully transfer and apply state of the art solutions from software-centric domains to such CPSs. Collaborations between industry and research institutes are a well-known construct to speed-up knowledge transfer from academia to practitioners [2]. Together, an overall plan for quality assurance activities and the appropriate tasks can be developed. Building knowledge in this area can thus be done more quickly to close the gap with other areas where test automation is already more widespread used and more mature today.

The aim of this paper is to present an example of a successful collaboration of research and industry in an application oriented setting. Therefore, one specific use case has been selected and is shown in more detail to grasp the advantages and special challenges of such collaborations more easily.

In the following section we describe the background of the project and the frame of our collaboration. We explain the presented use case and our approach afterwards in Sect. 3. Section 4 highlights the results and lists the challenges and success factors. The paper concludes with a summary of our work and findings.

2 Background

Palfinger is an international company well known for its loader cranes. Loader cranes help to load and unload trucks and other vehicles. The hydraulic knuckle boom crane is Palfigner's core product. As the name implies, a hydraulic system moves the mechanical arm of the crane. A programmable logic controller (PLC) drives the hydraulic system. Basically, a PLC is an industrial computer used to control assembly lines, machines, or robotic devices. In the past, Palfinger already intensified the efforts in system test automation, so they built a test system based on a hardware-in-the-loop (HIL) system offered by an international supplier.

The Software Competence Center Hagenberg (SCCH) provides application oriented research in the field of software and data science. Palfinger and SCCH had a close partnership in application oriented research in the field of test automation for many years. Currently, Palfinger and SCCH are partners within the nationally funded research project SmarTest (Smart Software Test) which addresses quality assurance of the ongoing evolution of large, complex software systems that are part of highly customizable and configurable products or product lines. The focus of the application oriented research program is to make steps towards a fully virtualized test environment for Palfinger's products.

To highlight the benefits and challenges of this cooperation best, we select one specific, easy to follow, research task. The team setup for this task was one researcher from SCCH and one practitioner from Palfinger. Besides the two active team members, four additional practitioners attended the regular meetings and workshops. Figure 1 shows the overview of the HIL based test environment. Lines and arrows depict the communication channels between the sub-parts. Dashed lines as well as the italic font style depict the entities affected by the previously mentioned research task.

The test environment consists of three major parts: the HIL-Test-System, the Test Interfaces and the Test Execution.

HIL-Test-System: All hardware parts (*HIL-System*, *PLC*, *Simulator-PC* and miscellaneous other hardware) are mounted into a rack and connected together properly. The HIL Simulation software (crane simulation) as well as any virtualization (e.g. virtual radio) runs on the Simulator-PC, which is a standard windows PC.

Test Interfaces: Each part in the HIL-Test-System has its own application programming interface (API). The *Relay-Server* simplifies the communication with the Test-System, by combining all APIs to a single JSON-RPC [5] like API.

Test Execution: *TESSY*, a commercial tool for testing embedded systems, is used for executing the tests [10]. It also supports to be triggered from a *Continuous Integration Server*. Integration as well as system tests use the *Test Library*, which is a collection of functions for high level communication with the *HIL-Test-System*.

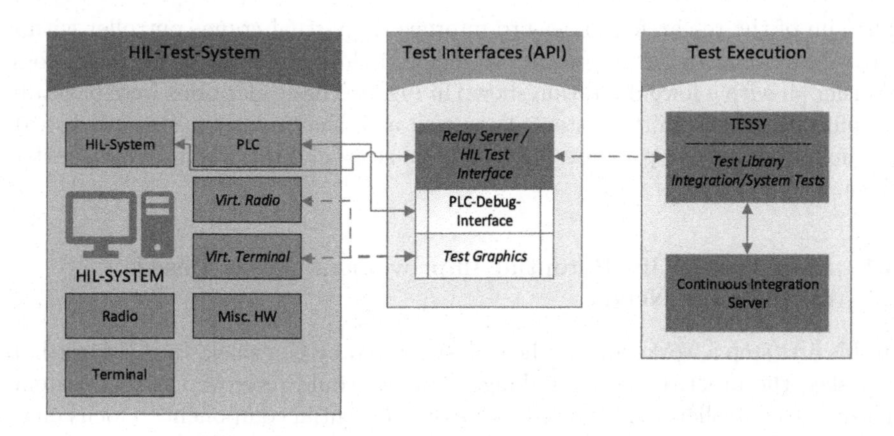

Fig. 1. Overview of the HIL based test environment.

In a nutshell, HIL testing is a technique where real signals from a PLC are connected to a test system that simulates the behaviour of the real hardware, so the PLC functions like the final product. It also enables virtualization of additional hardware components, like a wired terminal or radio controller, to control the crane's arm movement. Figure 2 shows a comparison between a photo of the real hardware Fig. 2a and 2b and its virtualized interface Fig. 2c.

(a) Picture of real radio controller hardware.

(b) Picture of real terminal hardware.

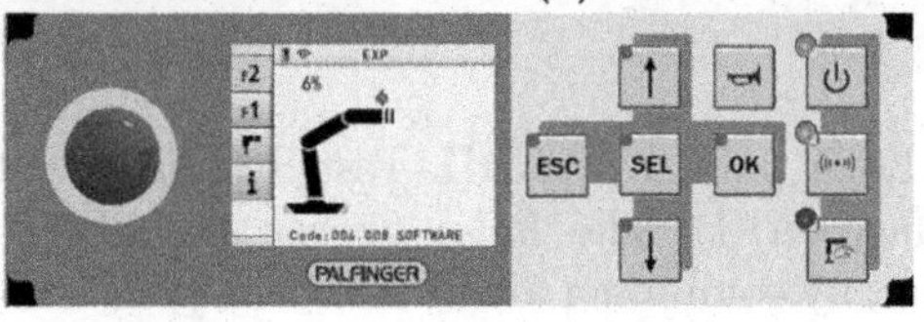

(c) Virtualized graphical representation.

Fig. 2. Crane controller: real versus virtualized presentations.

3 Approach

The aim of the research task was to improve the virtual crane controller within HIL test system. In our approach, we followed Gorschek's model for technology transfer [4] with a few adaptation shown in Fig. 3. These adaptions were basically simplifications by omitting step (4) as well as a shortcut from step (3) to (6). The following section presents the individual steps and the way we implemented them.

3.1 Step 1: Identify Potential Improvement Areas Based on Industry Needs

In this first step a workshop was held as suggested by Gorschek. In this workshop we asked the practitioners at Palfinger how we could improve the test system. The crane controller had been selected as one potential component to focus quality assurance activities, since there had been some hard to find issues regarding its graphical user interface (GUI) in the past.

The crane controller allows, as the name suggests, to control the loader crane. As depicted in Fig. 2a and 2b, it consists of hardware keys (right) to navigate through menus shown on the display (middle) and an emergency stop (left). The

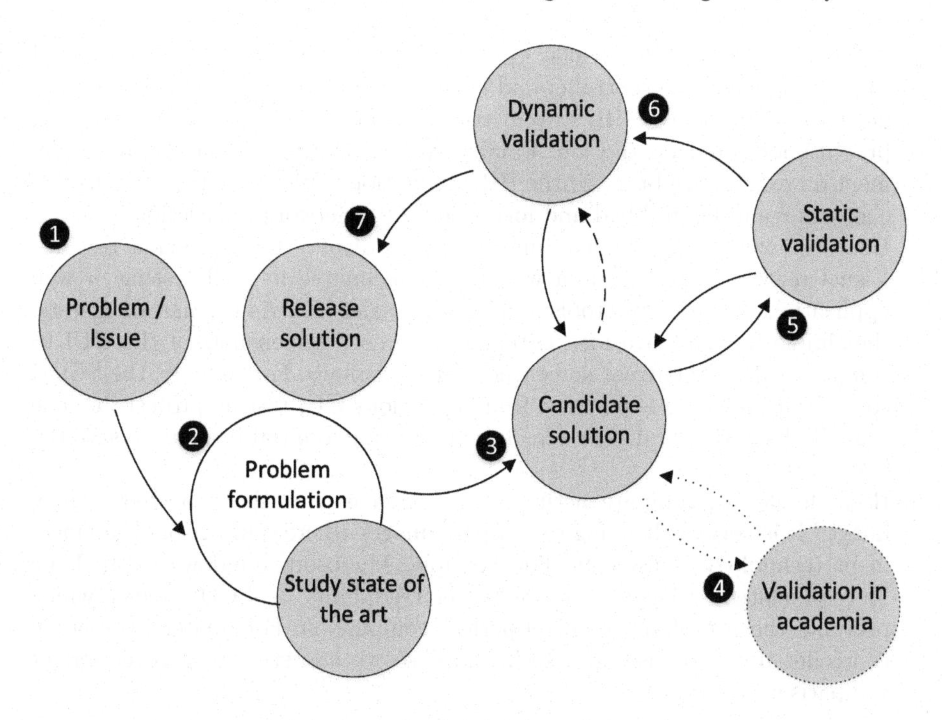

Fig. 3. Technology transfer model.

display shows the position of the crane's arm as well as additional status information about the crane. In the past, display errors have gone unnoticed because the visual representation of the GUI was not taken into account during automatic tests. Hence, the project team agreed to focus on testability improvement of the crane controller by also checking the display content.

3.2 Step 2: Formulate a Research Agenda

According to the identified needs in the previous step we identified a research agenda on the following questions:

1. How can we access the display contents?
2. How can we create reference images of the display for regression testing?
3. How should we compare actual display image with the expected display image?

3.3 Step 3: Formulate a Candidate Solution

In close cooperation (in form of several workshops) with the practitioners we created candidate solutions for each question above:

1. How can we access the display contents? To access the display contents of the virtualized crane controller and make it available during test execution, a deep knowledge of the HIL system was necessary. So only a well experienced practitioner could do the job. Finally, the proposed solution was to implement an extension library to the HIL simulation software so that the display contents could be grabbed and forwarded via a network connection.
2. How can we create reference images of the display for regression testing? Visual regression testing is a well-known technique in GUI testing of web applications where many tools and libraries exist [12]. As the name suggests, the simple but effective idea is to save reference screenshots of the GUI to compare with the current screenshots after changes. Fortunately, the SCCH could bring research experience from a previous GUI testing project [9]. The suggested solution grabs the screen buffer during a initial test run to save the base images.
3. How should we compare actual display image with the expected display image? It is very challenging to compare images where some display enhancement technology is used [9]. For example, Microsoft Windows Clear Type blurs pixelated or jagged text for better readability on LCD screens, which prevents the use of simple pixel-perfect comparison. Fortunately, the crane controller does not use such technology, so we can stick to a pixel-perfect comparison.

3.4 Step 4: Conduct Lab Validation

We agreed with the practitioners not to do explicit laboratory validation because we did not expect any major problems in implementing the integration. The dotted arrows depict that we skipped this step.

3.5 Step 5: Perform Static Validation

We presented the candidate solution from step 3 to the test experts and collected feedback from them. The main issue that was raised by them, was that the display presents different information at once, including dynamic content like the current time and date. In such cases, comparing the entire screen would always produce a difference. Therefore, we refined the candidate solution, to compare rectangular regions of the display instead. This has the additional advantage that these picture fragments can be reused in several test cases.

3.6 Step 6: Perform Dynamic Validation

For dynamic validation we implemented a prototype. Image grabbing and forwarding as pixel array via a network connection was done by a well experienced practitioner from Palfinger. Researchers took care of relaying the image to the test execution side as well as image processing (load, save, crop). Changes are emphasized in Fig. 1 as dashed lines as well as italic fonts. After creating a few test cases with image checks, it was obvious, that some structured storage of images is required, especially the separation of commonly used images from test case specific images. For minor changes we suggest a shortcut from step (3) to (6) in the model (see dashed arrow in Fig. 3 omitting steps (4) and (5)).

3.7 Step 7: Release the Solution

After adding the features regarding image management, we deployed the first release on the test system. Since then, the test engineers have evolved their test cases by adding image checks. Consequently, failed test runs were better and faster analyzable.

4 Results

Besides the successful implementation of the GUI testing approach of the virtualized crane controller and its integration into the testing pipeline, we gathered the main challenges and most important success factors for industry-academia cooperations in the context of application oriented software research in the field of CPSs.

4.1 Challenges

During this small specific task we did not face many challenges, which were caused by collaboration issues. However, throughout the complete collaboration, we have identified five particularly important challenges that you need to consider for successful collaboration, especially in the context of CPSs:

- **Hardware-dependency:** The most obvious challenge in this context of CPSs is that such systems can not be made available easily for external collaborators. As soon as hardware is involved, the effort to transfer a system to another location increases a lot. Therefore, whether hardware access is needed for a specific task should be considered early.
- **Access to specialized tools:** Test environment depends on specialized tools, which are often incurring corresponding license costs for every additional user. Hence, such industry-academia collaboration might cause unexpected costs.
- **Regulatory limitations:** Loader cranes have to fulfill safety regulations, which require the use of dedicated test procedures and certified tools. This may limit the introduction of new testing approaches and technologies, or at least limit its effectiveness. In addition, required changes to improve the testability of the system under test might be difficult to argue, if these changes require a cost-intensive recertification. Therefore, it is necessary to ask about the existence of such regulations at an early stage.

- **Binding to existing interfaces:** Many tools in the development/test chain of Palfinger have been developed a long time ago and have over years. Unfortunately, these internally used tools have not been developed like a professional software product and documentation is missing. To be able to extend or adapt these tools requires a substantial time investment or leveraging the knowledge of the company's tool maintainer. Hence, the responsible developers of such tools should be part of the team or at least available for consultation.
- **Involvement of stakeholders:** Working on the software production chain means that there are many different needs that must be met. As an external research institute, it is difficult to assess the impact of a particular change on the entire system and on specific user groups, as they are often not even known to the company members themselves. As a consequence, it is important to include representative stakeholders in the project team.

4.2 Success Factors

In [11] Wholin et al. analyzed success factors of industry-academia collaborations. We agree that *buy-in and support from company management and industry collaborators* are the most important success factors for such collaborations. The authors of the article argue that possibly the context of the project might have a big impact on the ranking of the most important success factors, what we can confirm. Although we were not able to conduct a survey and prioritize all of the success factors as they did, based on our many years of experience conducting applied research, we believe that *system and expertise knowledge* is more important to the success of such a collaboration in the field of CPSs than in systems without hardware dependencies. This knowledge is necessary for safety and cost reasons to minimize the probability of damage to physical components.

Software build chains are often very complex and the required test environments are difficult to set up due to the often long product history where testability was not one of the main concerns. The company partner brings in its system and product knowledge and the scientific expert provides its test expertise. So the need for regular discussion and knowledge transfer is very important. Hence it was very beneficial, that the project team was very stable over the last few years and that the integration task has been done by an experienced senior researcher. *On-site collaboration* by a researcher might also be a good way, to get to know the system faster. If this is not possible, more structured communication, knowledge transfer, and explicit responsibilities are required for successful cooperation.

We see fast and efficient communication channels at the industry partner and good networking skills of the company's project members as another important success factor. This is confirmed in the specific case by the fact that the results were already taken up by other departments during development and that the software part of the test environment will be rolled out to other departments in the near future. Furthermore, a prototype of a much smaller, mobile HIL-System was developed. The mobility of this environment improves and eases field tests as well as trainings for the service personnel.

5 Conclusion

In this paper we described the successful introduction of visual regression testing into a complex build and test environment of a cyber-physical system in the context of an application oriented funded research project. The research task has been setup as collaboration between the company Palfinger and the external research center SCCH. We reported about the successful cooperation and single steps following Gorschek's model for technology transfer [4]. In addition, we identified some unique challenges in industry-academia collaboration if a CPS is the core of the research objective: hardware dependency, access to specialized tools, regulatory limitations, binding to existing interfaces, and involvement of stakeholders. If these points are taken into account from the beginning of the cooperation, a successful collaboration becomes more likely. We also discussed the relevance of some factors known from the literature for successful collaboration, emphasizing that in the context of CPSs, appropriate *system and expert knowledge* are of particular importance. We want to evaluate this in the future in a broader context.

This work is intended to encourage researchers and practitioners from industry to also report on and analyze experiences with successful and unsuccessful collaborations. These reports contribute to a better understanding of the respective requirements and limiting factors and should increase the number of successful collaborations in the future by avoiding reported pitfalls.

Acknowledgements. The research reported in this paper has been funded by the Federal Ministry for Climate Action, Environment, Energy, Mobility, Innovation and Technology (BMK), the Federal Ministry for Digital and Economic Affairs (BMDW), and the Province of Upper Austria in the frame of the COMET - Competence Centers for Excellent Technologies Programme managed by Austrian Research Promotion Agency FFG.

References

1. Abbaspour Asadollah, S., Inam, R., Hansson, H.: A survey on testing for cyber physical system. In: El-Fakih, K., Barlas, G., Yevtushenko, N. (eds.) ICTSS 2015. LNCS, vol. 9447, pp. 194–207. Springer, Cham (2015). https://doi.org/10.1007/978-3-319-25945-1_12
2. Boardman, C., Gray, D.: The new science and engineering management: cooperative research centers as government policies, industry strategies, and organizations. J. Technol. Transf. **35**(5), 445–459 (2010). https://doi.org/10.1007/s10961-010-9162-y
3. Fischer, S., Ramler, R., Klammer, C., Rabiser, R.: Testing of highly configurable cyber-physical systems - a multiple case study. In: 15th International Working Conference on Variability Modelling of Software-Intensive Systems, Krems, Austria, pp. 1–10. ACM, February 2021. https://doi.org/10.1145/3442391.3442411
4. Gorschek, T., Garre, P., Larsson, S., Wohlin, C.: A model for technology transfer in practice. IEEE Softw. **23**(6), 88–95 (2006). https://doi.org/10.1109/MS.2006.147

5. jsonrpc.org: JSON-RPC 2.0 Specification. https://www.jsonrpc.org/specification. Accessed 31 Mar 2022
6. Lyu, M.R., et al.: Handbook of Software Reliability Engineering, vol. 222. IEEE Computer Society Press, Los Alamitos (1996)
7. Ogheneovo, E.E.: On the relationship between software complexity and maintenance costs. J. Comput. Commun. **02**(14), 1–16 (2014). https://doi.org/10.4236/jcc.2014.214001
8. Porter, M.E., Heppelmann, J.E.: How smart, connected products are transforming companies. Harv. Bus. Rev. **93**(10), 96–114 (2015)
9. Ramler, R., Wetzlmaier, T., Hoschek, R.: GUI scalability issues of windows desktop applications and how to find them. In: Companion Proceedings for the ISSTA/ECOOP 2018 Workshops, Amsterdam, Netherlands, pp. 63–67. ACM, July 2018. https://doi.org/10.1145/3236454.3236491
10. razorcat.com: TESSY - Test System. https://www.razorcat.com/en/product-tessy.html. Accessed 25 Mar 2022
11. Wohlin, C., et al.: The success factors powering industry-academia collaboration. IEEE Softw. **29**(2), 67–73 (2012). https://doi.org/10.1109/MS.2011.92
12. Wunschik, A.: Awesome Visual Regression Testing. https://github.com/mojoaxel/awesome-regression-testing. Accessed 22 Mar 2022

Vibration Analysis for Rotatory Elements Wear Detection in Paper Mill Machine

Amaia Arregi[1(✉)], Iñaki Inza[2], and Iñigo Bediaga[1]

[1] IDEKO S.COOP, Member of Basque Research and Technology Alliance, Elgoibar, Spain
{aarregui,ibediaga}@ideko.es

[2] Intelligent Systems Group, Computer Science Faculty, University of the Basque Country, Donostia-San Sebastián, Spain
inaki.inza@ehu.es

Abstract. Vibration analysis (VA) techniques have aroused great interest in the industrial sector during the last decades. In particular, VA is widely used for rotatory components failure detection, such as rolling bearings, gears, etc. In the present work, we propose a novel data-driven methodology to process vibration-related data, in order to detect rotatory components failure in advance, using spectral data. Vibration related data is first transformed to the frequency domain. Then, a feature called severity is calculated from the spectra. Based on the relation of this feature with respect to the production condition variables, a specific prediction model is trained. These models are used to estimate the thresholds for the severity values. If the real value of the severity exceeds the estimated threshold, the spectra associated to the severity is analyzed thoroughly, in order to determine whether this data shows any failure evidence or not. The proposed data processing system is validated in a real failure context, using data monitored in a paper mill machine. We conclude that a maintenance plan based on the proposed would enable to predict a failure of a rotatory component in advance.

Keywords: Vibration analysis · Predictive maintenance · Condition monitoring

1 Introduction

Vibration is the periodic motion or oscillation of the particles of an elastic body from the position of equilibrium [11]. Vibration analysis (VA) has received great interest in the industrial sector in the recent years due to the promising results that can be obtained by its application, especially in the case of rotatory elements. VA techniques provide to the analysts the possibility to design a correct maintenance plan based on the accurate knowledge of the health of the studied elements, such as bearings, gears, induction motors, etc. A large amount of failures can be detected in early stages, such as misalignment, eccentricity, imbalance, looseness, etc. If VA is performed over the time, it is possible to track the

G. Kotsis et al. (Eds.): DEXA 2022 Workshops, CCIS 1633, pp. 199–209, 2022.
https://doi.org/10.1007/978-3-031-14343-4_19

deteriorating process of the different failures associated to the industrial goods. By determining the health of these components, it would be possible to schedule an appropriate maintenance plan. This would result in a Predictive Maintenance (PM) assessment.

In the present work, we propose a novel data-driven methodology to process vibration-related data, in order to detect rotatory components failure in advance, using spectral data. The proposed system takes into account the possible relationships between the production related variables with respect to the features extracted from the vibration frequency spectra. The system is then tested in data monitored in a real industrial scenario. The analyzed rotatory components are part of a paper mill machine. In the analyzed use case, vibration data is monitored while the machine is producing paper i.e. in variant conditions. We prove that it is possible to detect failures in advance. The remaining of this paper is structured as follows. Section 2 presents the related work and the application in industrial scenarios of different VA approaches. In Sect. 3 the novel methodology is explained. In Sect. 4 the results of the case study are showed. Finally, Sect. 5 reports the conclusions.

2 Related Work

Industrial maintenance activity is a crucial part of the production process, so it is important to set up a proper maintenance plan for each plant. The absence of an adequate plan can lead to unwanted or unscheduled stops in machines/industrial goods, stopping production, increasing the risk of operators or generating failures or deficiencies in the quality of the products [9]. The most commonly used techniques are the corrective maintenance, preventive maintenance and predictive maintenance. The last type of maintenance task is the one addressed in this work.

As the consequences of a correct predictive maintenance plan can bring large benefits, it is undoubtedly the method that has attracted the most attraction in the literature [12]. Its aim is to schedule the maintenance actions only when a functional failure is detected [13]. To be able to carry out this type of action, it is necessary to have information about the condition of the machine, its components or industrial goods. By Condition Monitoring (CM) approaches, the health condition of the process/element analyzed is measured by sensors. Apart from that, predictive tools are used to determine when it is necessary to carry out maintenance tasks. These tools measure the state of the machine and by means of predictive models (statistical models, artificial intelligence, etc.) they try to predict a possible failure that may occur. Applying data analytic techniques to data flow can have a number of benefits for companies. Valuable information can be extracted from the processing of this data, making possible a better understanding of the operation of an industrial process. Various benefits that can be achieved are cited in the literature [2]. Among the most cited are the benefits that can be obtained in relation to maintenance work. Economic cost reductions, reduction of downtime (non-production time), possibility of planning maintenance tasks, and so on. The most used CM technique, among others, is the VA. On the contrary, it should be noted that designing a specific asset monitoring plan is complex, since it must be done for critical components or elements that indicate deterioration. On the other hand, it is necessary to make a significant initial investment, as well as to train specialized workers for this type of task. Finally, it is still necessary to improve the security and privacy in smart factories, as a large amount of private data is vulnerable to attacks [14].

2.1 Vibration Analysis in Industry

VA is the most widely used technique for CM for a simple reason: the possibility of determining a large number of defects, in a wide range of rotating elements (bearings, gears, motors...) with a reasonable initial investment.

Vibrations are generally measured using accelerometers. In the analysis of machine vibrations, the spectrum of vibrations is studied, so the raw data from these sensors is processed. One option is to transform the signal to the frequency domain, by computing the Fast Fourier Transform (FFT). Thus, the characteristic vibration of each of the components can be identified, giving the possibility of detecting the defects that these may have: play, imbalance, misalignment, loosening, bearing lubrication problems, damage to bearings, gears, engines, etc. It is also possible to extract features from the processed data.

Regarding the oil & gas sector, there are several proposals for monitoring failures of gas turbines, based on vibration analysis. These turbines are used in the oil industry, and are one of the most important elements in engines for generating mechanical energy (in the form of shaft rotation) from the kinetic energy of the gases produced in the combustion chamber. In the studies carried out in [3], authors propose failure monitoring systems learned using spectral analysis tools.

On the other hand, VA as a solution for the detection of failures in induction motors has been widely applied. These kind of motors are alternating current electric motors, and are currently widely used in industrial plants. In [15], the author proposeto use VA techniques for detecting electromagnetic faults in these kind of motors. Author shows several use cases demonstrating the capability of VA for fault diagnosis. On the other hand, authors in [1] make a comparison between three techniques: vibration analysis, motor current signal analysis and the discrete transform of the wave (DTW). They emphasize that by using these techniques it is possible to detect different types of failures in induction motors. It is also possible to detect failures caused by unbalanced loads, as shown in the study carried out in [7]. Authors used the Power Spectral Density (PSD) as the critical feature to detect these kind of failures. They studied the PSD for three different failure stages, and see how when the degree of unbalance increases the amplitude of the side-band harmonics also increases.

Regarding the aerospace industry, authors in [6] perform VA to evaluate the behavior of an airplane wing structure. Several vibration tests were carried out on the wing, to determine what its natural frequencies were and to understand what its damping characteristics were.

These work have studied the diagnosis capacity of different approaches. There are also some works regarding the deterioration evolution of the equipment good along time. Authors in [10] defend the use of vibration analysis techniques using time-frequency transforms for a specific failure. This type of failure happens in the bearings due to the electric currents induced by the motors. Thus, the authors suggest a component degradation analysis technique as a solution to identifying potential problems. Authors in [5] propose the use of autoregressive models to estimate the feature values extracted from vibration data. On the other hand, in [16] authors present a solution based on unsupervised machine learning techniques, analyzing the time series from the vibrations of a robotic arm used to paint the bodywork of an automobile manufacturer. The author proposes a technique for defining the state of health of the machine, based on the evolution of its vibrations, using Gaussian mixture models. Last, authors in [8] propose a pipeline for extracting the relevant features of the vibration related data to determine the remaining useful life (RUL). A Particle Filter (PM) algorithm is used to predict the RUL.

To the best of our knowledge, no work has been addressed for the detection of faults in rotating equipment goods, in real industrial scenarios where the data is obtained in variable working conditions.

3 Methodology

In the present work, we propose a novel data-driven methodology to process vibration-related data, in order to detect rotatory components failure in advance, using spectral data. The proposed system deals with vibration data monitored in variant conditions, making possible to analyze the evolution of the features extracted from the frequency spectra obtained from the vibration signal. Therefore, this system is applicable in production scenarios, with no need of stopping the machine or to performing periodic checkups in prefixed conditions. Thresholds are defined based on the working variables in the case that these affect to the severity values. On the contrary, if the values are not correlated, thresholds are calculated based on the evolution of the severity values in the last observations. The aim of the definition of the thresholds is to be able to select the spectrum to be analyzed knowing that its associated severity value is indeed anomalous. Besides, as the conditions for obtaining the data are not equal, this criteria aims to ensure that the analyzed spectrum are actually the ones that show valuable information.

First, data is collected for analyzing and training the models offline. Once the dataset is completed, the correlations between the variables related to the working condition and the severity values obtained from the vibration data are computed. Two correlation coefficients are calculated: the Pearson correlation coefficient (ρ_P) and the Spearman correlation coefficient (ρ_S).

There are several combinations where the correlation degrees between the variables are relevant. A criteria is designed for determining the models to learn to estimate the severity values based on the working conditions. Based on the empirical experience, three different categories are set up for the correlation values.

- $\rho_P, \rho_S \geqslant 0.8$, linear models are learned with the severity variable as the response variable.
- $\rho_P, \rho_S \in [0.6, 0.8)$, linear models are learned, after adding interactions between the variables, in order to add complexity to the models.
- $\rho_P, \rho_S < 0.6$, a model based on the temporary evolution of the severity is used for categorizing the values of the severity.

For both the linear and custom models, a threshold is established to be able to categorize the severity values, based on the difference of the estimated and real values. For the linear model case, the threshold is determined by the confidence interval of 95%. For the custom algorithm case, the threshold is computed based on the rolling mean and variance of the severity in previous observations.

Linear Models. The constraints for the correlation values justify the use of linear models. Using them make the interpretability of the model to be immediate [4]. If the dimensionality is too high (i.e. too many production related variables) it would be possible to learn penalized linear models. Examples of these type of models include Lasso, Ridge, ElasticNet, etc.

Custom Algorithm. For those scenarios where the correlations between the variables are not considered sufficiently strong, a custom algorithm is proposed to categorize the severity values. For these cases, only the threshold is estimated. The value of the threshold is based on the evolution of the severity (defined by the moving average and standard deviation) over time. The length of the temporary windows is selected based on the use case. The threshold is calculated using Eq. 1.

$$t_m = \mu_m + c\sigma_m, \tag{1}$$

where μ_m and σ_m are the moving average and standard deviation of the temporary window m respectively. The constant value is determined by empirical evidence for each use case.

By the implementation of this algorithm, the thresholds are dynamically defined and adapted based on the latest evolution of the severity values.

3.1 General Approach for Analyzing Anomalous Severity Values

For the cases where the severity value is categorized as anomalous i.e. the true value of the severity exceeds the estimated threshold, the spectrum related to the anomalous severity is analyzed thoroughly, in order to find amplitude peaks that identify possible failures occurring in the components of the different parts of the machine. It is necessary to know the rotary speed of the components, in order to determine the natural frequencies of the different failures that may be suffering the rotary components of the kinematic chain (rolling bearings, gears, etc.).

4 Case Study

The analyzed machine is located in a company that produces tissue paper, among other products. Paper mill machines incorporate a large number of rollers, all of them equipped with medium-large size bearings. Generally, a paper machine is composed of several sections. In this use case, monitoring system is focused on the data obtained from three critical parts of the machine: the forming section, the pressure section and the drier section. In each of them, a motor, gearbox and a bearing is analyzed.

4.1 Data

The dataset is constituted of production condition variables and vibration variables. No external sensor has been installed for monitoring the first type of variables. On the other hand, the monitoring of the vibrations is done by 18 accelerometers installed in the forming, suction and dryer sections. The sensors are installed in pairs: two on the motor, two on the gearbox and two on the bearings of each section. Through the processing of this type of data, VA is performed in acceleration, velocity and demodulation. These different types of processing the vibration signals provide a different insight of the health of the components. The data monitoring is done once a day. The training dataset dates from 02/01/2020 to 09/13/2020. On the other hand, the validation set ranges from 09/21/2020 to 10/16/2020.

4.2 Variables

Two types of variables are determined: in one hand, the variables that define the production condition, and in the other hand, the variables related with the vibration signal.

Production Condition Variables. Two condition variables are monitored for each section: forming, suction and drying sections. These variables are obtained from each of the 3 motors. On the one hand, load, which is the power at which the machine is working with respect to the maximum power (%). On the other hand, the rotary speed of the motor (RPM).

Vibration Variables. Only one feature is extracted from the different spectral data, the severity. Once the frequency spectra of the vibration signal is obtained, the value for the severity is obtained using the following equation:

$$Sev = \sqrt{\sum_{i=f_{min}}^{f_{max}} {A_i}^2 \frac{c_e}{c_a}}, \quad (2)$$

where A_i is the amplitude of the signal in the ith frequency, and the constants c_e and c_a are predefined by domain experts.

4.3 Training

Linear Models. Linear models are learned for the cases where the correlation is strong enough. As in this case there are only 2 independent variables, there are not dimensionality problems. So, no penalization has been applied to the linear models.

Custom Algorithm. For the custom algorithm, the values for m and c in Eq. 1 are set to 8 and 1.2, respectively. These values are based on empirical evidence.

4.4 Results

A validation set is used for evaluating the performance of the proposed algorithm. The objective is to analyze whether the method is able to detect a incipient failure of a rotatory element. For the present work it has been possible to obtain data related to an unwanted stop, due to a failure in a rotatory element of the machine. The failure was caused by a defect on a rolling bearing of the gearbox in the dryer section. This stop happened on 10/04/2020

As the failure happened in the dryer section, the models learned for this section had been analyzed, and more specifically the related to the gearbox of the motor. In Fig. 1 the evolution of the severity values calculated for the two signals obtained of the gearbox are shown. The thresholds for both cases are obtained by applying the algorithm explained in Subsect. 3. It is also possible to see how the threshold for the values of the severity has exceeded 3 days earlier than the day of the incident. Furthermore, if the spectrum associated to those anomalous values is analyzed, it is possible to see how the amplitude peaks correspond to possible failures of the defective rolling bearing. In the Fig. 2 is shown an example. For those days where the true severity values don't exceed

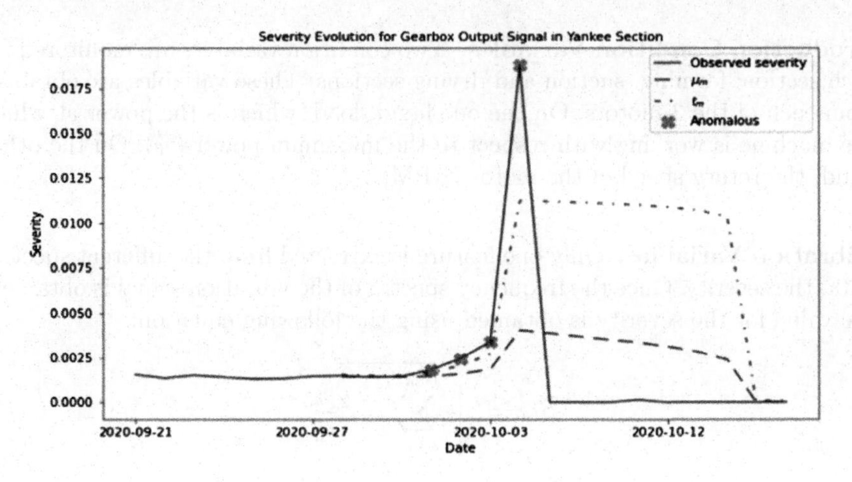

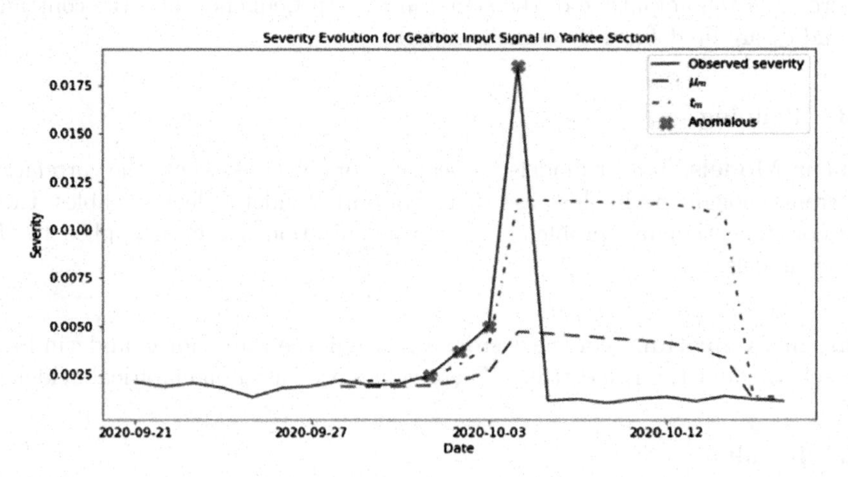

Fig. 1. Evolution of the severity values for the velocity analysis of the signals obtained from the output (upper figure) and input (lower figure) sensors installed in the gearbox of the dryer section. It is possible to see how the threshold for the severity values is exceeded from 10/01/2020 until the critical day, which was on 10/04/2020.

the estimated threshold, it can be seen that the first harmonics for the Ball Spin Frequency (BSF, a type of failure that occurs in rolling bearings) don't match with the amplitude peaks. On the other hand, for those spectra associated to the severities that exceed the estimated threshold, the amplitude peaks correspond to the BSF.

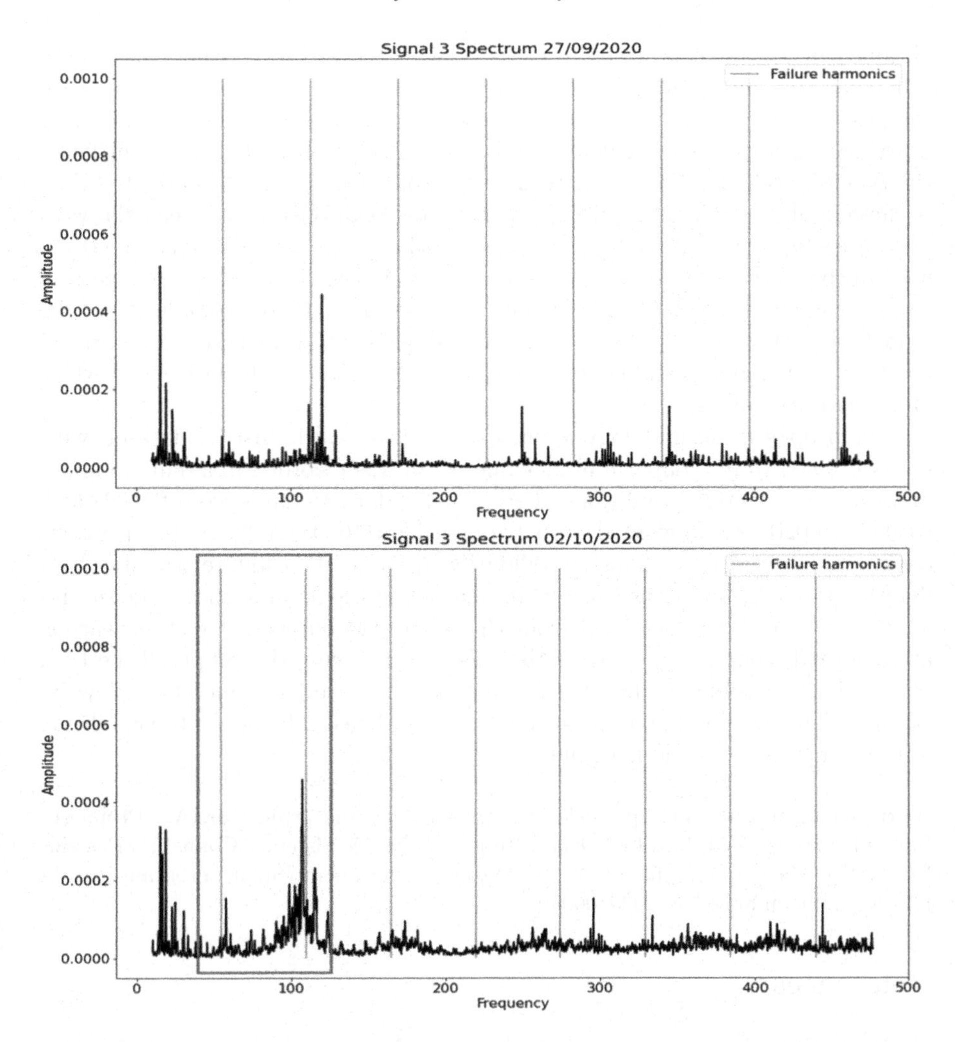

Fig. 2. Vibration spectra for the gearbox output signals, obtained on 25/09/2020 (upper figure) and 02/10/2020 (lower figure). The harmonics correspond to the BSF. The upper figure correspond to a spectra associated to a severity value that does not exceed the threshold value. The lower figure correspond to a spectra associated to a severity value that exceeds the estimated threshold value. The rectangle highlights the amplitude peaks that coincide with the failure harmonics.

5 Conclusions

In the present work, a methodology for analyzing vibration related data has been designed for detecting rotatory elements failures in advance. This method establishes the criteria for evaluating the frequency spectra of the vibration data obtained from the critical parts of the machine. As a preliminary study of the correlations is done before learning any model, this method can be used in data obtained in variant conditions, such as production scenarios, where the conditions may vary along time. For quantifying these possible relationships between the production related variables with respect of the severity values, Pearson and Spearman coefficients are calculated. Based on the values of these coefficients, the type of model used to estimate the thresholds for the vibration related data is chosen. The models learned for the strongest relationships between the variables are simple linear models. For the cases where the relationship is not strong enough (but still not that weak), interactions between the production condition variables are added to the linear models, in order to add more complexity to the models. For those cases where the relationships are the weakest, an algorithm has been designed for establishing the thresholds, based on the last observations of the severity values.

The proposed methodology is been tested in a real industrial dataset, with data obtained from a paper mill machine. Data was recorded in a daily manner, in variant production conditions. The training dataset dates from 01/02/2020 to 09/13/2020. An unwanted stop happened on 10/04/2020, so the previous days data is used to compose a validation dataset. The failure was due to a deterioration of a rolling bearing of the gearbox of the dryer section. The several analysis of the signal obtained from the inner and outer part of the gearbox are analyzed, concluding that for this case, by applying the proposed method, it would be possible to detect anomalies in the severity values four days in advance. This would enable the scheduling of PM tasks, based on the condition of the components of the machine.

Acknowledgement. This project was supported by the project 3KIA - Propuesta Integral y Transversal Para El Diseño E Implantación De Sistemas Confiables Basados En Inteligencia Artificial, funded by the Basque Government (Spain) program ELKARTEK, grant number KK-2020/00049.

References

1. Bessous, N., Zouzou, S., Sbaa, S., Bentrah, W.: A comparative study between the MCSA, DWT and the vibration analysis methods to diagnose the dynamic eccentricity fault in induction motors. In: 2017 6th International Conference on Systems and Control (ICSC), pp. 414–421. IEEE (2017). https://doi.org/10.1109/ICoSC.2017.7958655
2. Carvalho, T.P., Soares, F.A., Vita, R., Francisco, R.d.P., Basto, J.P., Alcalá, S.G.: A systematic literature review of machine learning methods applied to predictive maintenance. Comput. Ind. Eng. **137**, 106024 (2019). https://doi.org/10.1016/j.cie.2019.106024
3. Djaidir, B., Hafaifa, A., Kouzou, A.: Faults detection in gas turbine rotor using vibration analysis under varying conditions. J. Theor. Appl. Mech. **55**(2), 393–406 (2017). https://doi.org/10.15632/jtam-pl.55.2.393
4. Došilović, F.K., Brčić, M., Hlupić, N.: Explainable artificial intelligence: a survey. In: 2018 41st International Convention on Information and Communication Technology, Electronics and Microelectronics (MIPRO), pp. 0210–0215. IEEE (2018). https://doi.org/10.23919/MIPRO.2018.8400040

5. Fernández, A., Bilbao, J., Bediaga, I., Gastón, A., Hernández, J.: Feasibility study on diagnostic methods for detection of bearing faults at an early stage. In: Proceedings of the 2005 WSEAS International Conference on Dynamical Systems and Control, pp. 113–118 (2005)
6. George, J.S., Vasudevan, A., Mohanavel, V.: Vibration analysis of interply hybrid composite for an aircraft wing structure. Mater. Today Proc. **37**, 2368–2374 (2021). https://doi.org/10.1016/j.matpr.2020.08.078
7. Güçlü, S., Ünsal, A., Ebeoğlu, M.A.: Vibration analysis of induction motors with unbalanced loads. In: 2017 10th International Conference on Electrical and Electronics Engineering (ELECO), pp. 365–369. IEEE (2017)
8. Lei, Y., Li, N., Gontarz, S., Lin, J., Radkowski, S., Dybala, J.: A model-based method for remaining useful life prediction of machinery. IEEE Trans. Reliab. **65**(3), 1314–1326 (2016). https://doi.org/10.1109/TR.2016.2570568
9. Myklestad, N.O.: Fundamentals of Vibration Analysis. Courier Dover Publications (2018)
10. Ooi, B.Y., Beh, W.L., Lee, W.K., Shirmohammadi, S.: A parameter-free vibration analysis solution for legacy manufacturing machines' operation tracking. IEEE Internet Things J. **7**(11), 11092–11102 (2020). https://doi.org/10.1109/JIOT.2020.2994395
11. Palazzolo, A.: Vibration Theory and Applications with Finite Elements and Active Vibration Control. Wiley, New York (2016)
12. Poór, P., Basl, J., Zenisek, D.: Predictive maintenance 4.0 as next evolution step in industrial maintenance development. In: 2019 International Research Conference on Smart Computing and Systems Engineering (SCSE), pp. 245–253. IEEE (2019). https://doi.org/10.23919/SCSE.2019.8842659
13. Scheffer, C., Girdhar, P.: Practical Machinery Vibration Analysis and Predictive Maintenance. Elsevier, San Diego (2004)
14. Shi, Z., Xie, Y., Xue, W., Chen, Y., Fu, L., Xu, X.: Smart factory in industry 4.0. Syst. Res. Behav. Sci. **37**(4), 607–617 (2020). https://doi.org/10.1002/sres.2704
15. Tsypkin, M.: Induction motor condition monitoring: vibration analysis technique-a practical implementation. In: 2011 IEEE International Electric Machines & Drives Conference (IEMDC), pp. 406–411. IEEE (2011). https://doi.org/10.1109/IEMDC.2011.5994629
16. Wescoat, E., Krugh, M., Henderson, A., Goodnough, J., Mears, L.: Vibration analysis utilizing unsupervised learning. Procedia Manuf. **34**, 876–884 (2019). https://doi.org/10.1016/j.promfg.2019.06.160

Introducing Data Science Techniques into a Company Producing Electrical Appliances

Tim Kreuzer and Andrea Janes(✉)

Faculty of Computer Science, Free University of Bozen-Bolzano, 39100 Bolzano, Italy
{tkreuzer,ajanes}@unibz.it
https://www.unibz.it

Abstract. Industry-academia collaboration in the field of software engineering is posing many challenges. In this paper, we describe our experience in introducing data science techniques into a company producing electrical appliances. During the collaboration, we worked on-site together with the engineers of the company, focusing on steady communication with the domain experts and setting up regular meetings with the stakeholders. The continuous exchange of expertise and domain knowledge was a key factor in our collaboration. This paper presents the adopted collaboration approach, the technology transfer process, the results of the collaboration, and discusses lessons learned.

Keywords: Technology transfer · Industry-Academia collaboration · Data science

1 Introduction

Collaboration between industry and academia is beneficial for both: for example, industry obtains an early access to academic innovations and academia can test and validate its results in realistic contexts.

Both in academia and in industry, software engineering is an important field with many researchers and practitioners being involved. However, collaboration between the two groups is low [1,2]. Garousi and Varma [2] found that 88% of practitioners never or seldom interact with the research community. This division between researchers and practitioners has a negative impact on further advancements in the field. This paper shows how to achieve a win-win situation in which applied research can become a medium of technology transfer.

Past studies have suggested that technology transfer in software engineering presents many challenges that need to be overcome. [3] lists 10 challenges that are commonly faced in industry-academia collaboration. Kaindl et al. illustrate collaboration challenges in the field of requirements engineering [4]. In [5], the research-practice gap in the domain of requirements engineering is bridged, pinpointing possible solutions. A frequently mentioned challenge is that the results produced through research are not relevant for the industry [6]. Collaborations can help academia to better understand research or training needs from industry.

G. Kotsis et al. (Eds.): DEXA 2022 Workshops, CCIS 1633, pp. 210–220, 2022.
https://doi.org/10.1007/978-3-031-14343-4_20

When collaborating, it is important to take such challenges into account to foster a productive collaboration environment. To contribute to an improved mutual understanding between the parties, in this paper, we report about the introduction of data science techniques, specifically aiming at the improvement of the software engineering process of a team that produces software for electrical appliances, providing an experience report about the conducted research, the followed collaboration approach, and the lessons learned in the collaboration.

2 State of the Art

Several models for technology transfer between academia and industry have been proposed in the past. Gorschek et al. [7] have proposed a model based on seven steps. Figure 1 shows the structure of their approach. In Gorschek et al.'s proposal, industry detects a problem or issue and provides an initial problem formulation. Academia studies the problem and formulates research questions, compares them with the state of the art, and develops a candidate solution[1]. The solution is built and validated, in academia, but also statically and dynamically in the industry. Finally, after the dynamic validation succeeds, the solution is released. Please note that tasks carried out by industry are positioned on top half and tasks carried out by academia are on the bottom half.

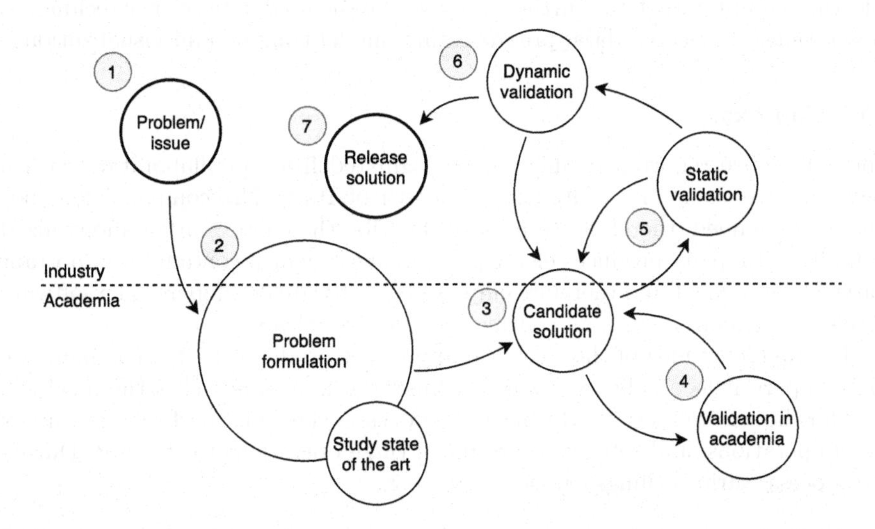

Fig. 1. Research approach and technology transfer process described as in [7].

[1] We report here the technology transfer process exactly as published in [7]. Activities are represented as nodes. The thicker lines around nodes represent starting and end nodes. The overlapping node "Study state of the art" over "Problem formulation" represents a "part-of" relationship between the two nodes. Directed edges represent transition possibilities between nodes.

Sandberg et al. [8] have proposed a list of key factors for industry-academia collaborations. They created this list based on their experience with such collaborations and proposed factors for an agile collaboration between researchers and practitioners.

Garousi et al. [6] have conducted a systematic literature review, analyzing challenges and best practices from industry-academia collaborations in the field of software engineering. They created a mapping between challenges and best practices to recommend how to overcome frequent obstacles. Their analysis shows a large number of challenges from different perspectives, making clear that technology transfer between academia and industry is a challenging task.

In [9], Garousi et al. conducted a survey with researchers from 101 industry-academia collaboration projects. Their survey reported which topics are most commonly pursued in industry-academia projects. They also focused on operational aspects of the projects, summarizing the impact of patterns and anti-patterns on project success.

3 Case Study

This section describes the industry-academia collaboration we carried out. First, we describe in which context our experience was made. Then, we describe the collaboration approach and the process to find an objective for the analysis and selecting appropriate data. Further, we describe the used data science techniques illustrating the selected data, preprocessing, model training and visualization.

3.1 Context

The experience reported in this paper originated from a collaboration with a medium-sized company located in the north of Italy. The company does not wish to be named, therefore we have to describe the context in an anonymized form. The company produces electrical appliances, which contain an important software component and therefore employs ten software developers. The software development team focuses on continuous software releases.

To provide an idea of the electrical appliances produced by the company, we can compare it to a coffee vending machine: the user approaches the machine, pays for the selected service, the machine processes the order, performs a complex set of operations, and returns the result of these operations to the user. During the process, various things can go wrong, e.g.:

- The user removed the cup too early from the machine: This is a failure that can be resolved by the user, as a consequence, the user is instructed what to do and needs to press a button to continue.
- Sugar is finished: This means that the service is degraded but it is still possible to provide coffee. The user is asked if the process should be continued without sugar or aborted and the money returned.
- Coffee is finished: This means that the service cannot be provided. The user is informed about the problem and the money is returned.

This system metaphor [10] can help to understand the context: The coffee vending machine has an important process component that involves a particular set of steps that have to be executed according to a given specification. The performed steps and their outcome are logged in detail to know exactly what happened in the event of a legal dispute. These log files are so large that they are not sent to the servers of the company, but remain within the electrical appliances for some time. A shorter version, a summary, is sent to the servers of the company.

Quality is one of the most important factors for the company as software faults have an immediate negative impact on their customers and, as a result, the reputation of the company. The goal of maintaining and improving quality was also the starting point of our collaboration: through the introduction of data science methods to analyze software logs, the team had the goal to improve the quality of the produced software. In this specific case, the model of Gorschek et al. [7] was a match: the company understood that they have a quality issue and defined an initial problem formulation.

Before our collaboration, the company used descriptive statistics techniques to examine log files coming from the electrical appliances in use by their clients (using our metaphor, this would mean from all the coffee vending machines around the world) and visualized it on a dashboard. For this, the data had not been processed in any way and the company had not employed any machine learning techniques or other automated analysis techniques before. Using this dashboard, e.g., problems that only impacted customers in Sweden could be quickly identified and countermeasures, i.e., investigating the log files of Swedish customers, could be taken.

3.2 Collaboration Approach

In previous experiences of industry-academia collaborations we observed that communication between the two parties, while being essential, occurs through very slow channels, e.g., e-mails or meetings. For this reason we opted for a model where one researcher was co-located and present on the industry site (as the suggested best practice number 61 in [6]). From an Agile point of view, this practice could be called "researcher on-site", as it serves the same purpose: facilitate communication.

This on-site presence has helped to shorten feedback loops, allowing us to gain direct feedback and also connect with domain experts present on-site without long lead times. This approach has sped up the collaboration, leading from an initial meeting to a dynamic system integration in less than six months. From an academic perspective, this approach corresponds to the "participant-as-observer" approach where the observer studies a social phenomenon as a participant, from inside [11].

The remaining researchers involved in the project communicated with the "researcher on-site" and the company through regular meetings.

3.3 Finding an Objective

As a first step, a common objective has to be found. This objective needs to be valuable from an industrial point of view and also feasible from the researchers' point of view. To achieve this, we discussed possibilities with all stakeholders over the course of three initial meetings. In those meetings, we focused on the following aspects:

- Understanding the domain: as researchers, we were not previously involved with the domain of the company, therefore, experts from the company were explaining the electrical appliances they are building and their key characteristics, starting from general information up to the company's internal systems and technical details. This was key to understand the logical correlations within the data and was a very important preparatory step for the following analysis.
- Presentation of data science techniques: neither the developers nor the management of the company were experienced with the topic of data science, its use-cases, limitations and possibilities. We were focusing on data science as a tool to gain insights from existing data, omitting technical details and providing a general overview to facilitate further communication between data scientists, developers, and managers.
- Problem analysis: in the discussion with the managers and the team lead of the IT department, several problems with the software, related to certain edge cases, came up. These problems helped us sketch an overview, identifying critical points in the system and the data these points are related to. With the domain knowledge of the engineers, the data science background of the researchers and the business standpoint of the managers, we brainstormed various approaches for the problems and possibilities for different analyses.

After the initial meetings, it became clear that the initial goal of quality assurance, more concretely, belonged to the area of anomaly detection. The electrical appliances produced by the company are complex and implement a specific process when a user interacts with it. Failures can occur at different points in this process and the company was interested in finding anomalies in their system and trace these anomalies, linking them to possible causes. Identifying the causes, so the idea, would help developers to remove errors to improve the resilience of their software and therefore improve its quality. If failures occur in known circumstances (as in the coffee vendor machine example above), it is possible to build a model of the machine and define measurements that help—through abductive and temporal reasoning—to identify the cause of the problem, i.e., to perform diagnosis [12]. In our case, the goal was to correlate unexpected failures with contextual information about the use of the electrical appliances. For example, in the case of our metaphoric coffee vending machine, a desirable problem to find would be that the machine often fails to make a cappuccino after making an espresso.

3.4 Data Selection and Preprocessing

A first challenge was to define what a failure exactly means and to create a model to recognize a failure. Therefore, before starting the analysis, the existing data has to be evaluated to understand which data represents a failure and which not. For this, domain knowledge is paramount, therefore we worked in close cooperation with engineers to outline details of the existing data structures. The data that is selected depends on the previously set objective, taking possible correlations between variables into account.

As a second step, various correlations were explored, based on the requirements of the company. In our case, the database contained many different variables that were related to the problem as well as textual log data, which represents the system-internal communication process. This log data contains crucial information, which was up until this point only used as a debugging tool for developers. For our analysis, the logs were included, as they represent a valuable source of information, greatly enhancing the potential of our analysis.

Having extracted the relevant data, preprocessing is the first step towards a further analysis. In our case, as we were working with textual log data, which is semi-structured data, structured information had to be extracted from the logs. This included log parsing, log structure analysis and time-series analysis, all resulting in further variables, representing the information contained in the log. During this process, we were in a steady exchange of information with domain experts from the company to reinforce our domain knowledge and elaborate the steps we took for preprocessing.

Preprocessing is a step, that varies depending on the dataset being used. Our dataset, including the log data, required the usage of custom algorithms to extract relevant variables. In the dataset we used, missing data was not a problem, therefore we did not implement a method to deal with missing data. The presence of missing data depends on the domain and communication with the responsible engineers is essential for an analysis.

3.5 Model Training and Visualization

With the preprocessed and cleaned data, machine learning models can be built. In our case, an anomaly detection model was designed, classifying logs into anomalous and non-anomalous categories. Depending on the goal of the analysis and the available data, different models need to be selected. In our case, the data was imbalanced, which required us to use a model that is resistant to imbalance. The library scikit-learn[2] was used for our machine learning implementation.

Building a model also includes an optimization process. In our use-case, a requirement from the company was that recall is more important than precision, which was a prerequisite that we took into account as we were optimizing the model. This means that is was more important to identify all cases in which a failure occurred, accepting the risk of incurring into false positives. This was one

[2] https://scikit-learn.org/stable/index.html.

aspect that needed to be clarified, since this depends on the company's business goals, which have an impact on the way how the analysis needs to be tuned and modified.

Visualization is a central part of data science, as it is a first step to grasp a large amount of data and to find anomalies and correlations. During our study, we conducted regular meetings with the stakeholders, visualizing the results of our models and our findings in dashboards. To create such dashboards we followed the recommendations described in [13].

4 Results and Discussion

In this section, we describe and discuss our results.

4.1 Outcome of the Collaboration

Our implementation, consisting of data science-focused technologies, detecting anomalies in the company's internal system, is the outcome of the research activity in collaboration with the company. This software is able to detect and label anomalous activities in the system, allowing the developers to work with the anomaly data to locate bugs and improve the system's reliability.

Our analysis was based on log data as well as other parameters from the company-internal database. To convey significant meaning, we connected our anomaly-labeled data with other types of information, showing how they correlate with each other. Having extracted labeled data points as anomalies, it is important to keep in mind that not all anomalies are necessarily caused by software errors, as the hardware involved in the process is another source of possible errors.

Using the coffee vending machine metaphor, some examples of analyses and visualizations illustrate the adopted approach:

- As different software versions cause the vending machine to behave differently, the software version is an important parameter for engineers. We visualized the anomaly rate related to the software version, showing which versions perform better and which perform worse. Additionally, grouped our analysis by error type to relate error frequencies between software versions. The engineers analyzed and reacted to the findings of these plots, comparing the software versions.
- The coffee vending machine has different user types that interact differently with the appliance, e.g. student and professor. This user type is logged every time an interaction with the coffee machine takes place. Since user behavior is a variable of key interest, this visualization is of particular interest to the user experience designers. We correlated the user type and the occurrence frequency of anomalies to see how both variables interact with each other.
- As temporal order plays a role in the interactions and behavior of the vending machine system, we also analyzed the order of user types interacting with a

vending machine, correlated to the anomaly rate. This allowed us to determine chains of user types, that are more likely to lead to an anomaly. This was crucial to analyze interactions within the system state, that lead to anomalies due to differences in user behavior.

After implementing our solution and validating it statically with company-internal data, we presented our results in a meeting to all stakeholders. During this meeting, the feedback was very positive, resulting in the request to implement the solution in the productive system. One of the main reasons for this was that our candidate solution was based on static data, which did not include the most recent software releases. Due to that, the question about generalizability and the possibility to work with dynamic data came up.

Having extracted dynamic data, our algorithms needed to be adapted to account for possible future changes and provide more stability with live data, which is constantly changing its structure, as new software versions are released. To integrate our analysis in the daily operations workflow of the company, we provided our code in the form of executable scripts, allowing operation engineers to include them in their system without necessarily understanding their inner workings. The generated visualizations were included in the dashboard that was already in use by the company, providing detailed information about specific parts of their system.

4.2 Collaboration Process

When we started our collaboration, we had the technology transfer process of Fig. 1 in mind. The *actual* process turned out to be different and is depicted—following the notation used in [7]—in Fig. 2: while the problem was identified by industry as in the original model (step 1), the problem formulation was more difficult than anticipated.

Since the electrical appliances and their related issues were unfamiliar to us, the company had to invest considerable time to explain the domain in which we were working. Moreover, while the industrial side may have this preconception, "academia" is not an abstract entity, knowledgeable in everything and capable to apply the state of the art in every context, but "academia", in the case of a concrete research project, consists of the researchers and students currently involved in the project. They have certain abilities and research interests that also influence the problem formulation. In academia, there is a strong fluctuation of students and researchers and technologies transferred last year might not be transferable anymore this year, due to key researchers changing work or studies. In industry, problems can become irrelevant. In summary, we observed the problem formulation not as a linear process where a problem leads to requirements, which are then converted into a specification and then to candidate solution (as Fig. 1 might suggest) but as a collaborative goal identification in which business goals have to be matched with the capabilities available in the project [14].

The candidate solution in our case was developed in an iterative manner, often leading back to the problem formulation since some solutions were not feasible due to hardware, networking, or storage constraints.

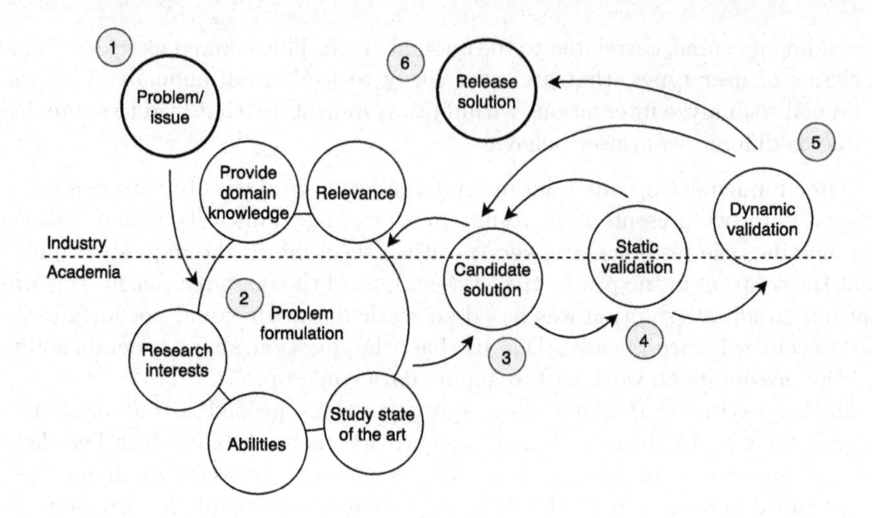

Fig. 2. Research approach and technology transfer process observed by us.

Since (due to the elevated costs) it was not feasible to have the electrical appliances produced by the company at the university, it was not possible to validate the solution in academia (step 4 in Fig. 1) and we validated the solution statically together with the company. At the dynamic validation step, we set up a DevOps [15] pipeline that executes the developed analyses and provides feedback to developers on a regular basis, based on the current software version. We conjecture that the approach of Fig. 2 more precisely (compared to Fig. 1) describes industry-academia collaborations since it adds different considerations that must be taken into account and, this is more specific to hardware related collaborations, it describes a situation in which it is difficult, if not impossible, to validate solutions within academia.

4.3 Lessons Learned

In this section, we summarize our lessons learned.

1. **About the problem definition**:
   - Defining the problem is an iterative task. Practical problems in industry are often a moving target, requiring problem formulation to be an iterative process. Through the continuous goal alignment between industry and academia, a higher match between target and solution can be achieved.
2. **About the collaboration approach**:
   - Understanding the business point of view is a necessary requirement for collaboration.
   - Face-to-face communication is key for a successful coordination.
   - A co-located researcher that is present on the industry site results in a more productive collaboration with less alienation of the researchers.

- Collaboration with domain experts is vital, as they provide the domain knowledge required to understand the use case and conduct a meaningful analysis.

3. **About the application of academic knowledge**:
   - Unfortunately adapting solutions from the literature is often not directly possible, as data used in academia is often sample data, varying from industry-data in structure and shape.
   - Use cases in the industry are unique, as is the data involved. This makes it important to analyze the existing dataset and perform an analysis specific to the use case.
   - Domain knowledge is essential, as it gives the data meaning, pointing an analysis in the desired direction.
   - Due to a continuous knowledge exchange between the researchers and developers, many developers acquired knowledge in the field of data science, allowing them to work with the topic and the results of our research.
4. **About the process of technology transfer**:
   - Practitioners often need to see the benefit of the application of scientific methods in their context to develop interest, which is a necessity to collaborate and to internalize the new knowledge.
   - Trust between industry and academia, especially concerning company-internal data, must be established to guarantee a positive working environment with a steady exchange between industry and academia.
   - Conveying the value of the research to management is critical, leading to the allocation of more resources to the topic.
   - Visualizations are paramount, they allow us to highlight key findings and transport a message to the stakeholders.
   - Conducting a study in collaboration with academia requires a company to commit resources. This might be in the form of developers supporting the researchers, managers taking part in meetings or tools being provided for the research. It is important that companies are aware of this before starting a research project, as research requires a constant interaction between academia and industry.

5 Conclusion

Collaboration between academia and industry in the field of software engineering is challenging but not impossible. During our study, we have learned to honor a humble approach towards the domain knowledge needed to learn to productively apply academic knowledge within an industrial context. Our industry partner has profited from the collaboration as we did, allowing them to change their software development process in a way that it is more data-driven and us to better understand how to develop and use innovative quality assurance techniques in industry.

References

1. Garousi, V., Felderer, M., Fernandes, J.M., Pfahl, D., Mäntylä, M.V.: Industry-academia collaborations in software engineering: an empirical analysis of challenges, patterns and anti-patterns in research projects. In: Proceedings of the 21st International Conference on Evaluation and Assessment in Software Engineering (2017)
2. Garousi, V., Varma, T.: A replicated survey of software testing practices in the Canadian province of Alberta: what has changed from 2004 to 2009? J. Syst. Softw. **83**(11), 2251–2262 (2010)
3. Wohlin, C.: Empirical software engineering research with industry: top 10 challenges. In: 2013 1st International Workshop on Conducting Empirical Studies in Industry (CESI). IEEE (2013)
4. Kaindl, H., et al.: Requirements engineering and technology transfer: obstacles, incentives and improvement agenda. Requirements Eng. **7**(3), 113–123 (2002)
5. Connor, A.M., Buchan, J., Petrova, K.: Bridging the research-practice gap in requirements engineering through effective teaching and peer learning. In: 2009 Sixth International Conference on Information Technology: New Generations. IEEE (2009)
6. Garousi, V., Petersen, K., Ozkan, B.: Challenges and best practices in industry-academia collaborations in software engineering: a systematic literature review. Inf. Softw. Technol. **79**, 106–127 (2016)
7. Gorschek, T., Garre, P., Larsson, S., Wohlin, C.: A model for technology transfer in practice. IEEE Softw. **23**, 12 (2006)
8. Sandberg, A., Pareto, L., Arts, T.: Agile collaborative research: action principles for industry-academia collaboration. IEEE Softw. **28**(4), 74–83 (2011)
9. Garousi, V., et al.: Characterizing industry-academia collaborations in software engineering: evidence from 101 projects. Empirical Softw. Eng. **24**(4), 2540–2602 (2019)
10. Beck, K., Andres, C.: Extreme Programming Explained: Embrace Change, 2nd edn. Addison-Wesley Professional, Boston (2004)
11. Gold, R.L.: Roles in sociological field observations. Soc. Forces **36**(3), 03 (1958)
12. Console, L., Torasso, P.: On the co-operation between abductive and temporal reasoning in medical diagnosis. Artif. Intell. Med. **3**(6), 291–311 (1991)
13. Few, S.: Information Dashboard Design: The Effective Visual Communication of Data. O'Reilly Series, O'Reilly Media Incorporated, Sebastopol (2006)
14. Schnabel, I., Pizka, M.: Goal-driven software development. In: 2006 30th Annual IEEE/NASA Software Engineering Workshop (2006)
15. Dyck, A., Penners, R., Lichter, H.: Towards definitions for release engineering and DevOps. In: 2015 IEEE/ACM 3rd International Workshop on Release Engineering (2015)

A Technology Transfer Portal to Promote Industry-Academia Collaboration in South-Tyrol

Roberto Confalonieri and Andrea Janes(✉)

Faculty of Computer Science, Free University of Bozen-Bolzano, 39100 Bolzano, Italy
{rconfalonieri,ajanes}@unibz.it
https://smart.inf.unibz.it

Abstract. Technology transfer is a complex and multifaceted activity whose main goal is to promote academic knowledge transfer from academia to industry. In this context, one of the most challenging parts of technology transfer activities is to inform stakeholders from the industry about the availability of academic results. Traditionally, this occurs through academic publications, and companies with a research department already use this knowledge source. Nonetheless, Small and Medium Enterprises (SMEs) do not often have the time or the resources to study and interpret results from academia. This paper describes a technology transfer Web portal that promotes technology transfer offers in a industry-friendly format. The portal aims at fostering innovation and collaboration between academia and industry.

Keywords: Industry-academia collaboration · Innovation model

1 Introduction

Technology transfer is the process of transfer and dissemination of in-situ academic developed technologies to industrial organisations [9]. Technology transfer has gained a lot interest in the last years. This is also due to the recently adopted R&D founding policies that promote industry-academic collaborations. For instance, the European Commission within the Horizon2020 and the new Horizon Europe frameworks have demanded academic and research institutions not only more applied research, but also a stronger commitment towards third-party founding and collaborations with industries. The term 'technology transfer' involves many activities such as, supporting academic practitioners in filing patents, helping them in the creation of spin-off and spin-out companies, and exploiting Intellectual Property through licensing, etc. Unfortunately, technology transfer is not easy to accomplish: the various stakeholders speak different languages, have different goals, and, therefore, they follow different approaches to achieve their objectives. It is often difficult to apply research results with ease, and, to enact processes of innovation [4]. Many companies need support to adopt the

G. Kotsis et al. (Eds.): DEXA 2022 Workshops, CCIS 1633, pp. 221–231, 2022.
https://doi.org/10.1007/978-3-031-14343-4_21

available research results, especially in the case of small and medium enterprises (SMEs).

Technology transfer activities do not only aim at promoting a close cooperation among the actors involved, but also at facilitating the co-design and co-development (including the design and participatory development) of IT applications whose innovation and complexity does not allow the independent realization by the various actors.

One of the most challenging parts of technology transfer activities is **to inform industry about the availability of academic results using an industry-friendly language**. Traditionally, this occurs through academic publications, and companies with a research department (capable of understanding academic papers) already use this knowledge source.

Unfortunately, the majority of the companies are SMEs (e.g., in Europe 99,8% [13]) that often do not have the time or the resources to study academic papers. It is in this setting that academia is called upon to do more to initiate collaborations with industry.

One way is to make academic research results more accessible, e.g., through the Web. Web portals for technology transfer enable technology providers to share their technologies and encourage their re-use, which would be of advantage both to the technology providers and the technology users who are looking for such a technology.

Addressing the challenge to initiate and conduct collaborations with industry is the main goal of the Smart Data Factory, a new applied research and technology transfer unit of the Faculty of Computer Science of the Free University of Bozen-Bolzano. In easing this process, a technology transfer portal of the Smart Data Factory was developed and will be presented in this paper.

The paper is organized as follows. In the next section, we describe the Smart Data Factory and its main goals. In Sect. 3, we overview related works and portals. Section 4 is devoted to the description of the portal's design and implementation. Section 5 presents the results of a usability test that we ran. Finally, Sect. 6 concludes the paper and outlines some improvements of the current version.

2 Smart Data Factory

The Smart Data Factory[1] (SDF) is an applied research and technology transfer unit of the Faculty of Computer Science[2] of the Free University of Bozen-Bolzano, located at the NOI Technology Park[3]. SDF's main goal is to foster industry-academia collaborations with an emphasis on the co-design and co-development of applications that focus on the intelligent use of data.

The mission of SDF is to promote technology transfer from academia to industry by implementing and adapting research results of the Faculty of Computer Science of the Free University of Bozen-Bolzano. This occurs through the

[1] https://smart.inf.unibz.it.
[2] https://www.inf.unibz.it.
[3] https://noi.bz.it.

transfer of advanced skills and competencies of researchers working at the Faculty of Computer Science, for providing innovative solutions to complex problems in the area of intelligent data management and in the co-design and co-development of IT applications, focusing on the intelligent use of data. Specifically, the goals of SDF are:

- To co-design and co-develop IT solutions focused on the intelligent use of data, whose innovation and complexity cannot be realised autonomously by the companies present in the territory.
- To collect experience in carrying out and facilitating industry-academia collaborations by collecting typical contracts, workflows, collaboration approaches, etc.
- To mediate between companies (located at the NOI Technology Park and outside) and the Faculty of Computer Science.

Whist technology transfer activities also encompass—as previously mentioned—patent filing, spin-off and spin-out creation etc., SDF only focuses on fostering industry-academia collaborations. Those activities will be in fact supported by other technology transfer managers at the university and at the NOI Technology Park.

3 State of the Art

There exist many papers about technology transfer models, strategies, or best practices, e.g., [1–3,9–11], but none of them discuss the possibility of such an exchange through on-line portals or platforms.

We also specifically searched for technology transfer and crowdfunding websites, looking for sites that illustrate concrete technologies that are the result of academic research or other forms of research cooperation in form of a market place, i.e., that go beyond a simple information page telling that cooperation is possible.

We identified five websites that we briefly describe: the web site of the Karlsruhe Institute of Technology, the platform "experiment", and the "technology offers" of the web sites of the Vienna University of Technology, the Graz University of Technology, and Wageningen University & Research.

The KIT - Research to Business[4] web site presents technologies and collaboration offers by the Karlsruhe Institute of Technology, Germany. The website uses a language that addresses businesses, e.g., the described technologies are not advertised as "research" but as"offers", which in German is even more surprising because it is unusual that a research institute is talking about "Angebote" (offers in German). The offers are grouped into categories and sub-categories, e.g., "Information" and "Automotive industry". Offers are described using one or more pictures, the description, categories, a contact person, related offers, related patents, related dissertations, and other additional material. Offers can

[4] https://www.kit-technology.de/.

be downloaded or added to a watch list. The page also offers the possibility to call a hotline for further information, follow the institute on various social media platforms and subscribe to a newsletter, which is published three times a year and informs about new offers and innovative projects. Similarly to the web site of the Karlsruhe Institute of Technology, the Vienna University of Technology[5], the Graz University of Technology[6], and Wageningen University & Research[7] created web sites to describe "technology offers". All three, but particularly the Vienna University of Technology, describe various technologies that might be interesting for companies describing the technology, potential applications, advantages, and contact information. The Vienna University of Technology also describes past collaborations[8], illustrating the project content, involved partners, the outcome, and contact information.

The web site "experiment"[9] is an online platform for "discovering, funding, and sharing scientific research [7]". Differently from the past platforms, "experiment" is a match-making web site to help researchers to describe research ideas and to find supporters that want to finance the research. The projects are organized into categories and described using pictures, videos, the involved scientists (that can be contacted directly), a description (describing context, significance and goals of the project), and the required budget (including the various cost categories). The web site describes itself as a "all-or-nothing funding platform" [6], meaning that backer's credit cards are only charged if a project reaches its funding target.

4 The Technology Transfer Portal of the SDF

During the instantiation of the laboratory of Smart Data Factory at the NOI TechPark, one of the first things we were concerned about was how the technology transfer should be enacted. As already mentioned, technology transfer is a complex activity, and we thought that having a series of reference activities—a sort of checkpoint list of what do to carry out technology transfer—could be useful for us. To this end, we designed a process according to which technology transfer can be enacted. The process is based on and extends the technology transfer model seen previously and foresees the following activities:

[5] https://www.tuwien.at/en/tu-wien/organisation/central-divisions/rti-support/research-and-transfer-support/technology-offers.
[6] https://www.wtz-sued.at/techoffer_tugraz/.
[7] https://www.wur.nl/en/value-creation-cooperation/collaborating-with-wur-1/entrepreneurship-at-wur/entrepreneurial-students/technology-offers-patents.htm.
[8] https://www.tuwien.at/en/tu-wien/organisation/central-divisions/rti-support/funding-support-and-industry-relations/industry-relations/support-for-businesses/best-practise-cooperations.
[9] https://experiment.com/.

1. **Contact:** SDF contacts the company or vice versa.
2. **Defining the problem to solve:** SDF presents the portfolio of transferable technologies to the company and/or the company presents projects for which it needs support. The outcome of this step is a problem description.
3. **Defining the team:** Based on the problem description, we will look for researchers in the Faculty of Computer Science that are interested to develop a solution to the presented problem. If no researcher is interested/able to collaborate, the project cannot be carried out. The outcome of this step (if the project is carried out) is the project team.
4. **Planning:** Once a project team is established, the company and faculty members define the technology transfer activities in detail. We foresee the following types of activities: *feasibility study*, *system design*, *prototyping*, or *training*. The outcome of this step is a project plan.
   - **Planning non-research activities:** Optionally, if non-research activities are part of the project, e.g., developing a web site, the faculty helps the company to find partner/s that carry out the activities.
   - **Apply for funding:** Optionally, if the company can obtain funding to carry out the project, the faculty might help the company to write a funding proposal.
5. **Preparation:** The collaboration contract is prepared by the university and signed by the company and university. This contract states the duration of the collaboration, the tasks, involved researchers, intellectual property rights, publication rights, and the compensation. The outcome of this step is the collaboration contract.
6. **Execution:** The technology transfer project is carried out by all partners. A typical approach for such a collaboration consists of the study of the state of the art by the faculty, the joint development of a candidate solution by the faculty and industry, and the validation in theory and in practice until a solution is found [9]. During the execution, the university issues one/partial invoice/s to the company, the company pays the invoice/s, and the money is used to cover the costs of the project.
7. **Closure:** The project is closed.

Since this process relies on a first contact with companies and stakeholders, we designed and implemented a technology transfer portal that introduces the activities carried out at SDF, and promotes a series of technology transfer activities that experts at the university can offer to companies and stakeholders. In the following, we describe the design requirements we stated and how these requirements were implemented.

4.1 Requirements

To decide the design of the technology transfer platform—incorporating the ideas we identified in the state of the art—we defined the following high-level requirements (following the categorization proposed by [12]).

- **Goal-level requirements:**
  1. The goal is to create a technology transfer platform to enable and promote collaboration between the Free University of Bozen-Bolzano and companies.
- **Domain-level requirements:**
  1. The Faculty of Computer Science shall have the possibility to present their technologies and thus help industry to find opportunities for collaboration.
  2. The Faculty of Computer Science shall have the possibility to present past projects to showcase examples of successful collaborations.
  3. The Faculty of Computer Science shall have the possibility to present information about faculty groups and their skills to guide companies to the right group if they have a specific problem in mind.
  4. The Smart Data Factory shall have the possibility to presents its process how industry can collaborate with academia to explain how such collaborations are carried out.
  5. Companies shall have the possibility to browse offers by category, type, or search them using text-based search to find the most adequate offer efficiently.
  6. Companies shall have the possibility to contact the Smart Data Factory to interact with its members or to send a new project proposal.

4.2 Implementation

This section explains the structure and the content of the created web site. Exports of the main pages of the web site are available at [5]. The main page of the *Smart Data Factory* website is subdivided into six sections: Offers, About, Our skills, Work with us, Team and Contact. These sections can be accessed either via the links on the top right corner of the page or by scrolling down. Given the multi-cultural characteristics of the Autonomous Province of Bozen-Bolzano, the content of the portal is offered in three languages, namely in English, German and Italian. The main page contains the following sections:

- **Offers and archive entries**: to view offers that are currently available or past projects, the visitor can either search by category or type, view a complete list of all offers, or search for a specific offer by clicking on the search button and entering a keyword into the pop-up window. Regardless of whether the offers were filtered or found through the search bar, they are displayed in a grid layout.
- The **About** section explains the idea behind the Smart Data Factory and its website.
- The section **Our Skills** lists the different groups within the Faculty of Computer Science. Detailed information about the group's mission, relevance and topics as well as key technologies, applications and contact details are provided when clicking on a group. Again, the layout of the group page remains the same as the offer page's layout.

- **Work with us** offers interested people the necessary information on how to collaborate with the Smart Data Factory. This section is split up into three subsections: information on collaborations for companies, job opportunities for data scientists and information on internships or diploma theses for students.
- **Team** is a short section about the employees currently working in the Smart Data Factory at the NOI Techpark.
- **Contact** contains contact details about the Smart Data Factory at the NOI Techpark.

To illustrate how an offer page is designed, the page for a *Recommender Widget*[10] will serve as an example (see Fig. 1).

On the left top side of the screen, a representative picture is displayed. This picture is either a picture of the project or a representative picture that should allow a visitor to memorize the offer. The first button beneath the picture enables the visitor to save this offer to a personal watch list or, if already added, remove it again. The second button evokes a pop-up window where the visitor can enter an email address to receive notifications of similar offers as well as newsletters. Below, two buttons allow visitors to directly jump back and forth between offers.

The first paragraph briefly explains what this offer is about. Along with the representative picture, this paragraph is key to a successful presentation of the offer. Both should help customers, regardless of their professional background, understand the process of this project.

Following the description paragraph, more detailed information is listed. It briefly explains what is needed from the customer in order to be able to perform the offered task along with what the customer will get. We consider these points as important and helpful, since the customer will know the requirements for each offer in advance.

Beneath the information and requirements paragraphs, contact details are listed. Name, email address and website of the professor working on this offer are displayed as well as a brief curriculum vitae. Where applicable, we created a diagram that displays the faculty member's most used keywords in their publications in the form of a bar chart. These were extracted from the most recent research papers written by the faculty member and indexed in *dblp*[11]. This diagram can be accessed through a link beneath the contact details.

Finally, the categories under which this offer can be found are listed, as well as its type. The type describes whether the offer is a feasibility study, a prototype, system design or training.

5 Evaluation

The evaluation was carried out through a questionnaire designed on the basis of the Unified Theory of Acceptance and Use of Technology (UTUAT3) model [8],

[10] https://smart.inf.unibz.it/en/offer/4.
[11] https://dblp.uni-trier.de/.

Offer: "Recommender Widget"

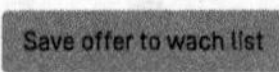

Description of this offer

We can integrate in the website of an e-Shop a widget for providing recommendations based on collaborative filtering (CF) and/or content-based (CB) filtering. On one hand, CF filtering can provide recommendations of the style "other people who have interacted with this product have also interacted with these products". On the other hand, CB filtering can provide recommendations of the style "similar products to those you have already interacted with, are these products". These recommendations can increase the website customers' visits and products' purchases.

What we need from you

The users, the products that they have interacted with and the timestamp of interaction.

What you will get

A widget that provides CF and CB recommendations.

Costs

The costs amount to 5.000€+VAT. You can also take advantage of the Lab bonus, which covers 50% to 65% of the incurred costs.

Contact for this offer

Panagiotis Symeonidis (✉ panagiotis.symeonidis@unibz.it, web site)
Panagiotis Symeonidis is a researcher with a fixed-term contract at the Faculty of Computer Science of the Free University of Bozen-Bolzano. Find out more about him checking out his website.

We have created a diagram that displays common keywords from articles written by Panagiotis Symeonidis. This is often useful to see what the respective researcher is doing: click here.

Categories

Artificial Intelligence, Machine learning

Type of offer

Prototyping

Fig. 1. Smart Data Factory - "Recommender Widget" offer.

particularly the "Effort expectancy" construct. This construct describes the degree of the technology's user-friendliness. Perceived ease of use and complexity are variables of this construct.

The questionnaire contained eight questions and seven tasks focusing on two specific points, namely, if the purpose of the site was understandable, and if the offers were easily accessible. Questions mainly targeted first impressions and overall feedback of respondents (e.g., "*What do you think is the purpose of this site?*"), while tasks aimed to understand if certain actions were easily solvable in the portal (e.g., "*Get more information about research groups and their offers.*"). We kept the duration of these interviews as short as possible, and we randomly selected five industry stakeholders at the NOI Techpark. According to Nielsen and Landauer [14,15] no more than five people are needed in a usability test. The questions were the following: 1) What do you think is the purpose of this site? 2) Who might this site be intended for? 3) Which options you see on the homepage and what do they do? 4) What catches your eye on the main offer page? 5) How would you describe the layout of a single offer page? 6) What do you think about the archive? 7) What do you think about the news section? 8) What other changes should be done to the website?

The tasks were the following: 1) Display all the offers that are currently on the website. 2) Find further information about people working on a specific offer. 3) Save an offer to your watch list. 4) Access your watch list. 5) Find the offer "Cloud migration". 6) Submit an collaboration idea to the Smart Data Factory team. 7) Get more information about research groups and their offers.

During the interviews, we took note of the answers to the questions and the reactions of the participants to the tasks. Overall, the responses received from the questionnaire were **positive** regarding the layout and features of the portal (e.g., "*Very nice and simple layout. The pictures are a great additive*"). Among major **issues**, respondents reported that the main page would benefit from a re-organization and reduction of the content that, at first glance, appeared overwhelming (e.g., "*There should be less text. Maybe redirect to other pages or structure it in a different way*"). Furthermore, the content in the main page should be more related to the scope of the portal (e.g., "*It would be better to have a couple of offers already visible on the homepage, rather than having to click around*"). As far as tasks are concerned, most of the respondents were able to solve the required task successfully. They also provided some feedback that will be taken into account in the design of the revised version of the portal. It was pointed out, for instance, that the search bar would be more efficient if it were possible to search by tags associated to existing offers and best practices (e.g., "*I tried to search for an offer by typing in its short form, but there were no results. Adding tags to offers that include their short forms would be helpful*").

6 Conclusion and Future Developments

Frequently, industries adhere to a closed innovation model, according to which they generate, develop, and market their own ideas in-house. In contrast, radical transformation and innovation models encourage a collaborative model involving industry and academia. Also from the point of view of academia, knowledge is not published in a way that makes it easy to be exploited to address societal challenges or transferred to industries. Technology transfer aims at blurring the borders between industries and academia, and facilitating the collaboration between these two stakeholders.

In this paper, we presented the technology transfer portal of the Smart Data Factory (SDF), a technology transfer centre promoting the transferable assets of the Faculty of Computer Science of the Free University of Bozen-Bolzano, and the knowledge transfer model that it is adopting.

In summary, the initial requirements that we had collected for the portal were successfully implemented. Due to its structure, the platform can easily be extended and modified. An example for the extension of the website would be the addition of other Faculties of the Free University of Bozen-Bolzano and their offers. Optionally, the platform could also be modified so that the website also allows crowdfunding.

The usability of the portal was evaluated through a questionnaire designed on the basis of the UTUAT3 model [8] "Effort expectancy" construct. With the help of the feedback received, it was possible to uncover shortcomings that will be revised in future versions. One of the major points is the amount of text throughout the website. On the main page, this could be resolved by redirecting the menu items to separate pages rather than leaving everything on the homepage. The amount of text inside the archive entries, group and offer pages could be visually reduced by creating collapsible content, where the text is displayed by clicking on the section title. Another point to be revised is the visibility of the offers. Since it is important that visitors understand the purpose of the portal, the offers need to be presented in the main page.

Furthermore, the concept of the archive and its distinction from the offers might have to be reconsidered in the future. It might not be clear to the visitors why we distinguished between past and future projects. This might complicate the search for a specific offer. On other web sites, e.g., the one of the Vienna University of Technology, archive items are called "best practice cooperations", which maybe conveys better the usefulness of the archive entries.

The portal is just one component in a wider strategy to support industry-academia collaborations. Even being only one aspect, it deserves attention because the portal achieves a good visibility. On average, the web site has 1300 visitors per month.

The major difficulty that occurred during the creation of the portal was not of technical nature, but lied in the formulation of offers for the industry. Academics perceive research in a different way than entrepreneurs: they prioritize publications rather than e.g., to develop a new product or to improve an existing one. We also noticed that some researchers are reluctant to formulate research in form of an activity with a defined duration and price since they are used to see research as a continuous exploration of a field. In this case, it revealed as helpful to define time-constrained activities, like a one-day workshop to brainstorm about a problem or a one-month feasibility study to investigate how a research project could be conducted, to elicit possible offers from researchers.

In the long term, we expect that SDF will increase the innovation potential and competitiveness of the region in the IC&T sector as well as the emerging "data science" through the creation of communities with diversified and innovative experiences at national and international level.

References

1. Arza, V., Carattoli, M.: Personal ties in university-industry linkages: a case-study from Argentina. J. Technol. Transf. **42**(4), 814–840 (2016). https://doi.org/10.1007/s10961-016-9544-x
2. Calcagnini, G., Favaretto, I.: Models of university technology transfer: analyses and policies. J. Technol. Transf. **41**(4), 655–660 (2015). https://doi.org/10.1007/s10961-015-9427-6
3. Cesaroni, F., Piccaluga, A.: The activities of university knowledge transfer offices: towards the third mission in Italy. J. Technol. Transf. **41**(4), 753–777 (2015). https://doi.org/10.1007/s10961-015-9401-3
4. Chesbrough, H.W.: The era of open innovation. MIT Sloan Manag. Rev. **44**(3), 35–41 (2003)
5. Confalonieri, R., Janes, A.: PDF exports of pages from the Smart Data Factory web site, April 2022. https://doi.org/10.5281/zenodo.6412535
6. experiment.com: How it works: The Public Funding Model. https://experiment.com/how-it-works. Accessed 3 Mar 2022
7. experiment.com: Website Description. https://experiment.com/. Accessed 3 Mar 2022
8. Farooq, M.S., et al.: Acceptance and use of lecture capture system (LCS) in executive business studies: extending UTAUT2. Interact. Technol. Smart Educ. **14**(4), 329–348 (2017)
9. Gorschek, T., Wohlin, C., Garre, P., Larsson, S.: A model for technology transfer in practice. IEEE Softw. **23**(6), 88–95 (2006)
10. Hobbs, K.G., Link, A.N., Scott, J.T.: Science and technology parks: an annotated and analytical literature review. J. Technol. Transf. **42**(4), 957–976 (2016). https://doi.org/10.1007/s10961-016-9522-3
11. Kumar, S., Luthra, S., Haleem, A., Mangla, S.K., Garg, D.: Identification and evaluation of critical factors to technology transfer using AHP approach. Int. Strateg. Manag. Rev. **3**(1), 24–42 (2015)
12. Lauesen, S.: Software Requirements: Styles and Techniques. Addison Wesley, Harlow (2002)
13. Muller, P., et al.: Annual report on European SMES 2020/2021 (2021). https://ec.europa.eu/docsroom/documents/46062. Accessed 4 Apr 2022
14. Nielsen, J.: Why You Only Need to Test with 5 Users, March 2000. https://www.nngroup.com/articles/why-you-only-need-to-test-with-5-users/. Accessed 4 Apr 2022
15. Nielsen, J., Landauer, T.K.: A mathematical model of the finding of usability problems. In: Proceedings of the INTERACT 1993 and CHI 1993 Conference on Human Factors in Computing Systems, pp. 206–213, April 1993

Fast and Automatic Object Registration for Human-Robot Collaboration in Industrial Manufacturing

Manuela Geiß[1(✉)], Martin Baresch[2], Georgios Chasparis[1], Edwin Schweiger[3], Nico Teringl[3], and Michael Zwick[1]

[1] Software Competence Center Hagenberg GmbH, Softwarepark 32a, 4232 Hagenberg, Austria
{manuela.geiss,georgios.chasparis,michael.zwick}@scch.at

[2] KEBA Group AG, Reindlstraße 51, 4040 Linz, Austria
bare@keba.com

[3] Danube Dynamics Embedded Solutions GmbH, Lastenstraße 38/12.OG, 4020 Linz, Austria
{edwin,nico}@danube-dynamics.at

Abstract. We present an end-to-end framework for fast retraining of object detection models in human-robot-collaboration. Our Faster R-CNN based setup covers the whole workflow of automatic image generation and labeling, model retraining on-site as well as inference on a FPGA edge device. The intervention of a human operator reduces to providing the new object together with its label and starting the training process. Moreover, we present a new loss, the intraspread-objectosphere loss, to tackle the problem of open world recognition. Though it fails to completely solve the problem, it significantly reduces the number of false positive detections of unknown objects.

Keywords: Automatic data labeling · Open world recognition · Human-robot collaboration · Object detection

1 Introduction

For small-lot industrial manufacturing and adaptability becomes a major factor in the era of Industry 4.0 [9]. In Smart Factories, individual production machines are equipped to an increasing degree with the ability to perform inference using machine learning models on low-resource edge devices [10]. To meet the promises of adaptability, there is a growing demand for (re-)training models on the edge in reasonable time without the need for a cloud infrastructure [2].

State-of-the-art (deep learning) models usually operate under a closed-world assumption, i.e., all classes required for prediction are known beforehand and included in the training data set. New classes not present during training are often falsely mapped to some class close to the unknown class in feature space. However, a robot that has to react to a changing working environment is often

G. Kotsis et al. (Eds.): DEXA 2022 Workshops, CCIS 1633, pp. 232–242, 2022.
https://doi.org/10.1007/978-3-031-14343-4_22

confronted with an open world setting (open world/open set [1,15]), where new classes not present in the data during model training need to be correctly recognized with as little delay as possible. In object detection (see the use case description in Sect. 2), unknown objects should not be recognized as some similar object class in the training set, instead the system should initiate an incremental training step to integrate the new object class into the model. The problem of open world object detection has only been tackled very recently in the scientific literature [6,17]. Moreover, during incremental training, automatic labeling of new training data is essential in order to avoid the time-consuming and expensive manual labeling process. In the current work, we present an end-to-end pipeline that addresses these objectives in the context of edge devices.

Section 2 describes the setting used by KEBA to demonstrate how to apply our pipeline in a real-world use case. The details about the used model architecture, data, and training setup are presented in Sect. 3. In Sect. 4, an overview of the necessary modifications to deploy the network on the FPGA-based AI module developed by Danube Dynamics is given. Section 5 introduces our approach for automatically generating training data for new objects, especially regarding automatic segmentation. In Sect. 6, we address the open world challenge in object detection by developing a new loss function that achieves some improvement.

2 Use Case

In human-robot collaborations (Cobots, [8]), a robot assisting a human operator has to adapt to a changing working environment constantly. In the context of object detection, this means to be able to quickly recognize new objects present in the shared working environment, without forgetting previous objects [11]. To ensure a seamless interaction between robot and human, the robot needs to learn objects quickly (i.e. integrate a new class into the model), ideally in less than one minute. In such a setting, the number and variety of images for training is usually limited. Gathering images for a new object class as well as the signal to start training on this object is initiated by the operator in our setup. In this work, we consider the use case of putting objects (e.g. fruits) into designated target baskets depending on the object class (Fig. 1). Initially objects of different classes are located in a single basket from which the robot picks individual objects.

In addition, the model has to run on low-power/low-cost edge-devices, which puts further constraints on the model architecture as well as the data set size.

3 Model and Training Workflow

3.1 Model Architecture

We compared different state-of-the-art object detection architectures, in particular one-stage vs. two-state architectures. In the context of learning new objects quickly, we found the one-stage approach Faster R-CNN [13] to be superior to the two-stage candidates SSD [7] and the YOLO family (e.g. YOLOv3 [12]). Though the one-stage architectures needed less training time per epoch than two-stage

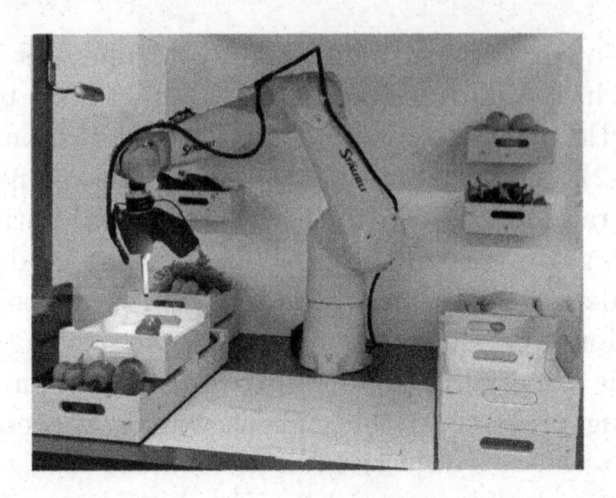

Fig. 1. Showcase of a robot sorting different types of fruits used to demonstrate fast retraining of new classes. The left basket holds a mix of fruits which are picked up by the robot and transferred to the correct basket on the right. For generation of training images, the new object is placed in the center of the white area (Source: KEBA).

methods, their overall training time was larger, while Faster R-CNN achieved good and robust results after a relatively small amount of training epochs. Due to the limitation in training time, we do not aim to train until the optimal point of high accuracy and low training loss but to stop training earlier without loosing too much in model performance. For this reason, we chose Faster R-CNN[1] with a VGG-16 backbone as the starting point for our initial framework, based on which the results of this paper have been derived. For more details on the architectures see also Fig. 2 and Sect. 4. Note, however, that in the meanwhile new YOLO architectures have been developed at rapid pace and latest versions such as YOLOX [4] are able to compete with Faster R-CNN in our application.

3.2 Data and Training Procedure

In our use case, we are given a collection of base objects on which we can train the model without time constraints. As a basis for this training, we use the pretrained weights from the Keras Applications Module[2] for the VGG-16 backbone. For the subsequent steps of incrementally learning new objects, that is learning the objects that are presented to the robot by a human operator, fast training is an important issue. It is essential that the training data of the new objects can be generated and labeled automatically, while training data for the base objects can also be labelled manually. The automatic generation of labeled training data is described in detail in Sect. 5. The base training data set consists of a collection of images, each containing either a single object or multiple objects, that contains each base object approximately 150–200 times. Images of the new object, in contrast, do only contain one instance of this new object each. We currently

[1] Code adapted from https://github.com/yhenon/keras-frcnn/ (last pulled 02/03/2020).

[2] https://github.com/fchollet/deep-learning-models/releases (pretrained on Imagenet).

use ~60 images (80%/20% training/validation) of the new object with different orientations but we found that also ~25 are sufficient to reach similar results. We currently use different collections of model fruits on a white background (Fig. 3 for an example), where, for training of the base model as well as testing, objects are presented in different combinations, angles, and distances but without being partly covered. Data augmentation methods such as flipping and rotation are used during training. Given these data sets, sufficient accuracy of the base training is achieved after 30–50 minutes on an NVIDIA GTX 1080 (8 GB RAM) graphics card. For the training of the new object, we merge the data set of the new and the "old" objects to avoid catastrophic forgetting [11]. We observed that, although the base objects are more frequent in the data set than the new objects, this imbalance does not negatively influence the training results. For learning new objects, we tried freezing different combinations of layers (mainly in the VGG-16 backbone but also other parts) but obtained worse results, both in accuracy and training time, compared to training the entire network architecture.

4 Inference on FPGA

Using deep neural networks on low-resource edge devices is one vision in the context of Industry 4.0. However, current frameworks only support inference but not training on the edge. In this project, we use the Deep Neural Network Development Kit (DNNDK)[3] from Xilinx to transfer our trained Faster R-CNN model to an FPGA. However, the DNNDK framework imposes technical limitations, e.g., it requires a 4-dimensional convolutional layer as input and cannot cope with Keras' Dropout and TimeDistributed layers. For this reason, the initial Faster R-CNN architecture had to be changed, mainly by replacing fully connected layers by convolutional layers. More precisely, the network's head now consists of one single convolutional layer with Softmax activation for classification and one convolutional layer with linear activation for bounding box regression (Fig. 2).

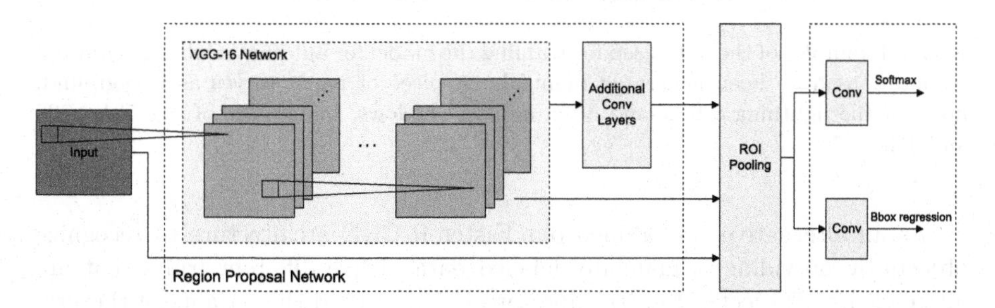

Fig. 2. Architecture of the used Faster R-CNN model with a VGG-16 backbone. The fully connected layers of the original Faster R-CNN architecture are replaced by convolutional layers to achieve compatibility with DNNDK. Figure adapted from [16].

[3] https://www.xilinx.com/support/documentation/user_guides/ug1327-dnndk-user-guide.pdf.

5 Automatic Generation of Images and Labels

For continuous and fast learning, it is crucial to automatically generate training data of the new objects to avoid time-consuming manual data labeling. This consists of two steps: taking pictures, and providing class labels as well as bounding box coordinates. In our setup, the human operator places the new object in a designated area and the robot moves around the object and takes about 30 pictures from different angles with a camera installed on the gripper. To also capture the object's back side, the human operator then flips the object and the robot takes a second round of pictures. The class label is provided by the human operator. For the extraction of the bounding box, we first tried traditional image segmentation tools such as a morphological transformation (opencv: morphologyEx()[4]) combined with an automatic threshold (scikit-image: local[5]). Despite the simplicity of our data with objects on white background, these methods did not succeed in sufficiently eliminating noise caused by lighting conditions (e.g. shadows). This heavily decreased the resulting accuracy of the bounding boxes.

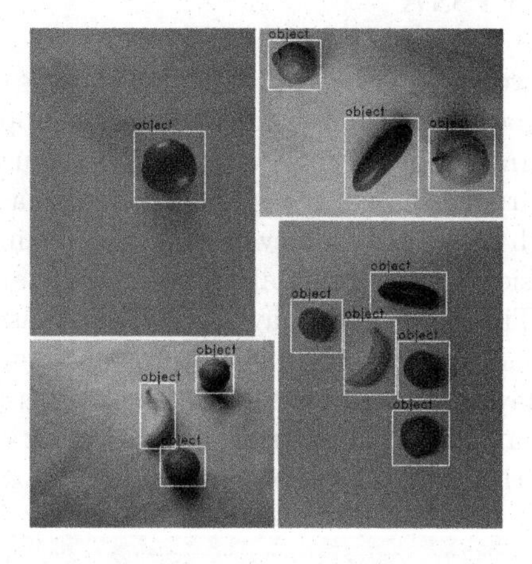

Fig. 3. Examples of the data used for training the model for automatic label generation. Annotations have been generated manually. A piece of paper serves as background, however the lightning differs and often induces shadows. Images are of size 531×708 and 432×576.

As an alternative, we trained our Faster R-CNN architecture to recognize objects by providing it manually labeled data of five different fruits that are all labeled as "object" (Fig. 3). More precisely, we used the weights of the pre-trained base model for object detection (Sect. 3.2) and only trained the classifier part of the Faster R-CNN architecture to identify objects. The model gave good results after only a few epochs of training, the results of the final model after

[4] https://docs.opencv.org/4.x/d9/d61/tutorial_py_morphological_ops.html.

[5] https://scikit-image.org/docs/dev/api/skimage.filters.html.

(a) **Inference results before and after post-processing**

(b) **Inference results on other backgrounds**

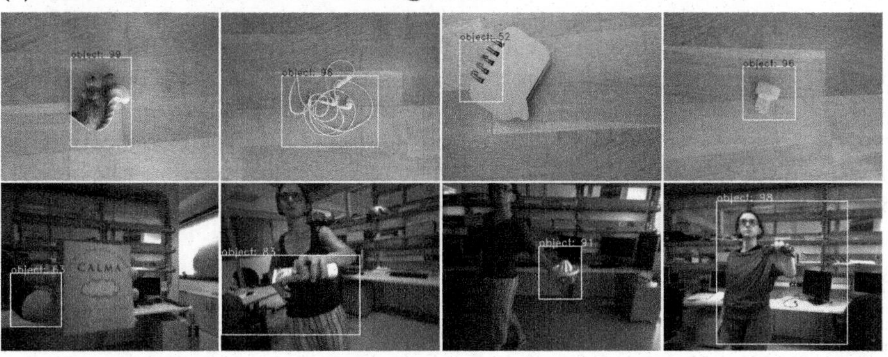

Fig. 4. Results of the automatic generation of labeled data. The model has been trained on five fruits on a white background. (a) The raw inference results are shown on the left, the post-processed results (merging bounding boxes and adding some slack) on the right. After post-processing, all objects are well detected with high scores. (b) Inference results (after post-processing) on data sets with different backgrounds. Images in the bottom row are taken from the iCubWorld data set [14].

200 epochs are shown in Fig. 4(a) on the left. The results are equally good for objects that have not been in the training data set. As can be seen in Fig. 4(a), in some cases (~50% in our setup) the bounding boxes are not ideal, that is, they either miss small parts of the object or multiple bounding boxes for different parts of the object are predicted. It is therefore beneficial to (i) merge multiple boxes into one and (ii) add some additional "slack", i.e., increase the box by a few pixels on each side. For our setup, the slightly enlarged bounding boxes do not pose a problem since the gripper takes care of fine-tuning when picking up single objects. As an interesting side effect, the model trained on white background is partly transferable to other homogeneous backgrounds (Fig. 4(b)). In summary, even though the Faster R-CNN based image labeling is slightly slower than the scikit-image based segmentation method (approx. 12%) and it initially needs manual data labeling and training of the segmentation model, it has the

significant advantage of being non-sensitive to reflections and therefore much more accurate.

Finally, we would like to emphasize that due to the Open World Problem (see Sect. 6) current state-of-the-art object detectors are not able to reliably distinguish between known and unknown objects. In our setup it is therefore only possible to automatically label images containing objects of a single class. In the specific case of our Faster R-CNN approach with the merge step during post-processing, this is even further restricted to one object per image. However, our experiments showed no significant performance difference between training images with one or multiple objects when learning a new class.

6 Open World Challenge: The Objectosphere Loss

One of the largest limitations of current deep learning architectures is the issue of open set recognition, more precisely the incorrect detection of unknown objects (Fig. 5, top right) that cannot be solved by simply thresholding the score functions. Inspired by the work of [3,5] in the context of open set classification, we developed the new *intraspread-objectosphere loss* function to reduce the number of false positive detections of unknown objects.

6.1 Methods

Given a set of known classes $\mathcal{C}$, we write $\mathcal{X}_k$ for the set of samples belonging to a known class and $\mathcal{X}_b$ for the samples of background resp. unknown classes. Following the observations in [3], we aim to enlarge the variety of the background class by including some unknown objects in the training data set and handling them as background. We emphasize, however, that this can barely cover the near infinite variety of the background space.

Our new intraspread-objectosphere loss consists of two parts. The first part corresponds to the *objectosphere loss*, developed by [3] for image classification. This loss aims at concentrating background/unknown classes in the "middle" of the embedding space (Eq. 1), which corresponds to a small magnitude of the feature vector, and put some distance margin between this region and known classes (Eq. 2). The compaction of unknown classes and background is achieved by the so-called *entropic open-set loss* that equally distributes the logit scores of an unknown input over all known classes as not to target one known class. This induces low feature magnitudes of these samples (see [3]). This loss is given by:

$$L_e(x) = \begin{cases} -\log S_c(x) & x \in \mathcal{X}_k \\ -\frac{1}{|\mathcal{C}|} \sum_{c \in \mathcal{C}} \log S_c(x) & x \in \mathcal{X}_b, \end{cases} \tag{1}$$

where $c \in \mathcal{C}$ is the class of x and $S_c(x) = \frac{e^{l_c(x)}}{\sum_{c' \in \mathcal{C}} e^{l_{c'}(x)}}$ is the softmax score with logit scores $l_c(x)$. To increase the distance of known classes and background, the entropic open-set loss is extended to the *objectosphere loss* which puts a so-called objectosphere of radius ξ around the background samples and penalizes all samples of known classes with feature magnitude inside this sphere:

$$L_o(x) = L_e(x) + \lambda_o \cdot \begin{cases} \max(\xi - ||F(x)||, 0)^2 & x \in \mathcal{X}_k \\ ||F(x)||^2 & x \in \mathcal{X}_b, \end{cases} \tag{2}$$

where $F(x)$ (with magnitude $||F(x)||$) is the feature representation of the last network layer's input. It has been shown in [3] that L_o is minimized for some input $x \in \mathcal{X}_b$ if and only if $||F(x)|| = 0$, which implies that the softmax scores are identical for all known classes if $x \in \mathcal{X}_b$.

For a better compaction of the known classes, we combine the objectosphere loss with a variant L_i of the *intraspread loss* (see [5]). This is tackled by using the mean feature vector of each class as the centroid of the cluster corresponding to this class. Given an input $x \in \mathcal{X}_k$ of class $c \in \mathcal{C}$, let μ_c be the mean feature vector of class c taken after the previous epoch. Then, we define L_i as

$$L_i(x) = \sum ||\mu_c - F(x)||. \tag{3}$$

The intraspread-objectosphere loss is then given as $L(x) = L_o(x) + \lambda_i \cdot L_i(x)$.

6.2 Results and Discussion

In these experiments, we focus on training the base model, again using pretrained weights from the Keras Application Module. Our training data set contains the three base classes apple, tomato, and lime as well as 20 different unknown objects (two images each) that are added to the background class. The test data set consists of 59 images containing known objects (tomato: 35x, apple: 24x, lemon: 31x) as well as two other unknown objects on white background (ficus: 18x, kiwi: 13x) that have not been in the training data set.

Using the classical Faster R-CNN loss, ficus and kiwi appear mostly as false positive detections with high score (¿80%), mainly classified as "apple" (Fig. 5, top left). Results of the whole test data set are visualized as t-SNE embedding in Fig. 5 (top right). Using the objectosphere loss reduces the number of false positives, especially for kiwi. This appears in the t-SNE embedding as less detected samples of ficus/kiwi (Fig. 5, bottom left). In addition, there is a tendency of better separation of the classes compared to the classical loss. The intraspread-objectsphere loss further improves this separation and further reduces the detection of false positives (Fig. 5, bottom right). Kiwi is detected only once and ficus is detected less often and if so, then the score is rarely above 90%.

Hence, the intraspread-objectosphere loss clearly improves the issue of false positive detection in the context of open set recognition. However, it fails in resolving it completely.

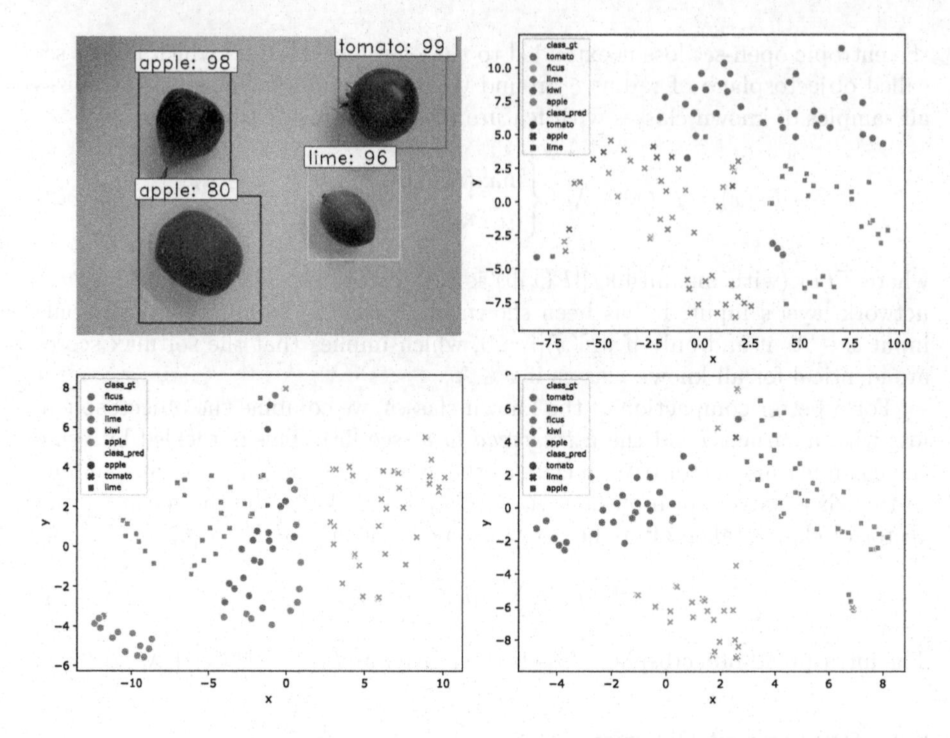

Fig. 5. *Top left:* Inference example after training on apple, tomato, and lime, and without the intraspread-objectosphere loss. Ficus and kiwi are mostly classified as "apple" with high score. *Top right:* t-SNE embedding showing the results of the whole test data set for the experiment on the left after 30 training epochs. Colors indicate the ground truth class, symbols show the predicted class. *Bottom left:* t-SNE embedding for the objectosphere loss after 30 epochs, $\xi = 300$, $\lambda_o = 10^{-4}$. *Bottom right:* t-SNE embedding for the intraspread-objectosphere loss after 30 epochs, $\xi = 300$, $\lambda_o = 10^{-4}$, $\lambda_i = 10^{-2}$.

7 Summary

We presented an end-to-end framework for fast retraining of object detection models in human-robot-collaboration. It is increasingly important for machine learning models to adapt to changing environments as the integration of prediction models into the workflow of human operators (human-in-the-loop, decision support) in industrial settings advances with rapid speed. Our setup covers the whole process of retraining a model on-site, including automatic data labeling and inference on FPGA edge-devices, however, some limitations still exist:

- In order to obtain acceptable results with respect to fast retraining as well as automatic data labeling, we limit ourselves to homogeneous backgrounds. Homogeneous backgrounds are often encountered in industrial manufacturing, however, it currently limits the applicability of our approach.

- For our automatic labeling approach to work, new classes have to be trained using single-instance images only, i.e., each training image can contain only one object of one class.
- One of the biggest challenges is learning in an open world, as closed world learning is still the standard for state-of-the-art object detection architectures using benchmark data sets. We introduced a new loss, the intraspread-objectosphere loss, that achieves a reduction of false positive detections but fails to completely solve the problem. It remains to compare the performance of our method to other simultaneously developed approaches (e.g. [6,17]). In particular, the intraspread loss shows similarities to the contrastive learning approach used by the `ORE` method [6].
- With our choice of architecture (Faster R-CNN), learning on edge devices is not yet feasible.

We are currently working on porting the whole framework to a new architecture (YOLOX [4]). Development of one-stage object detection architectures has been very fast paced, especially the YOLO family. By using a less resource-intensive architecture, training on low-power edge devices should become more feasible in the future. Due to time constraints during training, we have not investigated model pruning yet. However, with our current work on newer, improved YOLO architectures as well as the recent development on pruning methods, this aspect is an interesting topic for further research to further increase inference performance w.r.t. time on edge devices. Finally, the optimal training time (i.e. number of epochs) is heavily dependent on the use case and its data. Moreover, a new indicator for early stopping with regard to new/old objects would increase the usefulness of retraining in a collaboration environment, as standard methods like validation sets are not robust enough when training only a few epochs.

Acknowledgements. The research reported in this paper has been funded by BMK, BMDW, and the State of Upper Austria in the frame of SCCH, part of the COMET Programme managed by FFG.

References

1. Bendale, A., Boult, T.: Towards open world recognition. In: Proceedings of IEEE Conference on Computer Vision Pattern Recognition, pp. 1893–1902 (2015)
2. Chen, J., Ran, X.: Deep learning with edge computing: a review. Proc. IEEE **107**(8), 1655–1674 (2019)
3. Dhamija, A.R., Günther, M., Boult, T.: Reducing network agnostophobia. Adv. Neural Inf. Process. Syst. **31**, 9157–9168 (2018)
4. Ge, Z., Liu, S., Wang, F., Li, Z., Sun, J.: YOLOX: exceeding YOLO series in 2021. arXiv preprint arXiv:2107.08430 (2021)
5. Hassen, M., Chan, P.K.: Learning a neural-network-based representation for open set recognition. In: Proceedings of SIAM International Conference on Data Mining, pp. 154–162. SIAM (2020)
6. Joseph, K., Khan, S., Khan, F.S., Balasubramanian, V.N.: Towards open world object detection. In: Proceedings of IEEE Conference on Computer Vision Pattern Recognition, pp. 5830–5840 (2021)

7. Liu, W., et al.: SSD: single shot MultiBox detector. In: Leibe, B., Matas, J., Sebe, N., Welling, M. (eds.) ECCV 2016. LNCS, vol. 9905, pp. 21–37. Springer, Cham (2016). https://doi.org/10.1007/978-3-319-46448-0_2
8. Matheson, E., Minto, R., Zampieri, E., Faccio, M., Rosati, G.: Human-robot collaboration in manufacturing applications: a review. Robotics **8**, 100 (2019)
9. Mehrabi, M., Ulsoy, A., Koren, Y.: Reconfigurable manufacturing systems: key to future manufacturing. J. Intell. Manuf. **11**, 403–419 (2000)
10. Murshed, M.G.S., Murphy, C., Hou, D., Khan, N., Ananthanarayanan, G., Hussain, F.: Machine learning at the network edge: a survey. ACM Comput. Surv. **54**(8), 1–37 (2021)
11. Ramasesh, V.V., Dyer, E., Raghu, M.: Anatomy of catastrophic forgetting: hidden representations and task semantics. In: International Conference on Learning Representation (2021)
12. Redmon, J., Farhadi, A.: YOLOV3: an incremental improvement. arXiv preprint arXiv:1804.02767 (2018)
13. Ren, S., He, K., Girshick, R., Sun, J.: Faster R-CNN: towards real-time object detection with region proposal networks. In: Advanced Neural Information Processing Systems, vol. 28 (2015)
14. Ryan Fanello, S., et al.: iCuB world: friendly robots help building good vision datasets. In: Proceedings of IEEE Conference on Computer Vision Pattern Recognition Workshops (2013)
15. Scheirer, W.J., de Rezende Rocha, A., Sapkota, A., Boult, T.E.: Toward open set recognition. IEEE Trans. Pattern Anal. Mach. Intell. **35**(7), 1757–1772 (2012)
16. Zhao, X., Li, W., Zhang, Y., Gulliver, T.A., Chang, S., Feng, Z.: A faster RCNN-based pedestrian detection system. In: VTC2016-Fall, pp. 1–5 (2016)
17. Zhao, X., Liu, X., Shen, Y., Ma, Y., Qiao, Y., Wang, D.: Revisiting open world object detection. arXiv preprint arXiv:2201.00471 (2022)

Distributed Ledgers and Related Technologies

Sending Spies as Insurance Against Bitcoin Pool Mining Block Withholding Attacks

Isamu Okada[1,2], Hannelore De Silva[1], and Krzysztof Paruch[1(✉)]

[1] Research Institute for Cryptoeconomics, Vienna University of Economics, Welthandelsplatz 1, 1020 Vienna, Austria
krzysztof.paruch@wu.ac.at

[2] Faculty of Business Administration, Soka University, 1-236 Tangi, Hachioji, Tokyo 192-8577, Japan
https://www.wu.ac.at/en/cryptoeconomics, https://www.soka.ac.jp/

Abstract. Theoretical studies show that a block withholding attack is a considerable weakness of pool mining in Proof-of-Work consensus networks. Several defense mechanisms against the attack have been proposed in the past with a novel approach of sending sensors suggested by Lee and Kim in 2019. In this work we extend their approach by including mutual attacks of multiple pools as well as a deposit system for miners forming a pool. In our analysis we show that block withholding attacks can be made economically irrational when miners joining a pool are required to provide deposits to participate which can be confiscated in case of malicious behavior. We investigate minimal thresholds and optimal deposit requirements for various scenarios and conclude that this defense mechanism is only successful, when collected deposits are not redistributed to the miners.

Keywords: Blockchain · Bitcoin · Proof-of-Work · Pool mining · Block withholding attack · Game theory · Agent-based simulation

1 Introduction

1.1 Bitcoin, Pool Mining and Block Withholding Attacks

The Bitcoin network relies on the Proof-of-Work (PoW) consensus mechanism for its security where computational power provided by miners which are economically incentivized to participate to achieve a collectively agreed state of the system is utilized. Since reward mechanisms are stochastic agents aim to reduce their payout volatility by forming mining pools orchestrated by pool managers. All pool earnings are equally distributed among miners depending on their contributions allowing them to create a stable source of income with identical expected payoff but much lower variance compared to the solo mining case.

G. Kotsis et al. (Eds.): DEXA 2022 Workshops, CCIS 1633, pp. 245–257, 2022.
https://doi.org/10.1007/978-3-031-14343-4_23

In a pool mining system multiple miners collaboratively find a *Nonce* number satisfying the following condition: $h(Merkle + h(PreBlock) + Nonce) < D$ where $h(\cdot)$ is a hash function, *Merkle* refers to the merkle root, *PreBlock* indicates the Block ID, and D is a threshold which indicates the computation difficulty. An eligible *Nonce* number is a full solution of Proof-of-Work (fPoW). Miners in a pool agree to share their fPoW reward with all other participating miners in the pool as the merkle root includes a coinbase transaction which funnels rewards to the pool manager who distributes them among all miners after taking a management fee.

An attack vector in form of the block withholding attack (BWA) (Rosenfeld (2011)), (Zhu et al. (2018)), (Eyal and Sirer (2018)) is a serious threat to the pool mining approach and with this the whole legitimacy of the PoW consensus. In this attack the miner who finds a fPoW does not report it to the pool manager - either wanting to monopolize the reward for themselves or operates within another pool and aims to harm their competitors. For provability of provided efforts a pool mining system adopts an indirect index to ensure miners do not withhold correct solutions. This is accomplished via reporting of partial solutions of Proof-of-Work (pPoW) which solve the hash equation above for a D' with $D' >> D$. Since pPoW are much easier to find they provide a statistically corresponding signature for the contributed efforts over time. A pool manager can reward his miners in relation to their pPoW solutions and keep the fPoW rewards for himself.

1.2 Literature Review and Previous Approaches

(Rosenfeld (2011)) is one of the first to analyse bitcoin pooled mining reward systems. He introduces terminology in the comparison of solo mining versus pool mining techniques and describes the main benefit of pool mining in reducing earnings variance. He treats different reward systems and associated score-based methods for reward distributions and outlines potential attack vectors *pool hopping* and *block withholding* for one malicious actor. Rosenfeld proposes two solutions which of one is a *pop quiz* that identifies not truthfully reporting agents and the other one is *oblivious shares* which necessitates a change in the protocol logic.

(Eyal (2013)) formally shows that the Bitcoin mining protocol is not incentive-compatible as it allows for colluding miners to obtain a bigger reward than their share of contribution to the network. He captures mathematically the logic behind the BWA and further shows this design will lead to the formation of centralized colluding entities: the selfish mining pools. Eyal suggested solution requires an alternation of the protocol to resolve these threats.

(Luu et al. (2015)) use game-theoretic methodology to assess BWA in the context of Bitcoin mining pools. Their model allows to treat mining decisions as computational power splitting games where agents distribute their resources among competing pools conditioned upon the rewards they offer. The authors are able to show that selfish mining is always favorable in the long run, that optimal behavior is a stochastic mixed strategy and thus the network in total always wastes resources for competition. In addition the authors introduce seven desired properties of countermeasures and evaluate a change in reward payoff under these considerations.

(Eyal (2015)) extends the previous research from one-miner attacks to collaborative attacks performed by malicious pool managers grouping their participants to infiltrate other pools. Eyal introduces a model of the pool game where a number of pools can attack each other for varying pool sizes and attack strategies. This results in the Miner's Dilemma where similar to the iterative prisoner's dilemma pool managers decide continuously whether to attack the competitor. The strategy not to attack is dominated but leads to long-term highest gains if cooperation is established. Eyal suggest two countermeasures which of one goes in line with Rosenfeld's proposal to change the protocol and establish a honeypot technique to expose attackers and the other one claims that in closed pools no block withholding is possible.

(Lee and Kim (2019)) discuss available countermeasures to BWA and categorize them according to properties introduced by (Luu et al. (2015)). Moreover they propose an additional countermeasure that does not require a change in protocol and performs better with respect to the other characteristics. Their suggested method includes sending sensors to competing pools to be able to detect infiltrations and punish attackers. The authors consider a model with one attacking pool and one defending pool and asymmetric behavior.

1.3 Contribution of This Paper

We theoretically explore effective protocol defensive methods against BWA when assuming rational pool managers by developing an agent based framework. We build our analysis on top of the model introduced by (Eyal (2015)) and extend the defending method suggested by (Lee and Kim (2019)) by incorporating both pools capabilities to attack and send sensors to their competitors. After our model confirms existing results additional improvements are suggested: the effects of punishment and deposit on the agent behaviors.

The model is developed in four versions. The first includes two mutual attackers (Model 1) and is able to replicate Eyal's miner's dilemma situation. This is extended to a model applying mutual punishment (Model 2) using Lee and Kim's (2019) idea. The next model introduces a deposit system for participation in a pool (Model 3). The final model shows the effects of collected deposits redistribution (Model 4). All models are supported with numerical experiments.

In all models two pool managers adopt the Pay Per Latest N Share method and maximize their expected payoffs by adjusting their strategies on how to infiltrate the other pool with spies. The analysis of more complex management compensation schemes which could also impact agent's decisions is set aside for future research. With this paper we contribute to the existing discussion on the topic of pool mining and BWA as outlined above.

At first we solve each model version analytically for Nash equilibria of the two managers by calculating their best responses to finally derive best response strategies from numerical simulations.

2 Results

2.1 Model Definition

We consider a model of two pools A and B and assume respective relative mining powers within the whole network α and β, where $0 < \alpha + \beta \leq 1$. With m the total number of mining tasks $m\alpha$ tasks are calculated in Pool A while $m\beta$ tasks in B. Each miner joining in either Pool A or B is assigned one task of same size resulting in *one task per miner*. We conceptualize the possibility of more powerful miners which occur in reality as if virtually dividing them into an accumulation of miners with the capability of calculating one task only. The consequences of this simplification is left for further research.

We further allow both pools to send spies to each other. We define x and y as the fraction of miners which are spies meaning that A sends $s_A \equiv m\alpha x$ spies to B while B sends $s_B \equiv m\beta y$ spies to A. The process of spying is as follows. Manager A generates s_A miners and sends them to B where they are being assigned s_A tasks by manager of B. Then the manager of A reassigns these s_A tasks to *non-spy* s_A miners in his pool. Note that the manager never calculates tasks directly as this is delegated to *non-spy* miners who report to him. Finally manager A reports selected results to manager B to receive mining rewards. Manager B mirrors the behavior of manager A.

It is each pool manager's choice which results to forward to the other pool, therefore with malicious intent no fPoW and all pPow solutions are submitted. This allows to receive mining rewards corresponding to pPoW solutions. The manager of B cannot receive any fPoW solution of s_A tasks which reduces his the probability to win. Such behavior is defined as a BWA by A to B.

Eyal (2015) showed that BWA bears the miner's dilemma. Individually the attack increases the relative winning probability by decreasing the victim pool's chances. However, if both managers attack each other both winning probabilities decrease which corresponds to the prisoners' dilemma.

In our approach several versions of a model and associated metrics were developed to measure the pool's performance and to quantify the manager's and miner's incentives.

2.2 Model 1: Confirming the Miner's Dilemma

Following Eyal (2015) we use the three metrics direct revenue, revenue density and efficiency. The direct revenue indicates the winning probability of a mining race in the whole network and the revenue density indicates the expected reward of each miner. With m as total network task number the revenue density is $1/m$ if there are no attacks. The efficiency defining the revenue density multiplied by m reflects the attractiveness of a pool as miners have stronger incentives to join a pool when the efficiency is high. An efficiency of one is considered to be normal attractive for miners to join. Therefore a pool manager aims to maximize its efficiency.

Definition 1. *The* ***direct revenues*** *of pool A and B denoted by D_A and D_B respectively are given as $D_A = \frac{m\alpha - m\alpha x}{m - m\alpha x - m\beta y}$ and $D_B = \frac{m\beta - m\beta y}{m - m\alpha x - m\beta y}$.*

Definition 2. *The* ***revenue density*** *of Pool A and B denoted by R_A and R_B respectively are given as $R_A = \frac{D_A + m\alpha x R_B}{m\alpha + m\beta y}$ and $R_B = \frac{D_B + m\beta y R_A}{m\beta + m\alpha x}$.*

Definition 3. *The* ***efficiency*** *of Pool A and B denoted by E_A and E_B respectively are given as $E_A = mR_A$ and $E_B = mR_B$.*

Using these metrics we prove two theorems that allow us to conclude that there are strong incentives to attack competing pools as a pool manager. This holds in both cases when the counter party attacks or not as an attack always increases the efficiency regardless of the other managers strategy. We show that the efficiencies of both pools are equal to one if there are no attacks. In case a pool attacks the other pool has the incentive to retaliate to increase its relative efficiency. Proofs of Theorems are provided in the Appendix.

Theorem 1. *In the case that there are no attacks in the network it follows that the efficiency of both pools equals to one: $x = y = 0 \Rightarrow E_A = E_B = 1$*

Theorem 2. *In the case that one pool attacks the other it follows that the pool being attacked has an incentive to attack back to prevent having a lower efficiency: $x > 0 \wedge y = 0$ implies $E_A > E_B \wedge E_B < 1$. The efficiency of pool A is greater than one if the fraction x of spies is below a threshold: $E_A > 1 \iff x < \frac{\beta}{1-\alpha}$*

This theorem implies pools efficiency suffers from attacks causing it to lose miners and an attacker can increase their attractiveness above 1. Since these implications influence each individual pool managers in aggregate the miner's dilemma will follow as proposed by Eyal (2015).

In a situation where both pool managers send spies to each other $x > 0$ and $y > 0$. To be more attractive for miners each pool manager wants to maximize their efficiency. Therefore w.l.o.g. manager A is looking for the optimal x which maximizes E_A for given α, β, y and m and manager B makes analogous decisions. This idea is based on the best response analysis where $x^{BR}(y) = x_{E_A(y|\alpha,\beta,m)}$ is the best response for manager A while $y^{BR}(x) = y_{E_B(x|\alpha,\beta,m)}$ is the best response for manager B. If $(x, y) = (x^*, y^*)$ satisfies $(x^*, y^*) = (x^{BR}(y^*), y^{BR}(x^*))$ then (x^*, y^*) is a Nash equilibrium. We defer from analytical treatment in this case and show the numerical results of best responses in Fig. 1 for all four versions of the model: standard two spy model, model with punishment, model with deposit and model with deposit redistribution.

Figure 1a shows that due to the miner's dilemma the efficiency of a pool manager is always below 1 even if best responses are selected. Furthermore the figure shows that there is an incentive to conduct a BWA, especially as a small pool - therefore managers of big pools have to consider suitable defences against BWA. If there is a pool with a notable share of mining power in the network the losses incurred by all pools due to the miner's dilemma are significant - and this effect is stronger the bigger the largest pool is. Consequently all managers observe the biggest pools of the network and aim to decrease their power.

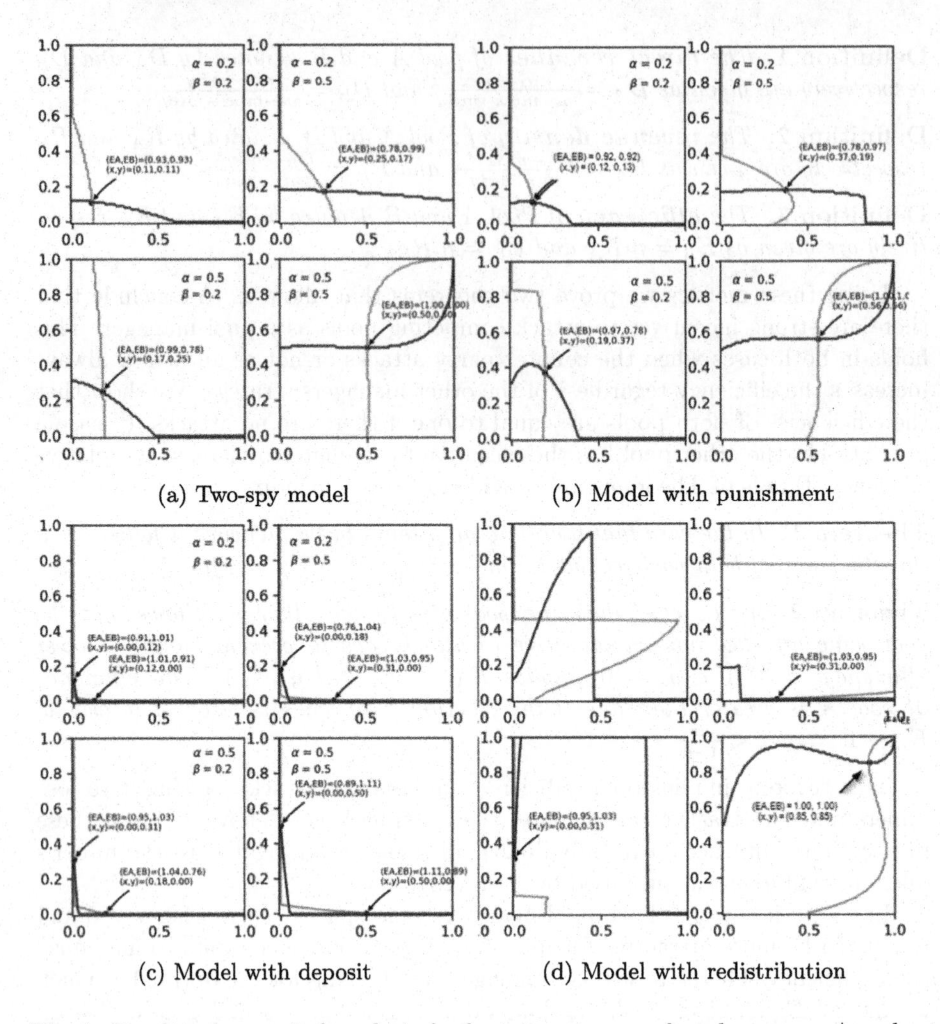

(a) Two-spy model

(b) Model with punishment

(c) Model with deposit

(d) Model with redistribution

Fig. 1. Results of numerical analysis for best responses x of pool manager A and y for pool manager B are shown in red and blue respectively where x is depicted on the horizontal axis and y on the vertical axis in each panel. The Nash equilibrium is the intersection of best response lines where both efficiency values and both efficiency values E_A and E_B are shown. We set $m = 10,000,000$ and $f = 10$. (Color figure online)

2.3 Model 2: Punishment System

To prevent the BWA, Lee and Kim (2019) proposed sending spies. In this strategy the manager of A send s_A miners as spies to B, collects some open tasks and assigns those s_A tasks to miners joining A. However these tasks are branded with information of the Coinbase Transaction which exposes their real affiliation to the respective pool and they can be therefore identified and correctly categorized by any miner. If spies are sent to A by manager B they would be able to detect the attack - if these individuals are assigned the tasks with an inappropriate Coinbase transaction. As a consequence manager B can sanction pool A by not awarding any rewards for pPoW solutions if he discovers to be attacked. The probability to discover the attack can be calculated by an application of a combination problem: For the expected value of d_{AB} of A's spies detected by the manager of pool B d_{AB} it holds $d_{AB} = \sum_{k=0}^{m\beta y} k \frac{{}_{m\alpha x}C_k \times {}_{m\beta}C_{m\beta y-k}}{{}_{m\beta+m\alpha x}C_{m\beta y}}$. The solution to this equation can be approximated by $d_{AB} = \frac{m\alpha\beta xy}{\alpha x+\beta}$ and in similar fashion $d_{BA} = \frac{m\alpha\beta xy}{\alpha+\beta y}$ to meet the requirements for this paper.

(Lee and Kim (2019)) proposed an approach which does provide improvements to previous models as it describes the one-attacker case but it does not cover the situation where both pools send spies to each other. It might very well be the case that pool manager A himself considers being attacked if his s_A miners find a task consisting of information on Coinbase Transactions including Pool B. The contribution of this paper is to extend Lee and Kim (2019)'s model to contain the two-attacker case and further to allow to assess both manager's strategies.

In comparison to the first model presented in the previous section the extended model of this section only differs in the definitions of equations for the revenue densities R_A and R_B. Their numerators and denominators now include the terms for the expected detection rates d_{AB} and d_{BA} respectively which are subtracted from the mining tasks attributed to spying $m\alpha x$ and $m\beta y$ accounting for the fact that the revenue density of one pool is reduced by these terms if tasks are assigned to spies: $R_A = \frac{D_A+(m\alpha x-d_{AB})R_B}{m\alpha+m\beta y-d_{BA}}$ and $R_B = \frac{D_B+(m\beta y-d_{BA})R_A}{m\beta+m\alpha x-d_{AB}}$

Figure 1b shows the results of numerical analysis of best responses and Nash equilibria replicating the experiments presented in Fig. 1a. A comparison of both figures allows to conclude that a punishment scheme as defined by the second model performs worse when confronted with a BWA. Contrasting the numerical results for e.g. parameter values of $(\alpha, \beta) = (20\%, 20\%)$ reveals that in the former model pool efficiencies E_A and E_B are reduced to 93% when a fraction of 11% spies are sent while in the latter model both managers send 13% spies reducing the efficiency to 92%. The effects of the miner's dilemma are magnified in the model with punishment even more when considering pools of different sizes so that for e.g. $(\alpha, \beta) = (20\%, 50\%)$ the fraction of spies increases from 25% to 37% when comparing both models. Consequently the punishment system of mutually sending spies does not provide a satisfactory countermeasure to prevent BWH attacks.

2.4 Model 3: Deposit System

To overcome the shortcomings of the model with punishment we propose a deposit system. Each new miner who wants to join a pool is required by the pool manager to pledge a certain amount that lapses in the case of attack detection. Ceased amounts are attributed to the manager's income and are not redistributed to other pool members. If no misbehavior is ever discovered the deposit is fully returned to the initial miner when leaving the pool. The model of this system includes an additional parameter f representing a deposit unit which is included in the definitions of revenue densities. This value satisfies that (deposit fee) $= f \times$ (expected reward unit) $= f/m$ and is equivalent to the expected reward of a task for $f = 1$ in the case of no attack: $R_A = \frac{D_A+(m\alpha x-d_{AB})R_B-d_{AB}f/m}{m\alpha+m\beta y-d_{BA}}$ and $R_B = \frac{D_B+(m\beta y-d_{BA})R_A-d_{BA}f/m}{m\beta+m\alpha x-d_{AB}}$

Numerical results of the model with deposit system are shown in Fig. 1c where the miner's dilemma appears to be resolved. A comparison of the panels for $(\alpha, \beta) = (50\%, 50\%)$ in Fig. 1a and Fig. 1c yields the conclusion that for the deposit system there no incentives to attack if the ratio of the other manager's spies exceeds a certain threshold. The best responses of both managers spy ratio indicated by the blue and red lines vanish almost everywhere in Fig. 1c in contrast to the model without punishment in Fig. 1a. Below a certain threshold of foreign spies however the incentive to attack becomes very large. Smaller pools are even less attractive to attack as the critical attacking threshold gets lower see e.g. $(\alpha, \beta) = (50\%, 20\%)$. This fact suggests that sending a small number of spies can serve as a form of insurance against potential attacks by the competing pool manager. In an environment with a deposit system mutual determent against BWA can be realized when all pools send a minimal amount of spies to each other. This result would not hold true without the deposit as this would even decrease the Pareto efficiency.

To find the optimal size of the required deposit for the insurance mechanism to work we refer to Fig. 2a where thresholds of the deposit unit are shown. A comparison of the various panels indicates that the threshold does not depend on the fraction of spies sent by the competing manager. This is consistent with the previous discussion because if a sufficient deposit unit is required, a pool manager who intends to do the BWA loses one's economic incentive to do so. When the calculating power of a pool is relatively large (the case of $\alpha = 30\%$), if the deposit unit sets $f = 6$, about $x = 5\%$ is needed for a pool with the same power. If the power of an opponent gets small, the threshold of f drastically decreases.

In reality a pool manager never knows where a joining miner comes from and thus cannot change the deposit unit depending on a miner, rather it must be set depending on the most powerful pool. Although the relative mining power shares fluctuate on a daily basis they never change drastically. Therefore the optimal deposit requirement should depend on the biggest pool share. With no deposit, Fig. 1b shows that two big pools have strong economic incentives to do the BWH attack with many spies. Figure 2b shows the suitable deposit for this case, e.g. when the biggest pool in the network has a relative share of 25% and the insurance spies are $x^* = 3\%$, then the suitable deposit requirement is about 4.

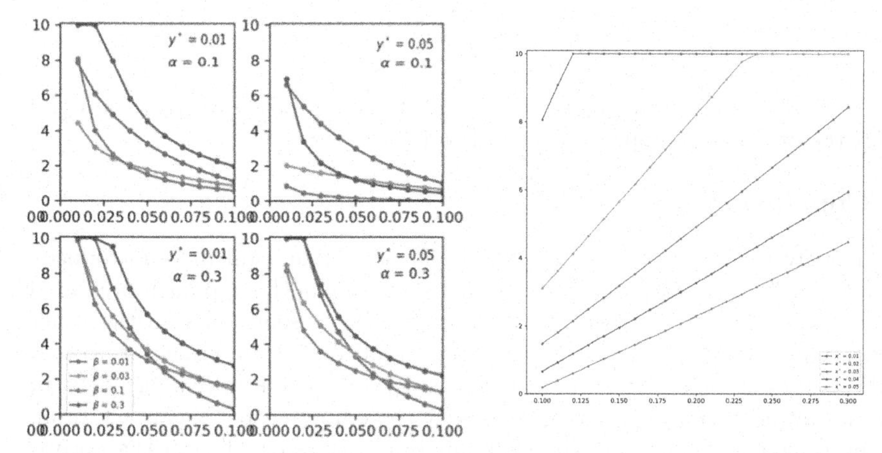

(a) x represented on the horizontal axis while threshold of f shown on the vertical axis. For given (α, β) and x the best response y is regarded as a function of f: $y(f)$. The threshold is the minimum value of f that satisfies $y(f) < y^*$.

(b) Suitable deposit requirement: The horizontal axis represents a relative mining power of the biggest pool in the whole network while the vertical axis represents the threshold of f where x^* is the fraction of spies sent to each other pool.

Fig. 2. Analysis of deposits: thresholds shown in Fig. 2a while optimal deposits shown in Fig. 2b. Thresholds above 10 are set to 10 in both Subfigures. We set $m = 10,000,000$.

Although the deposit system resolves the second issue of the miner's dilemma, the first issue is not resolved. How should a pool manager owning a small pool do against the BWA by a big pool? If the deposit unit is set to an extreme large value (for example, $f = 100$), this issue can possibly be resolved. This approach is feasible in practice as the deposit system bears no costs for honest miners besides opportunity costs of their deposits.

2.5 Model 4: Distributing the Lost Deposit

In this version of the model we consider the case of redistribution of claimed deposits. In Model 3 they are attributed to the pool manager, and thus are never redistributed to the miners in the pool. Here, we consider two arguments when dealing with the deposit.

In the first, a BWA is done by a pool manager and not a miner and the losses are incurred by the manager. Therefore the manager should receive the income from deposits to compensate his losses. However a pool manager is also very interested in maximizing its efficiency to attract more miners. Thus the income from collected deposits from attacks should be redistributed to the joining miners to raise the pools efficiency. Model 4 considers this argument.

For a manager of Pool A resp. B the collected deposit is $d_{BA}f/m$ resp. $d_{AB}f/m$ and thus the equations of R_A and R_B are revised to

$R_A = \frac{D_A+(m\alpha x-d_{AB})R_B-d_{AB}f/m+d_{BA}f/m}{m\alpha+m\beta y-d_{BA}}$ and
$R_B = \frac{D_B+(m\beta y-d_{BA})R_A-d_{BA}f/m+d_{AB}f/m}{m\beta+m\alpha x-d_{AB}}$

The result is shown in Fig. 1d. We see that the incentives to send spies and attack the opponent reappear compared to Fig. 1c.

3 Conclusion

In our analysis we have shown that results for the Miner's dilemma introduced by (Eyal (2015)) can be replicated in a model extending the approach presented in (Lee and Kim (2019)) where both pool managers send spies to the other system. Our numerical results indicate that an effective countermeasure to the BWH attack can be posed by introducing a deposit system in mining pools. In this case even a small number of spies sent to competing pools can act as an insurance against being attacked. This analysis is supported with the derivation of optimal deposit units and thresholds for various scenarios. Contrary to previously proposed countermeasures, this method seems to be effective even without the requirement to change protocol code and can be therefore implemented directly into already operating systems. Our analysis further shows that slashed deposits from malicious miners cannot be redistributed to the mining pool to increase it's efficiency since this would again reintroduce the problem of the Miner's dilemma and render the deposit insurance ineffective.

Acknowledgements. We would like to thank Michael Zargham (BlockScience, USA), Jamsheed Shorish (ShorishResearch, Belgium), and Hitoshi Yamamoto (Rissho University, Japan) for their useful comments. We thank Alfred Taudes (former Director of the interdisciplinary Research Institute for Cryptoeconomics at the Vienna University of Economics) for his management of our project.

A Proof of Theorems

A.1 Proof of Theorem 1

Proof (Theorem 1). If no attacks occur in the network both $x = y = 0$ by definition. Plugging both in into Definition 1 we get

$$D_A = \frac{m\alpha - m\alpha x}{m - m\alpha x - m\beta y} = \frac{m\alpha}{m} = \alpha$$

and

$$D_B = \frac{m\beta - m\beta y}{m - m\alpha x - m\beta y} = \frac{m\beta}{m} = \beta$$

which respecting Definition 2 yields

$$R_A = \frac{D_A + m\alpha x R_B}{m\alpha + m\beta y} = \frac{\alpha}{m\alpha} = \frac{1}{m}$$

and

$$R_B = \frac{D_B + m\beta y R_A}{m\beta + m\alpha x} = \frac{\beta}{m\beta} = \frac{1}{m}.$$

Due to Definition 3 we get $E_A = E_B = 1$.

A.2 Proof of Theorem 2

Proof (Theorem 2). For $x > 0$ and $y = 0$ plugging in into 1 yields

$$D_A = \frac{m\alpha - m\alpha x}{m - m\alpha x - m\beta y} = \frac{m\alpha - m\alpha x}{m - m\alpha x} = \frac{m(\alpha - \alpha x)}{m(1 - \alpha x)} = \frac{\alpha(1 - x)}{1 - \alpha x}$$

and

$$D_B = \frac{m\beta - m\beta y}{m - m\alpha x - m\beta y} = \frac{m\beta}{m - m\alpha x} = \frac{m\beta}{m(1 - \alpha x)} = \frac{\beta}{1 - \alpha x}$$

which again can be inserted into 2 and thus since $y = 0$ then R_B reduces to

$$R_B = \frac{D_B + m\beta y R_A}{m\beta + m\alpha x} = \frac{\frac{\beta}{1-\alpha x}}{m(\beta + \alpha x)} = \frac{\beta}{1 - \alpha x} \cdot \frac{1}{m(\beta + \alpha x)} = \frac{\beta}{m\beta + m\alpha x - m\beta\alpha x - m\alpha^2 x^2}$$

which can be plugged into R_A

$$\begin{aligned}
R_A = \frac{D_A + m\alpha x R_B}{m\alpha + m\beta y} &= \frac{\frac{\alpha(1-x)}{1-\alpha x} + m\alpha x \frac{\beta}{1-\alpha x} \cdot \frac{1}{m(\beta+\alpha x)}}{m\alpha} \\
&= (\frac{\alpha(1 - x)}{1 - \alpha x} + m\alpha x \frac{\beta}{1 - \alpha x} \cdot \frac{1}{m(\beta + \alpha x)}) \cdot \frac{1}{m\alpha} \\
&= (\frac{\alpha(1 - x)}{1 - \alpha x} \cdot \frac{m(\beta + \alpha x)}{m(\beta + \alpha x)} + \frac{m\alpha x\beta}{1 - \alpha x} \cdot \frac{1}{m(\beta + \alpha x)}) \cdot \frac{1}{m\alpha} \\
&= (\frac{\alpha(1 - x) \cdot m(\beta + \alpha x) + m\alpha x\beta}{(1 - \alpha x) \cdot m(\beta + \alpha x)}) \cdot \frac{1}{m\alpha} \\
&= (\frac{(\alpha - \alpha x) \cdot (m\beta + m\alpha x) + m\alpha x\beta}{(1 - \alpha x) \cdot m(\beta + \alpha x)}) \cdot \frac{1}{m\alpha} \\
&= (\frac{m\beta\alpha - m\beta\alpha x - m\alpha^2 x^2 + m\alpha^2 x + m\alpha x\beta}{(1 - \alpha x) \cdot m(\beta + \alpha x)}) \cdot \frac{1}{m\alpha} \\
&= (\frac{m\alpha(\beta - \alpha x^2 + \alpha x)}{(1 - \alpha x) \cdot m(\beta + \alpha x)}) \cdot \frac{1}{m\alpha} \\
&= \frac{\beta - \alpha x^2 + \alpha x}{(1 - \alpha x) \cdot m(\beta + \alpha x)}
\end{aligned}$$

and therefore both efficiencies compute as:

$$E_A = mR_A = m \cdot \frac{\beta - \alpha x^2 + \alpha x}{(1 - \alpha x) \cdot m(\beta + \alpha x)} = \frac{\beta - \alpha x^2 + \alpha x}{(1 - \alpha x)(\beta + \alpha x)}$$

and

$$E_B = mR_B = m \cdot \frac{\beta}{(1 - \alpha x) \cdot m(\beta + \alpha x)} = \frac{\beta}{(1 - \alpha x)(\beta + \alpha x)}$$

This shows that $E_A > E_B$ because

$$E_A - E_B > 0$$

$$\frac{\beta - \alpha x^2 + \alpha x}{(1-\alpha x)(\beta + \alpha x)} - \frac{\beta}{(1-\alpha x)(\beta + \alpha x)} =$$

$$\frac{\alpha(x - x^2)}{(1-\alpha x)(\beta + \alpha x)} > 0$$

This is true because $(x - x^2) > 0$ as $x < 1$. This further shows that $E_B < 1$ because:

$$\begin{aligned} E_B &= \frac{\beta}{(1-\alpha x)(\beta + \alpha x)} \\ &= \frac{\beta}{\beta + \alpha x - \beta \alpha x + \alpha^2 x^2} \\ &= \frac{\beta}{\beta + \underbrace{\alpha x}_{>0} \underbrace{(1 - \beta + \alpha x)}_{>0}} < 1 \end{aligned}$$

since the denominator is greater than the numerator. Further we show when $E_A > 1$:

$$\begin{aligned} E_A &> 1 \\ \frac{\beta - \alpha x^2 + \alpha x}{(1-\alpha x)(\beta + \alpha x)} &> 1 \\ \frac{\beta - \alpha x^2 + \alpha x}{\beta - \beta \alpha x + \alpha x - \alpha^2 x^2} &> 1 \\ \beta - \alpha x^2 + \alpha x &> \beta - \beta \alpha x + \alpha x - \alpha^2 x^2 \\ -\alpha x^2 &> -\beta \alpha x - \alpha^2 x^2 \\ 0 &> -\beta \alpha x - \alpha^2 x^2 + \alpha x^2 \\ 0 &> x(-\beta \alpha - \alpha^2 x + \alpha x) \end{aligned}$$

which is the case when

$$\begin{aligned} 0 &> -\beta \alpha - \alpha^2 x + \alpha x \\ \beta \alpha &> x(-\alpha^2 + \alpha) \\ \frac{\beta \alpha}{-\alpha^2 + \alpha} &> x \\ x &< \frac{\beta \alpha}{\alpha(1-\alpha)} = \frac{\beta}{(1-\alpha)} \end{aligned}$$

References

Eyal, I., Sirer, E.G.: Majority is not enough: bitcoin mining is vulnerable. In: Conference Paper (2013)

Eyal, I.: The Miner's dilemma. In: IEEE Symposium, Security and Privacy, pp. 89–103 (2015)

Eyal, I., Sirer, E.G.: Majority is not enough: bitcoin mining is vulnerable. Commun. ACM **61**, 95–102 (2018)

Lee, S., Kim, S. Countering block withholding attack efficiently. In: IEEE INFOCOM (2019)

Luu, L., Saha, R., Parameshwaran I., Saxena P., Hobor A.: On power splitting games in distributed computation: the case of bitcoin pooled mining. In: 2015 IEEE 28th Computer Security Foundations Symposium (2015)

Rosenfeld, M.: Analysis of bitcoin pooled mining reward systems. arXiv, 1112.4980 (2011)

Zhu, S., Li, W., Li, H., Hu, C., Cai, Z.: A survey: reward distribution mechanisms and withholding attacks in Bitcoin pool mining. Math. Found. Comput. **1**(4), 393–414 (2018)

Risks in DeFi-Lending Protocols - An Exploratory Categorization and Analysis of Interest Rate Differences

Marco Huber[1] and Vinzenz Treytl[2](✉)

[1] Wirtschaftsuniversität Wien, Vienna, Austria
[2] ABC Research, Favoritenstraße 111, 1100 Vienna, Austria
vinzenz.treytl@abc-research.at

Abstract. According to well-established principles of risk management, interest rates reflect different levels of risk. Understanding this level of risk is crucial for investors when making investment decisions. In this paper, risk factors in DeFi-lending are identified and categorized within a framework which is not only focused on technical but also on financial aspects. In a subsequent step, the influence of unsystematic risk factors on interest rates are explored to tentatively assess the validity of the literature-derived framework. Our observations indicate that operational risks emanating from the underlying layer-1-solution do seem to have a strong influence. Furthermore, first indications for the validity of scalability challenges and smart contract risks were found. For oracle risks as well as governance risks our approach yielded no results.

Keywords: Decentralized finance · DeFi · Risks · Interest rates

1 Introduction

Built on blockchain technology, decentralized finance (or DeFi) protocols have managed to constantly expand their offering of financial services, ranging from decentralized exchanges, lending & borrowing services over insurance services to trading platforms [1]. The market size of DeFi – as measured by the total value locked – has increased substantially over the last four years. By the end of 2017, the DeFi market size was hovering slightly above zero levels. Since 2020, however, DeFi picked up momentum and increased its popularity among crypto investors reaching a market size of more than $100B as of November 2021 [2].

Although the DeFi-space is very diverse, one can observe the counter-intuitive fact that different protocols, that offer seemingly identical services, do not offer similar interest rates. For example, in the period from Sept. 2021 to Dec. 2021, the AAVE v2 lending protocol offered lower interest rates then its precursor AAVE v1. These differences ranged between 1.5 and 2.5%-points. Likewise, AAVE Polygon and AAVE Avalanche yielded between 0.5 and 1%-point less than AAVE v2. This seems to be

This work is based on the first author's thesis which was co-supervised by the second author.

G. Kotsis et al. (Eds.): DEXA 2022 Workshops, CCIS 1633, pp. 258–269, 2022.
https://doi.org/10.1007/978-3-031-14343-4_24

in contradiction with established theories in traditional finance. The arbitrage pricing theory and the law of one price state that two or more investment strategies with the same level of risk should always have the same price [3, 30]. In other words, if two investment opportunities offer different investment returns, the return differences should be explainable by different levels of underlying risks.

In traditional finance, the relationship between risk and return is well documented and the underlying risk factors causing interest rate differences are well described [3, 13, 30]. In contrast, risks of DeFi-protocols that enable functionality comparable to traditional financial instruments are yet poorly understood. The main goal of this paper is to tentatively explore these interest rate differences among DeFi-lending-platforms. From a practitioners' perspective, it provides indications on why the interest rates differ among DeFi-platforms. Understanding the reasons for these differences in DeFi-protocols, helps investors to manage their risk exposure. From an academic perspective, exploring and analyzing interest rate differences among DeFi-protocols expands our understanding of this new form of financial services.

2 DeFi-Lending Protocols

2.1 Key Concepts and Components

Decentralized finance protocols, or DeFi protocols, are applications that provide a great variety of financial services without relying on central institutions like banks to act as an intermediary [1]. This section gives and overview of key concepts and components as well as metrics in relation to DeFi-lending protocols.

Key Participants. DeFi-lending systems typically involve three key parties: lenders, borrowers, and so-called liquidators [4].

Lenders aim to generate capital income from their excess liquidity. In return for interest payments, lenders transfer tokens to so-called lending pools and in exchange receive IOU tokens. IOU tokens allow lenders to withdraw the deposited funds as well as to collect accrued interest [5].

Borrowers are taking out loans in exchange for interest payments. Normally, borrowers in DeFi are required to deposit cryptocurrencies as a collateral for a loan [4]. As such, DeFi-loans typically take the form of collateralized loans. Normally the value of the collateral is higher than the amount borrowed [4].

Liquidators are responsible for monitoring debt positions and for liquidating "unhealthy" loans [4]. Usually, liquidators are computer programs automatically seeking out under-collateralized loans. If an individual debt becomes under-collateralized, the collateral is automatically sold to liquidators at a discount. Liquidators buying the collateral then settle the loan plus accrued interest. They thus serve as a safety mechanism for lenders to reduce risks from a default. The profit liquidators can make by this, serves as a strong incentive to detect and liquidate "unhealthy" loans.

Key Elements. DeFi-lending systems typically have some unique components that are important for the functioning of the protocol.

Liquidity pool/Lending pool: A fund of cryptocurrencies automatically managed by the protocol and its underlying smart contracts. The pool constantly provides liquidity to the market. In DeFi-lending, liquidity pools are also referred to as lending pools [5].

IOU token: A token that is issued by a liquidity pool when funds are deposited into it by a borrower. IOU tokens serve two main purposes. First, they can be redeemed to the pool by lenders to withdraw the original deposits [6]. Second, the exchange rate between IOU tokens and deposited funds is increasing over time. The longer funds are deposited, the more can be redeemed. From a lender's perspective, this is equivalent to an interest payment. IOU tokens can normally be traded in secondary markets.

Key Metrics. Several key metrics can be analyzed in connection with DeFi-lending protocols. Two of the most important ones are the health factor and the utilization rate.

Health factor: The health factor quantifies the safety of an individual debt position. It is based on two factors: a) the total amount of cryptocurrencies a borrower is allowed to take out and b) the outstanding loan. The former depends on the value of the collateral whereas the latter depends on accrued interest and on the amount repaid. The health factor is normally expressed as a ratio of these two factors [4]. If the health factor of an individual debt position falls below a certain threshold, the position is sold to liquidators at a discount in order to settle the loan plus accrued interest as quickly as possible.

Utilization rate: The utilization rate indicates the percentage amount of a pool's liquidity that is currently borrowed by users. A high utilization rate means that the asset is highly demanded, whereas a low utilization rate stands for lower popularity.

2.2 DeFi-Lending Process

This section discusses the basic lending process in a DeFi-protocol. The process starts with lenders depositing cryptocurrencies in the DeFi protocol. In exchange, platform-specific IOU tokens are issued. These tokens can be redeemed at any time to withdraw the amount locked in the smart contract as well as accrued interest. Once there is liquidity in the lending pool a borrower can take out a loan from the pool. In order to take out liquidity, borrowers are required to deposit collateral into the protocol, which is used by the protocol to grant further loans to more users. The borrower receives platform-specific tokens representing the collateral. After the entire debt and accrued interest have been repaid, these tokens can be used to withdraw the collateral. If the health factor of the individual debt position drops below a pre-defined threshold (usually below 1), the protocol automatically sells the collateral to liquidators to cover any outstanding loan amount.

From the description above some differences between a DeFi lending and lending in traditional finance become clear. In DeFi lending, there is no intermediary such as a bank which normally services the loan and performs important functions such as term transformation and risk management. Another major difference lies with the handling of risks. DeFi-lending protocols normally do not conduct an evaluation of credit risks. Risk is addressed via the collateralization of debt positions. Except for flash loans, DeFi-loans are always collateralized.

2.3 Risk Factors in DeFi-Lending

A first holistic categorization of DeFi-risks was provided by [7]. They investigated the security of DeFi-protocols and classified the risks in technical and economic security. Werner et al. [7] used the expression "security" over "risk" as they focus on how protocols can be manipulated and exploited, rather than focusing on a financial aspect.

Technical security includes all aspects that allow an exploitation of the protocol via technical issues [7]. *Smart contract risks* and vulnerabilities such as exploitable coding errors represent the main technical insecurity. Smart contract vulnerabilities, for instance, allow attackers to withdraw and steal funds from a lending pool [8]. Furthermore, *oracle manipulations* can also be considered a technical risk. The purpose of oracles is to "feed" a smart contract with off-chain data, such as market prices. A manipulation could falsely trigger liquidations and open arbitrage opportunities for attackers [9]. According to [8] oracles are the greatest threat to the DeFi-ecosystem to date.

From an economic perspective, a DeFi-protocol is insecure if its equilibrium and *incentive system* can be manipulated for a profit [7]. This risk can be increased by the high volatility of cryptocurrencies and illiquidity. In such a scenario the incentive for liquidators to liquidate unhealthy positions might be low, which can result in unmet debt positions and an increased risk for lenders [7, 10].

Other economic insecurities emanate from *operational risks* which are dependent on the underlying blockchain. These risks include, for example, threats resulting from unethical miner behavior such as prioritizing or frontrunning users [7, 9]. Scalability can also be a risk in this section. Schär [11] defines scalability as a risk where the underlying blockchain cannot meet the demand for transactions required by users. This risk could be observed over the past as the Ethereum blockchain was frequently congested. Carter and Jeng [9] extend the term scalability risks to include high volatility of transaction prices.

Lastly, *protocol governance* also represents a major economic insecurity [7]. Since DeFi is built on a blockchain network, decisions are taken democratically. This can lead to situations where governance tokens are acquired strategically. In the extreme case, if one individual or a group of individuals manage to control more than 51% of all governance tokens, a protocol can be controlled entirely by one actor [7–9].

In addition to the risks discussed, the increasing size of DeFi-ecosystems can lead to systematic risks. These include regulatory uncertainty, interdependencies between protocols and macroeconomic factors, which can influence entire DeFi markets [11, 12].

3 Methodology

3.1 Framework Development

The development of the risk framework in this project was based on existing academic as well as grey literature. Literature search focused on two strands of research: traditional finance and DeFi. First, traditional finance offers a wide, well-researched body of knowledge with regards to risk [cf. 3, 13, 30, 31]. Second, the literature search focused on the nascent field of DeFi-risk. The keyword-based literature search relied on established academic databases (e.g. EBSCO, Google Scholar) and used backward reference searching to identify relevant literature. As the number of publications addressing risks in DeFi is still limited, selected websites (e.g. https://defiscore.io/), Social Media (e.g. #DeFi on Twitter, Reddit) and market risk reports [e.g. 14–16] were also screened in order to enrich the information found in the literature.

Subsequently, risks identified in the literature review were assigned to categories based on their similarities. Helping in this categorization, it could be observed that even though this field of study is still at a very early stage, there is already strong accordance among scholars in labeling DeFi-risks [e.g. 7, 8, 11, 17]. The major categories identified were interoperability & interdependencies between protocols, market illiquidity risk, regulatory uncertainty, operational risk, scalability challenges, smart contract risk, oracle risk and governance risk.

3.2 Multiple Case Study

For this paper, a qualitative multiple case study approach was also conducted [cf. 18]. Based on the framework formulated by [18] and enriched by the guidelines defined by [19], this research followed a five-step process. This process consisted of 1) Defining the scope, 2) setting propositions and selecting appropriate cases, 3) collecting and 4) analyzing the necessary data and finally 5) describing the results.

Case Selection. Miles and Huberman [19] outline that the selection of appropriate cases is crucial for the study's outcome. For this analysis, the case selection focused on isolating unsystematic risk factors to assess their influence on the interest rate differences. There are multiple risks related to DeFi-protocols and – based on their features – protocols are affected by them differently. To isolate specific risks, cases were selected to be similar in most aspects. Ideally, a protocol only differs in one selected aspect when compared to a benchmark enabling the isolation and observation of single risk factors.

Because of its market size, popularity, and high degree of transparency at the time of writing AAVE v2 has been defined as the benchmark-case to which the other cases were compared. Additional protocols where then selected with the goal to isolate specific risks. These included: AAVE v1, AAVE Polygon, and AAVE Avalanche [29] as well as MakerDAO [27].

The protocols from the AAVE Ecosystem were chosen for their similarity. As these protocols are all part of the AAVE ecosystem, they are built on similar principles, offer identical services, and show no major differences in their functionalities. What differentiates them, however, are some technical aspects.

Data Collection and Analysis. Data was mainly gathered from primary data sources. The main source was the protocols' documentation on GitHub and the protocol's website. (e.g., aave.com). Specifically, the whitepapers of the selected cases were reviewed and analyzed. Additionally, published whitepapers of third-party projects such as DeFi Score were also considered. In addition, if available, audit reports concerning vulnerabilities and changes in the code were reviewed. To identify risks of selected protocols, the materials were analyzed with an inductive coding approach following established procedures [18, 19]. For the analysis of interest rates, comparisons focused on each protocol's DAI liquidity pool. DAI is a stable coin, pegged to a fiat currency with the purpose of reducing price volatility [28]. This eliminated the volatility aspect of cryptocurrencies. Furthermore, deposit rates and utilization rates for selected protocols from Oct. 15th, 2021 to Mar. 31st, 2022 were collected from aavescan.com.

Selected Cases. A brief overview of the protocols analyzed is given below.

AAVE v1 is a decentralized, Ethereum blockchain-based, open-source, platform that enables users to lend and borrow crypto-tokens [20]. The project was launched in 2017 as a first version of a lending protocol part of the AAVE ecosystem.

AAVE v2 (benchmark) is an updated version of the original AAVE v1 protocol which includes several security upgrades. AAVE v2 was launched in December 2020 and is now one of the largest lending protocols in the entire DeFi-ecosystem.

AAVE Polygon: Similar to AAVE v1 and v2, AAVE Polygon is a DeFi-lending project built on the AAVE ecosystem. In contrast to the Ethereum-based AAVE 2, this protocol is built on the more scalable Polygon sidechain.

AAVE Avalanche is built on the same principles as the three protocols above. However, it operates on the Ethereum competitor Avalanche. Compared to Ethereum, Avalanche is highly scalable and interoperable with other blockchains. According to [29], this network is the first platform designed to sustain the scale of global finance.

4 Results

4.1 DeFi-Risk Framework

In the broadest sense, two major categories of risk can be distinguished [3, 30, 31]. The first one is ***systematic risk*** which includes risks that influence an entire financial market equally. Systematic risks are determined, for instance, by macroeconomic factors such as regulations, market news, and technological changes. The second category is defined as ***unsystematic risk*** which includes all risks that are specific to a single firm, or in this case a specific protocol [21, 30]. The main difference between these two types of risks is that systematic risk cannot be diversified away, as it affects an entire market. In contrast, unsystematic risk can be diversified away, as it is specific to a single entity. These two risks form the basis of the framework to which the main risk categories found in literature were assigned to. The resulting framework is shown in Table 1.

Systematic Risks

The first three DeFi-risks were classified as systematic risks since they depend on the macro environment and affect an entire market equally [30]. Firstly, *interoperability risk* refers to risks arising to a protocol for being interoperable with other protocols. With the rise of multiple Layer-1 protocols, interoperability has been gaining importance for the DeFi-ecosystem [11, 22]. Protocols have started to become deeply interconnected with each other. This, however, also increased the risk of a focal protocol being influenced by problems in other protocols. Jensen and Ross [17] also refer to this as a "systemic dependence between applications". Secondly, *market illiquidity risk* is the risk of users being hindered to liquidate their position as the circulating supply of cryptocurrencies is insufficient. A frequent reason for market illiquidity is the relatively high volatility of cryptocurrencies [7]. Illiquid markets lead to illiquid lending pools, which complicates the withdrawal of deposited funds for lenders as well as collateral for borrowers. Since market illiquidity does not depend on a single protocol but on several macroeconomic factors such as volatility [7, 23], this risk is categorized as a systematic risk. Lastly, regulatory risk refers to the risk of future governmental regulations influencing the whole crypto ecosystem negatively. These potential regulations would not just affect one individual protocol but the entire ecosystem.

Table 1. DeFi-risk framework.

Risk category	Type of risk
Systematic risk	Interoperability and interdependencies between protocols
	Market illiquidity risk
	Regulatory uncertainty
Unsystematic risk	Operational risk
	- Scalability challenges
	Smart contract risk
	Oracle risk
	Governance risk

Unsystematic Risks

The remaining risks (operational risk, scalability challenges, smart contract risk, oracle risk and governance risk) can be classified as unsystematic risks since they depend on the individual protocol. Operational risks generally refer to a failure of internal operations [12]. In the context of DeFi-lending, operational risks can be seen as all risks stemming from the blockchain underlying a DeFi-application. This risk includes outages, consensus failures as well as opportunistic miner extractable value [9].

One important sub-category of operational risk is scalability challenges. Growing DeFi-protocols at some point struggle as the underlying blockchain does not support the increasing demand for computing power and block space [11]. This is the case when the demand for transactions exceeds the computing capacity of a given blockchain [9, 11]. As a result, transaction prices rise in order to reduce the number of pending transactions. In such a case, for instance, smaller users might not be able to afford transaction fees forcing them to leave funds locked in smart contracts or excluding them from participating in the system [9]. As scalability is dependent on the underlying network it is considered unsystematic as part of the operational risk.

The risk factor which has been given most attention in literature is *smart contract risks* [7, 8, 11]. It refers to the risk that errors in the code can be exploited for attacks. Such vulnerabilities can lead to intentional manipulations and illegal withdrawals of funds from lending pools [8]. Smart contract vulnerabilities can be classified as unsystematic since exploitable errors in the code are specific to a DeFi-protocol.

Oracle risk captures the risk of a technical failure or manipulation of oracles providing off-chain data to a protocol. Oracles are automated algorithms that provide protocols with data which is not stored on the blockchain. Given their importance for executing transactions correctly, their safety is a very important component for a functioning DeFi-ecosystem. A potential failure as well as a case of off-chain data manipulation (e.g., of market prices) can lead to severe consequences for users [7, 11]. Oracle risk is classified as unsystematic, as it is independent of a protocol's macroenvironment.

Lastly, a DeFi-protocol partly depends on human interaction and decisions which constantly adjust parameters and keep the protocol operative [7]. These decisions are made by holders of governance tokens who are allowed to vote on decisions regarding the protocol's future developments. As a result, governance risk is the risk that decisions which do not benefit all users are made if the majority of governance tokes is controlled by only a few stakeholders [7]. Since every protocol has its own governance system, this risk can be classified as unsystematic.

4.2 Observations on DeFi Risk Categories

Having developed a framework for categorizing risks in DeFi-protocols, next, we tentatively evaluate if the unsystematic risk categories within the framework actually influence interest rates. According to the arbitrage pricing theory and the law of one price, if two or more investment strategies have the same risk exposure, they should always have the same price [3, 30]. To observe specific risk categories, it is necessary to isolate these risks. This was done by choosing protocols based on their characteristics. These protocols were subsequently compared to the benchmark protocol AAVE v2.

Operational Risk. In order to observe effects of operational risk on the interest differences, Ethereum-based AAVE v2 was compared with Avalanche-based AAVE Avalanche. Because these two protocols don't rely on the same blockchain, this comparison should isolate possible operational risks such as consensus failures, outages, and congestions as these depend on the underlying blockchain [9].

Interest rates in the AAVE family partly depend on the utilization of a lending pool. The more a lending pool is used, the higher the interest rate. Therefore, for a comparison it is important to examine data points at approximately the same utilization rate. Within the observation period (Oct. 15th 2021 to March 31st 2022) a total of 49 observations were made where the utilization rate of the two protocols were within a 1,5%-points range. For the total observation period, interest rates offered by AAVE Avalanche are on average 26 bps lower. For the 49 observations with comparable utilization rates, this difference increased to an average discount of 46 bps on AAVE Avalanche.

Our observations seem to indicate that risks to the underlying infrastructure of a DeFi-protocol are reflected in lending rates. Literature provides some leads to the underlying factors. For example, [9] point out that Ethereum is more fragile to outages as centralized third parties such as Infura are responsible for keeping the blockchain operable. Additionally, due to its optimized consensus mechanism [24] Avalanche is more efficient than Ethereum and supports a greater number of transactions. Consequently, transaction fees are kept low and blockchain congestions are less frequent.

Scalability Challenges. In order to analyze the effect of efficiency and scalability challenges on the interest rate differences, AAVE v2 was compared with AAVE Polygon. As pointed out by [9], Layer 2 solutions – also referred to as sidechains – are more efficient than the mainnet and can process a larger number of transactions. Thus, scalability risk should be reduced by using these solutions.

Since Polygon is a sidechain of the Ethereum mainnet, this comparison should isolate the scalability risk by comparing DAI lending pools running on both AAVE v2 as well as AAVE Polygon.

Within the observation period (Oct. 15th 2021 to March 31st 2022) a total of 26 observations were made where the utilization rates of the two protocols were within a 1,5%-points range. The analysis showed that interest rates offered by AAVE Polygon are on average 75 bps higher in the observation period. For the 26 observations with comparable utilization rates, this difference decreased to an average premium of 61bps on AAVE Polygon.

This observation is difficult to interpret. Overall, a relatively high premium was observed in the samples used. This observation runs counter to the importance that scalability risk is given in literature [8, 9, 11]. One potential explanation is that utilization rates on AAVE Polygon were for most of the observation period higher than the utilization rates on AAVE v2 and hovered around the "optimal utilization rate" of 80%. As a safety mechanism the protocol increases interest rates exponentially if the optimal utilization rate threshold is surpassed. Another explanation could be that sidechains as a technology are at a very early stage and therefore considered more risky.

Smart Contract Risk. In order to analyze the effect of smart contract risk on the interest rates, AAVE v2 was compared with its prior version AAVE v1. As an updated version of v1, AAVE v2 was adapted to make the protocol more efficient and secure [15, 16, 20, 25, 26]. Since both protocols run on the same blockchain network, operational risk, scalability challenges as well as governance and oracle risks should be equal. Therefore, the selection potentially allows the isolation of smart contract risk.

We observed that in the observation period interest rates offered by AAVE v1 were on average somewhat higher than the interest rates offered by AAVE v2. Furthermore, the differences between these two protocols cannot just be found in the DAI lending pool which is the focus of this research but are consistent throughout all lending pools available on both protocols. Various reports [15, 16, 20, 26] show the increased smart contract security of AAVE v2 when compared to AAVE v1. This increased security could be an explanation for this phenomenon.

However, our observations have two important limitations. First, the interest model of AAVE v2 has slightly different parameter settings (AAVE v2 vs. v1: Optimal utilization rate: 80% vs. 80%, base: 0% vs. 1%; slope (utilization < 80%): 4% vs. 7%; slope (utilization < 80%): 75% vs. 150%). Second, the utilization rates recorded during the observation period are considerably higher for AAVEv2 (Avg: 72,8% vs. 46,2%). These limitations as well as the fact that the development of AAVE v1 has been discontinued make clear interpretations impossible.

Oracle Risk. The whole AAVE ecosystem relies on the Chainlink network for oracle services. The role of Chainlink is to validate and aggregate off-chain data such as market prices from several sources and make it accessible on the blockchain. In order to isolate the effect of oracle risk on the interest rate differences, two protocols with different oracle solutions but equal other features need to be compared.

Despite an extensive search, only one major DeFi-protocol (MakerDAO) which is not relying on Chainlink data could be identified. While this makes an isolation of this risk possible in principle, an analysis of MakerDAO showed that other features of MakerDAO are not comparable to the AAVE protocol family. With Compound, a second, more comparable, major protocol was identified. While both AAVE and Compound primarily rely on the Chainlink oracle network, AAVE additionally operates an inhouse "fallback price oracle" which constantly doublechecks the data provided by Chainlink. This adds another layer of security. Thus, the oracle risk should be slightly lower on AAVE. However, no meaningful interest rate differences could be observed.

Governance Risk. On the AAVE platform either AAVE tokens or Staked AAVE tokens (stkAAVE) are used for governance purposes. The amount of both gives users a proportional voting power, as AAVE operates a majority voting-scheme. As the AAVE protocol family are part of the same ecosystem, they rely on the same governance model. Thus, similar to oracle risk, AAVE needs to be compared with a protocol with different governance features but equal other features. However, a review of literature as well as internet sources showed that at the time of writing, majority voting-scheme governance systems were the predominant industry standard. Therefore, this single risk could not be isolated.

5 Conclusion

The starting point of this research was the observation of different interest rate levels among DeFi-lending protocols. According to well-established principles of risk management, interest rates reflect different levels of risk. Understanding this level of risk is crucial for investors when making investment decisions. In this project, we developed and started to assess a framework that helps explaining the observed interest rate differences. Based on a review of existing literature, several categories of systematic as well as unsystematic risks could be identified. The framework provides a first categorization of DeFi-risks which is not only focused on technical but also on financial aspects.

In a subsequent step, the influence of unsystematic risk factors on interest rates was explored in order to tentatively assess the validity of the literature-derived risk categories. This analysis was conducted via a pair-wise comparison of selected protocols that tried to isolate risk factors. The observations indicate that operational risks emanating from the underlying layer-1-solution do seem to have a considerable influence. Furthermore, first indications for the validity of scalability challenges and smart contract risks were found. For oracle risks as well as governance risks, an isolation by pairwise comparison was not possible and our approach yielded no results. For both risks standard solutions (i.e. Chainlink network and majority voting) seem to have become dominant in the DeFi-Space.

References

1. Kaal, W.A.: Digital asset market evolution. J. Corporation Law **46**(4), 910–963 (2021)
2. defipulse.com (Website). https://defipulse.com. Accessed 14 Apr 2022
3. Berk, J., DeMarzo, P.: Corporate Finance, 3rd edn. Pearson Education Inc., London (2014)
4. Qin, K., Zhou, L., Gamito, P., Jovanovic, P., Gervais, A.: An empirical study of DeFi liquidations: incentives, risks, and instabilities (2021). https://doi.org/10.1145/3487552.3487811
5. Bartoletti, M., Chiang, J.H., Lluch-Lafuente, A.: SoK: lending pools in decentralized finance (2020). http://arxiv.org/abs/2012.13230
6. Xu, J., Vadgama, N.: From banks to DeFi: the evolution of the lending market (2021). http://arxiv.org/abs/2104.00970
7. Werner, S.M., Perez, D., Gudgeon, L., Klages-Mundt, A., Harz, D., Knottenbelt, W.J.: SoK: decentralized finance (DeFi) (2021). http://arxiv.org/abs/2101.08778
8. Harvey, C.R., Ramachandran, A., Santoro, J.: DeFi and the Future of Finance, 1st edn. Wiley, Hoboken (2021)
9. Carter, N., Jeng, L.: DeFi protocol risks: the paradox of DeFi. In: Coen, B., Maurice, D. (eds.) Regtech, Suptech and Beyond: Innovation and Technology in Financial Services. RiskBooks (2021)
10. Castro-Iragorri, C., Ramirez, J., Velez, S.: Financial intermediation and risk in decentralized lending protocols (2021). http://arxiv.org/abs/2107.14678
11. Schär, F.: Decentralized finance: on blockchain-and smart contract-based financial markets. Fed. Reserve Bank St. Louis Rev. **103**(2), 153–174 (2021). https://doi.org/10.20955/r.103.153-74
12. Committee on Payment and Settlement Systems, & Technical Committee of the International Organization of Securities Commissions: Principles for financial market infrastructures. Bank for International Settlements (2012)
13. Geyer, A., Hanke, M., Littich, E., Nettekoven, M.: Grundlagen der Finanzierung: verstehen – berechnen – entscheiden, 5th edn. Linde Verlag, Vienna (2015)
14. Gauntlet: Market Risk Assessment: an analysis of the financial risk to participants in the Aave V1 and V2 protocols (2021)
15. Grieco, G., Czarnota, D., Colburn, M.: Aave Protocol Security Assessment (2019). https://github.com/aave/aaveprotocol/blob/master/docs/ToB_aave_protocol_final_report.pdf
16. Openzeppelin Security: Aave Protocol Audit, 15 January 2020 (2020). https://blog.openzeppelin.com/aave-protocol-audit/
17. Jensen, J.R., Ross, O.: Managing Risk in DeFi Position Paper (2020)
18. Yin, R.K.: Case Study Research – Design and Methods, 3rd edn, vol. 5. Sage Publications, Thousand Oaks (2003)

19. Miles, M.B., Huberman, A.M.: Qualitative Data Analysis, 2nd edn. Sage Publications, Thousand Oaks (1994)
20. SigmaPrime: Aave Protocol v2.0 – Smart Contract Security Assessment (2021). https://github.com/aave/protocol-v2/blob/master/audits/SigmaPrime-aave-v2-01-2021.pdf
21. Chaibi, H., Ftiti, Z.: Credit risk determinants: evidence from a cross-country study. Res. Int. Bus. Finan. **33**, 1–16 (2015). https://doi.org/10.1016/j.ribaf.2014.06.001
22. Sober, M., Scaffino, G., Spanring, C., Schulte, S.: A voting-based blockchain interoperability Oracle (2021). http://arxiv.org/abs/2111.10091
23. Gudgeon, L., Werner, S.M., Perez, D., Knottenbelt, W.J.: DeFi protocols for loanable funds: interest rates, liquidity and market efficiency (2020). http://arxiv.org/abs/2006.13922
24. Sekniqi, K., Laine, D., Buttolph, S., Gün Sirer, E.: Avalanche Platform Whitepaper (2020)
25. AAVE: AAVE Protocol Whitepaper V2.0 (2020)
26. Certora: Formal Verification of Aave's protocol-V2 (2020). https://github.com/aave/protocolv2/blob/master/audits/Certora-FV-aave-v2-03-12-2020.pdf
27. MakerDAO: The Maker Protocol White Paper (2020). https://makerdao.com/en/whitepaper#thedai-savings-rate
28. Van der Merwe, A.: A Taxonomy of Cryptocurrencies and Other Digital Assets (2021). https://bravenewcoin.com/general-taxonomy-forcryptographic-
29. Avalanche: What is Avalanche?—Avalanche Docs (n.d.). https://docs.avax.network/. Accessed 1 Feb 2022
30. Sharpe, W.: Capital asset prices: a theory of market equilibrium under conditions of risk. J. Finan. **XIX**(3), 425–442 (1964)
31. Koutmos, G., Knif, J.: Estimating systematic risk using time varying distribution. Eur. Financ. Manag. **8**(1), 59–73 (2002)

Battling the Bullwhip Effect with Cryptography

Martin Hrušovský(✉) and Alfred Taudes

Institute for Production Management, WU Vienna University of Economics and Business, Vienna, Austria
martin.hrusovsky@wu.ac.at
https://www.wu.ac.at/en/prodmanengl

Abstract. In real-world supply chains it is often observed that orders placed with suppliers tend to fluctuate more than sales to customers and that this deviation builds up in the upstream direction of the supply chain. This bullwhip effect arises because local decision-making based on orders of the immediate customer leads to overreaction. Literature shows that supply chain wide sharing of order or inventory information can help to stabilize the system and reduce inventories and stockouts. However, sharing this information can make a stakeholder vulnerable in other areas like the bargaining over prices. To overcome this dilemma we propose the usage of cryptographic methods like secure multiparty computation or homomorphic encryption to compute and share average order/inventory levels without leaking of sensitive data of individual actors. Integrating this information into the stylized beer game supply chain model, we show that the bullwhip effect is reduced also under this limited information sharing. Besides presenting results regarding the savings in supply chain costs achieved, we describe how blockchain technology can be used to implement such a novel supply chain management system.

Keywords: Supply chain management · Secure multiparty computation · Homomorphic encryption · Bullwhip effect

1 Introduction

A supply chain (SC) is the entire process from the customer's order to the delivery and payment of a product or service. It involves different companies that collectively manage the flows of materials, information and funds. A key challenge of supply chain management (SCM) is the tension between the need for cooperation and the competition for the share of the total surplus. For instance, an end customer facing entity needs to know the inventory and capacities of upstream partners in order to be able to check the feasibility of a demand plan via supply network planning. However, a manufacturer might be reluctant to share this information, as his customer might demand lower prices when learning that the inventory has increased, signaling low sales and a weaker bargaining position.

G. Kotsis et al. (Eds.): DEXA 2022 Workshops, CCIS 1633, pp. 270–281, 2022.
https://doi.org/10.1007/978-3-031-14343-4_25

Thus, often the communication within the SC is reduced to the necessary minimum, i.e. passing of orders between different stages. However, this limited information gives rise to the so-called bullwhip effect: small fluctuations in retail demand cause progressively larger fluctuations in derived demand upstream at the other SC actors. This is similar to the effect of cracking a whip, where a small movement of the wrist causes the whip's wave patterns to amplify [1].

The reason for this phenomenon is that actors have only a partial understanding of demand but through their local actions influence the entire chain. One behavior causing the bullwhip effect is the extrapolation of demand trends and safety stock adjustment. To give an example, suppose a retailer sells 20 units of a product and keeps one period of demand as safety stock. If a permanent increase in demand occurs, now he sells 30 units. Then, instead of ordering 20 units from the wholesaler, he orders 40 units, 30 units for the newly forecasted demand, and 10 units to raise the safety stock to 30 units. Thus, a 50% increase in customer demand has led to a 100% increase of the derived demand. If the other actors employ the same policy, wholesaler orders 60 units, distributor orders 100 and manufacturer 180 units. If the retailer would share his sales figures, the other actors could separate the inventory adjustment effect and learn that they only have to increase their safety stocks to 30. However, the retailer might not be willing to share this data, e.g. fearing that he might not be serviced when capacities are tight and his suppliers infer that he has enough inventory.

Instead of sharing local data, a SC member might be willing to participate in the computation of average values if he can keep his local information private, though. Such methods are available in Secure Multiparty Computation (SMPC) or Homomorphic Encryption (HE). SMPC [2] provides protocols, where, by exchanging messages with each other, participants can learn the value of a function without revealing their inputs and relying on a trustworthy party. In a simplistic setting, the participants are organized in a ring, and a leader starts by adding a private random number to his input. Then he passes the result to his neighbor, who adds his local value to the input and passes the result to his neighbor. The process continues until the leader is contacted, who then subtracts his private random number and announces the average value based on the sum calculated.

HE [3] is an encryption scheme that allows a third party to compute with encrypted inputs while preserving the features of the function and format of the encrypted data. For instance, in the additively homomorphic scheme of [4], each of the stakeholders involved in the information sharing would encrypt their local value using a public key and send the cipher text to a coordinating entity. This entity would multiply these inputs and each entity could decrypt the resulting value using the private key to recover the sum and determine the average value.

Implementing such schemes is costly and should only be considered if they can significantly reduce the bullwhip effect. To study this question we adapt the beer game developed by the MIT Sloan School of Management to introduce students of SCM to the bullwhip effect [5]. This role-play simulation game simulates a beer SC using retailer, wholesaler, distributor and brewer, who manage their inventories, production and backlogs, and communicate orders with the goal of minimizing operating cost. We will adapt this setting by incorporating information sharing of average orders and inventory levels to infer whether SMPC/HE should be employed to mitigate the bullwhip effect in a data parsimonious way.

2 Literature Review

[6] has first analyzed the bullwhip effect in the context of systems dynamics. A seminal work describing the phenomenon is [1], who identify demand signal processing, rationing game, order batching and price variations as sources of the bullwhip effect. [7] find that smoother forecasts and shorter lead times decrease the bullwhip effect. [8] find that inventory information helps upstream chain members to better anticipate and prepare for fluctuations in inventory needs downstream. [9] show that order-up-to replenishment policies lead to fluctuations and develop a decision rule that avoids amplifications.

Coordination of multiple actors in networks is a widely discussed topic in the literature with different approaches used to solve this problem. As an example, [10] use a mean field approach to coordinate charging of electric vehicles. In cases where independent agents need to be coordinated by sharing only limited information, decentralized techniques, such as the dynamic average consensus, can be used [11,12]. Another option is to use stochastic models to solve inventory network control problems, such as the queue clearing algorithm used by [13].

In SCM context, papers derived from the SecureSCM project [14] apply SMPC to supply network planning via linear programming (LP). They develop algorithms for solving LPs using a variant of the simplex algorithm and secure computation with fixed-point rational numbers based on secret sharing [15]. [16] describe a scheme for coordinating decentralized parties that share central resources but hold private information about their decision problems modeled as LPs, and [17] develop SMPC protocols for the Joint Economic Lot Size Model.

A number of papers describe ways of implementing information sharing via blockchains: [18] develop a blockchain architecture for general SC information sharing based on the codification of contractual relations and receiver-specific encryption, while [19] describe a shared ledger for a particular SC through which all relevant information is broadcasted. [20] enhance this design by adding a reputation system for tracking the truthfulness of the information sharing of a stakeholder and by considering these inputs when determining the validators in a PoA consensus mechanism.

In the case of HE, a permissioned blockchain operated by the SC members can execute a smart contract that collects the encrypted local values, performs the computation of the encrypted sum, checks whether all participants in the computation have taken part in the protocol and stores the decrypted average as common knowledge. Similarly, as described in [21], blockchain can be employed to enhance the efficiency and fairness of SMPC.

3 Modelling Approach

In our approach, we consider a SC consisting of several actors that are participating on a production and delivery of a product to the final customer. Although usually the bullwhip effect is modelled using two (see, e.g., [9,22]) or four actors (see, e.g., [8]), we present a SC containing six actors in order to be able to calculate averages and analyze the effect of proposed policies. Our SC starts with a retailer, who receives orders from the customer and is supplied by the distributor. The distributor receives goods from the manufacturer and the rest of the SC consists of three levels of suppliers: Tier 1, Tier 2 and Tier 3 suppliers, who deliver parts needed for the production. SC actors and the flows between them are shown in Fig. 1. Whereas the retailer receives exact demand from the final customer, all other actors only have the information about the orders from their direct predecessor in the SC which they need to fulfil. The information about orders on other stages or about the customer demand is not shared.

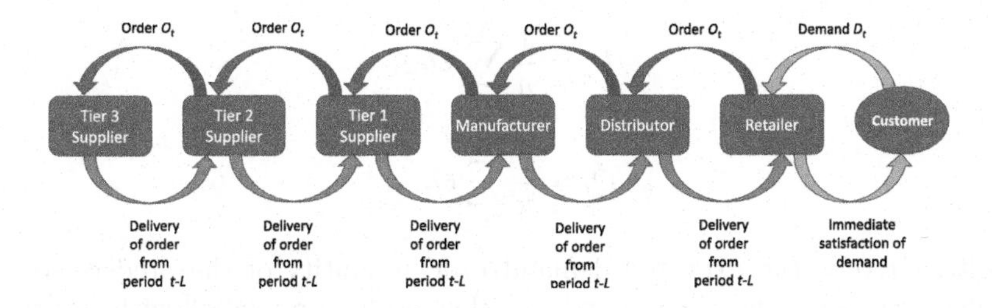

Fig. 1. Supply chain actors and flows between them

3.1 Basic Model

In the basic scenario, we assume that there is no information sharing and therefore only the orders of the predecessor are visible for each actor. Each SC actor uses the order-up-to inventory model, in which the following sequence of actions is observed in every ordering period: (1) at the beginning, a new order is placed, afterwards (2) the goods ordered in previous periods are received, (3) the demand is observed and satisfied, and (4) the end period inventory is counted. Similar to [23], the order quantity O_t in period t for each actor is calculated as:

$$O_t = S_t - S_{t-1} + D_{t-1} \tag{1}$$

for the retailer and

$$O_t = S_t - S_{t-1} + Q_{t-1} \tag{2}$$

for all other stages, where S_t denotes the order-up-to level in period t and D_{t-1}/Q_{t-1} represents the demand of the customer in case of the retailer or the quantity ordered by the predecessor from the last period for all the other stages.

The ordered quantities are delivered after a certain delivery time consisting of L periods. In order to cover the demand between the point when the order is placed and the moment when the goods are delivered, the respective order-up-to level should be chosen so that $L+1$ periods are covered [22], therefore,

$$S_t = (L+1) * \hat{D}_t \tag{3}$$

for the retailer and

$$S_t = (L+1) * \hat{Q}_t \tag{4}$$

for all other stages, where $\hat{D}_t/\hat{Q}_t$ is the expected customer's demand for the retailer or the expected order quantity from the predecessor for all the other stages. We consider two models for forecasting the expected demand/order quantity. Firstly, we use the moving average method [23], where the expected demand/order quantity is calculated as the average of observed values over the last n periods:

$$\hat{D}_t = \frac{1}{n}\sum_{i=1}^{n} D_{t-i} \tag{5}$$

$$\hat{Q}_t = \frac{1}{n}\sum_{i=1}^{n} Q_{t-i} \tag{6}$$

Alternatively, the expected demand/order quantity of the predecessor can be calculated using the simple exponential smoothing model [9] with smoothing parameter α (values between 0 and 1) where

$$\hat{D}_t = \alpha * D_{t-1} + (1-\alpha) * \hat{D}_{t-1} \tag{7}$$

$$\hat{Q}_t = \alpha * Q_{t-1} + (1-\alpha) * \hat{Q}_{t-1} \tag{8}$$

The inventory level is adjusted based on the received goods and the goods shipped to the preceeding stage. Unfulfilled demands/orders are considered as backlogs and will be fulfilled in the next period when sufficient goods are available. We also accept negative orders which represent returns of goods [24].

To summarize the logic of the model, each actor only receives information about the orders of its direct predecessor, which is the basis for calculating the expected demand/order quantity. Based on the expected demand/order, the new order-up-to level is calculated, which is together with the previous order-up-to level and the current order of the predecessor relevant for deciding about the order quantity in the current period that is then given to the next actor in the chain as input for his decision about the order-up-to level and the order quantity.

3.2 Considered Policies

In order to investigate the possibilities of using SMPC/HE in this setting, we extend the demand/order signal received by each actor by additional information about the orders of other preceeding SC actors. For this, we consider two policies that might help to mitigate the bullwhip effect: Firstly, we adjust the expected demand by the changes in the average SC inventory and, secondly, we combine the expected demand with the average order quantities in the SC. In this way, each SC stage receives additional information about changes in the demand that might be already present in the preceeding stages, but, in the basic case, they would be observed at the given stage only later due to the delays caused by delivery times. In this context, it is necessary to point out that we assume that all parties are acting honestly according to the assumptions of the model so that correct signal about the average order quantity is received by each SC actor.

Policy 1 - Average Inventory Changes: In this case, we extend the average expected demand/order quantity used for the calculation of S_t by the information about the changes in the average inventory level of all predecessors in the SC. Since the actors are usually reluctant to share the exact information, we use SMPC/HE to build an average inventory level at the end of each period. As there are at least two values needed to build an average, this policy is only applied to the manufacturer and suppliers. Therefore, the sum of inventory levels of retailer and distributor divided by two is reported as the average inventory $\bar{I}_t$ in period t to the manufacturer, Tier 1 supplier gets the inventory information based on the inventory levels of retailer, distributor and manufacturer and this continues up to Tier 3 supplier, who receives inventory information as the average of inventory levels of all five predecessors. Then the expected order quantity and the inventory level are combined in the following way:

$$\hat{UQ}_t = \beta * \hat{Q}_t + (1 - \beta) * (\bar{I}_{t-1} - \bar{I}_{t-2}) \tag{9}$$

where $\hat{UQ}_t$ is the updated expected order quantity that replaces $\hat{Q}_t$ in Eq. 4 and β is a smoothing factor with values between 0 and 1. We decided to use the changes in inventory level instead of the inventory level itself since the inventory level might be much higher than the relevant demand or orders.

Policy 2 - Average Orders: This policy is very similar to the previous one, but the difference of inventory levels is replaced by average orders of all predecessors. We again start with the manufacturer, who has direct access to the orders of the distributor, but does not know what the real demand of the customer and the order of the retailer were. Therefore these two values are again summed together and divided by two to build the average order quantity $\bar{Q}_t$ in period t. Tier 1 supplier receives the average of the order values from the customer, distributor and manufacturer and this again continues up to Tier 3 supplier who receives the exact order of Tier 2 supplier and the average orders based on all five previous SC stages. The updated expected order quantity is therefore calculated as

$$\hat{UQ}_t = \beta * \hat{Q}_t + (1 - \beta) * (\bar{Q}_{t-1}) \tag{10}$$

3.3 Model Implementation

The SC is built in form of a simulation model which allows to easily compare different demand patterns. Since each actor makes decisions independently based on the inputs coming from its predecessor (order quantity) as well as its successor (received goods), we chose to model each SC stage as an own agent in an agent-based simulation approach [25].

The model was created using Anylogic 7.2.0 University [26]. This simulation software allows to build agent-based models as well as to define different scenarios and experiments and to build graphs and export the results. Each actor is defined as an agent that follows its own logic, which is illustrated in Fig. 2 that shows the behavior of manufacturer in form of statecharts and transitions.

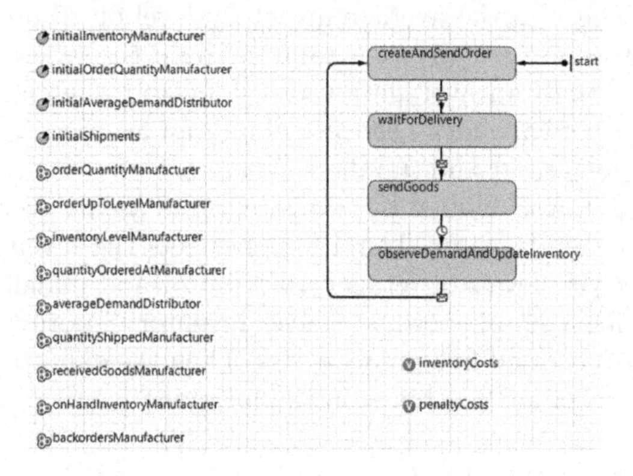

Fig. 2. Behaviour of manufacturer in Anylogic

As it can be observed, the manufacturer starts with creating an order that is then sent to Tier 1 supplier. Afterwards he waits for the delivery of goods ordered L periods before and together with receiving the goods from the supplier he also sends the goods to the distributor. At the end, he observes the demand from the distributor and updates the inventory level. The transitions are triggered by messages sent between the different agents. In addition to that, various parameters showing initial values for, e.g., inventory or average demand, can be defined. Moreover, multiple variables recording the values for, e.g., inventory level, received and shipped goods, demand, backlogs etc. were defined to be able to observe differences between models. The performance of the models is based on total SC costs which consist of inventory costs for each unit left in inventory and penalty costs for each unit of backlogs at the end of each period.

4 Case Study and Results

The case study used for the analysis of the proposed policies is based on the popular "beer game", described by e.g. [5]. In the original version of this game, each player represents one actor, including retailer, wholesaler, distributor and manufacturer. The retailer observes demand from the customer and orders goods from the wholesaler, who only receives the order information from the retailer and orders from the distributor without knowing the real demand of the customer. The delivery of the goods is delayed due to the time needed to prepare and ship the order. This works similar for the other stages. Due to the delays and missing information sharing, a small change in the demand of the customer creates the bullwhip effect within the whole SC.

In our case we use the initial data from [5]: we start with initial inventory of 12 units and planned deliveries of 4 units at each SC stage. The demand in the first periods is constant and amounts to 4 units. Based on the order-up-to model presented in Sect. 3.1, the order quantity is also stabilized in the first periods at 4 units. Delivery time L is 2 periods, inventory costs are 0.50 EUR/unit per period and penalty costs are 1 EUR/unit per period at each stage. We run the model for 50 periods and observe the total costs at the end of this time horizon.

After stabilizing the system in the first five periods, customer's demand increases to 8 units starting with period 6 and stays the same until the end of the game. This change in demand causes high fluctuations in orders and inventory levels on the different SC stages. This is illustrated in Fig. 3, where the inventory levels of all stages are shown when the basic model with moving average method from Eqs. 5 and 6 with $n = 5$ is used. As Fig. 3 shows, the inventories react with a slow decrease followed by a sharp increase and another sharp decrease. At the end there are some minor fluctuations until the inventory stabilizes after ca. 30 periods. The total costs after 50 periods are 3,984.5 EUR.

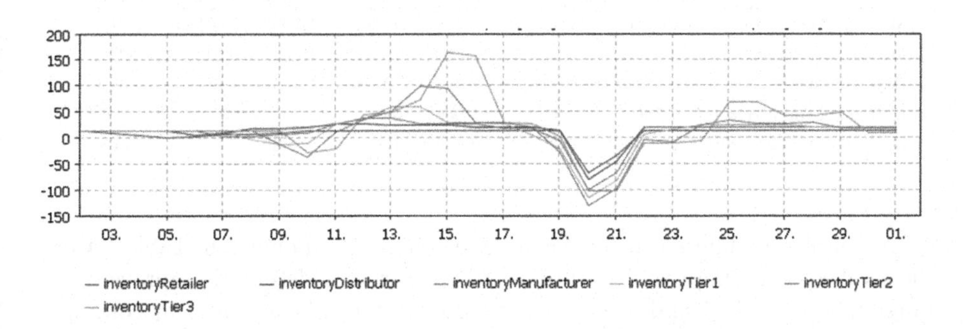

Fig. 3. Inventory levels for basic model with moving average ($n = 5$)

In comparison to that, if the basic model is combined with, e.g., Policy 2 (Average orders) with $\beta = 0.5$, the picture substantially changes, as it is shown in Fig. 4. Due to the additional information used, the fluctuations are flattened and the resulting total costs decrease by 33% to 2,661.5 EUR.

However, the flattening of the curves is not always the case, as our initial results for the two applied policies show. Especially in case of Policy 1 (see Table 1) the inclusion of the inventory differences does not help, but has a rather

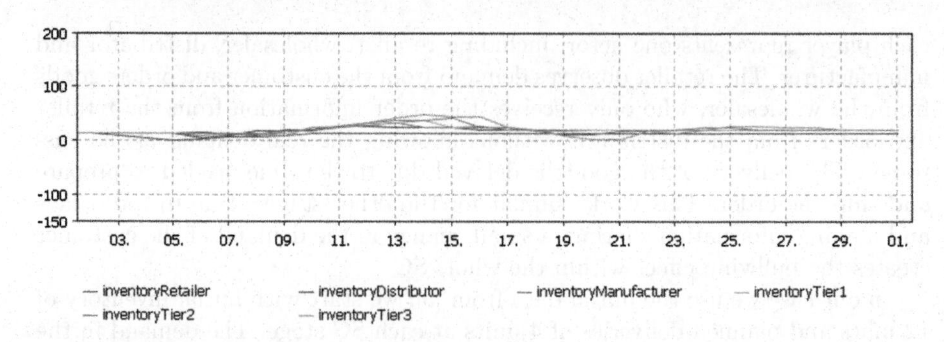

Fig. 4. Inventory levels for basic model with moving average ($n = 5$) combined with Policy 2 where $\beta = 0.5$

negative effect - due to the fact that the fluctuations of the inventories are very high, they even amplify their effect on the average expected demand/orders and therefore the resulting total costs are soaring. Only in cases where β is high and, thus, the inventories have a very low weight in the calculation, it might have positive impact on the total costs for some of the instances.

Table 1. Comparison of total costs for Policy 1

Basic model with	Basic total costs	Policy 1 - total costs and change with					
		$\beta = 0.1$	Change	$\beta = 0.5$	Change	$\beta = 0.9$	Change
Moving average $n = 3$	14,166.5	3.4e7	2.3e5%	2.6e5	1,700%	341,391	2,310%
Moving average $n = 5$	3,984.5	2.3e7	5.9e5%	10,211	156%	3,915	−2%
Exp. smoothing $\alpha = 0.1$	2,417.5	3.3e7	1.4e6%	33,485.5	1,285%	1,684.5	−30%
Exp. smoothing $\alpha = 0.5$	6,796	6.9e7	1.0e6%	3.6e8	5.3e6%	955,144.5	13,954%
Exp. smoothing $\alpha = 0.9$	192,425	4.1e8	2.1e5%	1.7e12	8.8e8%	3.5e11	1.8e8%

A completely different picture can be seen in case of Policy 2 in Table 2. Here the total costs are improved for 4 of the 5 considered models, with improvements ranging between 17% and 99%. These high savings can be achieved by considering the average orders in SC which offer an additional signal about the changes that might be happening at the preceeding stages. The reason why the costs are slightly higher in case of the exponential smoothing model with $\alpha = 0.1$ might be that due to the low α-value a lot of demand smoothing already happens in the basic model and the average orders do not bring much additional value.

Table 2. Comparison of total costs for Policy 2

Basic model with	Basic total costs	Policy 2 - total costs and change with					
		$\beta = 0.1$	Change	$\beta = 0.5$	Change	$\beta = 0.9$	Change
Moving average $n = 3$	14,166.5	2,634.5	−81%	3,291	−77%	10,500.5	−26%
Moving average $n = 5$	3,984.5	2,618	−34%	2,661.5	−33%	3,326.5	−17%
Exp. smoothing $\alpha = 0.1$	2,417.5	2,526.5	5%	2,529.5	5%	2,484.5	3%
Exp. smoothing $\alpha = 0.5$	6,796	2,625.5	−61%	2,832	−58%	5,017.5	−26%
Exp. smoothing $\alpha = 0.9$	192,425	2,691.5	−99%	25,533	−87%	140,919.5	−27%

When looking at the changes in costs between the basic case and Policy 2, it can be seen that including the additional information significantly reduces the penalty costs at all stages. For Policy 2 with β-factors 0.1 and 0.5, penalty costs are reduced by 90–100%, whereas in case of $\beta = 0.9$ the reductions are lower, but still significant, ranging between 27% and 52%. This results from the fact that the upstream SC actors can observe the increasing average demand in the downstream stages and prepare earlier for the increasing order quantities. Once the higher order is placed by the predecessor, there is already sufficient inventory available to satisfy that order and, thus, backlogs are eliminated. As a consequence, inventory costs at the downstream stages, especially for the retailer, distributor, manufacturer and Tier 1 supplier, are slightly increasing, on average by 4–18%. However, this increase is overcompensated by the savings in penalty costs and therefore every SC actor can reduce its overall costs, as illustrated in Table 3. Here the average savings in total costs over all demand forecasting models for each actor are shown if Policy 2 is employed. The highest savings are achieved by the Tier 2 and Tier 3 suppliers, who also had the highest total costs, and then by the retailer, who now has sufficient inventory and can therefore better serve customers. Relative savings of the actors in the middle of the SC are slightly lower, but still significant. However, the presented results only represent first investigations of the model and policies and therefore more detailed analysis about the influence of the different factors will be needed in the future.

Table 3. Average costs savings for each SC actor for Policy 2

Policy 2 with	Average total cost change over all demand forecasting models for					
	Retailer	Distributor	Manufacturer	Tier 1 sup.	Tier 2 sup.	Tier 3 sup.
$\beta = 0.1$	−52.35%	−46.73%	−46.89%	−51.29%	−56.81%	−61.16%
$\beta = 0.5$	−49.81%	−44.28%	−43.87%	−47.12%	−51.84%	−55.54%
$\beta = 0.9$	−41.61%	−36.67%	−37.09%	−39.87%	−43.98%	−46.41%

5 Conclusion

Sharing information between the different SC actors is important to avoid fluctuations in inventory levels defined as bullwhip effect. However, practice shows that the actors are reluctant to share any additional information due to various reasons and, therefore, only the minimum necessary information is shared. In order to contribute to the solution of this problem, we proposed to avoid direct information sharing and only calculate average values for inventory or orders using SMPC/HE techniques. As our first results show, this method might have some potential in mitigating the bullwhip effect, but further investigations regarding the best policies are necessary. Moreover, extensions of the model to more complex supply chain networks as well as investigations of additional demand models and scenarios are necessary to increase the robustness of the results. These points should be tackled in further research.

References

1. Lee, H.L., Padmanabhan, V., Whang, S.: Information distortion in a supply chain: the bullwhip effect. Manage. Sci. **43**(4), 546–558 (1997)
2. Cramer, R., Damgard, I.B., Nielsen, J.B.: Secure Multiparty Computation and Secret Sharing. Cambridge University Press, New York (2015)
3. Acar, A., Aksu, H., Uluagac, A.S., Conti, M.: A survey on homomorphic encryption schemes: theory and implementation. ACM Comput. Surv. **51**(4), 1–35 (2018)
4. Paillier, P.: Public-key cryptosystems based on composite degree residuosity classes. In: Stern, J. (ed.) EUROCRYPT 1999. LNCS, vol. 1592, pp. 223–238. Springer, Heidelberg (1999). https://doi.org/10.1007/3-540-48910-X_16
5. Sterman, J.D.: Teaching takes off - flight simulators for management education - "The Beer Game". OR/MS Today, pp. 40–44 (October 1992)
6. Forrester, J.W.: Industrial dynamics - a major breakthrough for decision makers. Harv. Bus. Rev. **36**(4), 37–66 (1958)
7. Chen, F., Drezner, Z., Ryan, J.K., Simchi-Levi, D.: Quantifying the bullwhip effect in a simple supply chain: the impact of forecasting, lead times, and information. Manage. Sci. **46**(3), 436–443 (2000)
8. Croson, R., Donohue, K.: Upstream versus downstream information and its impact on the bullwhip effect. Syst. Dyn. Rev. **21**(3), 249–260 (2005)
9. Dejonckheere, J., Disney, S.M., Lambrecht, M.R., Towill, D.R.: Measuring and avoiding the bullwhip effect: a control theoretic approach. Eur. J. Oper. Res. **147**, 567–590 (2003)
10. Zhu, Z., Lambotharan, S., Chin, W.H., Fan, Z.: A mean field game theoretic approach to electric vehicles charging. IEEE Access **4**, 3501–3510 (2016)
11. Yang, P., Freeman, R.A., Lynch, K.M.: Multi-agent coordination by decentralized estimation and control. IEEE Trans. Autom. Control **53**(11), 2480–2496 (2008)
12. Kia, S.S., Van Scoy, B., Cortes, J., Freeman, R.A., Lynch, K.M., Martinez, S.: Tutorial on dynamic average consensus: the problem, its applications, and the algorithms. IEEE Control Syst. Mag. **39**(3), 40–72 (2019)
13. Zargham, M., Ribeiro, A., Jadbabaie, A.: Discounted integral priority routing for data networks. In: 2014 IEEE Global Communications Conference (2014)

14. Kerschbaum, F., et al.: Secure collaborative supply-chain management. Computer **44**(9), 38–43 (2011)
15. Catrina, O., de Hoogh, S.: Secure multiparty linear programming using fixed-point arithmetic. In: Gritzalis, D., Preneel, B., Theoharidou, M. (eds.) ESORICS 2010. LNCS, vol. 6345, pp. 134–150. Springer, Heidelberg (2010). https://doi.org/10.1007/978-3-642-15497-3_9
16. Albrecht, M., Stadtler, H.: Coordinating decentralized linear programs by exchange of primal information. Eur. J. Oper. Res. **247**(3), 788–796 (2015)
17. Pibernik, R., Zhang, Y., Kerschbaum, F., Schroepfer, A.: Secure collaborative supply chain planning and inverse optimization - the JELS model. Eur. J. Oper. Res. **208**(1), 75–85 (2011)
18. van Engelenburg, S., Janssen, M., Klievink, B.: A blockchain architecture for reducing the bullwhip effect. In: Shishkov, B. (ed.) BMSD 2018. LNBIP, vol. 319, pp. 69–82. Springer, Cham (2018). https://doi.org/10.1007/978-3-319-94214-8_5
19. Ghode, D.J., Yadav, V., Jain, R., Soni, G.: Lassoing the bullwhip effect by applying blockchain to supply chains. J. Glob. Oper. Strateg. Sourcing **15**(1), 96–114 (2022)
20. Sarfaraz, A., Chakrabortty, R.K., Essam, D.L.: A blockchain-coordinated supply chain to minimize bullwhip effect with an enhanced trust consensus algorithm. Preprints (2021)
21. Zhong, H., Sang, Y., Zhang, Y., Xi, Z.: Secure multi-party computation on blockchain: an overview. In: Shen, H., Sang, Y. (eds.) PAAP 2019. CCIS, vol. 1163, pp. 452–460. Springer, Singapore (2020). https://doi.org/10.1007/978-981-15-2767-8_40
22. Lee, H.L., Kut, C.S., Tang, C.S.: The value of information sharing in a two-level supply chain. Manage. Sci. **46**(5), 626–643 (2000)
23. Damiani, E., Frati, F., Tchokpon, R.: The role of information sharing in supply chain management: the secure SCM approach. Int. J. Innov. Technol. Manag. **8**(3), 455–467 (2013)
24. Chatfield, D.C., Kim, J.G., Harrison, T.P., Hayy, J.C.: The bullwhip effect - impact of stochastic lead time, information quality, and information sharing: a simulation study. Prod. Oper. Manage. **13**(4), 340–353 (2004)
25. Borshchev, A.: The Big Book of Simulation Modeling: Multimethod Modeling with AnyLogic 6. AnyLogic North America, Chicago (2013)
26. AnyLogic: AnyLogic Timeline - Anylogic 7.2. https://www.anylogic.com/features/timeline/. Accessed 13 Apr 2022

Reporting of Cross-Border Transactions for Tax Purposes via DLT

Ivan Lazarov[1(✉)], Quentin Botha[2], Nathalia Oliveira Costa[1], and Jakob Hackel[2]

[1] Institute for Austrian and International Tax Law, Vienna University of Economics and Business, Vienna, Austria
ivan.lazarov@wu.ac.at

[2] Research Institute for Cryptoeconomics, Vienna University of Economics and Business, Vienna, Austria

Abstract. Finding the right balance between effective cross-border exchange of tax information and limiting tax authorities access to sensitive data of foreign taxpayers is among the key issues for international tax policy. The legal protection of private and commercially sensitive information, as well as the need to demonstrate that data is foreseeably relevant before requesting it are amongst the main backstops to having a symmetrical data flow between purely domestic and cross-border tax information. Our paper suggests a technological solution that strikes a balance between privacy and cross-border transparency. All taxpayers involved in a cross-border transaction need to report the transactional data to their domestic tax authorities. The tax authorities transform the standardized transactional data with a hashing algorithm and upload the resulting hash to a shared, permissioned blockchain platform. If both parties to a transaction reported it to their domestic authorities, two identical hashes would appear on the blockchain, raising no concern. If one of the parties fails to report, only one hash would appear, demonstrating non-reporting on one side of the border. This would give sufficient grounds to consider that the sensitive information underpinning the transaction is foreseeably relevant for establishing tax liability leading to traditional exchange of information between the authorities. Based on this, a failure to report would be detected both for income and VAT purposes thereby substantially reducing the possibility for tax evasion.

Keywords: Tax policy · Tax law · VAT · Tax evasion · DLT · Blockchain · Cryptoeconomics · Exchange of information

1 Introduction

Contemporary tax systems rely predominantly on self-declaration for establishing tax liability. Hence, the accurate assessment of taxes depends upon the ability of tax authorities to verify whether the information declared by taxpayers corresponds to reality. When the tax authorities lack this ability - e.g. because they

G. Kotsis et al. (Eds.): DEXA 2022 Workshops, CCIS 1633, pp. 282–293, 2022.
https://doi.org/10.1007/978-3-031-14343-4_26

have no access to accurate and timely information - there is a significant margin for tax fraud caused either by misreporting altogether or by the time gap between the event and its discovery. Generally, tax audits are performed by matching the information available in the tax accounting of a given taxpayer with the one available from other parties that have had dealings with this taxpayer, as well as against the available assets such as cash or stock. Hence, efficient auditing requires that tax authorities have access to transactional and financial data held by third parties. In this way, what has been declared by the audited party can be cross-checked. In a domestic context, gaining access to information held by third parties is relatively straightforward as everyone involved falls under the jurisdiction of one country (that can generally enforce its laws domestically to all private parties). In a cross-border context on the other hand, establishing effective exchange of tax information (EoI) between countries is imperative. Yet, cross-border EoI still significantly lags behind what is available domestically. There are two reasons for this: the limited amount of information that is automatically exchanged and the prohibition of so-called 'fishing expeditions' - i.e. a generalized request by a country for information that might have some relevance for establishing tax liability on its territory without first demonstrating any foreseeable relevance of the information. These limitations are enhanced in an analogue world due to the administrative burden that automatic EoI entails for tax administration (in terms of processing the information and in terms of gathering it for the purposes of sharing) and the privacy of taxpayers, which should not be subject to a generalized interference in their personal sphere without any preexisting foreseeable relevance of the information that is being exchanged for establishing their tax liability. Thus, the question that arises is how to make the international exchange of tax information more efficient as regards cross-border transactional data whilst simultaneously ensuring privacy, compliance with EoI standards and keeping the administrative burden low.

2 Aim and Concept

The goal of this paper is to suggest a technological solution to that question. This solution should have the following features: 1) deployable on a large scale, encompassing all taxpayers engaged in cross-border B2B activity; 2) preserve the privacy of taxpayers while being able to red-flag instances which need further inquiry; 3) allow for (near) real time data flow and; 4) be capable of automation and integration with existing processes. In an oversimplification, existing business systems for e-invoicing allow for issuance of electronic invoices in a standardized, structured format that is automatically incorporated in the seller and buyer's accounting systems. Some tax authorities create a framework enabling them access to the information flowing on e-invoicing systems for the reason of receiving real time transactional data that they can later match with what is being reported by the taxpayers (or possibly even fully automate the tax compliance process). While such systems function well for purely domestic situations (where both the buyer and the seller are domestic entities), they do not in cross-border contexts due to the aforementioned limitations of existing EoI solutions, taxpayers' right to privacy, and the lack of interoperability between the different domestic systems. We therefore suggest an alternative system for cross-border reporting of transactions. This system is based upon the idea of 'same utility of the information flow, without the actual information flow'. It requires the establishment of a platform standardizing the data on transactions, independent of the natural language used at the domestic level. Such standardized information is then transmitted to the tax authorities of each party to the transaction (both by the buyer and the seller to their respective tax authorities). At the level of the tax authorities, the standardized information is hashed, with the hash being uploaded on a permissioned blockchain system run by participating tax authorities. If both parties to the transaction duly reported it domestically, two identical hashes would appear on the blockchain system. If one of the parties failed to report or reported different information, the system would red-flag that only one hash is uploaded and further investigation would be needed. It should be noted that establishing an e-invoicing system prior to the implementation of this solution is desirable, but not necessary as the system concerns only B2G communication, thereby not preventing the B2B utilization of paper invoices.

2.1 State of Existing EOI System

Contemporary international EoI systems are predominantly based on the framework established by the OECD under the Global Forum on Transparency and Exchange of Information for Tax Purposes and a network of international agreement that is in force between the different jurisdictions around the globe.[1] However, as work on the global level relies on a complicated interrelationship between soft law (non-binding) instruments issued by international standardizing bodies such as the OECD and their implementation in hard law (binding) instruments agreed upon by countries, we focus on one of the most developed systems for cross-border cooperation that is coordinated by hard law uniform rules, namely that which is established within the European Union (EU). If our findings regarding the deficiencies stand in an EU context, they would be a fortiori valid for the other existing forms of international EoI coordination. The main legal instrument that governs the rules on EoI in an EU context is the Directive on Administrative Cooperation (DAC) [6]. The system for cooperation established under DAC contains three main forms of exchange of information: EoI on *request*, [6, Sect. 1] *automatic* EoI, [6, Sect. 2] and *spontaneous* EoI [6, Sect. 3]. EoI on request relates to the tax authorities of one Member State exchanging tax information that they have in their possession or should collect upon the request of the tax authorities of another Member State. Automatic EoI concerns an exhaustive list of tax relevant information that is exchanged on regular intervals between the tax authorities of Member States. Finally, spontaneous EoI concerns the sharing of information between tax authorities that is considered relevant for the assessment of tax liability in another Member State. However, there are key limitations underpinning the intra-EU EoI legal framework. The first is the requirement

[1] See oecd.org/tax/transparency/.

established under Art. 1(1) DAC that information exchanged must be already "foreseeably relevant" for the establishment of tax liability at the moment of exchange. Thus, there must be already a reasonable suspicion that the information might be relevant for the assessment of the tax liability of a particular taxpayer, excluding therefore generalized inquiries. Hence, "Member States are not at liberty to engage in 'fishing expeditions' or to request information that is unlikely to be relevant to the tax affairs of a given taxpayer" [3]. A request would not lack foreseeable relevance "inasmuch as it concerns contracts, invoices and payments that were concluded or carried out, during the period covered by the investigation, by the person holding the information concerning those contracts, invoices and payments, and that are connected with the taxpayer concerned by that investigation" [4]. A similar standard exists under the framework established by the OECD [13, Article 26(1)]. An exception to this rule is the systemic exchange of predefined information under the rules of automatic EoI. This predefined information is irrebuttably presumed to be foreseeably relevant but is limited to a specific set of circumstances thereby preventing generalized cross-border data flows of unlimited taxpayers' information. Automatic EoI covers, for instance, employment income and financial accounts of non-residents but is far from including all cross-border activities that likely impact the tax position of taxpayers. Moreover, the intervals in which information is currently exchanged vary from a year to several months, bringing about substantial time lag before information is available. Hence, the system for EoI proposed in this paper does not substitute but rather complements the system of automatic EoI. It allows for encompassing all cross-border activities but provides for actual exchange of taxpayers' information only when a transaction is red-flagged by the system.

Data Protection Standards. Certain data protection standards must be followed by all jurisdictions taking part in the Convention on Mutual Administrative Assistance in Tax Matters,[2] a joint initiative by the OECD and the Council of Europe since 1988 and amended in 2010, which prohibits arbitrary or unlawful interference in the privacy of an individual [11]. When applying the mutual assistance framework, there must be a balance between exchanging information and protecting the confidentiality of personal data. Articles 21 and 22 of the Multilateral Convention on Mutual Administrative Assistance in Tax Matters state that the jurisdiction receiving the information [14]:

- must maintain the confidentiality and protection of the data;
- is limited on how to use the information and only allowed to disclose it to persons or authorities *"concerned to the assessment, collection or recovery of, the enforcement or prosecution in respect of, or the determination of appeals in relation to, taxes of that Party, or the oversight of the above"*;
- is required to seek authorization of the supplying state for onward transfer of the data to a third state.

[2] Convention on Mutual Administrative Assistance in Tax Matters - OECD.

The EU data protection framework goes even further in protecting the processing of personal data and the free movement of such data (due to the DAC on administrative cooperation in the field of taxation and the General Data Protection Regulation, GDPR) [8]. Under the GDPR, authorities must have the ability to modify or delete data to comply with legal requirements, since individuals have the right to demand that their personal data is rectified or deleted ('right to be forgotten') under certain circumstances [8, Chap. 3]. In addition, individuals have the right to obtain confirmation as to whether or not his/her personal data are being processed. Where that is the case, access to the personal data is ensured [8, Article 15]. Furthermore, authorities must adhere to data minimisation to limit the quantity of private information being held.

2.2 State of the Distributed Ledger Technology

This proposal makes use of Distributed Ledger Technology (DLT), which caught public attention with its implementation in Bitcoin [12], a public permissionless blockchain facilitating the exchange of value in the form of cryptocurrency. DLT enables users to share, use, and agree upon a common set of information without the need of a trusted party or intermediary. At its core, DLT provides the technical capability for maintaining a decentralized, time-stamped database. This characteristic means its applications extend beyond just cryptocurrency. There are two main subcategories of blockchains, a specific application of DLT, relevant to our discussion: public and permissioned. Public blockchains allow anyone to read and store all transactions as well as participate in building new blocks, a bundle of transactions recorded on a distributed ledger and sequentially linked through timestamps. While public blockchains exhibit strong resilience and immutability for already settled transactions, they do come with significant overhead of coordinating new changes among an unknown set of participants. This is known as consensus, and in the current most popular public blockchains (Bitcoin, Ethereum) is handled through a process called Proof-of-Work (POW) [2]. Permissioned blockchains on the other hand require coordination only among their trusted and known participants, reducing overhead and energy expenditure dramatically but at the expense of resilience and immutability. In this proposal, we employ a permissioned blockchain used for storing real-time transaction data for the purpose of tax compliance. While this is not the first proposal of its kind, [9] this is the first concept that reflects a comprehensive system capable of simultaneous direct and indirect tax compliance. In practice, this entails the sharing of information cross-border on a need-to-know basis and makes existing methods of VAT fraud essentially impossible.

Hashing. Initially developed as a technique to efficiently search and validate data, hashes are now essential in many modern applications such as cloud services, data structures, and password verification, among others [15]. In blockchain applications, such as Bitcoin, hashing is used to easily identify transaction and ledger legitimacy. Without hashing, the censorship-resistance of public blockchains would not be possible. Hashing algorithms have favorable qualities for our application. Cryptographic hash functions accept any data of arbitrary length as input (in our case basic information extracted from an invoice) and efficiently produce a fixed length output (called a "hash") in a process that is infeasible to reverse. No information can be gained from the hash without having the original data used as input (i.e., it is not possible to obtain the original information when starting from the hash). The same input data always produces the same output hash. Any change to the input (invoice information) results in a drastically different output. Any two parties (in our case tax authorities) can extract a pre-defined part of the information contained in an invoice and use a hashing algorithm to yield an identical hash, without any coordination or data exchange. This allows for the exchange of hashes without privacy concerns. Exchanging hashes instead of actual invoice data enables efficient and secure matching of transactions by those who already have the original data, without revealing any private information.

3 Implementation of Blockchain in the EOI System (POC)

We propose a basic system, which on a higher level can be functionally separated into two subsystems. The first is accessed and used separately by each tax authority and deals with the operations for receiving invoices and transforming them to hashes, all happening on the tax authority's servers. The second is shared equally between all tax authorities and serves as our distributed ledger for red-flagging unreported tax transactions. Our focus in this paper lies on the shared system, but the first system is included to show the basic functionality and potential improvement on current cross-border tax reporting. This proposal could be extended with larger scope, various complexities and technical creativity. We want to show that even a highly simplified and permissioned blockchain can offer clear benefits for the described use-case.

3.1 Processing Invoices

To better explain the system, we will follow a standard process where invoices are either matched, without sharing any private information, or red-flagged, where only the information relevant to this exact invoice is shared, avoiding fishing expeditions. Our proposal makes use of the SHA-3 family of hashing algorithms [7]. However, practically any hashing algorithm considered secure and efficient should be usable for implementations. Note that while we describe participants following these processes, everything we describe can be automated and does not introduce any significant additional work to tax authorities. An Austrian company (Aco) sells goods to a French company (Bco) for EUR 1000. Aco issues an invoice[3] containing a set of pre-determined information which

[3] In an optimal case, this is a standardized e-invoice.

they send to the domestic tax authority in Austria and to the buyer in France. Bco incorporates the invoice in its accounts and also transmits it to the French tax authority. At this point, both the tax authorities in Austria and France have possession over an invoice describing the same purchase of goods, including identical details on the buyer, seller and invoice number. The full invoice is stored by the tax authorities independently, who separate the relevant data from the invoice. Following a standardized process, this data is used as input for the hashing algorithm, resulting in a hash. Both tax authorities, if following the process correctly, arrive at the exact same hash. Then, the Austrian tax authority adds to their hash an identifier for France, while the French tax authority adds an identifier for Austria. Both tax authorities then each propose this appended hash to the blockchain. The same hash will now appear twice on the blockchain - once submitted by Austria, with a French identifier, and once submitted by France, with an Austrian identifier. If both hashes appear, each side knows that the same invoice was reported in the respective other country too. If the hash appears only once, either only one company reported the invoice or an error in the process led to a different hash (in which case two lone hashes should appear). Each tax authority can actively scan for hashes with their country identifier. Unmatched hashes are red-flagged and investigated further by requesting more information on the specific invoice that resulted in the lone hash. Each tax authority submits the country identifiable hashes to the shared ledger, in a process described further below. As each tax authority only submits hashes of invoices that they received from companies in their own jurisdiction, and signs them with their private key, the submissions can be tracked to both concerned countries, even though no actual invoice information was disclosed. The only information available to all parties in the system is that a transaction occurred between two countries. No information about the companies, the amounts, the types of goods or any other private information can be extracted from this hash. This allows every tax authority to efficiently verify whether any transactions need red-flagging and further EoI. The only requirement is that at least one of the countries received an invoice to create the published hash, in which case at least one country has the details to exchange. Naturally, therefore, if both the payor and the payee do not report a transaction to their local tax authorities, the system is of no use. Thus, it cannot detect income tax evasion stemming from the fact that both parties to the transaction are parts of the informal economy. However, for example, VAT fraud is generally dependent on at least one of the parties reporting the transaction so that a VAT refund can be claimed.

Necessary Data. While electronic invoices today are not yet standardized on a worldwide scale, most economic blocks have some form of standardization. These standards usually include information such as machine-readable unique invoice numbers, unique identifiers for each respective party, timestamps, but also natural language descriptions of goods and services, names and addresses for the respective parties, and more [5, Articles 217-240]. For the purpose of our proposed system, only minimal data should enter the system. From an invoice we can easily extract the machine readable invoice ID number, seller ID number and buyer ID number. This information does not differ depending on who extracts this information, as natural language might. This minimum set of information is the only input to the hashing function, ensuring that each hash is unique to each transaction, while also limiting potential for errors and privacy breaches. This also results in transactions looking quite different from regular blockchain transactions. Since there is no exchange of value, a transaction consists only of the country identifier and the respective hash. Since tax authorities construct blocks only with transactions they themselves created, they only need to sign the full block instead of every transaction separately. The full information, stored separately by the individual tax authorities, will only be used in case there is a specific request for EoI and for specific dispute resolution, which avoids fishing expeditions.

Permissioned Setup. While a public blockchain network would bring higher censorship resistance and immutability, these characteristics are not strictly necessary for the purpose of this system. Instead, a permissioned network allows for tighter access control, strongly increased scalability, reduced complexity, fewer privacy concerns and very limited energy expenditure. As only the tax authorities have access, no outside user sees any data exchanged, nor would they gain any meaningful information from the hashes exchanged. The tax authorities share and co-own the infrastructure on equal footing. For this basic system, we propose a simple Proof-of-Authority (PoA) consensus mechanism, where all changes are submitted by trusted entities, known generally as "authority nodes". They share responsibility and duties and operate on a "one entity one vote" basis. Tax authorities are selected in a round-robin format to propose a new block. The chosen tax authority picks a set of transactions from its own storage, constructs a block according to a predefined format and proposes it to the other tax authorities. All tax authorities will then vote on the finalization of the proposed block[4]. As soon as a block is accepted by a majority of the tax authorities, all tax authorities store this block as an update to the ledger, which finalizes all transactions contained within the block. Then, a new round begins and the next tax authority is selected to create a block from their collection of transactions. If the block is not accepted, no new block is added to the ledger. A new tax authority is then selected by the PoA mechanism to propose a block. The proposing tax authority will have a new chance to construct a valid block in the next round. This simple model allows for high efficiency, gives tax authorities equal power, and requires neither transaction fees nor high energy expenditure.

[4] As mentioned before, these processes are automated to the point where "voting" means that software validates the proposed entries against a set of rules.

3.2 Implications of Blockchain in the EOI System

As explained above, the current EOI system could be improved in a cross-border scenario by increasing the speed of processing and sharing information between authorities, reducing the administrative burden to fulfill the mutual assistance obligations, and shortening (or even eliminating) the time-lag in which the information is exchanged, also taking into account the limitations imposed by data protection regulations. An illustration of how the inefficiencies in EoI facilitates tax fraud is evident in the EU VAT system, which is vulnerable to sophisticated and systematic fraud schemes, resulting in a significant gap in collection for EU Member States (around EUR 140bln in 2018) [10]. A key issue is that the right to deduct input VAT is, generally, detached from reporting or actual VAT payment by the supplier to the tax authorities, and intra-Community supplies are zero rated. This is the basis for carousel fraud, illustrated in Fig. 2 in the appendix. The carousel fraud constitutes a B2B circular-based scheme involving at least 3 parties - in our example below Aco, Bco and Cco. Aco is located in Austria, while Bco and Cco are in France. Aco sells products to Bco and charges 0% VAT as this constitutes an intra-Community supply of goods.[5] Bco then sells the products to Cco, which is considered a domestic supply and VAT applies. Bco does not remit this VAT to the tax authorities of France, but Cco, who sells the products back to Aco, files a claim for refund of VAT invoiced to it by Bco. Cco itself incurs no VAT liability as the intra-Community supply is with a VAT zero-rate. As a result Cco receives a VAT refund in France without VAT ever being paid in the country. The goods start and end at the same trader (Aco) and usually no actual delivery of goods takes place (only a paper trail is created) [1]. Also usually, there would be a number of other entities in France that are interposed between Bco and Cco creating a buffer and diversion between the failure to pay VAT (Bco) and the VAT refund (Cco). By the time the information is received and processed, all companies involved are left with little or no assets, making it difficult for authorities to collect any VAT due.

The solution that we propose in this paper reduces greatly the possibility to engage in VAT fraud as the one described above. If Aco, Bco and Cco have to all report the transactions to the authorities, then at the moment when Cco claims a refund, it would be visible that the 'missing trader' Bco failed to report the intra-Community transaction related to acquiring the goods (Aco-Bco) or that there is VAT that it still owes under the domestic transaction (Bco-Cco). A further layer of sophistication would require that the system is capable of matching not only the reporting of invoices but also give information whether the respective VAT under these invoices is paid to the tax authorities. In case the amount due does not correspond to the amount paid, the system can red-flag the given transaction as suspicious and require further investigation before a corresponding VAT deduction is granted. In the context of corporate or personal income tax, cross-border tax underreporting is relatively straightforward. Aco is an Austrian company (could also be an individual) that provides services to Bco,

[5] However, this is not an exempted transaction and therefore Aco is entitled to receive a refund on its own input VAT.

tax resident in France (see Fig. 3 in the appendix). The corresponding payment constitutes a taxable income for Aco and a deductible expense for Bco. If the transaction was purely domestic, a failure to report the income on the side of Aco could be caught when auditing the books of Aco and Bco and seeing that the tax deduction of Bco leads to no corresponding inclusion of income in Aco. However, in a cross-border transaction, unless suspicion arises in Austria and a reasonable ground for asking for information exchange with France is established, the Austrian authorities would have no way of finding out that their resident taxpayer generated taxable income.

The solution that we are proposing would make such failure to report cross-border income by Aco easily discoverable. It would highlight all instances when Bco reports the transaction (and it has the incentive to do so, since only in this way it could deduct the expense from its tax base). Hence, in such instances France and Austria would be notified that a transaction between these two countries has occured while it was reported only in France.

4 Conclusion

Our proposal improves on EoI efficiency without increasing administrative burden at the institutional level by retaining the same authorities as are characteristic of contemporary systems. Our solution can be adapted to various jurisdictions by standardizing the set of information extracted from invoices. It makes both failure to report cross-border transactions for VAT and income tax purposes easily discoverable. Additionally, new members can always be onboarded through rights management, extending the set of authority nodes. To comply with privacy obligations, authorities must have the ability to modify or delete the collected data. While public blockchains are considered immutable, our proposal does not suffer the same hindrance as authorities can vote to make modifications to the ledger. Given the nature of participants in this EoI scheme, censorship is not a concern, as each tax authority has an incentive to match their hashes, which requires the other tax authorities to submit their hashes too. The benefits of reduced overhead and simplified consensus outweigh threats to resilience and immutability. This also enables further internal conflict resolution and adheres to privacy principles. Moreover, the limited information necessary to store results in limited overhead and easily manageable processes, while complying with legal requirements.

Appendix

(See Fig. 1)

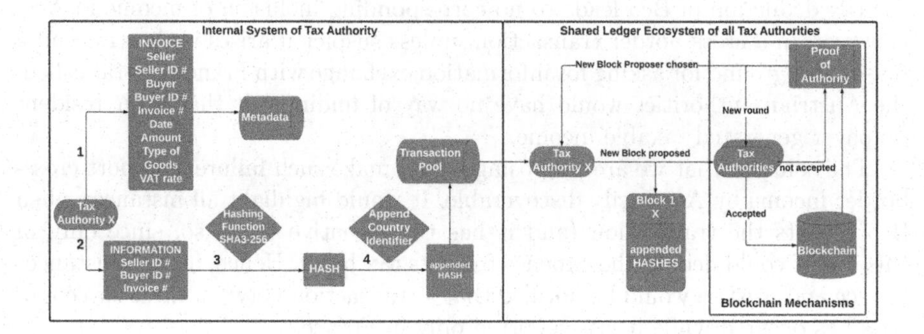

Fig. 1. Overview of the system

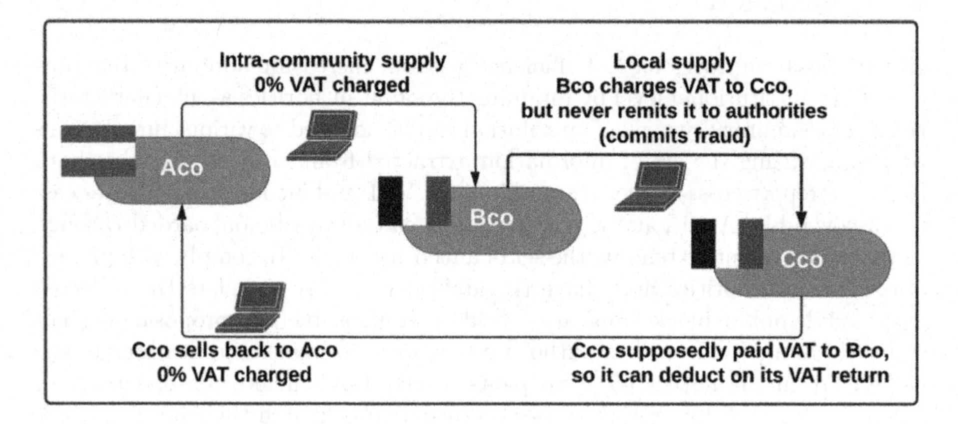

Fig. 2. VAT fraud

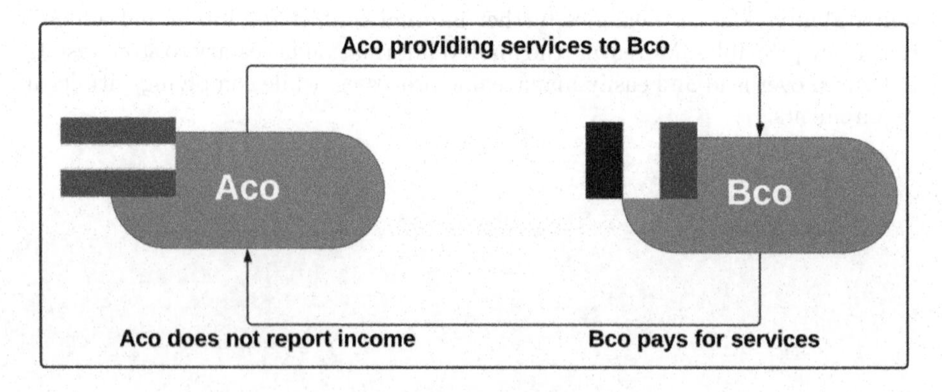

Fig. 3. Income Tax

References

1. van Brederode, R.: Combating carousel fraud: the general reverse charge vat, p. 146. International VAT Monitor (May 2015)
2. Buterin, V.: Ethereum: a next-generation smart contract and decentralized application platform (2014). https://github.com/ethereum/wiki/wiki/White-Paper. Accessed 22 Aug 2016
3. CJEU: Case C-682/15, Berlioz Investment Fund, EU:C:2017:373, para. 66 (2017)
4. CJEU: Case C-245/19, État luxembourgeois (droit de recours contre une demande d'information en matière fiscale), para. 120 (2020)
5. Council of European Union: Council directive 2006/112/EC on the common system of value added tax (2006). https://eur-lex.europa.eu/legal-content/EN/ALL/?uri=CELEX:32006L0112
6. Council of European Union: Council directive 2011/16/EU on administrative cooperation in the field of taxation and repealing directive 77/799/EEC (2011). https://eur-lex.europa.eu/legal-content/EN/TXT/?uri=CELEX%3A02011L0016-20200701
7. Dworkin, M.: SHA-3 standard: permutation-based hash and extendable-output functions (04 August 2015). https://doi.org/10.6028/NIST.FIPS.202
8. European Parliament and Council of the European Union: Regulation (EU) 2016/679 on the protection of natural persons with regard to the processing of personal data and on the free movement of such data, and repealing directive 95/46/EC (general data protection regulation) (2016)
9. Jafari, S.: Combining modern technology and real-time invoice reporting to combat VAT fraud: no revolution, but a technological evolution (2020). https://books.google.at/books?id=QFPSzQEACAAJ
10. Karaboytcheva, M.K.: Addressing the vat gap in the EU (2020). https://www.europarl.europa.eu/RegData/etudes/BRIE/2020/659423/EPRS_BRI(2020)659423_EN.pdf
11. Krähenbühl, B.: Personal data protection rights within the framework of international automatic exchange of financial account information (2018). https://books.google.at/books?id=WemezgEACAAJ
12. Nakamoto, S.: Bitcoin: a peer-to-peer electronic cash system (December 2008). https://bitcoin.org/bitcoin.pdf. Accessed 18 Mar 2022
13. OECD: model tax convention on income and on capital: condensed version 2017 (2017). https://doi.org/10.1787/mtc_cond-2017-en. https://www.oecd-ilibrary.org/content/publication/mtc_cond-2017-en
14. OECD, Council of Europe: The multilateral convention on mutual administrative assistance in tax matters (2011). https://doi.org/10.1787/9789264115606-en. https://www.oecd-ilibrary.org/content/publication/9789264115606-en
15. Stevens, H.: Hans Peter Luhn and the birth of the hashing algorithm. IEEE Spect. **55**(2), 44–49 (2018). https://doi.org/10.1109/MSPEC.2018.8278136

Securing File System Integrity and Version History Via Directory Merkle Trees and Blockchains

Andreas Lackner(✉), Seyed Amid Moeinzadeh Mirhosseini, and Stefan Craß

ABC Research – Austrian Blockchain Center, Favoritenstr. 111, 1100 Wien, Austria
{andreas.lackner,amid.moeinzadeh,stefan.crass}@abc-research.at

Abstract. In our data-driven world, the secure storage of information becomes more and more important. Digital data is especially affected by this topic, as digital records can be simply manipulated in the absence of special securing mechanisms. Hence, for digital archives, a verifiable mechanism to guarantee data integrity is of great importance. While it must be able to rule out manipulation, in many scenarios data updates are desirable. In this case, the version history of data must be traceable. In this paper, we propose an approach based on blockchains and Merkle Trees that fulfills both criteria: It provides a verifier with a proof of data integrity while allowing traceability of changes in the stored data.

Keywords: Digital archives · Blockchain · Integrity · Versioning

1 Introduction

The main task of dedicated storage systems like digital archives is to provide long-term protection for stored data and make it accessible for authorized users. For many practical applications, it is crucial that users can trust the validity of the returned information. Archives may contain different kinds of data, like historical records, scientific data, construction plans, or administrative documents. Certain parties might have commercial, legal, or political incentives to manipulate documents after they have been archived, e.g., to cover up a scandal or to create fake evidence in a legal dispute. Therefore, an archive has to provide suitable measures to ensure data integrity and prevent illicit modifications.

Using suitable IT security measures, archives may prevent attackers from breaking into the system. However, full trust can only be established if the archive can indisputably prove to any user that the requested data has not been manipulated after storage. Otherwise, late modifications by system administrators or authorized (but malicious) users would remain undetected. To prevent

This work is funded within the framework of the COMET centre ABC – Austrian Blockchain Center by BMK, BMDW and the provinces of Vienna, Lower Austria and Vorarlberg. The COMET programme (Competence Centers for Excellent Technologies) is managed by the FFG.

G. Kotsis et al. (Eds.): DEXA 2022 Workshops, CCIS 1633, pp. 294–304, 2022.
https://doi.org/10.1007/978-3-031-14343-4_27

this, an archive can provide a so-called *Proof of Existence (PoE)* that determines since when a document has existed [12]. A common approach to solve this task is via cryptographic hash functions in combination with secure timestamps.

Cryptographic hash functions like SHA-3 [5] are one-way functions that map input data with arbitrary length (e.g., a file) into a distinct hash value with a fixed length. Their mathematical features ensure that it is practically impossible to find two inputs that produce the same output. The hash of a file can be used as its digital fingerprint because it is practically impossible for attackers to modify a file without changing its hash. The integrity of a retrieved document can be verified by comparing its hash value with the original hash that was created when the file entered the storage system. The crucial point is the secure storage of this hash value for each document. If it were stored in the archive together with the file, an attacker could simply modify both. Therefore, an additional mechanism is required for securing hash values.

A common way to tackle this problem is by creating a PoE via cryptographically signed timestamps [3]. Hash values are transmitted to a trusted *Time Stamping Authority (TSA)*, which creates a response that includes the hash, the current time, and a digital signature. This record is then stored by the archive together with the data and returned upon request. The TSA signature ensures the integrity of this PoE, as neither the hash nor the timestamp could be changed without the collaboration of the third-party TSA. As an optimization, the hash values of multiple documents can be combined using a *Merkle Tree* [9], where document hashes form the leaf nodes and the inner node values are computed by executing the hash function over the respective children. In this case, only a timestamp for the value of the root node (i.e., the root hash) is required to provide a PoE for each included document. In addition to the signed timestamp, the archive has to include some inner node values (a so-called *Merkle proof*) that enable users to verify that the document hash was indeed a part of the corresponding Merkle Tree. Additional information on the usage of signed timestamps and Merkle Trees for integrity verification in digital archives can be found in [12].

The main problem with this approach is the need for a trusted third party (i.e., the TSA). If it gets compromised, all issued timestamps become invalid. Therefore, a decentralized mechanism using *blockchain* technology is beneficial. A blockchain is a distributed ledger that is operated by a peer-to-peer network of decentralized nodes. The integrity of the stored data is ensured via a combination of hash functions, digital signatures, and a consensus mechanism that incentivizes honest participants and prevents manipulations by a malicious minority. Although blockchains were originally established as cryptocurrencies [10], they can store arbitrary data, including Merkle Tree hashes. As blockchain transactions are associated with the timestamp of their enclosing block, a (public or permissioned) blockchain can fulfill a similar role as a centralized TSA for providing a PoE. For verification, a user needs access to the corresponding blockchain transaction (and—if applicable—the Merkle proof from the archive). There are already several examples that follow this approach [4,6,11,13].

However, current approaches focus on providing integrity mostly on a per-file basis. A change in the directory structure or the deletion of a file from a folder cannot be detected easily. A naive solution would be the creation of separate proofs for each directory at any level of the file system by considering directories as archive files themselves, but this would require a lot of redundant hashing operations (as files may be included in multiple folders from the direct parent until the file system root) and additional overhead for PoE storage. This paper describes an efficient mechanism for a blockchain-based PoE mechanism that protects the entire file system using an adapted Merkle Tree version termed *Directory Merkle Tree (DMT)*.

Integrity proofs are also related to the protection of the version history of a file system. Common version control systems like Git [2] already provide mechanisms to store the complete revision history while detecting (accidental) inconsistencies, but manipulations by insiders cannot be ruled out. Users may not only want to have a PoE for a specific file or directory, but there might also be a need to know if they have received the most recent version of the data or if the file or directory has been modified or deleted since the specified timestamp. Using concepts from Git and similar approaches, we therefore extend our approach to enable proofs of existence (or deletion) for any version in the archive's history.

Our contribution consists of a description for a *verifiable versioned archive* that makes it possible for a verifier to check the integrity of single files or whole directory structures and trace the version history of the data stored in the archive. In Sect. 2, a detailed explanation of the technical concepts is given. A discussion of the approach considering also related work can be found in Sect. 3. Section 4 gives a summary of the content as well as an outlook on future work.

2 Towards a Verifiable Versioned Archive

We assume two different roles that interact with each other: the archive and the verifier. The archive is responsible for storing the data and keeping track of all the occurring changes. The verifier possesses data that is allegedly included in the archive. The main objective of the verifier is to validate the data's integrity. That is, to verify the membership of the data in the archive's database. As updates to data occur, the verifier may want to check if her data is up-to-date.

In general, the archive is not trusted by the verifier. However, we do assume that the archive is cooperative. That is, the archive is eager to prove the integrity of data. For that purpose, the archive stores additional metadata needed for verifying data integrity and, upon request, delivers this metadata to the verifier.

Furthermore, we assume that the archive stores its data inside a file system supporting all the commonly available operations. In particular, we assume that data is stored in files that can be grouped together by folders and that there are no circular dependencies. More formally, we define the structure of the data stored in the archive over a finite set D of directories and a finite set F of files. A file $f \in F$ has properties such as a name and content denoted by f_n and f_c, respectively. The name of the file is simply a string, while the content is a finite binary sequence.

Likewise, a directory $d \in D$ has a name and a content denoted by d_n and d_c, respectively. The name of the directory is again a string, while the content of the directory is a finite sequence $\langle e_i \rangle_{1 \le i \le n}$ where $e_i \in D \cup F$ and $n \in \mathbb{N}$. The structure of the archive's data can be described by a *Directory Tree*:

Definition 1 (Directory Tree). *Given a set of files and directories F and D respectively, let $\mathcal{DT} = (V, E)$ be a directed graph such that $V = D \cup F$ and $E = \{(d, e) \in D \times V \mid e \in d_c\}$. Then $\mathcal{DT}$ is a* ***Directory Tree*** *if and only if $\mathcal{DT}$ is a DAG such that there exists a directory $r \in D$ with in-degree $\delta^+(r) := |\{(u, r) \in E \mid u \in V\}| = 0$ and for all $v \in V \setminus \{r\}$ we have $\delta^+(v) = 1$. We refer to r as the root directory of $\mathcal{DT}$.*

2.1 Directory Merkle Tree

A Directory Merkle Tree is—as the name suggests—based on Merkle Trees [9] and can be used to cryptographically secure a Directory Tree. Similar approaches are already known and are for example used by the versioning system Git [2].

Definition 2 (Directory Merkle Tree). *Given a Directory Tree $\mathcal{DT} = (V, E)$ with $V = D \cup F$ and a cryptographic hash function $H(\cdot)$, we define the* ***Directory Merkle Tree*** *of $\mathcal{DT}$ to be a vertex labeled graph $\mathcal{DMT}(\mathcal{DT}, H) = (V', E')$ such that $V' := \{\mathcal{H}(v) \mid v \in V\}$, $E' := \{(\mathcal{H}(v_1), \mathcal{H}(v_2)) \mid (v_1, v_2) \in E\}$ and for all $v' := \mathcal{H}(v) \in V'$ the label v'_n of v' is defined as $v'_n := v_n$. The function $\mathcal{H}(\cdot)$ is a cryptographic hash function over V based on H and defined as follows:*

$$\mathcal{H}(v) := \begin{cases} H(v_n \,||\, H(v_c)) & \text{if } v \in F, \\ H(v_n \,||\, \mathcal{H}(e_1) \,||\, \dots \,||\, \mathcal{H}(e_n)) & \text{if } v \in D \text{ and } v_c = \langle e_i \rangle_{1 \le i \le n}, \end{cases}$$

where $||$ denotes a binary concatenation.

If the particular hash function H chosen, is not of any interest, we might also write $\mathcal{DMT}(\mathcal{DT})$ to denote the Directory Merkle Tree of $\mathcal{DT}$. Given $\mathcal{DT}$ and its root directory r, we refer to the vertex $\mathcal{H}(r)$ as the *root hash* of $\mathcal{DMT}(\mathcal{DT})$.

Definition 3 (Proof of Membership). *Let $\mathcal{T} := \mathcal{DMT}(\mathcal{DT}) = (V, E)$ be a Directory Merkle Tree, of the Directory Tree $\mathcal{DT}$, a* ***Proof of Membership*** *of a vertex $v \in V$ in $\mathcal{T}$ is a sequence $P_{\mathcal{T}}(v)$ defined as follows:*

$$P_{\mathcal{T}}(v) := \begin{cases} \emptyset & \text{if } v \text{ is the root hash of } \mathcal{T}, \\ P_{\mathcal{T}}(u) \,||\, \langle (u_n; \{w \in V \setminus \{v\} \mid (u, w) \in E\}) \rangle & \text{where } (u, v) \in E. \end{cases}$$

Given a Directory Tree $\mathcal{DT} = (V, E)$, for a vertex $v \in V$ we may also write $\mathcal{P}_{\mathcal{T}}(v)$ to denote the Proof of Membership $\mathcal{P}_{\mathcal{T}}(\mathcal{H}(v))$ of the vertex $\mathcal{H}(v)$ in the corresponding Directory Merkle Tree $\mathcal{T} = \mathcal{DMT}(\mathcal{DT})$. In Fig. 1 we see a graphical representation of a Directory Tree (colored in blue) and its corresponding Directory Merkle Tree (colored in red). Considering this example, the Proofs of Membership of the vertex h_{a1}, which represents *file*$_{a1}$, is given by the sequence

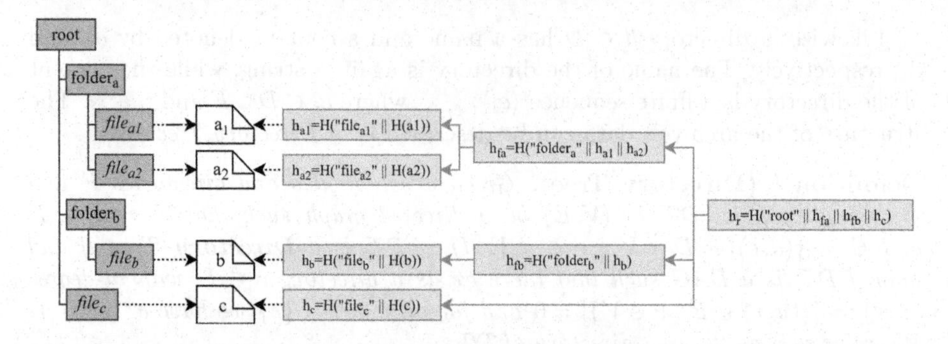

Fig. 1. A Directory Tree (blue) and its corresponding Directory Merkle Tree (red) (Color figure online)

$P_{\mathcal{T}}(h_{a1}) := \langle(\text{"}root\text{"}; \{h_{fb}, h_c\}), (\text{"}folder_a\text{"}; \{h_{a2}\})\rangle$, where $\mathcal{T}$ denotes the Directory Merkle Tree. Now given only the root hash h_r, we can prove the membership of *file*$_{a1}$ given the proof sequence $P_{\mathcal{T}}(h_{a1})$ as follows: First, we assume that we have the file *file*$_{a1}$ and know that it is named that way. We then can recompute the hash h_{a1}. Together with h_{a2} and the name of *folder*$_a$ we can then recompute h_{fa} which, together with h_{fb}, h_c and the name of the root folder, gives us the root hash. We now have not only proven that *file*$_{a1}$ is actually contained in the directory, but also its location within the directory structure.

2.2 Versioning

The version history of the data stored in the archive can be represented by a sequence $\langle \mathcal{DT}_i \rangle_{0 \leq i < n}$ of Directory Trees. Typically, two consecutive versions $\mathcal{DT}_i$ and $\mathcal{DT}_{i+1}$ differ on multiple vertices, hence we assume that multiple changes are summarized into one transaction that leads to a new version.

On a Directory Tree, different changes can occur. In particular, we are considering **i)** creation of a file or a folder, **ii)** modification of a file and **iii)** deletion of a file or a folder. Note that we do not consider renaming of files or folders, but instead handle them as deletion and recreation. Updates to the Director Tree are handled the following way:

(i) If a file or a folder is created, we simply add a new node to the tree.
(ii) If a file is modified, we replace the old node by the new one in the tree.
(iii) If a file or a folder is deleted, we replace it with a special `DELETED` entry.

2.3 The Archiving Process

For archiving, we assume that both the verifier and the archive have access to a trusted, unforgeable storage. The storage is designed as an append-only public bulletin board, where data is stored in storage cells. We assume that each such cell can be addressed by its index. The first cell has the index 0. Moreover, we expect the data storage to support the following methods:

- `write(data)` Stores the data by appending it at the end of the storage.
- `read(i)::(data;time)` Returns the i–th data cell and a timestamp.
- `last()::i` Returns the index of the last cell stored. Note that indices start at 0. Thus, if we have n data cells stored, `last()` will return $n-1$.

A simple approach to realize such a storage is using a blockchain and store the data inside of transactions together with a reference to the transaction containing the previous entry. Conceptually, this yields a linked list. However, data look-ups are rather inefficient using this approach. For example, the successor of a data entry cannot be found directly because of the lack of forward pointers. Thus, we suggest using smart contracts [14], which are basically programs executed on top of a blockchain (e.g., maps that allow random access in $\mathcal{O}(1)$ time).

The archive itself keeps track of the sequence $\langle \mathcal{DT}_i \rangle_{0 \le i < n}$ of versions of Directory Trees. Additionally, the archive stores the sequence $\langle \mathcal{DMT}(\mathcal{DT}_i) \rangle_{0 \le i < n}$. Here we want to point out that typically, two consecutive versions $\mathcal{DT}_i$ and $\mathcal{DT}_{i+1}$ have many vertices in common. Therefore, also the corresponding Directory Merkle Trees $\mathcal{DMT}(\mathcal{DT}_i)$ and $\mathcal{DMT}(\mathcal{DT}_{i+1})$ have the same number of vertices in common. In practice, it is therefore possible and also sensible to reuse common vertices in order to save storage. Note that the version history of the Directory Tree as well as the corresponding history of the Directory Merkle Tree is stored only at the archive and not in the smart contract.

Now, for the first version $\mathcal{DT}_0$, the archive computes the Directory Merkle Tree $\mathcal{DMT}(\mathcal{DT}_0)$ and stores the root hash in the smart contract: $\texttt{write}(\mathcal{H}(r_0))$. Note that since it is the first version, the data will be written at index 0. Hence, if we start numbering the versions from 0, the version number equals the index of the storage cell where the version was anchored.

When the archive creates a new version $\mathcal{DT}_i$ with $i > 0$ of the Directory Tree, it first computes $\mathcal{DMT}(\mathcal{DT}_i)$ as before. For optimization, hash values for unchanged files and directories can be reused from $\mathcal{DMT}(\mathcal{DT}_{i-1})$. Then, the new version can be secured by performing the following command: $\texttt{write}(\mathcal{H}(r_i))$.

Note that we are assuming that the archive behaves honestly when creating the Merkle Directory Tree and storing the root hash in the smart contract. This assumption can be justified, as the original creator of the data can verify the data's integrity after uploading it to the archive.

2.4 The Verification Process

For an archive supporting versioning, there are various properties of the stored data and its version history that a verifier may want to validate. We, however, have identified three main use cases that are interesting for a verifier to check:

- The verifier wants to check if data she possesses was in the archive at a given point of time (Proof of Existence).
- The verifier wants to check if data she possesses is up-to-date (Proof of Topicality).
- The verifier wants to check if data has been deleted (Proof of Deletion).

Proof of Existence (PoE). The PoE refers to a certain point in time t for which a verifier wants to prove that some data has been stored in the archive. Let d denote the data the verifier wants to validate. This can be either a single file or a directory. First, the verifier needs the Proof of Membership of d. Since d can be included in several versions, there also exist multiple proofs for d. Let $\mathcal{DT}_i$ be any of those versions and let $\mathcal{T}_i := \mathcal{DMT}(\mathcal{DT}_i)$, then, in order to check membership of d in $\mathcal{DT}_i$, the verifier needs the proof $\mathcal{P}_{\mathcal{T}_i}(d)$.

For the verifier to obtain the proof, there are different options. We could for example assume that by possessing d, the verifier also has the Proof of Membership. This is reasonable as the archive can store the proof in the metadata of d, once d has been added to the archive. Anyway, by assuming a cooperative archive, the verifier can request the proof from the archive. After that the verifier requests the entry that is stored at index i: $(h_r; t') \leftarrow \texttt{read}(i)$.

Given the proof sequence $\mathcal{P}_{\mathcal{T}_i}(d)$ together with d, the verifier can recompute the root hash of the Directory Merkle Tree. Note that this assumes that the verifier has certain knowledge of the structure of the Directory Merkle Tree (e.g., the order of the files in a directory). However, we can assume that such information is included in the proof.

At the end of the computation, the verifier obtains a root hash h'_r. If d is a member of the Directory Merkle Tree represented by h_r, then h'_r must be equal to h_r. Otherwise, the verifier rejects d. If the two hash values match, it remains to compare the two timestamps t and t'. If $t' > t$, the proof $\mathcal{P}_{\mathcal{T}_i}(d)$ is too new to prove existence at time t'. Then the verifier has to ask the archive for a proof referring to an older version. If $t' \leq t$ then d was already stored in the archive at time t and the PoE succeeds. However, it still could be the case that d has been overwritten by a newer version or a `DELETE` entry at time t. To make sure d was in the latest version at time t, we need a Proof of Topicality at time t.

Proof of Topicality (PoT). For the PoT the verifier wants to check if d was stored in the current version at time t. In the simplest case, t refers to the current time. In that case, a PoT shows that the corresponding data is up-to-date.

If a verifier has data d and wants to check whether d is up-to-date, the verifier executes $i \leftarrow \mathit{last}()$ to obtain the latest index and then requests the stored data at that index: $(h_r; t) \leftarrow \texttt{read}(i)$. After that, the verifier requests the proof $\mathcal{P}_{\mathcal{T}_i}(d)$ where $\mathcal{T}_i := \mathcal{DMT}(\mathcal{DT}_i)$ is the Directory Merkle Tree of the i–th version. Then again, given d and $\mathcal{P}_{\mathcal{T}_i}(d)$, the verifier can recompute the root hash h'_r and compare it with the root hash h_r stored in the smart contract. If the two hash values match, the verifier has shown that d is included in the latest version.

To check the PoT at an arbitrary time t, the verifier must first find an index i with $(h_{r_i}; s) \leftarrow \texttt{read}(i)$ and $(h_{r_{i+1}}; p) \leftarrow \texttt{read}(i+1)$ such that $s \leq t \leq p$. If i does not exist, t refers to the latest version, and the verifier proceeds as before. Otherwise, the verifier proves membership of d in the i–th version.

Proof of Deletion (PoD). As we have already mentioned before, we handle deletion of files and folders by overwriting it with a special `DELETE` entry. So if a document or a file d has been deleted at version i, there must exist a proof $\mathcal{P}_{\mathcal{T}_i}(\texttt{DELETE}_d)$ where $\mathcal{T}_i$ is again the Directory Merkle Tree of the i–th version.

Note that the proof contains the path of the proven data. Hence, the verifier can actually check if the PoD is referring to the data she has requested the proof for. We are using the notion $\texttt{DELETE}_d$ to emphasize that it is the PoD of a specific d and not just a Proof of Membership of any DELETE entry in $\mathcal{T}_i$.

Given $\mathcal{P}_{\mathcal{T}_i}(\texttt{DELETE}_d)$, the actual proof works as before. Note that this is by no means a Proof of Non-Membership, as we obviously can only show deletion of data that was stored in the archive in previous versions.

2.5 Performance Evaluation

In the following, we analyze relevant performance properties of our approach from a theoretical viewpoint.

Additional to the archived data, the proof files have to be stored for each item in the directory, which leads to additional space consumption. For a Merkle Tree, the size of a leaf node's proof requires $\mathcal{O}(\log n)$ space, where n is the number of leaves stored in the Merkle Tree. However, we cannot balance the Directory Merkle Tree the same way a Merkle Tree can be balanced: The structure of the Directory Merkle Tree is obviously restricted by the structure of the corresponding Directory Tree. Regarding the proof size, the worst-case structure would be a single folder which contains all the files of the directory. In that case, the size of the proof is in $\mathcal{O}(n)$ for a single file. By using a naive approach where we store the complete proof for each file, this would lead to an overall space complexity of $\mathcal{O}(n^2)$ just for storing the proof files of the archived data. Before coping with this problem, one should consider that the proof files are usually quite small compared to the potentially large files stored in an archive. Still, a quadratic growth in space could lead to problems in practice. In order to improve the worst-case space complexity of our solution, there are different starting points.

For folders that contain numerous files and subfolders, optimizations are possible. For example, a combination using Merkle Trees to combine the hash values of the items of a folder seems like a promising approach. The quadratic growth in disk space, however, is also related to a second problem: the redundant information stored in the proof files. For example, the proofs of files inside the same folder only differ in one hash value. Conceptually, by storing the proof files, we implicitly store the Directory Merkle Tree. However, by this implicit representation of the tree we also introduce redundancy as the same hash values are stored in multiple proofs. A simple solution for this problem is for the archive to store the Directory Merkle Tree explicitly and generate the proof files each time it is requested based on the stored tree. Generating a proof obviously takes additional computation time, but the worst-case run time can be bound by $\mathcal{O}(n)$.

Besides the additional space needed for storing the proof files, another critical part to consider is the interaction with the smart contract. Typically, the use of such smart contracts involves additional costs (e.g., transaction fees). Therefore, it may also be important to reduce the number of write operations and the amount of data written per write operation. In our approach, for each version, the root hash of the Directory Merkle Tree has to be stored. Here, optimizations are possible by using a hash function mapping to shorter hash values. However, one has to be careful when using this approach, as the length of the hash values typically corresponds to the security parameter of the hash function.

The number of data writes strongly depends on the versioning strategy used by the archive. Clearly, there is a trade-off between traceability and costs: if we summarize multiple updates to a single version, we may lose traceability of the changes. On the other hand, if we create a new version on each update, this may lead to a lot of write operations, which increase the costs. For this problem, there does not exist a one-fits-all solution. Instead, the chosen strategy depends on the requirements of the archive.

3 Discussion and Related Work

Securing a directory structure using Merkle Trees is an approach that is well known. Maybe most noteworthy in this context is the versioning system Git [2]. The data structure used there to represent the content of a repository at a point in time is conceptually a Merkle Tree that is quite similar to our Directory Merkle Tree. In fact, a simple solution would be for the archive to use Git to track the changes inside a directory and secure the hashes of each commit using the unforgeable storage. By following this approach, we get a similar level of integrity protection. However, there is one disadvantage of that solution, which makes it practically unusable in our scenario: In order for the verifier to validate the integrity of a file using Git, she would need access to the whole repository.

In theory, Git already produces all the data necessary for creating proofs for single files and folders, similar to the proofs we have introduced previously. Hence, in order to implement a first prototype of our solution, one approach would be to adapt Git to support the creation and validation of such proofs.

Regarding other approaches for securing the integrity of digital archives, Haber et al. have described a system based on timestamps [7]. They use a hash-link method that batches the requests into Merkle Trees and builds up a linked list using their root hashes. The root hashes, witness hash values, are stored by a TSA or published as a widely witnessed event. However, this method does not capture the hierarchical relationship between files and folders.

Burns et al. [1] suggested a verifiable versioning file system that can be used to secure the content of a directory structure at a given time, as well as the whole version history of a file or a directory. They use a Merkle Tree-like data structure that is similar to Directory Merkle Trees, but more complex. Their approach is also rather suited for internal audit trails where the verifier has access to the file system itself. For securing the hash values, they assume a trusted third party.

Using a blockchain to timestamp a document is also described by Gipp et al. [6]. Their approach uses Bitcoin transactions to timestamp documents. To reduce the costs, multiple hash values are combined using Merkle Trees. However, their approach describes timestamping of individual documents. Furthermore, they do not describe a method for securing the version history of a document.

A similar approach is described by Magrahi et al. [8]. They also suggest securing the integrity of a document using a blockchain. Different to Gipp et al., they do not aggregate multiple hash values, but store the hash value of each document individually. Hence, for large archives, their approach does not scale. Also, versioning is not covered explicitly.

Nizamuddin et al. described an approach of keeping track of a document's version history by making use of smart contracts [11]. The smart contract does not store the actual document, but only the hash of each version of the document. However, their approach does not scale very well for large archives, as it creates a smart contract for each document.

Finally, we also want to mention ARCHANGEL [4] and the National Archive of Korea [13] as two examples for digital archives with blockchain-based integrity protection mechanisms that have been evaluated in practical scenarios.

Compared to the other approaches shown in this section, our approach in particular focuses on the verification of individual files and folders inside an archive. In practice, a verifier can be an external person or system (with respect to the archive) that does not have access to the whole data stored in the archive. Our approach attempts to ease the verification process for the verifier while preserving data protection. Also, all the other approaches discussed in this section either do not handle versioning, do not secure the whole directory structure but only single documents, or are relying on a trusted third party.

4 Conclusion and Future Work

Summing it up, our approach combines well-known solutions for versioning of directory structures (e.g., Git) and timestamping using distributed ledger technologies (i.e., blockchains) in order to yield a simple solution for a verifiable versioned archive. The usage of Directory Merkle Trees allows us to prove various properties on an arbitrary level inside a directory tree (file-level or folder-level).

We plan to implement a runnable prototype of our approach in order to analyze performance and compare it with other solutions. Furthermore, for long-term protection of data, further topics like format changes or breaking crypto routines (e.g., hash functions can become outdated) have to be covered.

During the work on this paper, we have explored further solutions and optimizations that we plan to investigate in the future. For now, the verifier is strongly dependent on the cooperation of the archive. In future work, we want to focus on reducing that dependency. Optimization of the Directory Merkle Tree and the required proofs with regard to scalability and storage requirements will also be investigated. Finally, we will analyze how additional security properties like authenticity can be protected based on our approach.

Acknowledgements. We would like to thank the Austrian Federal Chancellery (BKA) and the Austrian Federal Computing Center (BRZ) for their support within this project.

References

1. Burns, R., Peterson, Z., Ateniese, G., Bono, S.: Verifiable audit trails for a versioning file system. In: Proceedings of the 2005 ACM Workshop on Storage Security and Survivability, pp. 44–50 (2005)
2. Chacon, S., Straub, B.: Pro Git. Apress (2014)
3. Chokhani, S., Wallace, C.: Trusted archiving. In: Proceedings of the 3rd Annual PKI R&D Workshop. NIST (2004)
4. Collomosse, J., et al.: ARCHANGEL: trusted archives of digital public documents. In: Proceedings of the ACM Symposium on Document Engineering 2018, pp. 1–4 (2018)
5. Dworkin, M.: SHA-3 standard: permutation-based hash and extendable-output functions (2015). https://doi.org/10.6028/NIST.FIPS.202
6. Gipp, B., Meuschke, N., Gernandt, A.: Decentralized trusted timestamping using the crypto currency Bitcoin. In: Proceedings of the iConference (2015)
7. Haber, S., Kamat, P.: A content integrity service for long-term digital archives. In: Archiving Conference, vol. 2006, pp. 159–164. Society for Imaging Science and Technology (2006)
8. Magrahi, H., Omrane, N., Senot, O., Jaziri, R.: NFB: a protocol for notarizing files over the blockchain. In: 2018 9th IFIP International Conference on New Technologies, Mobility and Security (NTMS), pp. 1–4. IEEE (2018)
9. Merkle, R.C.: A digital signature based on a conventional encryption function. In: Pomerance, C. (ed.) CRYPTO 1987. LNCS, vol. 293, pp. 369–378. Springer, Heidelberg (1988). https://doi.org/10.1007/3-540-48184-2_32
10. Nakamoto, S.: Bitcoin: a peer-to-peer electronic cash system (2008). https://bitcoin.org/bitcoin.pdf
11. Nizamuddin, N., Salah, K., Azad, M.A., Arshad, J., Rehman, M.: Decentralized document version control using Ethereum blockchain and IPFS. Comput. Electr. Eng. **76**, 183–197 (2019)
12. Vigil, M., Buchmann, J., Cabarcas, D., Weinert, C., Wiesmaier, A.: Integrity, authenticity, non-repudiation and proof of existence for long-term archiving: a survey. Comput. Secur. **50**, 16–32 (2015)
13. Wang, H., Yang, D.: Research and development of blockchain recordkeeping at the National Archives of Korea. Computers **10**(8), 90 (2021)
14. Wood, G.: Ethereum: a secure decentralised generalised transaction ledger. Ethereum Project Yellow Paper (2014). https://ethereum.github.io/yellowpaper/paper.pdf

Taxation of Blockchain Staking Rewards: Propositions Based on a Comparative Legal Analysis

Pascal René Marcel Kubin(✉)

German University of Administrative Sciences Speyer, Freiherr-vom-Stein-Straße 2, 67346 Speyer, Germany
pascalkubin@protonmail.com

Abstract. Blockchain technology is seen as an essential part of a decentralized society. Given the relatively high energy consumption of proof-of-work consensus algorithms, proof-of-stake blockchains are becoming increasingly popular to reach consensus on the current state of the blockchain. With proof-of-stake, a randomized process determines which node is allowed to validate the next block of transactions. In order to participate, coin holders must stake their coins and receive staking rewards in return. Yet, there still is high uncertainty around the taxation of these staking rewards. This analysis compares the taxation of staking rewards in Germany, Austria, and Switzerland and derives four propositions under the premise of reaching a higher degree of tax neutrality. The legal comparison illustrates the heterogeneity in terms of taxing staking rewards. To achieve a more neutral taxation, (1) staking should only qualify as a business activity under clearly defined and restrictive circumstances. (2) Whenever staking does not represent a main business activity, staking rewards should not be taxed upon receipt. (3) Instead, they should be taxed upon disposal. (4) Moreover, staking should not cause further tax consequences for staked coins.

Keywords: Distributed ledger technology · Blockchain · Staking rewards · Taxation · Proof-of-stake

1 Introduction

In times of rising inflation and hyper centralization, digital assets and the underlying blockchain technology have gained widespread recognition [15]. Moreover, the revitalization of libertarian values and self-determination have also contributed to the diffusion of blockchain technology. This technology enables use cases such as peer-to-peer decentralized financial services ('DeFi') and powers the so-called Web3, which represents a decentralized and token-based version of the Internet. A blockchain is a distributed data structure for the immutable, tamper-resistant, and transparent storage of transaction data in a chronological order [1]. For example, this data can provide information on the ownership of cryptocurrencies, digital art, or gaming items. When network participants begin transacting with each other, the blockchain-based network must reach consensus

G. Kotsis et al. (Eds.): DEXA 2022 Workshops, CCIS 1633, pp. 305–315, 2022.
https://doi.org/10.1007/978-3-031-14343-4_28

on the post-transaction state of the ledger. It must be determined which network node is allowed to verify new transactions and produce the next block containing these transactions. In general, there are two dominant consensus algorithms to achieve network-wide agreement on the current state of the ledger, which are namely 'proof-of-work' and 'proof-of-stake' [2]. In proof-of-work, so-called mining nodes use their CPU power to try out nonce values until they come up with a value that provides the necessary number of zero bits [18]. Nonce values are arbitrary numbers, which are only valid for single use [16]. The nonce value together with the hash data of the previous block must produce a hash value with the necessary number of zero bits [5, 18]. The miner who finds this nonce value first is allowed to validate the next block of transactions and receives the block reward in the form of native coins of the respective network. The two largest blockchain-based networks by market capitalization Bitcoin and Ethereum both rely on a proof-of-work algorithm. However, the environmental impact caused by the high energy consumption of proof-of-work is often criticized [23]. Against this background, proof-of-stake blockchains such as Cardano, Cosmos, Algorand, or NEAR Protocol are increasingly gaining attention. Even Ethereum intends to move from a proof-of-work to a proof-of-stake consensus algorithm. With proof-of-stake, a randomized process determines which node is allowed to validate the next block of transactions. In this regard, nodes that want to validate blocks must prove their stake in the network in the form of native coins. Compared to proof-of-work, a proof-of-stake consensus algorithm requires less computing power, is considered more efficient, and consumes less energy. Proof-of-stake blockchains also economically incentivize validators by rewarding them with block rewards in the form of native coins. The entry barriers to participate in staking are becoming increasingly lower. On the one hand, many proof-of-stake blockchain protocols allow staking without operating a separate validator node (e.g., through delegated proof-of-stake, nominated proof-of-stake, or staking pools) [10]. This opens up staking for technically less experienced network participants. On the other hand, an increasing number of centralized exchanges offer staking opportunities directly to their customers [14]. Given these lower entry barriers, it is becoming increasingly attractive for investors to hold coins for the long term and to hedge themselves against crypto volatility by passively earning staking rewards.

However, there is considerable uncertainty surrounding the taxation of these staking rewards, as most tax regimes were not prepared for this new form of yield. Thus, staking rewards are treated very differently across countries [24]. For instance, the taxation of staking rewards in Germany, Austria, and Switzerland differs substantially, thereby providing a suitable basis for a legal comparison. Against this background, this study investigates (1) how Germany, Austria, and Switzerland differ in taxing staking rewards and (2) how the taxation of staking rewards could be designed under the premise of reaching a higher degree of tax neutrality? Tax neutrality refers to a tax system that does not synthetically distort decision making [11].

By comparing different concepts of taxing staking rewards and by analyzing how they incentivize investor behavior, I intend to contribute to the field of regulatory and legal research around digital assets and blockchain technology. In practical terms, this analysis can provide a selective overview on the taxation of staking rewards for investors and may serve as an indicative framework for policy makers and tax authorities. The analysis begins with a more detailed explanation of the proof-of-stake consensus algorithm. Subsequently, I compare the taxation concepts in Germany, Austria, and Switzerland, before I derive propositions for a more neutral taxation of staking rewards. I only investigate the taxation of staking rewards received by natural persons, which means that the corporate income taxation of staking rewards received by legal persons is not within the scope of this analysis. As a major value proposition of blockchain technology is the possibility of individual self-custody of digital assets, the taxation of natural persons should be of particular practical relevance. I also neglect implications related to value-added taxes. Moreover, it must be noted that this analysis neither covers country-specific details nor provides tax advice. Even though I solely refer to blockchain technology, the findings can also be transferred to other distributed ledger technologies that also rely on proof-of-stake consensus algorithms (e.g., Fantom).

2 Proof-of-Stake Consensus Algorithms

In order to reach an agreement on the current state of the ledger among the various network nodes, blockchain-based networks depend on decentralized consensus algorithms [15]. These consensus algorithms enable transactions between network participants by providing a governance mechanism to achieve network-wide consensus on the post-transaction state of the ledger. In technical terms, consensus algorithms determine which network node is allowed to add the next block to the blockchain. Without such an algorithm, each node could arbitrarily add its own blocks to its copy of the blockchain. This would undermine the fundamental principles of a distributed and decentralized system, as it could result in multiple entries of single transactions on the blockchain, thereby causing the problem of 'double spending'. Before the node that has been selected through the consensus algorithm can add its new block, it has to validate whether the transactions within this block are accurate and comply with the protocol's principles [13]. Subsequently, the transactions within this new block are once again verified by the other network nodes, before they also add the new block to their copy of the blockchain [13].

With proof-of-stake consensus algorithms, randomized selection processes determine which node is allowed to create the next block. Depending on the respective blockchain protocol, nodes with a higher share of staked coins or a longer staking duration may increase their probability of being selected as block validators [2]. The underlying idea is that nodes with higher stakes in the network have greater interest in securing the network as well as maintaining its robustness and decentralization [10]. Coin holders are economically incentivized with staking rewards to participate in the proof-of-stake consensus. Depending on the blockchain protocol, the staking rewards consist of newly minted coins (inflation) and/or transaction fees that have been paid by the network participants, when initiating transactions. In some blockchain protocols, nodes that behave maliciously can lose part of their staked coins as a punishment for their misbehavior (also known as slashing) [20]. The risk of losing coins in combination with the prospect of receiving staking rewards can discipline validator nodes. After all, validators would devalue their own stake in the network by behaving maliciously. Some blockchain protocols additionally require lock-up periods, during which coin holders cannot dispose of their coins [22]. Advanced forms of proof-of-stake such as staking pools, delegated, or nominated proof-of-stake enable coin holders to participate in staking without having the technical knowledge to independently operate a separate validator node. Hence, while validator nodes are validating new blocks, 'regular' coin holders solely delegate their stake in the network to trusted validator nodes [10].

3 Comparative Legal Analysis

The following section provides a legal comparison between the taxation of staking rewards in Germany, Austria, and Switzerland. Table 1 illustrates the central tax differences between these three tax regimes.

Table 1. Taxation of staking rewards in Germany, Austria, and Switzerland.

Country	Taxation upon receipt	Taxation upon disposal
Germany	**Yes** (up to 45% + further levies + possibly trade tax for businesses ~ 14%)	**Yes** (up to 45% + further levies + possibly trade tax for businesses ~ 14%)
Austria	**No** (but up to 55% if the business activity is centered around cryptocurrencies)	**Yes** (27.5% if traded for fiat, goods, or services; up to 55% if the business activity is centered around cryptocurrencies)
Switzerland	**Yes** (33.73% on average)	**No** (but 33.73% on average if staking rewards qualify as independent business income)

3.1 Germany

As in most countries, the tax treatment of staking rewards is not yet directly regulated in German tax law. However, the German Federal Ministry of Finance has published a draft BMF circular and a subsequent final BMF circular on the taxation of virtual currencies [4, 5]. Such BMF circulars are intended to provide legal clarity on certain tax issues for tax authorities as well as taxpayers.

Taxation Upon Receipt. According to these BMF circulars, staking rewards are taxable upon receipt [4, 5]. For this purpose, they must be valued at their current market value at the time of receipt [4, 5]. Thus, the progressive income tax rate is applicable, resulting in a marginal tax rate of up to 45% (plus further levies). Expenses related to the generation of staking rewards such as transaction fees are deductible [24]. Regarding the qualification of staking for tax purposes, the BMF circular distinguishes between the actual block creation by a validator node and the sole delegation of coins to a trusted validator node [5]. The block creation by a validator node qualifies as a business insofar as a sustainable repetition of this activity is intended and insofar as the validator node has the intention to make a profit [5]. Thus, especially professionally operated validator nodes may qualify as a business [4], thereby being subject to trade tax (depending on the location: on average around 14%) [19]. If the block creation of a validator node does not qualify as a business activity because of the lack of repetitive intent or profit-making intent, then the staking rewards are taxable as 'miscellaneous income' (sec. 22 no. 3 German Income Tax Act) [5]. This miscellaneous income is not subject to trade tax and solely taxed at the progressive income tax rate. The same taxation applies if staking is done by delegating coins to a validator node without participating in the block creation itself (e.g., through a staking pool) [5]. However, if the delegated coins are held as business assets, the received staking rewards are deemed as income from trade or business due to the subsidiarity principle. This makes them subject to both progressive income tax as well as trade tax.

Taxation Upon Disposal. If the staking rewards are later sold at a profit, these realized gains are also taxed at the progressive income tax rate in Germany. Thus, staking rewards are subject to taxation upon receipt as well as upon profitable disposal. Depending on whether the staked coins are held as private assets or as business assets, staking rewards are either subject to income tax (private disposal) or subject to income as well as trade tax (business disposal). Regarding the private disposal of staked coins or staking rewards, there is an additional feature in Germany. Capital gains on cryptocurrencies are tax-exempt after a holding period of one year. Thus, there is no tax on staking rewards that are sold at a profit one year after the acquisition [5].

Accordingly, capital gains from the disposal of coins that have been used for staking are also tax-exempt after holding them for at least one year. However, the Federal Ministry of Finance initially assumed that by using coins for staking, the staked coins become a source of income, which results in an extension of the holding period from one to ten years [4]. Following this view, previously staked coins that are sold at a profit within ten years after their acquisition would be taxable. However, this extension of the holding period due to staking has been controversial, as the underlying tax regulation was created for the very specific case of container leasing [12]. Hence, a contrary opinion assumes that the staked coins themselves do not represent a source of income. Instead, the act of providing coins for securing and stabilizing the network constitutes the actual source of income [12]. In the meantime, the Federal Ministry of Finance has endorsed this view and eliminated the extension of the holding period in its BMF circular from May 2022 [5].

3.2 Austria

In Austria, the taxation of staking rewards was originally handled in a similar way as in Germany. However, the so-called eco-social tax reform has led to specific provisions with regard to the taxation of cryptocurrencies [7].

Taxation Upon Receipt. In general, continuous income from cryptocurrencies is still taxed upon receipt (sec. 27b para. 1 Austrian Income Tax Act). Continuous income includes block rewards that are received through proof-of-work mining (sec. 27b para. 2 sentence 1 no. 2 Austrian Income Tax Act) [7]. However, there is an exception for staking rewards (sec. 27b para. 2 sentence 2 alternative 1 Austrian Income Tax Act). Under this exception, staking rewards received by coin holders for processing transactions are not deemed as continuous income, thus not taxable upon receipt. Hence, in contrast to Germany, staking rewards are exclusively taxed upon disposal in Austria. Furthermore, it becomes apparent that proof-of-stake receives a preferential tax treatment compared to proof-of-work in Austria.

Taxation Upon Disposal. In Austria, staking rewards are only taxed, when the coin holder leaves the crypto sphere by trading staking rewards for fiat money, goods, or services (sec. 27b para. 3 Austrian Income Tax Act). In contrast, there is no taxation, when staking rewards are converted into other cryptocurrencies. While Germany taxes staking rewards upon receipt as well as upon profitable disposal, Austrian tax law allows an indefinite postponement of taxation, as long as the value generated through staking remains in crypto. However, as soon as coin holders decide to trade their staking rewards for fiat money, goods, or services, they are subject to a flat-rate tax of 27.5% [7]. Related expenses such as transaction fees are deductible [7]. The taxpayer may opt for the lower tax rate if the personal progressive income tax rate is lower than 27.5% [6]. In case staked coins are held as business assets, they are also subject to the flat-rate tax of 27.5% upon disposal [7]. However, if the business activity is centered around cryptocurrencies and/or staking, staking rewards fall under the regular progressive income taxation, which means that they are taxed upon receipt as well as upon profitable disposal with a marginal tax rate of up to 55% [7].

3.3 Switzerland

The tax system in Switzerland is characterized by a federalist structure. The Federal Constitution explicitly determines, which taxes the Federal Government is allowed to levy. Each of the 26 cantons may decide within the scope of the Federal Constitution which taxes are levied [8]. The municipalities can also levy taxes, but their competence to tax is constrained by cantonal laws [8].

Taxation Upon Receipt. Similar to Germany, staking rewards are also taxed upon receipt in Switzerland. According to a working paper of the Swiss Federal Tax Administration, staking rewards received by natural persons are considered as 'income from movable assets' (art. 20 para. 1 Swiss Federal Law on Direct Federal Tax) [9]. They are valued in Swiss francs at the time of receipt [9]. This income is subject to the direct federal, cantonal, and municipal income tax. The structure of the cantonal and municipal income tax is similar to that of the direct federal tax on income [3]. Depending on the canton and municipality, the average income tax rate is approximately 33.73% including the direct federal tax [21]. Expenses related to the generation of staking rewards are deductible [8].

Taxation Upon Disposal. If the staking rewards are privately sold at a profit, these capital gains are tax-exempt according to Swiss tax law (art. 16 para. 3 Swiss Federal Law on Direct Federal Tax; art. 7 para. 4 lit. b Swiss Tax Harmonization Act) [3]. Thus, unlike Austria, Switzerland taxes staking rewards upon receipt. However, if the staking rewards are generated by setting up a separate validator node, it must be assessed whether they are taxable as 'independent business income' (art. 18 para. 1 Swiss Federal Law on Direct Federal Tax) [9]. In this regard, they are taxed upon receipt as well as upon profitable disposal [3].

4 Propositions

Based on the approaches of taxing staking rewards in Germany, Austria, and Switzerland, I derive four propositions for a more neutral taxation of staking rewards.

4.1 Proposition 1: Staking Should Only Qualify as a Business Activity Under Clearly Defined and Restrictive Circumstances

In all three tax regimes, the participation in a proof-of-stake consensus may qualify as a business activity under certain circumstances. As business activities typically result in more extensive taxation (e.g., taxation upon receipt and upon disposal; trade tax), this could discourage individuals from participating in staking or from setting up their own validator node. Such a discouragement may reduce the decentralization of the network, as coin holders who would actually stake their coins refrain from staking or from setting up a validator node because they want to avoid the consequences of being taxed as a business. Against this background, there is a need for clearly defined provisions that determine under which circumstances staking qualifies as a business activity. Regarding this, the classification of staking as a business activity should be handled restrictively. Staking should only qualify as a business activity if it substantially exceeds the scope of private asset management. Moreover, it should be noted that staking inherently depends on chance [12]. Thus, unlike typical business activities, the success of staking cannot be directly managed or controlled through entrepreneurial strategies [12]. This characteristic of staking underlines the proposition of a cautious and restrictive qualification of staking as a business activity.

4.2 Proposition 2: Staking Rewards Should Not Be Taxed Upon Receipt If They Do Not Originate from a Main Business Activity

In Germany and Switzerland, staking rewards are taxed upon receipt, even though they do not originate from a main business activity. However, taxation upon receipt can substantially distort the decision making of coin holders. As there is a significant risk of volatility in the crypto sphere, taxation upon receipt may encourage coin holders to immediately sell their staking rewards in order to reduce the volatility risk and prevent an excessive tax burden as soon as the tax liability is due. For instance, the received staking rewards could substantially decrease in value. If coin holders do not realize the value upon receipt, they nonetheless pay taxes on the value of the coin at the time of receipt, even though the coin may be worth significantly less in the meantime. Thus, taxation upon receipt could incentivize coin holders to realize the value of their received staking rewards immediately in order to obtain liquidity for paying their taxes later on. Consequently, they may sell their coins despite not wanting to. Hence, whenever staking does not represent a main business activity, staking rewards should not be taxed upon receipt.

Moreover, it should be noted that staking rewards often consist of newly minted coins on the one hand and retained transaction fees on the other hand. Morton (2021) compares newly minted coins with a newly baked cake by a baker [17]. This cake is only taxed when the baker sells it. Following this logic, the newly minted coins should not be taxed upon receipt. A similar view is held by a plaintiff in a U.S. lawsuit, who considers it unjustified that newly minted coins created through staking are taxed upon receipt [17]. Moreover, the creation of new coins through staking creates inflation. Unlike the normal fiat monetary system, coin holders can compensate for the devaluation of their coin holdings by staking and receiving rewards. This means that the network share of coin holders does not diminish, as long as they stake their coins. However, if the staking rewards are immediately taxed upon receipt, the coin holders lose parts of their original share in the network because of taxation [17]. This may incentivize investors to move to blockchains whose staking rewards are solely composed of retained transaction fees instead of newly minted coins. In fact, moving to proof-of-stake blockchains without inflation enables coin holders to prevent a tax-induced erosion of their network share. Thus, taxation upon receipt may favor blockchains that have tokenomics without staking-induced inflation.

4.3 Proposition 3: Staking Rewards Should Be Taxed Upon Disposal

In Austria, staking rewards are not taxed until they are traded for fiat money, goods, or services. Therefore, the coin holder only triggers a taxable event when eventually realizing the value from the staking rewards. Obviously, the taxation upon disposal can also incentivize the coin holder to hold the staking rewards longer than actually intended. However, as any form of taxation somehow affects decision making, taxation of staking rewards upon disposal seems reasonably close to the idea of a neutral tax system. The problem with the Austrian form of taxing staking rewards upon disposal is that taxation can be shifted almost indefinitely into the future, as long as the value generated through staking remains in the crypto sphere. For example, taxpayers could simply trade their staking rewards for a stable coin pegged to the U.S. Dollar in order to avoid taxation. This enables them to lock-in their staking value without having to hold a highly volatile native crypto asset obtained through staking for a longer period of time. Thus, coin holders may rather trade their staking rewards for other cryptocurrencies than convert them into fiat money, goods, or services. This affects their decision as to whether they extract generated value from the crypto sphere. Against the background of tax neutrality, staking rewards should be taxed upon disposal regardless of whether they are traded for other cryptocurrencies, fiat money, goods, or services. Otherwise, cryptocurrencies receive a preferential tax treatment compared to other more 'traditional' assets.

4.4 Proposition 4: Staking Should Not Cause Further Tax Consequences for Staked Coins

Prior to the publication of the final BMF circular, the German Ministry of Finance suggested that participating in staking extends the holding period of staked coins from one to ten years. According to this opinion, potential profits from the sale of coins that have been used for staking only become tax-exempt after ten years instead of being tax-exempt after the standard holding period of one year. This is a substantial tax-related disadvantage that only arises from the decision to stake. Even though this opinion has been revoked in the meantime, it illustrates that adverse tax consequences caused by staking may substantially influence the initial decision to stake. In fact, such tax consequences for the staked coins may eventually discourage coin holders from participating in staking. Thus, in accordance with a more neutral tax system, participating in staking should not cause further tax consequences for the underlying staked coins.

5 Conclusion

This analysis illustrates that the taxation of staking rewards is very heterogeneous across different countries. As a result, there is considerable uncertainty especially among retail investors, which could eventually prevent them from participating in staking. This may harm the decentralization of networks with proof-of-stake consensus algorithms. More specifically, far-reaching and complex taxation of staking rewards could reduce the share of staked coins in a network and potentially result in a concentration of validators in certain tax-beneficial countries. To ensure that taxation does not undermine a central value proposition of blockchain technology, there is a need for a neutral taxation of staking rewards that acknowledges specific blockchain characteristics such as the volatility of digital assets as well as the unique functionality of staking. Otherwise, coin holders may not participate in staking and decentralization will be undermined by government intervention in the form of taxation. Against this background, the propositions of this analysis can incrementally contribute to an innovation-friendly and technology-specific regulatory environment. However, it should be noted that the analysis is limited to only three countries, thereby covering a variety, but by no means all possibilities of taxing staking rewards. In addition, this analysis excludes the corporate income taxation of legal persons as well as value-added taxes related to staking. Furthermore, it only considers the taxation of staking rewards, thereby neglecting other tax issues related to digital assets and blockchain technology. Against this background, future research should have a closer look at the tax implications of airdrops, forks, or certain forms of financing such as initial stake pool offerings (e.g., Cardano) or parachain auctions (e.g., Polkadot). All things considered, taxation and other regulatory initiatives should take into account the decentralized nature of blockchain technology.

References

1. Alt, R.: Blockchain. In: Gramlich, L., Gluchowski, P., Horsch, A., Schäfer, K., Waschbusch, G. (eds.) Gabler Banklexikon (A – J), pp. 348–349. Springer Fachmedien Wiesbaden, Wiesbaden (2020)

2. Beck, R., Müller-Bloch, C., King, J.L.: Governance in the blockchain economy: a framework and research agenda. JAIS **19**(10), 1020–1034 (2018). https://doi.org/10.17705/1jais.00518
3. Blockpit: Bitcoin und Steuern in der Schweiz (2021). https://cryptotax.io/de-ch/bitcoin-und-steuern-in-der-schweiz/
4. Bundesministerium der Finanzen: Entwurf eines BMF-Schreibens: Einzelfragen zur ertragsteuerrechtlichen Behandlung von virtuellen Währungen und von Token (2021). https://www.bundesfinanzministerium.de/Content/DE/Downloads/BMF_Schreiben/Steuerarten/Einkommensteuer/2021-06-17-est-kryptowaehrungen.pdf?__blob=publicationFile&v=2
5. Bundesministerium der Finanzen: Einzelfragen zur ertragsteuerrechtlichen Behandlung von virtuellen Währungen und von sonstigen Token (2022). https://www.bundesfinanzministerium.de/Content/DE/Downloads/BMF_Schreiben/Steuerarten/Einkommensteuer/2022-05-09-einzelfragen-zur-ertragsteuerrechtlichen-behandlung-von-virtuellen-waehrungen-und-von-sonstigen-token.pdf?__blob=publicationFile&v=1
6. Bundesministerium für Finanzen: Besteuerung inländischer sowie im Inland bezogener Kapitalerträge (2022). https://www.bmf.gv.at/themen/steuern/sparen-veranlagen/besteuerung-kapitalertraege-inland.html
7. Bundesministerium für Finanzen: Steuerliche Behandlung von Kryptowährungen (2022). https://www.bmf.gv.at/themen/steuern/sparen-veranlagen/steuerliche-behandlung-von-kryptowaehrungen.html
8. Eidgenössische Steuerverwaltung ESTV: Das schweizerische Steuersystem (2019). https://www.zh.ch/content/dam/zhweb/bilder-dokumente/themen/steuern-finanzen/steuern/CH-Steuersystem_d.pdf
9. Eidgenössische Steuerverwaltung ESTV: Arbeitspapier. Kryptowährungen und Initial Coin/Token Offerings (ICOs/ITOs) als Gegenstand der Vermögens-, Einkommens- und Gewinnsteuer, der Verrechnungssteuer und der Stempelabgaben (2021). https://www.estv.admin.ch/dam/estv/de/dokumente/dbst/kryptowaehrungen/arbeitspapier-kryptowaehrungen.pdf.download.pdf/arbeitspapier-kryptowaehrungen-de.pdf
10. Emurgo: Explaining Cardano's Proof-of-Stake (PoS) vs. Delegated Proof-of-Stake (DPoS) Blockchain (2020). https://emurgo.io/ja/blog/explain-proof-of-stake-pos-dpos
11. Endres, D., Spengel, C.: International Company Taxation and Tax Planning. Kluwer Law International, Alphen aan Den Rijn (2015)
12. Hornung, P.: Staking und Steuern in Deutschland – Gewinne, Haltefristen, BMF-Schreiben (2021). https://hub.accointing.com/crypto-tax-regulations/germany/staking-und-steuern-in-deutschland-gewinne-haltefristen-bmf-schreiben
13. HUPAYX: How Are Blockchain Transactions Validated? Consensus VS Validation (2020). https://medium.com/hupayx/how-are-blockchain-transactions-validated-consensus-vs-validation-ada9c001fd0a
14. Kramer, B.: Top Staking Platforms for Massive Income (2021). https://medium.com/general_knowledge/top-staking-platforms-for-massive-income-8003b820c04d
15. Lumineau, F., Wang, W., Schilke, O.: Blockchain governance—A new way of organizing collaborations? Org. Sci. **32**(2), 500–521 (2021). https://doi.org/10.1287/orsc.2020.1379
16. Lutkevich, B.: Definition cryptographic nonce (2021). https://www.techtarget.com/searchsecurity/definition/nonce
17. Morton, J.: Crypto staking rewards and their unfair taxation in the US (2021). https://cointelegraph.com/news/crypto-staking-rewards-and-their-unfair-taxation-in-the-us
18. Nakamoto, S.: Bitcoin: A Peer-to-Peer Electronic Cash System (2008). https://bitcoin.org/bitcoin.pdf
19. Pielke, W.: Besteuerung von Kryptowährungen. Springer Fachmedien Wiesbaden, Wiesbaden (2018). https://doi.org/10.1007/978-3-658-23256-6

20. Polkadot: What does it mean to "get slashed"? (2021). https://support.polkadot.network/support/solutions/articles/65000110858-what-does-it-mean-to-get-slashed-#:~:text=Slashing%20will%20happen%20if%20a,the%20severity%20of%20the%20transgression
21. Statista, KPMG: Einkommenssteuersätze in der Schweiz nach Kantonen im Jahr 2021 (2021). https://de.statista.com/statistik/daten/studie/513589/umfrage/einkommenssteuersaetze-in-der-schweiz-nach-kantonen/
22. Terra Docs: About the Terra Protocol. Unbonding (2022). https://docs.terra.money/docs/learn/protocol.html#terra-and-luna
23. Truby, J.: Decarbonizing bitcoin: law and policy choices for reducing the energy consumption of Blockchain technologies and digital currencies. Energy Res. Soc. Sci. **44**, 399–410 (2018). https://doi.org/10.1016/j.erss.2018.06.009
24. Zhang, L., et al.: Taxation of Cryptocurrency Block Rewards in Selected Jurisdictions. The Law Library of Congresss, Global Legal Research Directorate, Washington, D.C. (2021). https://www.loc.gov/item/2021666100/

Comparison Framework for Blockchain Interoperability Implementations

Alexander Neulinger(✉)

Vienna University of Economics and Business, Welthandelsplatz 1, 1020 Wien, Austria
alexander.neulinger@s.wu.ac.at

Abstract. Blockchain interoperability has gained importance in practice, is increasingly discussed in literature, and serves as basis for new use cases such as manufacturing and financial services. However, many of the blockchain interoperability solutions discussed in literature are still in the design phase, are unpopular or have a small developer community. Therefore, this study proposes a comparison framework and examines implemented public blockchain interoperability solutions, focusing on data from published GitHub repositories. The results show that these implementations vary significantly in terms of popularity, their developer communities as well as their source code, indicating differences in quality. The insights gained in this work facilitate the selection of an appropriate implementation to enable blockchain interoperability use cases.

Keywords: Blockchain interoperability · Comparison framework · Empirical study · GitHub · Implementations

1 Introduction

Blockchain interoperability (BCI) [1] is increasingly discussed in literature [2], provides a wide range of potential applications [3] and can add value through network effects enabled by an extended number of participants [4]. The widespread adoption of blockchain technology in various sectors, such as manufacturing or financial services [5], also makes BCI increasingly important from a practical perspective [6]. For some blockchain-based use cases, BCI is even considered a prerequisite, for instance, in the financial services sector [7].

These BCI solutions are design science research artefacts [8,9]. Such artefacts aim to be relevant and practical by solving existing real-world problems [8]. However, there is a gap between theory and practice, as some of the identified BCI solutions [2] are still only theoretical concepts, lacking popularity and publicly

This study was conducted as part of a co-operation project between ABC Research and Raiffeisenbank International, funded within the framework of the COMET centre ABC, by BMK, BMDW and the provinces of Vienna, Lower Austria and Vorarlberg. The COMET programme is managed by the FFG.

G. Kotsis et al. (Eds.): DEXA 2022 Workshops, CCIS 1633, pp. 316–327, 2022.
https://doi.org/10.1007/978-3-031-14343-4_29

available source code. Even if a public implementation exists, it may be written in an undesirable language, may have an inactive developer community or may be of low quality, making the implementation of new features as well as maintenance difficult. There is no scientific comparison framework for BCI solutions. As the expectations of the designers of an artefact differ from the requirements of practitioners [10], such a comparison framework is needed to make informed decisions. To address this research gap, the following research questions (RQs) are asked in this study:

- RQ1: **Which BCI solutions have a public *implementation*?**
- RQ2: **How can public BCI implementations be *compared*?**
- RQ3: **How can the comparison framework for BCI implementations be *applied*?**

The purpose of this study is to bridge the gap between theory and practice by extending the *Blockchain Interoperability Framework* presented in [2] with an empirical component that enables a systematized comparison of BCI implementations. This work follows an empirical approach based on the data science life cycle [11]. First, BCI solutions are identified in literature [2]. Then, retrieved solutions are filtered for public implementations. This is accomplished by searching for published source code on GitHub[1], thereby allowing BCI implementations to be distinguished from conceptional or unpublished solutions. The identified implementations are further examined based on the comparison framework proposed in this work.

2 Background

In this section the concept of BCI is briefly described. In particular, it is depicted, which BCI functions can be enabled by BCI solutions. Furthermore, the classification of BCI solutions in connector types and their respective BCI mechanisms is outlined.

Cross-blockchain communication is enabled by BCI mechanisms [1]. These mechanisms are utilized by BCI solutions, as Polkadot or Cosmos, for instance, to enable BCI functions [9] as atomic swaps [12], token portability [13], blockchain oracles [9,14] and cross-blockchain smart contracts [15]. These functions serve as the foundation for BCI use cases.

BCI mechanisms can be categorized into three connector types, namely Public Connectors, Blockchains of Blockchains (BoB) and Hybrid Connectors [2]. Public Connectors connect public blockchains and include the BCI mechanisms sidechains & relays, notary schemes and hashed time-lock contracts (HTLCs). While communication between two blockchains is established in a more decentralized fashion using sidechains & relays [4] and in a more centralized fashion using notary schemes [1], [9], HTCLs enable direct exchange of assets between two participants on two different blockchains [1,9].

[1] https://github.com/.

In contrast to the other connector types, a BoB collects cross-blockchain transactions and uses them to form its own blocks and eventually its own blockchain [2]. Finally, Hybrid Connectors aim to connect public and private blockchains and include more novel mechanisms as Trusted Relays, Blockchain Agnostic Protocols and Blockchain Migrators [2].

3 Methods and Data Sources

In this section, the methods and data sources used in this work are described. First, the approach of retrieving BCI solutions from literature is outlined. Second, the process of searching and selecting BCI implementations delineated.

This work follows an empirical approach based on the *data science life cycle* [11], see Fig. 1. The approach starts with determining BCI solutions suitable for further analysis. For that purpose, a list of BCI solutions is taken from literature. Reference [2] was published in November 2021 in ACM Computing Surveys and serves as main source. The aim of [2] was to conduct a thorough literature review on developments in the field of BCI by examining a total of 404 different documents, consisting of about two-thirds scientific literature and about one-third gray literature. Of these 404 sources, 67 were considered by the authors as relevant BCI solutions. Since [2] provides a comprehensive overview, its 67 identified solutions form the initial sample of BCI solutions for the study at hand.

GitHub is the largest open source software platform [16] with more than 63 million users[2] and more than 36 million public repositories[3]. Hence, publicly available *GitHub repositories*[4] are the primary data source for BCI solutions examined in this work. Each GitHub repository is associated with a *GitHub project* or a *contributor*. A project or a contributor can have multiple repositories. Most identified BCI implementations belong to a project that has multiple repositories. While some projects are largely engaged with interoperability, such as *Uniswap*, in other projects interoperability is only one of several aspects, such as in *Cosmos*. Hence, the most suitable repository has to be selected for each BCI implementation to achieve an accurate analysis.

For the purpose of selecting the *most suitable GitHub-repository*, if several repositories are pinned in a project, the first pinned repository with interoperability reference was chosen. Since some GitHub projects do not have pinned repositories, its repositories were sorted in descending order by stars and the first repository with an interoperability reference was selected. While this process enables the identification of the repositories with the highest relevance in most cases, there were exceptions to this scheme. First, *Polkadot* is uniquely associated with a particular repository within the Parity Technologies project[5]. Second, the first pinned repository of *0x* was *0x-mesh*[6], at the time of

[2] https://github.com/search?q=type:user&type=Users, retrieved on May 11, 2022.
[3] https://github.com/search?q=is:public, retrieved on May 11, 2022.
[4] https://docs.github.com/en/rest/reference/repos.
[5] https://github.com/paritytech.
[6] https://github.com/0xProject/0x-mesh.

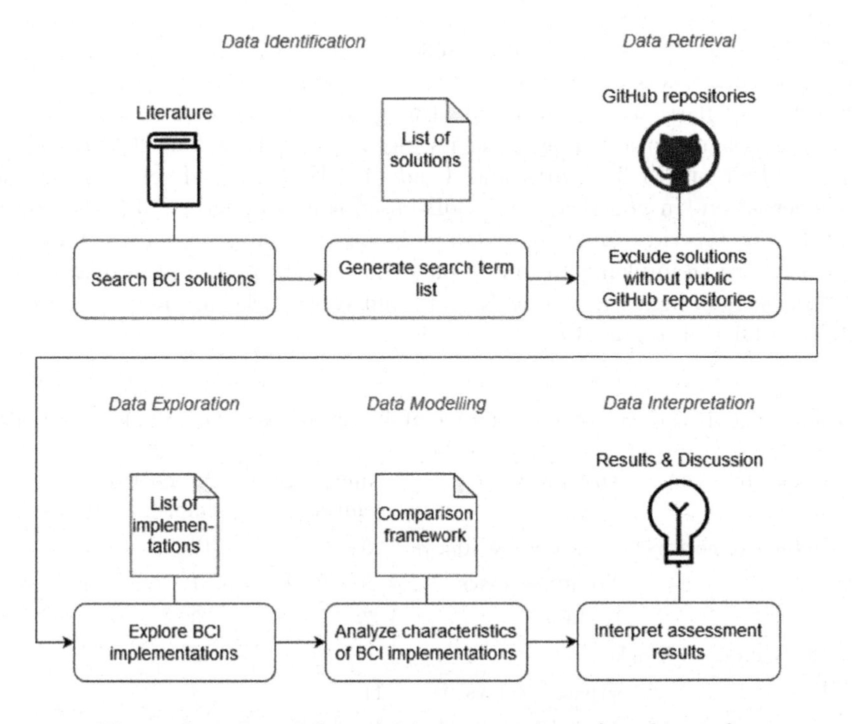

Fig. 1. Overview of the empirical approach used in this study.

writing. However, this repository was no longer being maintained and was therefore considered as irrelevant.

4 Results

In this section, the RQs asked in the Introduction are answered. In Sect. 4.1, BCI implementations are identified. In Sect. 4.2, a comparison framework for BCI implementations is suggested. Finally, the proposed comparison framework is applied on the retrieved BCI implementations in the Sect. 4.3.

4.1 Identified BCI Implementations

In this subsection, the first research question (RQ1), **which blockchain interoperability solutions have an *implementation***, is answered. To this end, BCI implementations were identified by examining which BCI solutions have published source code. Since this study follows the *blockchain interoperability framework* [2], the results were categorized into Public Connectors, BoBs and Hybrid Connectors. Table 1 compares the number of BCI solutions [2] with BCI implementations of each mechanism. In total, from 67 BCI solutions listed in [2], 37 solutions could be identified that have an implementation published on GitHub.

While a majority of the Public Connectors and all BoB-based solutions have an implementation, only a minority of Hybrid Connectors have one, as indicated in Table 1. In particular, from 36 identified Public Connector solutions [2], 22 have an implementation. Regarding its mechanisms, 17 out of 26 sidechain & relay-, only 1 out of 3 notary- and 4 out of 7 HTLC-based solution have an implementation. In contrast, all 6 BoB-based solutions have published source code. Lastly, Hybrid Connectors contain 25 solutions in [2], from which only 10 solutions have an implementation. In particular, 3 out of 11 trusted relays, 5 out of 10 blockchain agnostic protocols and 2 out of 4 blockchain migrators have a published GitHub repository.

Table 1. Number of blockchains interoperability solutions [2] and implementations

Connector type	Mechanism	Number of solutions	Number of implementations
Public connectors	Sidechains & Relays	26	17
	Notary schemes	3	1
	HTLC	7	3
Blockchains of Blockchains		6	6
Hybrid connectors	Trusted Relays	11	3
	Agnostic protocols	10	5
	Blockchain migrators	4	2

4.2 Comparison Framework for BCI Implementations

In this subsection, the second research question (RQ2), **how can public BCI implementations be *compared***, is answered by proposing a BCI comparison framework. Public GitHub repositories allow the analysis of multiple comparison criteria. For instance, the *number of stars* and the *number of forks* are the most and second most important measures of popularity [16]. Stars are a user-based rating retrieved from other developers, which can even indicate the quality of a GitHub repository [17]. Moreover, the number of stars correlates positively with the market capitalization of a token-based project [18].

Next to the popularity of a GitHub repository, its developer community can also be analyzed. The *number of contributors* determines the size of the developer community. Its activity can be measured as well and can additionally serve as an indicator of quality [19]. The *last commit in the repository* detects inactivity in the respective GitHub repository of the BCI implementation. Since repositories at an advanced stage of development may not need to be changed as regularly, the *last commit in the superior project* provides insights into the activity of the entire project to which the repository belongs. The total number of *commits during the last month* should give further indications about the developing activities within a repository.

Finally, the source code of a published BCI implementation can be analyzed as well. The *top language used* may be an important criteria for developers to decide which implementation to adopt. Furthermore, the *lines of code (LoC)*[7] of the source code within the GitHub repository can be counted, indicating the scope and developing effort of the implementation.

4.3 Assessment of Identified BCI Implementations

In this subsection, the third research question (RQ3), **how can the comparison framework for BCI implementations be *applied***, is answered by comparing the characteristics of the previously identified BCI implementations using the previously proposed comparison framework. The results are summarized in Table 2 and categorized as in [2], since their underlying technical mechanisms themselves already serve as a first comparison criteria for BCI implementations. Since the *number of stars* is the most important measure of popularity, indicates the quality of a repository, and further correlates with the market capitalization of token-based projects, the BCI implementations are ordered descending by their stars within a mechanism group. Additional comparison criteria allow assessing the popularity, developer community and source code of the BCI implementations.

As indicated in Table 2, BCI implementations differ greatly from each other on several comparison criteria, such as the number of stars, forks, contributors or LoC, sometimes by a factor of thousands. Thereby, the two popularity measures of number of stars and forks seem to correlate, with *Horizon* being a clear outlier in this regard. Only the repositories of 13 out of 37 BCI implementations registered commits within the last month. However, all BCI implementations belong to GitHub projects that registered recent commits, with *HyperService* as an exception. For instance, while the repository of *BTC Relay* is inactive since 2017, its superior project has a large and active developer community.

In terms of source code, the 18 BCI implementations with the highest number of stars have around 682 K LoC in average, while the 18 BCI implementations with the lowest number of stars only have 204 K LoC in average. *Uniswap* stands out in this respect, as it is one of the most popular implementations, while having one of the lowest LoC. Last but not least, in 3 out of 7 groups of mechanisms the most popular BCI implementations is written in the language *Go*.

[7] To retrieve the lines of code (LoC) the web tool *codetabs* was used; see https://codetabs.com/count-loc/count-loc-online.html.

Table 2. Comparison framework for blockchain interoperability implementations

Blockchain interoperability implementations listed by connector type & mechanism [2] and sorted descending by number of stars			Popularity measures		Developer Community & Activity Measures				Source Code Measures	
Connector type	Mechanism	Implementation	Number of Stars ▼	Number of Forks	Number of Contributors	Last commit in Repo	Last commit in superior project	Commits (last month)	Top language used	Lines of code*
Public Connectors	Sidechains & Relays	RSK [20]	643	241	52	Mar 21, 2022	Mar 31, 2022	159	Java 99.5%	2,401,692
		BTC Relay [21]	546	171	6	Oct 25, 2017	Mar 31, 2022	0	Python 99.8%	26,191
		BitXhub [22]	515	73	19	Mar 31, 2022	Mar 31, 2022	55	Go 87.6%	55,495
		Interledger [23]	373	110	34	Jul 6, 2021	Mar 31, 2022	0	HTML 47.3%	8,804
		Kyber Network [24]	346	161	19	Apr 9, 2021	Mar 31, 2022	0	HTML 37.5%	220,591
		Hyperledger Quilt [25]	220	78	16	Feb 23, 2021	Mar 31, 2022	0	Java 100.0%	49,535
		POA Network [26]	184	157	12	Mar 31, 2022	Mar 28, 2022	8	JavaScript 53.1%	22,815
		Ethereum Plasma [27]	147	33	2	13 Jun 2018	13 Jun 2018	0	JavaScript 100%	6,240
		XMR-BTC [28]	67	9	6	Nov 18, 2020	Mar 23, 2022	0	TeX 98.9%	3,398
		Block Collider [29]	44	20	7	Apr 23, 2019	Mar 29, 2022	0	JavaScript 78.9%	255,929
		ConsenSys [30]	36	13	4	Mar 31, 2022	Mar 31, 2022	25	Java 51.6%	70,655
		JugglingSwap [31]	27	5	2	Jan 19, 2021	Mar 29, 2022	0	TypeScript 100.0%	2,116
		Pantos [32]	26	7	5	Jul 4, 2021	Mar 26, 2022	0	JavaScript 88.4%	3,487
		Zendoo (Horizen) [33]	24	12	11	Feb 14, 2022	Mar 31, 2022	0	Rust 63.7%	14,875
		Plasma [34]	15	3	2	Dez 8, 2020	Mar 31, 2022	0	Solidity 81.2%	361
		NOCUST [35]	14	11	0	Jul 15, 2020	Mar 26, 2022	0	JavaScript 100.0%	21,265
		Horizon [36]	2	428	0	Dez 19, 2019	Mar 31, 2022	0	JavaScript 74.3%	132,982
	Notary Schemes	0x [37]	378	64	3	Dez 24, 2020	Mar 31, 2022	0	Rust 63.0%	116,476
	HTLC	Wanchain [38]	278	86	18	Feb 14, 2022	Mar 31, 2022	0	Go 85.8%	1,347,689
		Blocknet [39]	211	93	333	Aug 20, 2021	Mar 31, 2022	0	C++ 74.6%	485,412
		Fusion [40]	29	31	13	Mar 5, 2022	Mar 5, 2022	1	Go 86.1%	203,760

(continued)

Table 2. (*continued*)

Blockchain interoperability implementations listed by connector type & mechanism [2] and sorted descending by number of stars			Popularity measures		Developer Community & Activity Measures				Source Code Measures	
Connector type	Mechanism	Implementation	Number of Stars ▼	Number of Forks	Number of Contributors	Last commit in Repo	Last commit in superior project	Commits (last month)	Top language used	Lines of code*
Blockchain of Blockchains		Polkadot [41]	5,100	1,200	168	Mar 31, 2022	Mar 31, 2022	597	Rust 98.5%	205,613
		Cosmos [42]	466	180	58	Mar 30, 2022	Mar 31, 2022	61	TeX 93.3%	8,880
		AION [43]	337	114	43	Jan 8, 2021	Mar 29, 2022	0	Java 92.5%	225,978
		ARK [44]	311	313	48	Mar 24, 2022	Mar 31, 2022	12	TypeScript 94.6%	558,216
		Komodo [45]	188	131	31	Mar 29, 2022	Mar 31, 2022	158	QML 45.2%	95,527
		Overledger (Quant) [46]	34	10	10	Apr 20, 2021	Mar 17, 2022	0	TypeScript 83.5%	14,478
Hybrid Connectors	Trusted Relays	Hyperledger Fabric [47]	13,000	7,900	297	Mar 30, 2022	Mar 31, 2022	25	Go 99.3%	1,179,870
		CBT [48]	5	5	2	Sep 23, 2020	NA	0	Java 99.3%	1,053,312
		SCIP [49]	1	0	2	Mar 4, 2020	NA	0	NA	178
	Blockchain Agnostic Protocols	Uniswap [50]	2,000	1,000	10	Jan 26, 2022	Mar 31, 2022	0	TypeScript 51.7%	6,758
		Hyperledger Cactus [51]	209	163	57	Mar 30, 2022	Mar 31, 2022	40	TypeScript 84.4%	641,783
		ION [52]	129	24	10	May 12, 2021	Mar 30, 2022	0	JavaScript 59.6%	27,917
		Canton [53]	35	4	3	Mar 28, 2022	Mar 31, 2022	6	Scala 99.1%	183,752
		HyperService [54]	15	2	3	Oct 21, 2020	Oct 21, 2020	0	Java 93.3%	28,718
	Blockchain Migrators	blockchain-interop [55]	2	0	1	Apr 19, 2019	NA	0	Java 100.0%	5,457
		VeriSmart [56]	0	5	4	Apr 8, 2021	NA	0	JavaScript 96.4%	12,339

* Excluding comments, including data formats and markup language, e.g., JSON or XML.

5 Discussion

The Results of this study can be used to bridge the gap between theory and practice by facilitating the identification of suitable BCI implementations. Table 2 can be used to identify implementations that are popular, have a large and active developer community and are written in an appropriate language. This saves search costs and enables a faster realization of BCI use cases.

None of the applied comparison criteria - presented as columns in Table 2 - in itself objectively reflects the quality of a BCI implementation. In their entirety, however, they allow an initial quality assessment. One possible approach to use the framework could be to search for a particular connector type, e.g., Hybrid Connector, for connecting public and private blockchains, then select the appropriate mechanism within that type using [2], and finally examine the implementations listed above first. In addition, the result of the assessment is also a matter of requirements and their weighting. Therefor, the adopter of the framework can come up with an individual formula that weights different comparison criteria according to their desired BCI use case.

The majority of Public Connectors and all BoBs have an implementation, while only the minority of Hybrid Connectors have one. In terms of number of stars, contributors and LoC, Hybrid Connectors are overrepresented among the 10 BCI implementations with the lowest values. This indicates that Hybrid Connectors are still mostly being conceptualized, while Public Connectors and, especially, BoBs are more suitable for production. Nevertheless, practitioners should keep up to date with the development of Hybrid Connectors, since they can enable use cases with a high business potential by reaching a large number of prospective clients, for instance, by connecting a private banking blockchain with Ethereum.

6 Limitations

This study focuses on BCI solutions identified in literature [2]. However, not all BCI solutions are covered in literature because of the fast speed of development in the field and the possibility to copy or *clone* a published GitHub repository, make minor changes in its source code and publish it as an independent or *forked* implementation. For instance, *SushiSwap* is a well-known fork of *Uniswap*, but it is not mentioned in literature [2].

In addition, it could be argued that some of the listed blockchain interoperability implementations are difficult to compare. While some implementations are designed for a specific purpose, others have a much broader range of applications where blockchain interoperability is just an additional feature. For example, *XMR-BTC* is tailored to enable the blockchain interoperability function *atomic swaps* between the *Bitcoin* and the *Monero* blockchain, whereas *Hyperledger Fabric* is a multifaceted project where blockchain interoperability functions serve to *complement* a broad range of general blockchain use cases.

7 Future Work

Future work that will improve the set of BCI solutions to be analyzed, broaden the comparison framework and allow more precise results is summarized below:

- Extend the initial *list of BCI solutions* retrieved from literature, for instance, by searching relevant key words on GitHub or crypto-related platforms.
- Refine the GitHub analysis, for instance, by analyzing the *average time an issue remains unresolved* to make more precise assumptions about the activity of the developer community.
- Calculate the *bus factor* [57], that is, the number of developers essential to the project and the risk of losing them.
- Analyze *developer discussions* in public forums as an additional user-based comparison criteria, for instance, using text mining methods to scan Stack-Overflow discussions [58] in order to identify popular topics, issues, emerging trends or adoption rates per BCI implementation.
- Analyze *network data* as in [59] as an objective measurement of the popularity with end users and effective usage of an BCI implementation.

8 Conclusion

This empirical study provides an overview of BCI implementations and supports the selection of an appropriate solution for enabling BCI use cases. Furthermore, the gap between theory and practice is bridged by identifying BCI implementations, proposing a comparison framework and applying it on the identified implementations. The 37 identified BCI implementations vary significantly in terms of their popularity, developer community and source code. In their entirety these measures allow an initial quality assessment of BCI implementations.

Within a given connector type, the number of BCI implementations per mechanism and the comparison values of the individual artifacts differ greatly. While the majority of Public Connectors and all BoBs have implementations, many Hybrid Connectors remain as concepts. Only a little number of Hybrid Connectors has a high popularity and a large and active developer community. Hence, there are only a few production-ready BCI solutions to implement a blockchain interoperability use case, where a public and a private blockchain are connected.

References

1. Buterin, V.: Chain interoperability. R3 Research Paper (2016)
2. Belchior, R., Vasconcelos, A., Guerreiro, S., Correia, M.: A survey on blockchain interoperability: past, present, and future trends. ACM Comput. Surv. (CSUR) **54**(8), 1–41 (2021)
3. Schulte, S., Sigwart, M., Frauenthaler, P., Borkowski, M.: Towards blockchain interoperability. In: International Conference on Business Process Management, pp. 3–10, September 2019

4. Sunyaev, A., et al.: Token economy. Bus. Inf. Syst. Eng. **63**(4), 457–478 (2021). https://doi.org/10.1007/s12599-021-00684-1
5. Schär, F.: Decentralized finance: On blockchain-and smart contract-based financial markets. FRB of St. Louis Review (2021)
6. Akram, S.V., Malik, P.K., Singh, R., Anita, G., Tanwar, S.: Adoption of blockchain technology in various realms: opportunities and challenges. Secur. Priv. **3**(5), e109 (2020)
7. Bechtel, A., Ferreira, A., Gross, J., Sandner, P.G.: The future of payments in a DLT-based European economy: a roadmap. Available at SSRN (2020)
8. Bichler, M.: Design science in information systems research. Wirtschaftsinformatik **48**(2), 133–135 (2006). https://doi.org/10.1007/s11576-006-0028-8
9. Kannengießer, N., Pfister, M., Lins, S., Sunyaev, A.: Bridges between Islands: cross-chain technology for distributed ledger technology. In: Proceedings of the 53rd Hawaii International Conference on System Sciences (2020)
10. Lukyanenko, R., Parsons, J.: Design theory indeterminacy: what is it, how can it be reduced, and why did the polar bear drown? J. Assoc. Inf. Syst. **21**(5), 1 (2020)
11. Fayyad, U., Piatetsky-Shapiro, G., Smyth, P.: From data mining to knowledge discovery in databases. AI Mag. **17**(3), 37 (1996)
12. Herlihy, M.: Atomic cross-chain swaps. In: Proceedings of the 2018 ACM Symposium on Principles of Distributed Computing, pp. 245–254, July 2018
13. Sigwart, M., Frauenthaler, P., Spanring, C., Sober, M., Schulte, S.:. Decentralized cross-blockchain asset transfers. In: 2021 Third International Conference on Blockchain Computing and Applications (BCCA), pp. 34–41. IEEE, November 2021
14. Koens, T., Poll, E.: Assessing interoperability solutions for distributed ledgers. Pervasive Mob. Comput. **59**, 101079 (2019)
15. Nissl, M., Sallinger, E., Schulte, S., Borkowski, M.: Towards cross-blockchain smart contracts. In: 2021 IEEE International Conference on Decentralized Applications and Infrastructures (DAPPS), pp. 85–94. IEEE, August 2021
16. Borges, H., Valente, M.T.: What's in a GitHub star? Understanding repository starring practices in a social coding platform. J. Syst. Softw. **146**, 112–129 (2018)
17. Jarczyk, O., Gruszka, B., Jaroszewicz, S., Bukowski, L., Wierzbicki, A.: GitHub projects. Quality analysis of open-source software. In: Aiello, L.M., McFarland, D. (eds.) SocInfo 2014. LNCS, vol. 8851, pp. 80–94. Springer, Cham (2014). https://doi.org/10.1007/978-3-319-13734-6_6
18. Trockman, A., Van Tonder, R., Vasilescu, B.: Striking gold in software repositories? An econometric study of cryptocurrencies on GitHub. In: 2019 IEEE/ACM 16th International Conference on Mining Software Repositories (MSR), pp. 181–185. IEEE, May 2019
19. Dabbish, L., Stuart, C., Tsay, J., Herbsleb, J.: Social coding in GitHub: transparency and collaboration in an open software repository. In: Proceedings of the ACM 2012 Conference on Computer Supported Cooperative Work, pp. 1277–1286, February 2012
20. RSK. https://github.com/rsksmart/rskj. Accessed 31 Mar 2022
21. BTC Relay. https://github.com/ethereum/btcrelay. Accessed 31 Mar 2022
22. BitXhub. https://github.com/meshplus/bitxhub. Accessed 31 Mar 2022
23. Interledger. https://github.com/interledger/rfcs. Accessed 31 Mar 2022
24. Kyber Network. https://github.com/KyberNetwork/KyberSwap. Accessed 31 Mar 2022
25. Hyperledger Quilt. https://github.com/hyperledger/quilt. Accessed 31 Mar 2022

26. POA Network. https://github.com/poanetwork/tokenbridge. Accessed 31 Mar 2022
27. Ethereum Plasma. https://github.com/ethereum-plasma/plasma. Accessed 16 May 2022
28. XMR-BTC. https://github.com/h4sh3d/xmr-btc-atomic-swap. Accessed 31 Mar 2022
29. Block Collider. https://github.com/blockcollider/bcnode. Accessed 31 Mar 2022
30. ConsenSys. https://github.com/ConsenSys/gpact. Accessed 31 Mar 2022
31. JugglingSwap. https://github.com/ZenGo-X/JugglingSwap. Accessed 31 Mar 2022
32. Pantos. https://github.com/pantos-io/ethrelay. Accessed 31 Mar 2022
33. Zendoo. https://github.com/HorizenOfficial/zendoo-sc-cryptolib. Accessed 31 Mar 2022
34. Plasma. https://github.com/plasmadlt/PPAY-Governance. Accessed 31 Mar 2022
35. NOCUST. https://github.com/liquidity-network/nocust-contracts-solidity. Accessed 31 Mar 2022
36. Horizon. https://github.com/Horizon-Protocol/Horizon-Smart-Contract. Accessed 31 Mar 2022
37. 0x. https://github.com/0xProject/OpenZKP. Accessed 31 Mar 2022
38. Wanchain. https://github.com/wanchain/go-wanchain. Accessed 31 Mar 2022
39. Blocknet. https://github.com/blocknetdx/blocknet. Accessed 31 Mar 2022
40. Fusion. https://github.com/FUSIONFoundation/efsn. Accessed 31 Mar 2022
41. Polkadot. https://github.com/paritytech/polkadot. Accessed 31 Mar 2022
42. Cosmos. https://github.com/cosmos/ibc. Accessed 31 Mar 2022
43. AION. https://github.com/aionnetwork/aion. Accessed 31 Mar 2022
44. ARK. https://github.com/ArkEcosystem/core. Accessed 31 Mar 2022
45. Komodo. https://github.com/KomodoPlatform/atomicDEX-Desktop. Accessed 31 Mar 2022
46. Overledger. https://github.com/quantnetwork/overledger-sdk-javascript. Accessed 31 Mar 2022
47. Hyperledger Fabric. https://github.com/hyperledger/fabric. Accessed 31 Mar 2022
48. CBT. https://github.com/hpdic/cbt. Accessed 31 Mar 2022
49. SCIP. https://github.com/lampajr/scip. Accessed 31 Mar 2022
50. Uniswap. https://github.com/Uniswap/v3-core. Accessed 31 Mar 2022
51. Hyperledger Cactus. https://github.com/hyperledger/cactus. Accessed 31 Mar 2022
52. ION. https://github.com/clearmatics/ion. Accessed 31 Mar 2022
53. Canton. https://github.com/digital-asset/canton. Accessed 31 Mar 2022
54. HyperService. https://github.com/HyperService-Consortium/HyperService-Language. Accessed 31 Mar 2022
55. Blockchain-interop. https://github.com/pf92/blockchain-interop. Accessed 31 Mar 2022
56. VeriSmart. https://github.com/informartin/VeriSmart. Accessed 31 Mar 2022
57. Cosentino, V., Izquierdo, J.L.C., Cabot, J.: Assessing the bus factor of git repositories. In: 2015 IEEE 22nd International Conference on Software Analysis, Evolution, and Reengineering (SANER), pp. 499–503. IEEE, March 2015
58. Barua, A., Thomas, S.W., Hassan, A.E.: What are developers talking about? An analysis of topics and trends in stack overflow. Empir. Softw. Eng. **19**(3), 619–654 (2014)
59. Nadini, M., Alessandretti, L., Di Giacinto, F., Martino, M., Aiello, L.M., Baronchelli, A.: Mapping the NFT revolution: market trends, trade networks, and visual features. Sci. Rep. **11**(1), 1–11 (2021)

Cyber-security and Functional Safety in Cyber-physical Systems

Towards Strategies for Secure Data Transfer of IoT Devices with Limited Resources

Nasser S. Albalawi[1(✉)], Michael Riegler[2], and Jerzy W. Rozenblit[1]

[1] Department of Electrical and Computer Engineering, University of Arizona, Tucson, AZ, USA
nasseralbalawi@email.arizona.edu, jerzyr@arizona.edu
[2] LIT Secure and Correct Systems Lab, Johannes Kepler University Linz, Linz, Austria
michael.riegler@jku.at

Abstract. Many *Cyber Physical Systems* (CPSs) and *Internet of Things* (IoT) devices are constrained in terms of computation speed, memory, power, area and bandwidth. As they interact with the physical world, various aspects such as safety, security, and privacy should be considered while processing personal data. Systems should continue operating even under harsh conditions and when the network connections (e.g., to the cloud) are lost. If that happens and the storage capacity is limited, sensor data may be overwritten irrevocably. This paper presents preliminary ideas and the planned research methodology to examine and define strategies to secure the data transfer from IoT devices which have limitations to edge devices and the cloud, and to overcome the situation when a device loses its connection, to mitigate data loss.

Keywords: IoT · Resources · Limitations · Security

1 Introduction

Cyber Physical System (CPS) and *Internet of Things* (IoT) devices fall under the embedded system umbrella. The terms are largely interchangeable and are used to describe the interconnection of the physical world with the digital world [3]. Examples are devices and systems that are used in smart automobiles, smart manufacturing, and many other industries. IoT is the concept of combining disparate objects to create a seamless interface between physical and virtual entities [15].

Already, IoT devices generate a lot of raw data which need to have sufficient resources to be processed efficiently. In general, the resources in IoT devices are expected to be constrained due to space and cost limitations [8]. Moreover, the constraints would impact different aspects of IoT components, such as memory and processor as well as the environment such as power, storage, and bandwidth [15]. Based on that, IoT devices would usually connect either to edge or

G. Kotsis et al. (Eds.): DEXA 2022 Workshops, CCIS 1633, pp. 331–336, 2022.
https://doi.org/10.1007/978-3-031-14343-4_30

cloud computing to provide sufficient resources that can keep the IoT devices working properly [6]. Therefore, both edge and cloud computing, would then provide the resources such as computation, memory, and storage to the devices. Likewise, the IoT device would consume less power because most of the computation and storage is done either in the edge or cloud; this is known as computation offloading.

We aim to develop a technique or mechanism and specific strategies to prevent data losses and overcome the effects of a lost connection in case it occurs. In this paper, we present the first findings and our proposed methodology. Section 2 describes the background of IoT in general, especially the challenges of ensuring security despite limited resources. We present the problem and our motivation in Sect. 3. In Sect. 5, we show our proposed case study design and draw our conclusions in Sect. 6.

2 Internet of Things

IoT resources have an essential role in terms of device performance. In general, these resources can be classified into two main types. The first one is the tangible resources such as memory, processor, energy and networking. The second one is the intangible resources such as operating systems, protocols, algorithms and other resources [15]. However, the resources tend to be limited by virtue of size, power limitations, etc. For example, medical and other wearable devices such as insulin pumps should be lightweight and non-intrusive for the patients. In addition, smart sensors (or other devices, e.g., smart speakers or smart locks) should not take up too much space so that they can be integrated into existing environments relatively easily. This limits the hardware and, consequently, software used in the final product.

2.1 Resources Management

The key, essential aspect of the IoT ecosystem is the management scheme needed to assign and efficiently manage the resources [4]. One technique of resource utilization uses edge or cloud computing to offload tasks [5]. This mechanism can increase their computing power and accelerate their completion. In that context, virtualization can be used to efficiently operate hardware and limited resources [14]. Furthermore, resource management considers load balancing and prioritization to ensure the quality of service. Additionally, scheduling and allocation are used to optimize runtime despite limited resources [12] However, sending all sensor data to a centralized cloud platform typically leads to latency. Therefore, some systems rely on on-site edge computing to fulfill real-time requirements and overcome the bandwidth problem. Nevertheless, in case of lost or ineffective communication to the edge or the cloud, IoT devices (for example, in medical or power grid applications) have to continue processing tasks independently, especially in safety-critical systems.

2.2 Security

The connectivity of IoT devices leads to a much larger attack surface. They face many threats such as Denial of Service (DoS), malware, wiretapping, injection, Man-in-The-Middle, or zero-day exploits. Common entry points of attackers are the supply chain (e.g., SolarWinds, Kaseya), ransomware (Colonial Pipeline, Ryuk, WannyCry), or malware (Stuxnet). After a successful attack, IoT devices are misused as a botnet (Mirai) and for lateral movement. Security is a constantly evolving area. New vulnerabilities and exploits become public over time. Therefore, IoT devices should include methods to identify, protect, detect, respond, and recover from threats according to the NIST Cybersecurity Framework [7]. However, enhancing security while being constrained by limited resources is challenging. Insufficient memory and computing power restrict commonly used methods from personal computers and servers. An interesting approach in that context is security modes [10]. When specific events occur, the system switches its modes to reduce the attack surface while providing a reduced set of functionality.

3 Problem and Motivation

An inherently desirable characteristic of IoT is to stay connected. However, connection failures are conceivable and could occur under many circumstances, for example, issues with hardware or software, network congestion, etc. External factors may include natural disasters such as earthquakes, eruptions, fires, floods, or solar winds, which disrupt the communication of IoT devices. In addition, intentional, unintentional, and man-made disasters can also lead to a loss of communication. For example, for more than 18 months, an old TV caused irregular broadband outages by electrical interference [2]. In February 2022, a satellite outage, possibly caused by hackers, affected about 5,800 wind turbines in Europe [9]. In this case, the data transfer was interrupted and the isolated IoT devices could not offload work to the edge or cloud.

As mentioned above, not all IoT devices have sufficient resources to continue working autonomously (due to computing power, storage, etc.). Consequently, this can lead to service discontinuity, especially if heavy computational tasks overload the device. Latency will occur and the increased local computational demands lead to higher power consumption that can reduce battery life. While sensors keep sending data to the IoT devices, the data cannot be forwarded to the edge or the cloud and must be stored in the local memory. If the memory is full, data overwriting can occur, which leads to data loss. That may negatively impact the entire IoT ecosystem for a particular application. Winderickx et al. [13] and Alharaby et al. [1] analyzed different microcontrollers and IoT platforms regarding the energy consumption of security algorithms and protocols, but they did not consider the connection loss. Therefore, we aim to fill the gap and develop a new methodology which minimizes data loss, optimizes performance, secures data transfer, and provides accurate data for cloud-based data utilization. With the proposed case study we will evaluate our strategies.

4 Approach

As this paper presents an initial, conceptual idea for how to handle the resource (especially memory and storage) in IoT devices, we will tackle the development of the management methodology in a bottom-up and top-down manner. The bottom-up approach will be to use a case study (or studies once the concepts mature). The top-down approach will require that we study existing abstractions for resource management in embedded/cyber-physical systems with limitations of memory, storage, and processing power. Such existing approaches include, for example, using virtual mechanisms to optimize resources [14]. Our idea is innovative in that we are not aware of any methods that manage memory and storage in the case of loss of connections to the cloud.

5 Case Study Design

With our research, we want to answer the following research questions:

- How can data loss be prevented and mitigated in the short and long-term for IoT devices with lost connections?
- What strategies are feasible to overcome the connection loss for IoT devices?
- Is a general solution feasible enough, or are specific strategies tailored to device classes and the time span of the outage needed?
- How can security algorithms and protocols influence performance and lifetime?

In a first step of our research, we will conduct a case study according to the guidelines described by Runeson and Hoest [11]. We want to collect data from several sensors and use four different commonly used IoT devices, see Table 1, for processing. The main objective is to explore, implement and test several strategies to handle the situation when an IoT device loses its connection. With the different IoT devices, we want to triangulate the results and compare specific challenges. Figure 1 shows our proposed system architecture. We select *IoT Weather Stations* for our use case. In that context multiple sensors such as temperature, humidity and atmospheric pressure are needed. We want to monitor and analyze real-world data to predict the weather for the coming days. Therefore, the sensor data is transferred to the cloud. Optionally, there is an edge device in between, which works as a gateway. The cloud application as well IoT devices should be able to detect a lost connection, e.g., by using heartbeat protocols. In this case, we want to keep the data for a period of time on the IoT device while the signal is lost. However, the limitation will touch on different hardware aspects of IoT devices, especially limited storage. Therefore, we propose several strategies to optimize resource utilization by first defining types of data and their importance. For example, the barometric pressure may be more important to detect changing weather conditions than humidity. The second strategy is to compress data to save space in the storage, but this also needs processing power. We plan to test different compression algorithms and use statistical functions, e.g., using mean, median, minimum, maximum and variance instead of all single values. In addition, we can free up space by deleting all odd or even data values when memory is almost full. In addition, in the third strategy, we want to reduce the data transmission rate from a sensor to the IoT device and the cloud by increasing the time between two values. For example, we gradually increase the transfer rate and send data values every 10, 20, or 30 s or only changed values. In order to extend the runtime of a battery-operated IoT device, we are investigating a method to keep some resources working by sacrificing other resources, like sensors, in our fourth strategy. Decreasing security measures may not be a good idea, but could be beneficial for one-way data transfer in some circumstances. We plan to measure memory utilization, performance, network as well as power consumption to compare the results.

Table 1. IoT devices

Device	Processor	Memory		Wi-Fi
		Flash	RAM	
Raspberry Pi4 Model B	1.5G Hz 64bit ARM Cortex A72	128 GB	4 GB	802.11ac
NVIDIA Jetson NANO	1.43 GHz 64bit ARM Cortex A57	16 GB	4 GB	External
Particle Photon	120 MHz 32bit ARM Cortex M3	1 MB	128 KB	802.11b/g/n
Arduino Nano 33	48 MHz 32bit ARM Cortex M0	256 KB	32 KB	802.11b/g/n

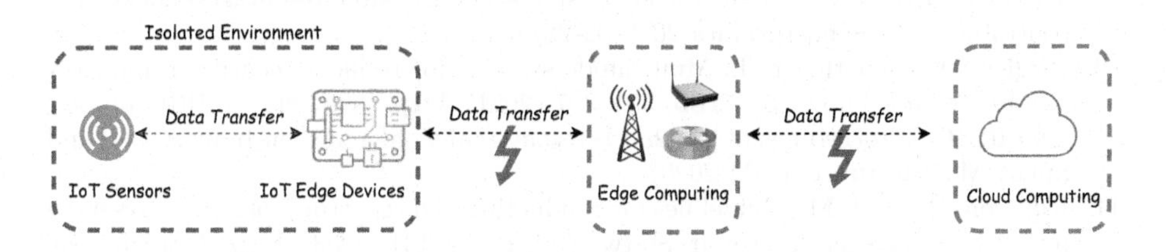

Fig. 1. Isolated IoT architecture

6 Conclusion

We presented our ideas about the planned case study research to secure the data transfer between IoT devices with limited resources and the cloud. Edge computing could help offload IoT devices' work and secure the communication between the IoT devices and the cloud. In addition, they can cache data if the connection drops or the IoT devices are exhausted. We plan to explore, implement and test strategies with different IoT devices in a weather station example in order to answer our initial research questions.

References

1. Alharby, S., Harris, N., Weddell, A., Reeve, J.: The security trade-offs in resource constrained nodes for IoT application. Int. J. Electr. Electron. Commun. Sci. (2018). https://doi.org/10.5281/zenodo.1315561
2. BBC: Internet: Old TV caused village broadband outages for 18 months. BBC News, September 2020. https://www.bbc.com/news/uk-wales-54239180
3. Greer, C., Burns, M.J., Wollman, D., Griffor, E.: Cyber-physical systems and Internet of Things. NIST SP-1900-202, March 2019. https://www.nist.gov/publications/cyber-physical-systems-and-internet-things
4. Jadhav, D., Nikam, S.: Need for resource management in IoT. Int. J. Comput. Appl. **134**(16) (2016)
5. Mijuskovic, A., Chiumento, A., Bemthuis, R., Aldea, A., Havinga, P.: Resource management techniques for cloud/fog and edge computing: an evaluation framework and classification. Sensors **21**(5), 1832 (2021)
6. Mohsin, S.M., Aslam, S., Akber, S.M.A., Iqbal, A., Waheed, A., Ikram, A.: Empowering cloud of things with edge computing: a comparative analysis. In: 2020 International Conference on Information Science and Communication Technology (ICISCT), pp. 1–6. IEEE (2020)
7. NIST: Framework for Improving Critical Infrastructure Cybersecurity, Version 1.1, April 2018. https://doi.org/10.6028/NIST.CSWP.04162018
8. Pisani, F., de Oliveira, F.M.C., Gama, E.S., Immich, R., Bittencourt, L.F., Borin, E.: Fog computing on constrained devices: paving the way for the future IoT. Adv. Edge Comput. Massive Parallel Process. Appl. **35**, 22–60 (2020)
9. Reuters: Satellite outage knocks out thousands of Enercon's wind turbines, February 2022. https://www.reuters.com/business/energy/satellite-outage-knocks-out-control-enercon-wind-turbines-2022-02-28/
10. Riegler, M., Sametinger, J.: Multi-mode systems for resilient security in industry 4.0. Procedia Comput. Sci. **180**, 301–307 (2021). https://doi.org/10.1016/j.procs.2021.01.167, Proceedings of the 2nd International Conference on Industry 4.0 and Smart Manufacturing (ISM 2020)
11. Runeson, P., Höst, M.: Guidelines for conducting and reporting case study research in software engineering. Empir. Softw. Eng. **14**(2), 131 (2008). https://doi.org/10.1007/s10664-008-9102-8
12. Sangaiah, A.K., Hosseinabadi, A.A.R., Shareh, M.B., Bozorgi Rad, S.Y., Zolfagharian, A., Chilamkurti, N.: IoT resource allocation and optimization based on heuristic algorithm. Sensors **20**(2), 539 (2020)
13. Winderickx, J., Braeken, A., Singelée, D., Mentens, N.: In-depth energy analysis of security algorithms and protocols for the Internet of Things. J. Cryptogr. Eng. (2021). https://doi.org/10.1007/s13389-021-00274-7
14. Zahoor, S., Mir, R.N.: Virtualization and IoT resource management: a survey. Int. J. Comput. Netw. Appl. **5**(4), 43–51 (2018). https://doi.org/10.22247/ijcna/2018/49435
15. Zahoor, S., Mir, R.N.: Resource management in pervasive Internet of Things: a survey. J. King Saud Univ. Comput. Inf. Sci. **33**(8), 921–935 (2021). https://doi.org/10.1016/j.jksuci.2018.08.014

Application of Validation Obligations to Security Concerns

Sebastian Stock(✉), Atif Mashkoor, and Alexander Egyed

Johannes Kepler University, Linz, Austria
{sebastian.stock,atif.mashkoor,alexander.egyed}@jku.at

Abstract. Our lives become increasingly dependent on safety- and security-critical systems, so formal techniques are advocated for engineering such systems. One of such techniques is validation obligations that enable formalizing requirements early in development to ensure their correctness. Furthermore, validation obligations help hold requirements consistent in an evolving model and create assurances about the model's completeness. Although initially proposed for safety properties, this paper shows how the technique of validation obligations enables us to also reason about security concerns through an example from the medical domain.

Keywords: Validation obligations · Security-critical systems · Model-driven engineering · Formal methods

1 Introduction

As software systems become more and more responsible for our daily life experiences, it is natural to discuss how to engineer them to ensure their safety and security. Both safety and security are already mature disciplines, and individual processes for dealing with the domain-specific problems are available such as attack trees [14] for security concerns or model checking [7] for safety properties. However, as pointed out in literature (e.g., see [3,11]), there are a limited amount of cross-cutting techniques available at our disposal, which are capable of being used effectively in both domains.

A validation obligation (VO) [10] is a logical formula associated with the correctness claim of a given validation property. This technique helps formalize and validate software systems, thus ensuring their overall correctness. While initially proposed for the safety domain, we believe VOs are equally beneficial for validating security concerns in software systems. The benefit of applying

The research presented in this paper has been conducted within the IVOIRE project, which is funded by "Deutsche Forschungsgemeinschaft" (DFG) and the Austrian Science Fund (FWF) grant # I 4744-N. The work of Sebastian Stock and Atif Mashkoor has been partly funded by the LIT Secure and Correct Systems Lab sponsored by the province of Upper Austria.

G. Kotsis et al. (Eds.): DEXA 2022 Workshops, CCIS 1633, pp. 337–346, 2022.
https://doi.org/10.1007/978-3-031-14343-4_31

VOs to assert the correctness of security concerns lies within the unified approach offered by VOs for modeling all sorts of proprieties, e.g., safety, security, and functional properties. In this fashion, we do not need individual correctness assurance approaches for each set of requirements, which can cause problems while keeping the model consistent. VOs, substantiated with multiple formal techniques, offer a property-agnostic approach for checking completeness and conflict freeness of models.

This paper aims to investigate the application of VOs for the correctness assurance of security concerns. The rest of the paper is structured as follows: Sect. 2 discusses the Event-B method – the formal method we have used in our approach. Next, Sect. 3 introduces the idea of VOs. Then, we exemplify the application of VOs to security concerns in Sect. 4 using an example from the medical domain. Next, Sect. 5 compares the current approaches with VOs for formal modeling of security-critical systems. Finally, we conclude the paper in Sect. 6 with an outlook on future work.

2 The Event-B Method

State-based formal methods [9] enable modeling systems with a strict formal syntax, thus allowing to establish correctness assurance with techniques like model checking and theorem proving. This establishes the model's consistency and shows that the model does not lead to a faulty state. Furthermore, state-based formal methods follow the correct-by-construction approach meaning that the model is incrementally enriched with behavior while the correctness of each step is ensured.

One of the well-known state-based formal methods is Event-B [1]. In Event-B states are made up of `variables` that are constrained by `invariants`. `Events` define transitions between the states. A model can be refined via the `refines` keyword. This means the original specification is advanced concerning variables or events. This refinement relationship, however, needs to be proven to ensure that existing model constraints are not violated in the process of refinement. This is done through proofs. `contexts` define the static elements of a model and are `seen` by `machines` defining the dynamic behavior.

3 Validation Obligations

A VO is composed of a model and a validation task (VT) that must be successfully executed on the model. Once a VT is successfully executed, it establishes the presence of the associated requirement. A VT comes in different forms, e.g. (LTL) model-checking, proof obligations (POs), or even manual inspection of the state space. The input parameters of each VT depend upon the associated requirement. For example, for the LTL model checking, the VT gets an LTL formula as an input parameter. It is the judgment of the designer which VT is best suited to the cause and what are its appropriate parameters. Following is the formal definition of a VO:

$$VO_{id} = VT_{id}/VT_{context}/VT_{technique} : VT_{parameters}$$

As a VO is assigned to a requirement, an `id` uniquely identfies it. The assigned VT has also an `id` to identify it. In the area of VOs `context` refers not to the `context` of machines as defined earlier but to the context the VO is applied in, i.e., the model we investigate. Second last is the applied `technique`, e.g., model checking, and last are the input `parameters`, e.g., the LTL formula.

The concept of VOs should be seen as a carrier technology. It is not bound to a particular formal method. In fact, VOs can be applied to all state-based formal methods alike. In this paper, however, we use Event-B only to exemplify the proposed approach. Following that, we argue that we can validate everything we explicitly model.

The power of VOs comes from three aspects.

1. By connecting requirements to VOs, designers can rely on the associated VTs for compliance. Once the VO composed of VTs is successfully discharged, it shows the presence of the associated requirement in the model. If the designers change the model subsequently, they can execute the VTs again and assure themselves of the requirement's presence in the evolved model.
2. Having all requirements written as VOs lets designers quickly spot conflicting requirements. For example, if requirements are correctly translated into VTs but contradict each other, one of the VOs will fail. In this case, stakeholders can reevaluate the associated requirements.
3. VOs open up the ability to view a model from perspectives outside the classical top-down refinement chain. These views can fall into the following categories: (1) abstract views with which we can safely drop information not necessary for reasoning about a VO, and (2) instantiation or scenario views, which transform the abstract model into a concrete example. The output of the associated validation task can give insight into how the view behaves, and the instantiation view can also provide multiple examples of the behavior. Emerging requirements stemming from this exercise can find their way back into the model.

4 Application of VOs to Security Concerns

We now show the application of VOs to establish the correctness of security concerns through the example of a hemodialysis machine model. As aforementioned, we believe that security concerns can be treated like other properties when developing models, which is an advantage as there is no need for their special treatment, thus making the overall modeling process simpler.

4.1 Illustrative Example

Hemodialysis is a medical treatment that uses a device to clean the blood. The hemodialysis device transports the blood from and to the patient, filters waste and salts from the blood, and regulates the fluid level of the blood. Due to the involved complexity of the dialysis process, the resulted medical treatment is monitored by a professional caregiver for treatment compliance and desired output. Traditionally hemodialysis is performed in a standalone mode, i.e., patients come to a medical facility, get connected to the device, and let dialysis be performed. However, this monitoring is also possible via remote access but demands additional security precautions. The basic architecture of hemodialysis machines is depicted in Fig. 1. The requirements specification of hemodialysis machines is discussed in detail in the work of Mashkoor [8].

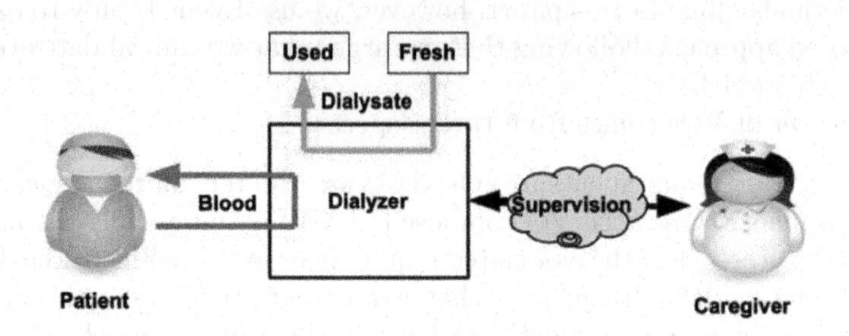

Fig. 1. Architecture of hemodialysis machines

We first design an abstract model and then refine it to show that VOs can be used to gain confidence in the soundness and conflict freeness of the requirements. We then extend the example a second time to show the advantages of creating views on a model. Creating a view means in our context that we keep the abstract model in the background but only focus on one detailed aspect of the model at a time. This helps better understand a model's behavior or debug it by only viewing what is necessary to satisfy a particular requirement.

4.2 Formal Model

We start the modeling process with a small subset of the hemodialysis machine requirements specification. We also add some additional security concerns as follows.

- `SAF1`: In order to start the treatment, the parameters must be within the permissible range.
- `FUN1`: There are three types of staff IDs: maintenance, nurses, and doctors.
- `SEC1`: The staff has to log in to start the treatment.
- `SEC2`: Only doctors and nurses are allowed to start the treatment.

Figure 2a shows the base model of the system. It models the `SAF1` safety requirement. In the original specification multiple parameters depending on the patients condition can be set thus ensuring a personalized treatment. `SAF1` is an abstract reference to that, we enter a treatment value which serves as a token for more complex parameters required in the original specification. We model

```
context ctx

sets login_status

constants yes no

axioms
@axm1 partition(login_status, {yes}, {no})


end
machine m1 sees ctx

variables treatmentValue loggedIn

invariants
  @inv1 treatmentValue ∈ ℕ
  @inv2 loggedIn ∈ login_status

events
  event INITIALISATION
    then
      @act1 treatmentValue ≔ 0
      @act2 loggedIn ≔ no
  end

  event login
    then
      @act loggedIn ≔ yes
  end

  event startSystem
    any parameters
    where
      @grd1 loggedIn = yes
      @grd2 parameters ∈ 1‥5
    then
      @act2 treatmentValue ≔ parameters
  end
end
```

(a) Abstract model

```
context ctx1 extends ctx

sets persons ids

constants doctors nurses maintenance

axioms
@axm1 partition(persons, {doctors}, {nurses}, {maintenance})

end
machine m2 refines m1  sees ctx1

variables treatmentValue loggedIn loggedInID treatmentAllowed

invariants
  @inv4 treatmentAllowed ∈ ids ⇸ persons ∖ {maintenance}
  @inv5 loggedInID ∈ ids

events
  event INITIALISATION extends INITIALISATION
    then
      @act4 treatmentAllowed :∈ ids ⇸ persons ∖ {maintenance}
      @act5 loggedInID :∈ ids
  end

  event login extends login
    any number
    where
      @grd3 number ∈ ids
    then
      @act2 loggedInID ≔ number
  end

  event startSystem extends startSystem
    where
      @grd3 loggedInID ∈ dom(treatmentAllowed)
  end
end
```

(b) First refinement

Fig. 2. The formal model

SAF1 along with the SEC1 security requirement. There is an Event-B context ctx modeling the login status. The corresponding machine m1 models two events: login and startSystem. Guards, namely @grd1 and @grd2, of the startSystem event prevent the event from being fired prematurely.

Figure 2b shows how the model looks after the refinement. It introduces FUN1 and SEC2 requirements into the model. We add roles as data type in the refined context ctx1 and configure the set of roles that are allowed to perform a treatment in @inv4. The login event ensures that the role logged into the system is now tracked via id. @grd3 in startSystem checks that only allowed ids are able to start the treatment. Also note that the startSystem event in m2 extends the event from m1 meaning that @grd1 and @grd2 are still present and active but are hidden form this view of the model to avoid confusion.

4.3 Ensuring Soundness and Conflict Freeness of Requirements

To each requirement, we associate a VT with which we ensure the presence of this requirement in the model. The VTs can be formulated as:

- SAF1/m1/PO: treatmentParameters $\in 1..5$
- SEC1/m1/LTL:G({loggedIn $=$ yes} $\Rightarrow$ e(startSystem))[1]

[1] e(startSystem) means that this event is enabled because its guard is true.

PO means we have a proof obligation that treatmentParameters $\in$ 1..5 is true in every state of the model, meaning the patients treatment parameters are within the allowed range. For the LTL formula G(lobally) means the condition in the brackets is always true. The term inside states that when loggedIn = yes the event startSystem is enabled. m1 is the model or in our case the machine we apply the VT to as laid out in Sect. 3.

The proof of the VT for SAF1 is relatively simple. We can use the Event-B tool support, i.e., the Rodin platform [2], to show that @inv1 holds in every state of the model. The proof can be discharged automatically by the tool. The VT regarding SEC1 was established using the LTL formula with the ProB [7] model checker, which we employed to do LTL model checking of the model. With both VOs discharged, we can be confident that the concerning requirements are correctly modeled.

Now we tackle FUN1 and SEC2. For those requirements we write the following VTs:

- FUN1/m2/PO/persons = {doctors, nurses, maintenance}
- SEC2/m2/LTL/G(e(startSystem) $\Rightarrow$ {treatmentAllowed(loggedInID) = doctors} $\vee$ {treatmentAllowed(loggedInID) = nurses})

We demand proof that the set of roles consists only of doctors, nurses, and maintenance for FUN1. For SEC2 we demand that globally the system can only be started if the logged-in role is registered in the set of roles that are allowed to perform treatment, i.e., doctors or nurses.

Again with the help of the corresponding tools, we see that both VTs on the model are executed and satisfied, thus ensuring the soundness of the model m2 regarding the requested requirements, i.e., FUN1 and SEC2. However, if we run our VTs for SEC1 and SAF1 previously established on m1 on m2 to ensure that these requirements are also present in the refinement, we spot a problem. The VT representing SEC1 fails. We can find an instance where we are logged into the system but not allowed to start the treatment. This is the case with the login of the maintenance role.

A failing VT helps us discover a flawed requirement/design in the model, provided that the task was correctly chosen and the parameter was correctly formulated. In our example the requirement encoded as the VT for SAF1 is no longer satisfiable as loggedIn = yes is no longer sufficient for starting the system. We could draw three consequences from this:

1. Requirements SEC1 and SEC2 may be contradicting each other. However, this is not the case here.
2. We could adjust the model by removing the event guard that controls the startSystem event allowing every loggedIn person to start treatment. However, this will lead SEC2 to fail; hence the formulation of the requirement as a VO prevented us from introducing new bugs when trying to fix the existing ones.

3. Requirements SEC1 and SEC2 are ambiguous in how they are formulated and need to be clarified. In our case, SEC1 is indeed very broad. We can solve the issue by either making SEC1 more explicit by stating that the allowed staff has to be logged in or merging the requirement entirely with SEC2.

In our example, we settle for option (3) to clarify requirements: we refactor the corresponding VO, and the fixed VO passes without any new conflict.

4.4 Creating Views

Different views help designers spot flaws in the model or help non-technical stakeholders better understand the model. Views are a unique form of refinement that aims not to introduce new behavior but to how the model is perceived. VOs help create views by showing their soundness and conflict freeness. However, views affect the associated VT, i.e., the model changes, and the output which decides VT's success or failure also changes accordingly. While the output of a VT is more concrete on an instantiation view, and thus, depending on the task, it can provide a concrete example of why the VT succeeds or fails. The stakeholders can then easily understand the scenario. The knowledge gained from the view can be feedback on the requirements of the model and may result in new VOs to ensure the presence of the emerging requirements.

Creating a scenario-view in Event-B is achieved by replacing deferred constructs with concrete ones, thus adapting the model for a specific scenario. Let us consider our hemodialysis example again. Figure 3 shows a created view. Instead of reusing a deferred function where we map arbitrary `ids` to roles as we did in `INITIALISATION/@act4` of `m2` we use concrete mappings as one can see in `@act4` of the`INITIALISATION` of `m2Concrete`. We successfully execute the VTs and establish conflict freeness and soundness of the requirements for this view.

The state-space of the model `m2Concrete` is visualized in Fig. 4[2]. It shows two states. One can switch between the states via logging in as a different user or looping back to them by logging in as the same user. From this view, the stakeholder may discover a flaw in the model, for example, the fact that `startSystem` in the upper state is a loop that always ends in the same state. The consequence of this is that after `startSystem` was fired by the logged-in role, the logged-in role can be changed. For further treatment, this might be unwanted as the responsible role should not be changed mid-session, nor should the maintenance be able to log in after treatment is started. From this we can formulate a new requirement, for example:

- FUN2 Once the treatment is started, the logged in ID can not change.

This requirement can then be encoded into a VO and run against the original model `m2`. A VT for this requirement could look like this:

- FUN2/m2/LTL:[`startSystem`]$X(G(not(login)))$

[2] We omitted parts of the graphic for space reasons.

This formula would check that after the firing of the `startSystem` event, the `login` event cannot occur, of course, until the treatment is over.

We can go even a step further. Suppose we find a property of a scenario desirable for all scenarios that are created in the same way, i.e., by using the concrete mapping of `ids` to roles we used in the example. In that case, we can translate this property into a VO that has to hold for all scenarios that are created, relying on this mapping for initializing the `treatmentAllowed` variable. Every time we create a scenario view the same way, it must comply with this VO.

```
context ctx2 extends ctx1
constants id1 id2 id3 noneID
axioms
@axm1 partition(ids, {id1}, {id2}, {id3}, {noneID})
end
machine m2Concrete refines m2  sees ctx2
variables treatmentValue loggedIn loggedInID treatmentAllowed
events
  event INITIALISATION
    then
      @act1 treatmentValue ≔ 1
      @act2 treatmentAllowed ≔ {id1 ↦ doctors, id2 ↦ nurses}
      @act3 loggedIn ≔ no
      @act4 loggedInID ≔ noneID
  end
  event login extends login
  end
  event startSystem extends startSystem
  end
end
```

Fig. 3. Concrete initialisation for stakeholders

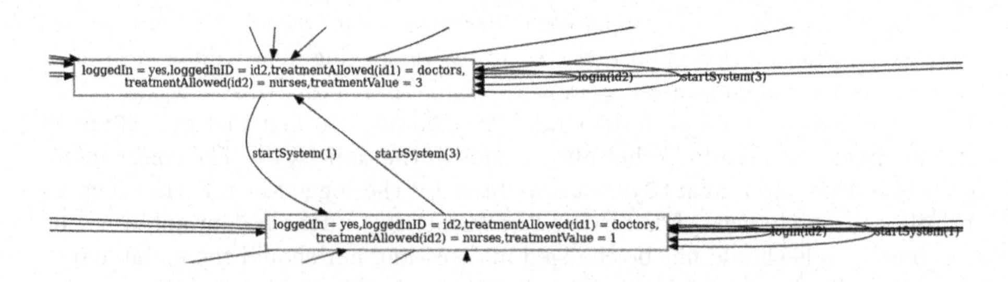

Fig. 4. Part of the concrete state-space created with ProB

5 Related Work

The Tokeneer case study [12] is a well-known example of the application of formal methods to the security domain. Tokeneer is an extensive system that provides safety for a network of workstations in a protected room, which can only be entered via verified bio-metric scans and id cards. In addition, the Tokeneer system has the task of verifying access for persons and checking certificates for credentials. The case study was modeled in different languages like the modeling language Z [4]. For this implementation, there exist several approaches to verify the system, e.g., Cristiá and Rossi [5]. The specification was also translated from Z to Event-B by Rivera et al. [13]. While these contributions show that formal methods can be successfully applied to help design and implement security-critical systems, they only tackle the problem from the verification side. The question of validation is mainly unaddressed.

VOs aim to complement by showing that the actual requirements specified by stakeholders are present in the model. Up to now, many techniques in formal methods have aimed to show the absence of faulty behavior. For example, model checking traverses the state space to check if a state violates a previously defined invariant. Another example is proofs that can show if the access to data structures is well defined. However, verification, i.e., checking whether faulty behavior is absent in the model, is not enough as a designer has no feedback on what the model is capable of. For this, we need validation, and while there exist validation approaches, e.g., proposed by Fitzgerald et al. [6], these are tool and language-specific. That is where VOs are a valuable addition, as they provide a tool and language-independent formalism. Requirements are formulated as VOs, which evolve with the model while remaining traceable. Additionally, different model views enable other stakeholders to understand the model and reason about the requirements relevant to them without going into unnecessary details.

6 Conclusion and Future Work

This paper shows how the VO approach can validate security concerns alongside safety and functional properties in a formal model. A uniform approach for validating different system properties is a big plus, thus making the overall development process more straightforward. Furthermore, the VO approach helps spot conflicting requirements and existing bugs and prevents from introducing the new ones. Additionally, we can create views on a model that facilitate the model's understanding by all stakeholders. From views, we can extract knowledge that we can feedback into the model. Finally, we can formulate VOs that describe the scenario's desired outcomes, thus ensuring compliance for the whole development cycle.

In the future, we want to apply the VO approach to large-scale case studies and show the multitude of VOs in different settings. In such a study, we would like to investigate the following research questions:

- What are the limitations of the VO approach?
- Is the VO approach equally beneficial to formal methods which are not state-based?
- How does the approach scale in large-scale case studies?

References

1. Abrial, J.R.: Modeling in Event-B: system and software engineering. Cambridge University Press (2010)
2. Abrial, J.R., Butler, M., Hallerstede, S., Hoang, T.S., Mehta, F., Voisin, L.: Rodin: an open toolset for modelling and reasoning in Event-B. Int. J. Softw. Tools Technol. Transfer **12**(6), 447–466 (2010)
3. Biró, M., Mashkoor, A., Sametinger, J., Seker, R.: Software safety and security risk mitigation in cyber-physical systems. IEEE Softw. **35**(1), 24–29 (2018)
4. Copper, D., Barnes, J.: Tokeneer id station eal5 demonstrator: Summary report. Tech. Rep., Augugst, Altran Praxis Limited (2008)
5. Cristiá, M., Rossi, G.: An automatically verified prototype of the tokeneer id station specification. J. Autom. Reason. **65**(8), 1125–1151 (2021)
6. Fitzgerald, J.S., Tjell, S., Larsen, P.G., Verhoef, M.: Validation support for distributed real-time embedded systems in vdm++. In: 10th IEEE High Assurance Systems Engineering Symposium (HASE 2007). pp. 331–340. IEEE (2007)
7. Leuschel, M., Butler, M.: ProB: a model checker for B. In: Araki, K., Gnesi, S., Mandrioli, D. (eds.) FME 2003. LNCS, vol. 2805, pp. 855–874. Springer, Heidelberg (2003). https://doi.org/10.1007/978-3-540-45236-2_46
8. Mashkoor, A.: The hemodialysis machine case study. In: Butler, M., Schewe, K.-D., Mashkoor, A., Biro, M. (eds.) ABZ 2016. LNCS, vol. 9675, pp. 329–343. Springer, Cham (2016). https://doi.org/10.1007/978-3-319-33600-8_29
9. Mashkoor, A., Kossak, F., Egyed, A.: Evaluating the suitability of state-based formal methods for industrial deployment. Softw. Pract. Exp. **48**(12), 2350–2379 (2018)
10. Mashkoor, A., Leuschel, M., Egyed, A.: Validation obligations: a novel approach to check compliance between requirements and their formal specification. In: 2021 IEEE/ACM 43rd International Conference on Software Engineering: New Ideas and Emerging Results (ICSE-NIER), pp. 1–5. IEEE (2021)
11. Mashkoor, A., Sametinger, J., Biro, M., Egyed, A.: Security- and safety-critical cyber-physical systems. J. Soft. Evol. Process **32**(2), e2239 (2020)
12. (NSA): The tokeneer case study. https://www.adacore.com/tokeneer, (Accessed 19 July 2022, 14:12:17)
13. Rivera, V., Bhattacharya, S., Cataño, N.: Undertaking the tokeneer challenge in event-b. In: Proceedings of the 4th FME Workshop on Formal Methods in Software Engineering, pp. 8–14 (2016)
14. Schneier, B.: Attack trees. Dr. Dobb's J. **24**(12), 21–29 (1999)

Mode Switching for Secure Edge Devices

Michael Riegler[1(✉)], Johannes Sametinger[1], and Christoph Schönegger[2]

[1] LIT Secure and Correct Systems Lab and Department of Business Informatics, Johannes Kepler University Linz, Linz, Austria
{michael.riegler,johannes.sametinger}@jku.at
[2] ENGEL AUSTRIA GmbH, Schwertberg, Austria
christoph.schoenegger@engel.at
https://www.jku.at/en/lit-secure-and-correct-systems-lab, https://www.se.jku.at, https://www.engelglobal.com/

Abstract. Many devices in various domains operate in different modes. We have suggested to use mode switching for security purposes to make systems more resilient when vulnerabilities are known or when attacks are performed. We will demonstrate the usefulness of mode switching in the context of industrial *edge devices*. These devices are used in the industry to connect industrial machines like cyber-physical systems to the Internet and/or the vendor's network to allow condition monitoring and big data analytics. The connection to the Internet poses security threats to *edge devices* and, thus, to the machines they connect to. In this paper (i) we suggest a multi-modal architecture for *edge devices*; (ii) we present an application scenario; and (iii) we show first reflections on how mode switching can reduce attack surfaces and, thus, increase resilience.

Keywords: Mode switching · Edge device · Security · Linux · Systemd · Ansible

1 Introduction

Manufacturers will increasingly become service providers and use condition monitoring, and big data analytics for predictive maintenance [25]. *Edge devices* connect physical devices like machines, robots, and sensors over various protocols like MQTT, OPC UA, and others to the virtual world. They are widely used in Cyber-physical systems (CPS) and Internet of Things (IoT) systems, and their use will increase in the next decade [24]. *Edge devices* provide computing power for lightweight devices and other facilities on-site and build a bridge between operational technology (OT) and information technology (IT), typically in the cloud. They allow machine-to-machine communication, remote monitoring and control across several locations and increase availability and efficiency. *Edge devices* provide load balancing, low latency, and service continuity in case of connection failures, optimize the network bandwidth to the cloud, and process the workload close to its occurrence [15]. Devices increasingly get interconnected,

G. Kotsis et al. (Eds.): DEXA 2022 Workshops, CCIS 1633, pp. 347–356, 2022.
https://doi.org/10.1007/978-3-031-14343-4_32

enhancing systems' complexity and unleashing threats and vulnerabilities. In the year 2021, vulnerabilities in *Log4J*, *Kaseya* and *Solar Winds* have affected thousands of companies. Cyberattacks like the one against *Colonial Pipeline* can lead to loss of productivity, business continuity problems, financial losses, and damage of reputation. According to the Allianz Risk Barometer [1], cyber incidents are the most important business risk in 2022.

Time is essential when a vulnerability becomes known. It typically takes a while until an update will be available. During that time, attackers can write exploits and attack systems. Workarounds are sometimes provided to mitigate known vulnerabilities. Nevertheless, manual changes are time-consuming and, if performed hastily, involve the danger of disrupting operations. In the worst case, services have to be completely stopped or shut down.

We have presented how multi-mode systems can provide resilient security in Industry 4.0 [18] and a web server case study using a multi-purpose mode switching solution to overcome vulnerabilities in [19]. Additionally, we have investigated how modes can secure deliberately insecure web applications [17]. In this paper, we will investigate the security of *edge devices* and reflect on how automatic and manual mode switches can reduce attack surfaces.

In Sect. 2, we present the idea of mode switching and its use in several domains. In Sect. 3, we describe common methods to secure *edge devices*. We present our considerations about mode switching in *edge devices* in Sect. 4. Automation of configuration management of modes and deployment are shown in Sect. 5. We discuss our results in Sect. 6 and draw conclusions in Sect. 8.

2 Mode Switching

We have provided first findings of a systematic literature review about mode switching from a security perspective in [16]. In general, modes are used to provide flexible adjustment of behavior, real-time adaptation, and complexity management. They have already been used in domains like automotive [2], aviation [22] and energy [26]. Modes divide and manage complexity, have specific configurations, and consist of specific behaviors. For example, nuclear power plants have multiple modes for power operation, startup, hot standby, hot shutdown, cold shutdown, and refueling [26]. They automatically change their modes and may even be completely switched off if parameters exceed or fall below critical thresholds [21]. Special accident and emergency systems and a degraded mode of operation [7] provide resilience and a better understanding of the system's state if there is any kind of malfunction. However, mode switching is not limited to safety precautions. Modes can also be used to control functionality and to increase security [20]. A web server case study [19] has shown that mode switching has reduced known risks in that specific scenario in 98.9% of the time and has shortened the window of exposure from 536 days to 8 days.

3 Securing Edge Devices

Making *edge devices* more secure begins with the key concepts of the security triad *CIA*: confidentiality, integrity, and availability. According to ISO/IEC 27000:2009 [11], authenticity, accountability, non-repudiation, and reliability are also important. Preventive, detective, and corrective security controls will help to minimize security risks and to recover faster whenever something bad happens. There should be multi levels of protection (defense in depth) for all system parts: cloud, network, device, application, and data. One compromised part should not affect the entire system. Following a *Secure Software Development Live Cycle* (sSDLC) will facilitate delivering effective and efficient systems during lifetime [4,10]. Several guidelines and frameworks [5,10,13,14] exist to make IoT devices more secure. They suggest physical controls like a *Trusted Platform Module* (TPM), mechanical locks, and minimized external ports. TPM goes hand in hand with a *Trusted Execution Environment* (TEE), secure boot, a protected *operating system* (OS), and secure updates. McCormack et al. provide an architecture for trusted edge IoT security gateways and recommend periodic remote attestation with TPM, to control that there are no fraudulent manipulations [12]. It is state-of-the-art to encrypt all communication to the outside of an IoT device. In addition to regulatory requirements and policies, technical and logical controls should also be implemented: authentication, access controls, a firewall, and antivirus software. Each device should have its own client certificate to communicate with manufacturers' cloud services. Even if compromised, only single devices are affected and can be blocked on the server-side, if necessary. Updates should be signed and verified, risky legacy protocols should be avoided, and open ports minimized.

Cejka et al. have compared *Amazon Web Services* (AWS) IoT Greengrass, Microsoft Azure IoT Edge, and a self-implemented framework for secure *edge devices* and distributed control of critical infrastructure [3]. They recommend filtering and monitoring communication for anomaly detection and reacting to deviations with countermeasures.

In our sample scenario, we use *edge devices* to connect industrial machines at the customers' site to the Internet and to allow communication with the machines' manufacturer. Typically, *edge devices* will be delivered to the customer's factory or machine. However, sometimes devices get stolen and even manipulated. Therefore, the hard disk is encrypted, and the device provides only limited initial functionality like configuring the network settings to onboard the device. Onboarding means that the device is registered with the manufacturer and gets a device certificate that activates the device's full functionality, which persists even if later the connection gets lost. Thus, once the device has successfully established a connection with the manufacturer's cloud service, it sends an onboarding request. Then the customer can authenticate at the manufacturer's cloud service to view the request and get a security token to assign the *edge devices* to the customer, similar to [27].

4 Edge Device Modes

In this section, we provide further details about the design and implementation of *edge device* modes. Figure 1 gives an impression about a few sample modes and their services and configurations. A system and its components can be in multiple modes. Initially, an *edge device* is in *factory mode*. After successful onboarding, it switches to the *onboarded mode*. In our sample scenario, *edge devices* are online most of the time. They check for updates in regular time intervals, typically once a day. If an update check was not possible for a specific time period, e.g., for x days, the device is considered to be outdated. A mode switch to *outdated mode* leads to limited functionality for security reasons. Thus, the onboarded mode has two sub-modes: *updated mode* and *outdated mode*. Modes can extend other modes similar inheritance in object-oriented programming (OOP).

We use modes to protect the system and its services from vulnerabilities found in the meantime. We consider it bearable that a system that has not been online for a predefined time period makes updates first. The system itself periodically checks if there is an update. After the successful update or when the last successful update check is younger than x days, the *updated mode* becomes active, which provides all services, like condition monitoring, big data analytics for predictive maintenance, remote support, and others. Additionally, we plan to analyze log files to react to specific events like denial-of-service (DoS) attacks.

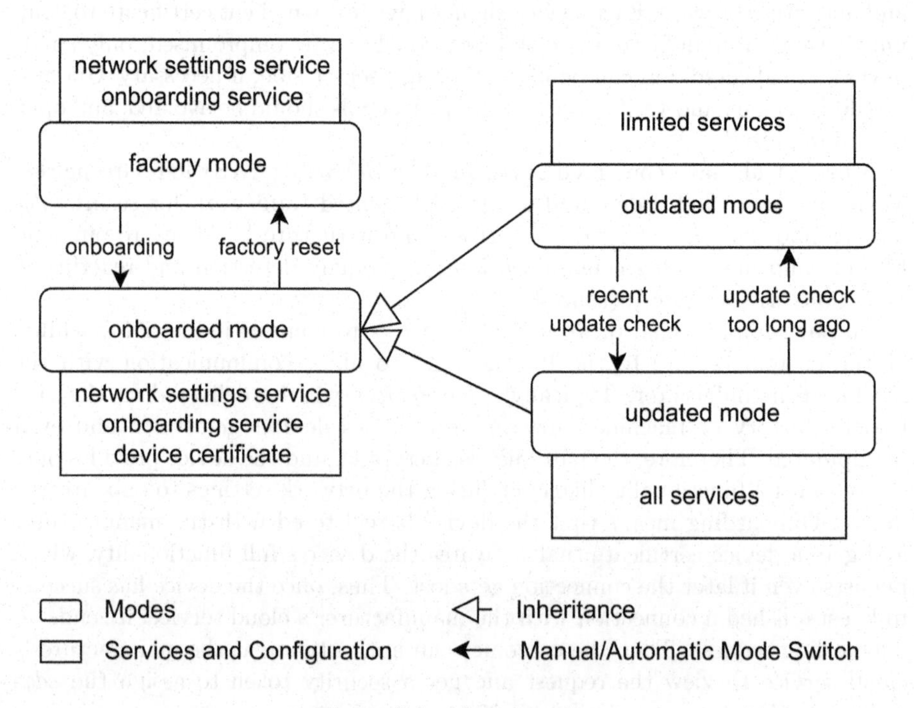

Fig. 1. Edge device modes

4.1 Implementation

For implementing the concept of modes on common *edge devices*, we have extended the system and service manager *systemd*[1], which is used in most Linux operating systems. We use *systemd* targets [8] to specify modes. They are similar to *SystemVinit* run levels [23], which were used in previous systems. A target is specified with a unit-file and combines several services. It is possible to define a default target as well as hard and soft dependencies to other modes, which should be started before. Usually the `multi-user.target` (on non-graphical systems) or the `graphical.target` are the default targets. We can switch to the mentioned mode by running `systemctl isolate mode.target`. Nevertheless, it only works if the mode meets the conditions. Services, like to define the network settings, for onboarding or for specific software containers, have their own unit-file and can be attached to one or multiple modes in their `[Install]` section with the command `WantedBy=mode.target`. Since release v250, the `factory-reset.target` has become available. Obviously, the developers think in a similar way. Listing 1.1 shows the `updated-mode.target`. It is started after the `onboarded-mode.target` only when a specified file exists. *Systemd* only supports some basic condition and assertion checks, before a target or service becomes active. For example, we can only check whether a file exists (`ConditionPathExists`), but we cannot check whether a certificate is valid, or whether the device is online. Therefore, we consider these parts in our *ModeSwitcher* shell script. As of now, all mode switches are executed by this shell script. A cron job runs periodically and checks if there are any changes. If an *edge device* is not onboarded, it tries to resolve that. If the device certificate becomes invalid, a mode switch to the *factory mode* is triggered. The cron job performs the updates for the device, software repositories, and the device certificate. If the last successful update check was as long ago as predefined, a mode switch is triggered from *updated mode* to *outdated mode*, and several containers are stopped in order to reduce the attack surface. If the update was successful, the system switches back to *updated mode*. Every mode switch will be notified to the manufacturer's cloud service.

Listing 1.1. Systemd updated-mode.target

```
#/etc/systemd/system/updated-mode.target
[Unit]
Description=Updated Mode
Requires=onboarded-mode.target
Conflicts=
After=onboarded-mode.target
AllowIsolate=yes
#Start target only when file does exists
ConditionPathExists=/var/lib/edgedevice/up-to-date
```

[1] https://systemd.io.

4.2 Fail2ban Sample Scenario

We are monitoring log files in our sample scenario with *fail2ban*[2], a Python-based *Intrusion Prevention Software* (IPS) that is mainly used to defend systems from brute-force attacks, e.g., by blocking IP addresses of attackers after several failed login attempts within a specific time frame. We can also specify multiple and less intrusive actions like sending an email to the administrator. It follows the rules of *event condition action* (ECA). New log entries get checked by a *filter* of regular expressions to extract interesting data. *Jail* configurations become relevant, if there is a match. These configurations specify services or attack scenarios to be monitored. They consist of the log file, the port or service, the *filter*, the time span and the maximum number of wrong attempts, the *actions* to be executed, and the ban time. For example, we can specify that after three failed SSH login attempts, the IP address of the possible attacker gets blocked, and the administrator will be informed by email. Then, after a ban time of 10 min, a new login attempt from that IP address will be allowed again. The *fail2ban* manual [6] and Hess [9] provide more details about these mechanisms. The standard configuration provides filters for *Apache*, *sshd*, *vsftpd*, *Postfix* and others. We have implemented additional filters, e.g., to detect (unsuccessful) terminal logins, (unsuccessful) SSH logins, port scans, HTTP errors, plugged-in LAN cables, attached USB devices, and lost connections. We imagine using *fail2ban* to send a notification to manufacturers' cloud services in case of abnormal behavior. In addition to that, we will use some of the *fail2ban* events as a source to trigger mode switches. For example, we envision switching from a normal to a denial-of-service (DoS) mode. The typical use case of *Fail2ban* is to block single IP addresses. However, if there is a DoS attack, it is not appropriate to block thousands of IP addresses. In this case, switching to a more secure mode is more practical, where this form of attack is blocked in general. Additionally, IP address ranges or countries could be blocked. Switching to another or degraded mode can reduce the attack surface and protect the system from further attacks. Another scenario is multiple wrong login attempts on the console. If that happens and the device is offline, we switch to the *factory mode* to prevent further attacks. We can react in the same way, if we detect anomalies regarding hardware attacks like sniffing, or port scans. In order to prevent the abusive use of insiders, administrative capabilities have to be limited, and all actions have to be logged and monitored. Procedural or administrative controls are needed to handle incidents and increase security awareness [13].

5 Automation

We have used our own package repository to provide modes for multiple devices and distribute changes. Software packages can be checked, adapted, and activated individually. Thereby, we can install modes as if they belonged to *systemd*. We have also examined system configuration management tools like *Ansible*, *Progress Chef* and *Puppet*. They are also used for provisioning, application

[2] https://www.fail2ban.org.

deployment, and infrastructure as a code. We had a closer look at *Ansible* and have developed a module for mode switching, i.e., switching *systemd* targets. With that implementation, we were able to use mode switching in *Ansible* tasks and playbooks. This feature is helpful to manage multiple *edge devices* and to switch the mode of either one or many devices depending on requirements. For example, Listing 1.2 shows how to switch all *edge devices* from group `devgroup1` to the `oudated-mode.target`.

Listing 1.2. Call developed Ansible module to change systemd target

```
ansible devgroup1 -m systemd_target -a "name=outdated-mode.
    target␣state=isolated"
```

6 Discussion

From a security perspective, it is an advantage to switch to a restricted mode as soon as possible, if attacks or abnormal behavior happen. From a customer perspective, it is highly unsatisfactorily if an *edge device* switches the mode automatically and seemingly unexpectedly denies services and interrupts operational processes. Therefore, full automation of mode switching is not desirable under all circumstances. We propose to inform a manufacturer's cloud service and put a human in the loop. Thus, online devices can be controlled semi-automatically or manually by experts. In addition, the devices can learn over time about false positives of possible attacks, e.g., with machine learning techniques. If *edge devices* were offline for a long time, they have to act autonomously and should be stricter about switching modes. However, even then, a specific threshold needs to be considered. Shutting down the device must be the last resort.

Safety is an issue for a CPS as well. Connected machines and robots can potentially harm people and damage property. In this context, the concepts of fail-secure and fail-safe contradict each other to some extent. Completely stopping an *edge device* in case of any issue is highly secure. Nevertheless, if a person is stuck in a machine, some basic functionality is needed to get her free. A differentiated approach is needed in such a scenario. From the outside, the system has to be in a fail-secure mode and may prohibit external access. A fail-safe mode can provide at least some basic on-site functionality.

We have also experimented with multiple modes with different arranged software containers on top of the *updated mode*. Further research is needed to define how many modes are necessary and are still manageable. However, we think it is more beneficial to stop single vulnerable services. *Ansible* and other configuration management tools are well usable to achieve that goal. They allow us to specify single, groups, or all hosts to make changes. We can go one step further and integrate predefined modes more deeply if a service supports different configurations or some kind of limited operation.

7 Limitations

Mode switching and our proposed *edge device* modes are not intended to replace other traditional hardware and software protection techniques. Secure system architecture and security by design are still necessary. Developers must not become less concerned about security and rely on the fact that they can later install an update or patch. However, modes can contribute to more secure and flexible system architectures. Our research has no empirical results yet. Experiments and more detailed comparisons against other protection techniques will have to underline the effectiveness of the approach.

8 Conclusion

We have given a first impression of how mode switching can increase the security of *edge devices* and enhance resilience. We have shown an example of how modes can be implemented in the Linux operating system and how system configuration management tools can help to manage modes of multiple *edge devices*.

Future work includes monitoring modes of multiple *edge devices* on the manufacturer's cloud service and considering the *edge device* modes of a customer and across different customers in a security information and event management (SIEM) system. Being able to change *edge device* modes manually is highly beneficial in case of known vulnerabilities or when stopping malware from spreading further. In addition, we plan to simulate attacks to demonstrate the effectiveness of our approach.

Acknowledgement. This work has partially been supported by the LIT Secure and Correct Systems Lab funded by the State of Upper Austria.

References

1. Allianz Global Corporate & Specialty SE: Allianz Risk Barometer (2022). https://www.agcs.allianz.com/content/dam/onemarketing/agcs/agcs/reports/Allianz-Risk-Barometer-2022-Appendix.pdf. Accessed 20 Feb 2022
2. AUTOSAR: Guide to Mode Management (2017). https://www.autosar.org/fileadmin/user_upload/standards/classic/4-3/AUTOSAR_EXP_ModeManagementGuide.pdf. Accessed 24 Feb 2022
3. Cejka, S., Knorr, F., Kintzler, F.: Edge device security for critical cyber-physical systems. In: 2nd Workshop on Cyber-Physical Systems Security and Resilience (CPS-SR), April 2019
4. European Union Agency for Cybersecurity (ENISA): Good Practices for Security of IoT - Secure Software Development Lifecycle, November 2019. https://www.enisa.europa.eu/publications/good-practices-for-security-of-iot-1
5. European Union Agency for Cybersecurity (ENISA): Guidelines for securing the Internet of Things: secure supply chain for IoT. Publications Office (2020). https://doi.org/10.2824/314452
6. Fail2ban: Manual Fail2ban 0.8. https://www.fail2ban.org/wiki/index.php/MANUAL_0_8. Accessed 20 Feb 2022

7. Firesmith, D.: System Resilience: What Exactly is it? (2019). https://insights.sei.cmu.edu/sei_blog/2019/11/system-resilience-what-exactly-is-it.html. Accessed 23 Feb 2022
8. freedesktop.org: systemd.target — Target unit configuration. https://www.freedesktop.org/software/systemd/man/systemd.target.html. Accessed 22 Feb 2022
9. Hess, K.: Linux security: Protect your systems with fail2ban. Red Hat, June 2020. https://www.redhat.com/sysadmin/protect-systems-fail2ban. Accessed 21 Feb 2022
10. International Electrotechnical Commission (IEC): IEC 62443-4-1:2018 — Security for industrial automation and control systems - Part 4-1: Secure product development lifecycle requirements. https://webstore.iec.ch/publication/33615
11. International Organization for Standardization (ISO): ISO/IEC 27000:2009 (2009). https://www.iso.org/standard/41933.html
12. McCormack, M., et al.: Towards an architecture for trusted edge IoT security gateways. In: 3rd USENIX Workshop on Hot Topics in Edge Computing (HotEdge 2020). USENIX Association, June 2020. https://www.usenix.org/system/files/hotedge20_paper_mccormack.pdf
13. National Institute of Standards and Technology (NIST): Framework for Improving Critical Infrastructure Cybersecurity, Version 1.1, April 2018. https://doi.org/10.6028/NIST.CSWP.04162018
14. National Institute of Standards and Technology (NIST): Security and Privacy Controls for Information Systems and Organizations, September 2020. https://doi.org/10.6028/NIST.SP.800-53r5
15. Noghabi, S., Kolb, J., Bodik, P., Cuervo, E.: Steel: simplified development and deployment of edge-cloud applications. In: 10th USENIX Workshop on Hot Topics in Cloud Computing (HotCloud 2018) (2018)
16. Riegler, M., Sametinger, J.: Mode switching from a security perspective: first findings of a systematic literature review. In: Kotsis, G., Tjoa, A.M., Khalil, I., Fischer, L., Moser, B., Mashkoor, A., Sametinger, J., Fensel, A., Martinez-Gil, J. (eds.) DEXA 2020. CCIS, vol. 1285, pp. 63–73. Springer, Cham (2020). https://doi.org/10.1007/978-3-030-59028-4_6
17. Riegler, M., Sametinger, J.: Mode switching for secure web applications – a juice shop case scenario. In: Kotsis, G., et al. (eds.) DEXA 2021. CCIS, vol. 1479, pp. 3–8. Springer, Cham (2021). https://doi.org/10.1007/978-3-030-87101-7_1
18. Riegler, M., Sametinger, J.: Multi-mode systems for resilient security in Industry 4.0. Procedia Comput. Sci. **180**, 301–307 (2021). https://doi.org/10.1016/j.procs.2021.01.167. Proceedings of the 2nd International Conference on Industry 4.0 and Smart Manufacturing (ISM 2020)
19. Riegler, M., Sametinger, J., Vierhauser, M., Wimmer, M.: Automatic mode switching based on security vulnerability scores (2022, submitted for publication)
20. Sametinger, J., Steinwender, C.: Resilient context-aware medical device security. In: International Conference on Computational Science and Computational Intelligence, Symposium on Health Informatics and Medical Systems (CSCI-ISHI), pp. 1775–1778 (2017). https://doi.org/10.1109/CSCI.2017.310
21. Shultis, J.K., Faw, R.E., McGregor, D.S.: Fundamentals of Nuclear Science and Engineering, 3rd edn. CRC Press (2016). https://cds.cern.ch/record/2245430. Accessed 24 Feb 2022
22. SmartCockpit: A330-A340 Flight Crew Training Manual, July 2004. https://www.smartcockpit.com/docs/A330-A340_Flight_Crew_Training_Manual_1.pdf. Accessed 24 Feb 2022

23. van Smoorenburg, M.: init, telinit - process control initialization. Debian, July 2004. https://manpages.debian.org/testing/sysvinit-core/init.8.en.html. Accessed 21 Feb 2022
24. Statista: Number of edge enabled internet of things (IoT) devices worldwide from 2020 to 2030, by market. https://www.statista.com/statistics/1259878/edge-enabled-iot-device-market-worldwide/. Accessed 20 Feb 2022
25. Statista: In-depth: Industry 4.0 2021, June 2021. https://www.statista.com/study/66974/in-depth-industry-40/. Accessed 20 Feb 2022
26. US Nuclear Regulatory Commission (NRC): Standard Technical Specifications – Operating and New Reactors – Current Versions (2019). https://www.nrc.gov/reactors/operating/licensing/techspecs/current-approved-sts.html. Accessed 24 Feb 2022
27. Zoualfaghari, M.H., Reeves, A.: Secure & zero touch device onboarding. In: Living in the Internet of Things (IoT 2019), pp. 1–3, May 2019. https://doi.org/10.1049/cp.2019.0133

Machine Learning and Knowledge Graphs

A Lifecycle Framework for Semantic Web Machine Learning Systems

Anna Breit[1], Laura Waltersdorfer[2(✉)], Fajar J. Ekaputra[2], Tomasz Miksa[2], and Marta Sabou[3]

[1] Semantic Web Company, Vienna, Austria
[2] Technical University Vienna, Vienna, Austria
laura.waltersdorfer@tuwien.ac.at
[3] Vienna University of Economics and Business, Vienna, Austria

Abstract. Semantic Web Machine Learning Systems (SWeMLS) characterise applications, which combine symbolic and subsymbolic components in innovative ways. Such hybrid systems are expected to benefit from both domains and reach new performance levels for complex tasks. While existing taxonomies in this field focus on building blocks and patterns for describing the interaction within the final systems, typical lifecycles describing the steps of the entire development process have not yet been introduced. Thus, we present our **SWeMLS lifecycle framework**, providing a unified view on Semantic Web, Machine Learning, and their interaction in a SWeMLS. We further apply the framework in a case study based on three systems, described in literature. This work should facilitate the understanding, planning, and communication of SWeMLS designs and process views.

Keywords: Semantic web · Machine Learning · Lifecycle framework

1 Introduction

Both the Machine Learning (ML) and Semantic Web (SW) domain have seen dynamic growth and application due to advancements in Deep Learning [14] and the emergence of large crowd-sourced Knowledge Graphs [12]. With growing popularity of the distinct approaches, also the hybrid application of both paradigms, coined as Semantic Web Machine Learning (SWeMLS) has increased [6,22] to benefit from complementary strengths. Examples include using embeddings for graph construction or completion [5] and leveraging SW resources for Question Answering [16]. However, as the two distinct approaches (SW and ML) have developed independently over the last decades, the disciplines have each emerged with different vocabularies, notations and methodologies. These developments led to the result that SWeMLS are often regarded from one limited viewpoint instead of an unified manner [11].

A first important step towards harmonisation was done by proposing taxonomies and design patterns, as known from software engineering: In the related

G. Kotsis et al. (Eds.): DEXA 2022 Workshops, CCIS 1633, pp. 359–368, 2022.
https://doi.org/10.1007/978-3-031-14343-4_33

area of neuro-symbolic systems (i.e., systems that combine ML and deductive systems), recent taxonomies, such as the neuro-symbolic taxonomy [13] and the boxology of hybrid AI [3] enable a categorisation according to patterns and provide initial building blocks to model components of neuro-symbolic AI. Inspired by these works, we recently conducted a systematic mapping study [4,26] which resulted in (among other outcomes) a collection of boxology-based processing-flow patterns and a classification system for SWeMLS.

While these tools facilitate the communication in the field by providing a unified way of documenting, describing and categorising *final* SWeMLS based on components, they lack the possibility to depict the holistic view on the steps included the entire creation and exploitation process of these systems, i.e. their *lifecycle.* Lifecycle representations come with many benefits. First, they provide a comprehensive view over the steps included in the entire development process of systems, and can therefore serve as guideline during planning, implementation, and deployment. Furthermore, given their simple representation and their universal understandability, they facilitate communication, especially to business representatives or people from outside the community.

To this end, in this work, we (i) introduce a *SWeMLS lifecycle framework* consisting of a lifecycle model for SW and ML, as well as their interaction paths in a SWeMLS, and (ii) we exemplify the applicability of this framework in a *case study*, analysing the lifecycle of three existing SWeMLS. The usage of lifecycles of different components -in this case ML and SW- can support the identification of ideal architectures regarding their combination by explicitly presenting possible interaction points and support communication on different architectures.

The remainder of this paper is structured as follows: We discuss related work in Sect. 2. In Sect. 3, we introduce our lifecycle models, while the case study applying our framework to three systems can be found in Sect. 4. Conclusions and future steps are presented in Sect. 5.

2 Related Work

2.1 Machine Learning Lifecycles

For the Machine Learning component, we analyse dedicated ML lifecycles, as well as those for Data Mining (DM) models, since this area is closely related.

Machine Learning. Most existing frameworks for ML usually focus on the development of models, from modelling and training to evaluation.

However, despite their popularity, most ML lifecycles do not account for the deployment of the developed models, despite being an important step with respect to applications exploiting ML models. Only a small amount of works identify the deployment phase of a ML solution as model-worthy, one of which is Garcia et al. [10], they also focus on subsequent steps like final training and inference, including the exploitation of the outcomes in an application, as well as a feedback loop. A detailed perspective on deep learning taking into account the various data artifacts created by such systems is described by Miao et al. [17].

Polyzotis et al. focus on the data being central in the ML lifcycle, specifically focusing on the first phases of the process (e.g. data preparation and understanding) [20]. A recent, detailed ML lifecycle describes quality assurance factors for each phase [1].

One aspect that is omitted in existing ML lifecycles is the exploration of the problem itself and oftentimes the origin of the data used to develop the model, even though these are big topics in real life ML development.

Data Mining. Opposed to existing ML lifecycles, Data Mining (DM) lifecycles lay great focus on domain understanding and data preparation. The most prominent definition of the DM lifecycle, is the well-established Cross Industry Standard Process for Data Mining (CRISP-DM) [18]. In this data and use-case-centric view, three out of the total of six steps focus on understanding the problem, and analysing and preparing the available data. Studer et al. combine CRISP-DM with the ML lifecycle (CRISP-ML) and add a Monitoring and Maintenance Phase [23].

In their survey, Ristoski and Paulheim [21] took a detailed look on data-related pre-processing steps. They provide insights on how SW approaches can be combined with the data mining and knowledge discovery process, but do not focus on ML approaches.

2.2 Semantic Web Lifecycles

At the center of the SW lifecycle is the development of a knowledge resource, often divided in an engineering and a usage phase, with problem definition being an integral part. For SW resources, the frameworks are tailored to specific types of resources, such as *Ontologies* [2,15], *Linked Open Data* [19] and *Knowledge Graphs* [8]. Reuse of SW resources is common and best practice, which leads to the crucial challenge of maintaining interoperability between resources: To this end, an evolution and matching phase might be introduced for ontologies [7], or multiple scenarios accounting for different needs [24]. While the steps for KG construction might be similar to other lifecycles, the purpose of a KG is often tailored to a specific service depending on it such as analytics or recommendation services, making the deployment phase also a critical phase [12].

Concluding, in contrast to ML, problem and scope definition are widely covered in SW lifecycle frameworks. However, the evolution of semantic resources, adaptions and active usage of semantic resources are an important topic, which seems to require more attention. Furthermore, there is a higher variety in described frameworks for different type of resources.

3 SWeMLS Lifecycle Framework

To our best knowledge, there is no other attempt to combine both lifecycles to a common framework. The closest work would be by Van Bekkum et al. [3] extending the boxology notation for hybrid AI with more sophisticated functional blocks to have improved modeling capabilities for processes.

A general overview of our framework is provided in Fig. 1. To better understand how ML models and SW resources can benefit from each other in a system, we are taking first a closer look on the scope of the framework and their respective lifecycles, ML and SW.

3.1 Scope of Framework

A SWeML system consists of (at least) three components, being a *ML* module, a *SW* module, and an *Application*. Each of these modules has their own lifecycle, where the goal of the SW and ML lifecycle is the deployment and the exploitation within an application. To do so, they can expose artefacts, –i.e., a *ML model* or a *SW resource*– to the application, where users can interact with these accessible resources. Both types of artefacts can either be accessed directly, or through provided interfaces, e.g., for the predictions of a ML model. The application on the other hand can provide *feedback* and *adoptions* to the ML and SW lifecycle.

The interconnection between the ML and the SW lifecycle can vary and happen on different levels: they can either directly intervene in each others lifecycle (e.g., SW generating training data for ML, ML generating triples for SW), or the two components interact only within the application context, via their artefacts. If the interconnection happens within the lifecycles, oftentimes, only one lifecycle exposes an artefact to the application.

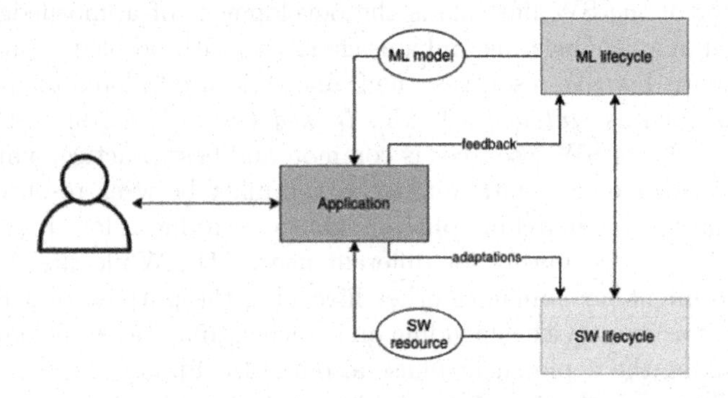

Fig. 1. Overview on SWeMLS lifecycle: it consists of a ML lifecycle, a SW lifecycle, and application lifecycle and interconnections between these lifecycle

In the further sections, we are taking a closer look on the ML and the SW lifecycle, both tightly linked to the CRISP-DM framework.[1] The decision to align both of them to CRISP-DM was done i) due to the wide adoption in data-related contexts and ii) after looking at existing lifecycle models in SW or ML with varying depths of detail: in order to facilitate drawing connections between

[1] The third lifecycle in a SWeMLS, being the Application lifecycle corresponds to the extensively discussed Software Development Lifecycle, which we will not discuss in the scope of this paper.

both components, we needed them to be on the same level. Thus, for both, we distinguish between a *Development Phase*, where the base artefact is modelled and created, and an *Operation Phase*, that provides a tangible and deployed outcome.

3.2 ML Lifecycle

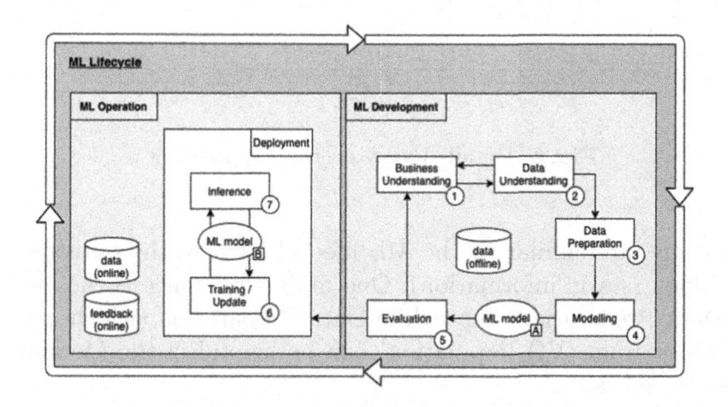

Fig. 2. Detailed view on the ML lifecycle

The developed ML lifecycle is tightly based on the CRISP-DM lifecycle, however, with a more detailed view on the deployment (cf. Fig. 2). In the *Development Phase*, first, under the presence of offline data, Business Understanding (1) as well as Data Understanding (2) must be established. Then, the data needs to be prepared (3) and used as basis for modelling a ML model (4). The modelling phase includes the decision for an architecture and parameters for the intended solution, as well as the training (if any). Therefore, the outcome of this step is a prototype ML model (A) that can be evaluated against the predefined needs in the Evaluation step (5).

If the ML model is satisfying, it can be used in the *Operation Phase*, by deploying it in an appropriate format. Herefore, the model first goes through a final training step (6), e.g., with all available data, to produce the production-ready model (B) that will be used for inference (7) on online data. The output of the inference step can be fed back to the model in a feedback loop to update the system (6). Occasionally, it might make sense to recreate or update the entire architecture of the ML model, in which case the entire lifecycle will be rerun (indicated by the outer arrows).

3.3 SW Lifecycle

The developed SW lifecycle is also based on the CRISP-DM framework (cf. Fig. 3): The first four steps in the *SW Development* phase –(I) Business Understanding (II) Data Understanding (III) Data Preparation (IV) Modelling, and

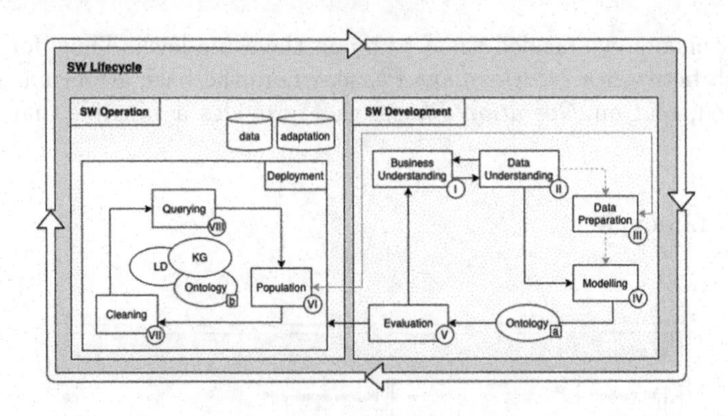

Fig. 3. Detailed view on the SW lifecycle

(V) Evaluation– are similar to the ML lifecycle, where the outcome of (IV) is an (a) ontology (static information). One major difference is the optionality of the (III) Data Preparation step: while Data Preparation (which can make use of methodology from (VI) Population) can be helpful in the Modelling step, it is not strictly necessary.

After successful Evaluation (V), *SW Operation* starts with (VI) Population with available data (VII) Cleaning erroneous data and (VIII) Querying to use the capabilities of the semantic resource, including also possible inference based on rules. The specific steps will also be dependent on the produced SW resource (b) which could be e.g., a Knowledge Graph (KG), Linked Data (LD), or the ontology itself. Similar to the ML lifecycle, it might also make sense to rerun the entire circle for larger changes, as indicated by the outer arrows.

4 Case Study

The following section provides three real-world Semantic Web Machine Learning Systems. The selected cases are taken from our systematic mapping study investigating the general characteristics of SWeMLS [4,26]. The goal is to show different interaction patterns between ML and SW components and the application of our framework to such systems. The first case, shows how the ML component is only used as a backup, in the second the SW component generates learning examples for the ML component and in the third case, we see multiple complex interactions between ML and SW.

ML as Backup: In this example [25], the authors build a system to perform sentiment analysis for restaurant reviews. Herefore, an ontology is predefined containing a number of classes (e.g., SentimentValue, AspectMention, and SentimentMention), relations between these classes, and axioms. Additionally, they deploy a Left-Center-Right Separated Neural Network with Rotatory Attention (LCR-Rot), that is trained on two SemEval datasets in a supervised fashion.

For their final system, the algorithm will first use the ontology to predict a positive or negative sentiment. If the ontology is not able to give a conclusive result, the system will use the ML method as a backup method.

As it can be seen in the corresponding lifecycle diagram (Fig. 4), the lifecycles of SW and ML do not influence each other directly, but the interconnection only takes place in the application, when the SW result is evaluated.

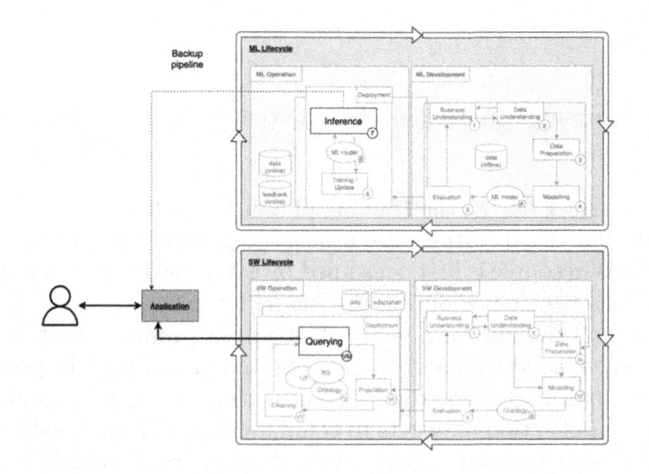

Fig. 4. ML as backup: interconnection in the application

SW Generates Learning Examples: The authors of [27] are developing a system that performs multi-instance entity typing from corpus. They deploy a system based on a Bi-LSTM model and an integer linear programming model (ILP), which they train in a distantly supervised manner. The training data is automatically generated by incorporating DBpedia as a Knowledge Base and DBpedia spotlight as an annotation tool. As it can be seen in Fig. 5, the interaction of the lifecycles takes place in the development phase of the ML model, more specifically, in the (3) Data Preparation step. Only the ML model is provided as artifact to the application.

SW Improves Training: The systems developed in [9] aim to perform a variety of analysis tasks on top of paintings, including author identification, type classification, and art retrieval. Therefore, they use use interacting ML models, where one (a Convolutional Neural Network or CNN) only uses the information obtained from the visual representation of the painting itself, while the other one (a Graph Embedding model) exploits contextual information obtained from an Artistic Knowledge Graph built with non-visual artistic metadata, as it can be seen in Fig. 6. In a first step (right ML lifeclcye), nodes from the KG are embedded using a Node2Vec model. These embeddings are used in the second training round (left ML lifecycle), where they influence the loss function of the

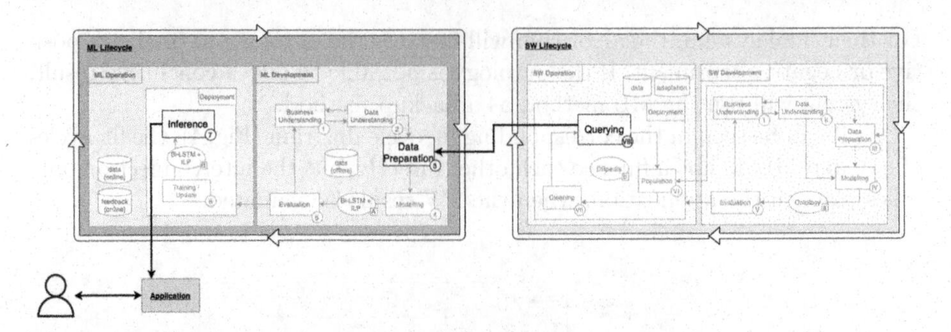

Fig. 5. SW generates learning examples: interconnection in ML data preparation

CNN-based classifier during the training phase. When the CNN is fully trained, however, only the visual features are utilized to generate predictions.

Here, the SW resource is first consumed by a ML model as an input, meaning, that interactions take place during the modelling phase, as well as during the final learning, and the inference phase. The output produced by this ML model (i.e., Node2Vec embeddings) is then incorporated in another ML process, which only uses this information for the loss calculation during the training process. Therefore, the application only gets to communicate with the predictions generated by the (CNN-based)ML model, and the incorporation of SW is completely hidden for the application, as image data only is used to generate the output.

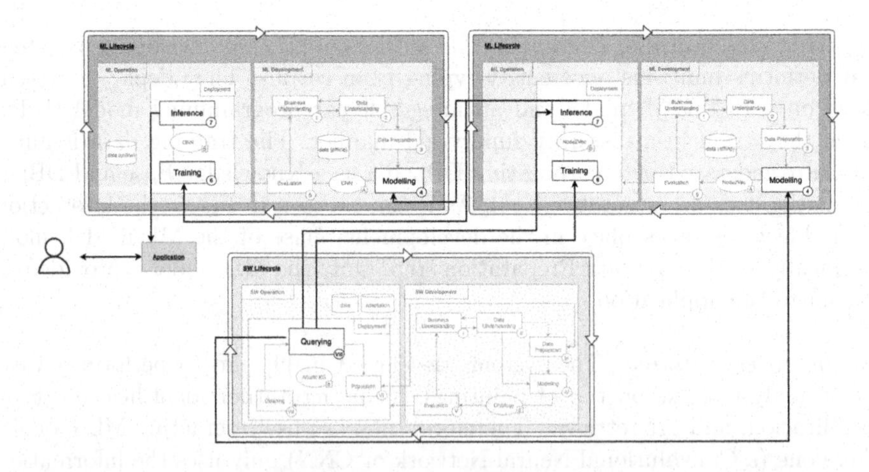

Fig. 6. SW improves training: complex interconnection in the ML training phase

5 Conclusion

Combining ML and SW techniques, advantages of both areas can be obtained, such as the ability to learn from sparse knowledge and increased levels of explainability. To the best of our knowledge, current literature only offers system lifecycle methods that focus on either SW or ML components. To make sense of complex SWeMLS architectures, we have presented the **SWeMLS lifecycle framework** to provide an overview. Furthermore, the lifecycle framework aims to support creators of such systems during design and operation of the system by identifying required key activities when creating SWeML systems.

Limitations. The SWeMLS lifecycle and case study is influenced by our system definition as described in [4,26]. Our system definition aims at systems with a task to ensure a certain maturity, which excludes theoretical approaches.

Future Work. With further growth of SWeMLS, we aim to cluster similar existing applications and derive best practices for certain tasks and domains.

Funding and Acknowledgement. This work has been funded by the project OBARIS (https://www.obaris.org/), which has received funding from the Austrian Research Promotion Agency (FFG) under grant 877389.

References

1. Ashmore, R., Calinescu, R., Paterson, C.: Assuring the machine learning lifecycle: desiderata, methods, and challenges. ACM CSUR **54**(5), 1–39 (2021)
2. Ashraf, J., Chang, E., Hussain, O.K., Hussain, F.K.: Ontology usage analysis in the ontology lifecycle: a state-of-the-art review. Knowl.-Based Syst. **80**, 34–47 (2015)
3. van Bekkum, M., de Boer, M., van Harmelen, F., Meyer-Vitali, A., Teije, A.: Modular design patterns for hybrid learning and reasoning systems. Appl. Intell. **51**(9), 6528–6546 (2021). https://doi.org/10.1007/s10489-021-02394-3
4. Breit, A., et al.: Combining machine learning and semantic web -a systematic mapping study (under review). ACM CSUR
5. Chen, P., Wang, Y., Yu, Q., Fan, Y., Feng, R.: Hamming distance encoding multi-hop relation knowledge graph completion. IEEE Access **8**, 117146–117158 (2020)
6. D'Amato, C.: Machine learning for the semantic web: lessons learnt and next research directions. Semant. Web **11**(1), 195–203 (2020)
7. Driouche, R.: Towards ontology lifecycle: building, matching and evolution to semantically integrate application ontologies. Int. J. Comput. Appli. Technol. Res. **6**, 109–116 (2017)
8. Fensel, D., et al.: Introduction: what is a knowledge graph? In: Knowledge Graphs, pp. 1–10. Springer, Cham (2020). https://doi.org/10.1007/978-3-030-37439-6_1
9. Garcia, N., Renoust, B., Nakashima, Y.: Context-aware embeddings for automatic art analysis. In: Proceedings of the 2019 on International Conference on Multimedia Retrieval, pp. 25–33 (2019)
10. Garcia, R., Sreekanti, V., Yadwadkar, N., Crankshaw, D., Gonzalez, J.E., Hellerstein, J.M.: Context: the missing piece in the machine learning lifecycle. In: KDD CMI Workshop, vol. 114 (2018)

11. Hitzler, P., Bianchi, F., Ebrahimi, M., Sarker, M.K.: Neural-symbolic integration and the Semantic Web. Semantic Web **11**(1), 3–11 (2020). https://doi.org/10.3233/SW-190368
12. Janev, V., Graux, D., Jabeen, H., Sallinger, E. (eds.): Knowledge Graphs and Big Data Processing. LNCS, vol. 12072. Springer, Cham (2020). https://doi.org/10.1007/978-3-030-53199-7
13. Kautz, H.: The Third AI Summer, AAAI Robert S. Engelmore Memorial Lecture, Thirty-fourth AAAI Conference on Artificial Intelligence (2020)
14. LeCun, Y., Bengio, Y., Hinton, G.: Deep learning. Nature **521**, 436–444 (2015)
15. Luczak-Rösch, M., Heese, R.: Managing ontology lifecycles in corporate settings. In: Networked Knowledge-Networked Media, pp. 235–248. Springer, Berlin (2009). https://doi.org/10.1007/978-3-642-02184-8_16
16. Lukovnikov, D., Fischer, A., Lehmann, J.: Pretrained transformers for simple question answering over knowledge graphs. In: Ghidini, C., et al. (eds.) ISWC 2019. LNCS, vol. 11778, pp. 470–486. Springer, Cham (2019). https://doi.org/10.1007/978-3-030-30793-6_27
17. Miao, H., Li, A., Davis, L.S., Deshpande, A.: Towards unified data and lifecycle management for deep learning. In: 2017 IEEE ICDE, pp. 571–582. IEEE (2017)
18. Ncr, P.C., et al.: Crisp-dm 1.0 (1999)
19. Ngomo, A.-C.N., Auer, S., Lehmann, J., Zaveri, A.: Introduction to linked data and its lifecycle on the web. In: Koubarakis, M., et al. (eds.) Reasoning Web 2014. LNCS, vol. 8714, pp. 1–99. Springer, Cham (2014). https://doi.org/10.1007/978-3-319-10587-1_1
20. Polyzotis, N., Roy, S., Whang, S.E., Zinkevich, M.: Data lifecycle challenges in production machine learning: a survey. ACM SIGMOD Rec. **47**(2), 17–28 (2018)
21. Ristoski, P., Paulheim, H.: Semantic web in data mining and knowledge discovery: a comprehensive survey. J. Web Semant. **36**, 1–22 (2016)
22. Seeliger, A., Pfaff, M., Krcmar, H.: Semantic web technologies for explainable machine learning models: a literature review. In: Proceedings of the 1st Workshop on Semantic Explainability co-located with the 18th International Semantic Web Conference (ISWC 2019), vol. 2465, pp. 30–45 (2019)
23. Studer, S., et al.: Towards crisp-ml (q): a machine learning process model with quality assurance methodology. Mach. Learn. Knowl. Extr. **3**(2), 392–413 (2021)
24. Suárez-Figueroa, M.C., Gómez-Pérez, A., Fernández-López, M.: The NeOn methodology for ontology engineering. In: Suárez-Figueroa, M.C., Gómez-Pérez, A., Motta, E., Gangemi, A. (eds.) Ontology Engineering in a Networked World, pp. 9–34. Springer, Heidelberg (2012). https://doi.org/10.1007/978-3-642-24794-1_2
25. Wallaart, O., Frasincar, F.: A hybrid approach for aspect-based sentiment analysis using a lexicalized domain ontology and attentional neural models. In: Hitzler, P., et al. (eds.) ESWC 2019. LNCS, vol. 11503, pp. 363–378. Springer, Cham (2019). https://doi.org/10.1007/978-3-030-21348-0_24
26. Waltersdorfer, L., Breit, A., Ekaputra, F.J., Sabou, M.: Bridging semantic web and machine learning: first results of a systematic mapping study. In: Kotsis, G., et al. (eds.) DEXA 2021. CCIS, vol. 1479, pp. 81–90. Springer, Cham (2021). https://doi.org/10.1007/978-3-030-87101-7_9
27. Xu, B., et al.: Metic: multi-instance entity typing from corpus. In: Proceedings of the 27th ACM International Conference on Information and Knowledge Management, pp. 903–912 (2018)

Enhancing TransE to Predict Process Behavior in Temporal Knowledge Graphs

Aleksei Karetnikov[1(✉)], Lisa Ehrlinger[1,2], and Verena Geist[1]

[1] Software Competence Center Hagenberg GmbH, Hagenberg, Austria
{aleksei.karetnikov,lisa.ehrlinger,verena.geist}@scch.at
[2] Johannes Kepler University Linz, Linz, Austria
lisa.ehrlinger@jku.at

Abstract. Temporal knowledge graphs allow to store process data in a natural way since they also model the time aspect. An example for such data are registration processes in the area of intellectual property protection. A common question in such settings is to predict the future behavior of a (yet unfinished) process. However, traditional process mining techniques require structured data, which is typically not available in this form in such communication-intensive domains. In addition, there exists a number of knowledge graph embedding methods based on neural networks, which are too performance-demanding for large real-world graphs. In this paper, we propose several extensions for preprocessing process data that will be embedded in the traditional triple-based TransE knowledge graph embedding model to predict process behavior in temporal knowledge graphs. We evaluate our approach by means of a real-world trademark registration process in a patent office and show its improved performance compared to the TransE base model.

Keywords: Graph embeddings · TransE · Process behavior prediction

1 Introduction

Problem and Motivation. The digitalization of knowledge work poses major challenges in communication-intensive areas. For example, in the legal sector the creativity of the individual experts is essential – despite clear structures through a generally applicable legal framework. In this context, creative knowledge work is understood as the systematic processing of interrelated information fragments into new knowledge by several people. Hübscher et al. [9] presented a graph-based solution that integrates process components (tasks) and knowledge work (data objects) by extending traditional graphs to temporal knowledge graphs [14]. In temporal knowledge graphs, facts evolve over time and each edge in the graph is tagged with temporal information [14].

However, a creative approach to knowledge work often generates complex interrelationships. Tracing and proposing continuous work processes based on exchanged data objects is crucial (not only for legal reasons) to enable learning of

G. Kotsis et al. (Eds.): DEXA 2022 Workshops, CCIS 1633, pp. 369–374, 2022.
https://doi.org/10.1007/978-3-031-14343-4_34

applicable knowledge in communication-intense processes [8]. Based on incoming and outgoing data for knowledge and communication tasks, temporal and logical relations from the underlying graph can be discovered and analyzed.

Related Work. Established data and process mining techniques (cf. [1,2]) require structured, denormalised data in the form of flat (event) logs or need simplification and abstraction methods for their application on large and dynamic graphs. Márquez-Chamorro et al. [13] present a survey on predictive monitoring of business processes. A predictive modeling approach by Breuker et al. in [4] proposes a way to visualize probabilistic process models using a Petri net visualization. However, these approaches usually support only simple, structured workflow patterns and require event logs of an apriori known process model. Since both requirements do not hold for the communication-intensive legal domain, we focus on knowledge embeddings for predicting process behavior in temporal knowledge graphs.

There is increasing interest in research on predicting process behavior using recurrent neural networks (RNN) [5] and long short-term memory (LSTM) neural networks [4]. However, all the studies have very high performance requirements, which limits their use with large graphs. Further, there is still a problem of working with paths of different length and its parallelism. An alternative is the extension of triple-based embedding models. For instance, Lin et al. [11] proposed PTransE, a method that combines a set of relations into a single relation, which describes the whole path. A further extension of PTransE is the C-PTransE model, which uses a different evaluation function that aims at reducing inequality between the high-connected and low-connected nodes [10]. Liu et al. [12] propose EPTransE, a model which also considers the associations of the related nodes. Zhang et al. [15] propose DPTransE (discriminative path-based embedding), which additionally uses statistical properties of nodes and relations. Another extension of TransE is the recurrent TransE (RTransE) model by García-Durán et al. [6], which modifies a training step by using similar composed relations between the graph nodes and an additionally introduced concept of the quadruples [6]. However, all existing extensions of TransE are tailored to a fixed path length, which is not suitable for application domains, where no fixed structure is provided.

Contribution. In this paper, we propose an approach to enhance TransE embeddings for predicting the next possible process step in a temporal knowledge graph by applying additional preprocessing steps suitable for process data. In contrast to existing extensions of TransE, our approach can be directly applied on dynamic graphs since it relies on instance data reflecting the different working styles of users and does not make assumptions about the path length. We evaluate our approach on a real-world scenario of a trademark registration process including administrative and knowledge-intensive work.

2 Method: Enhance TransE to Predict Process Behavior

Our approach is divided into several steps, where in each step, additional data is introduced. In the first step, the experiment requires only basic triples, whereas in the further steps, we will also exploit the retrieved paths. In all of the experimental settings, we add artificial relations between the graph nodes. The details of the single models trained in each step will be described in the following subsections. To evaluate the quality of the different extensions of TransE, we apply the models on a dataset containing process data for predicting the respective next step. Regardless of the fact that TransE is a rather old model, we choose it because it is more flexible for our preprocessing operations. All the experiments are executed with the python library PyKEEN [3]. This library provides easy access to most of the embedding models, such that we could focus on the required preprocessing steps.

Dataset. A process is considered as a sequence of task nodes, which are logically (i.e. indirectly) connected in the graph by data consumption and production. Each individual task is classified by a domain-specific type and has further attributes to describe how the component is specified, such as its id or the start and end time. The dataset includes 100 instances of a trademark registration process, with each containing approximately 30 task nodes and 65 data nodes. The trademark dataset is privately owned by an Austrian patent office and contains sensitive data, thus, it cannot be publicly referenced.

Base Model. The starting point for further experiments is a base model that represents the original dataset in triples in the form of task→nextTask→task (see Fig. 1). We introduced new relations for this purpose, which were derived from the logical data relations and temporal dependencies between tasks. The model should serve to predict the type of the following task depending on the previous one. This basic approach is expected to be limited in terms of performance and expressiveness, because the result only depends on a single step.

Model Extension 1. Next we extend the base model by adding artificial relations called *futureTask* (see Fig. 2). The idea is that each task node is additionally linked with all preceding ones within the same process instance – not only to its direct predecessor or successor (i.e., via *nextTask*). This helps to enrich the model with knowledge about previous actions taken in a process. Nevertheless, this approach is still limited by the required set of the edges.

Model Extension 2. Finally, the previous model can be further extended by a set of artificial relations called *task(N)Step*, where N is the respective number of steps before the targeted task. So we are able to also perceive the varying sequence of the tasks in a process instance. The approach is shown in Fig. 3. We expect this model to fit best for our requirements to predict process behavior.

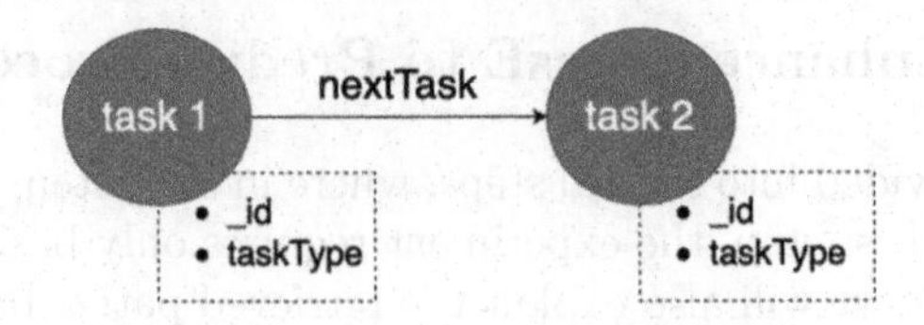

Fig. 1. Design of the base model. Each task is directly linked to its successor.

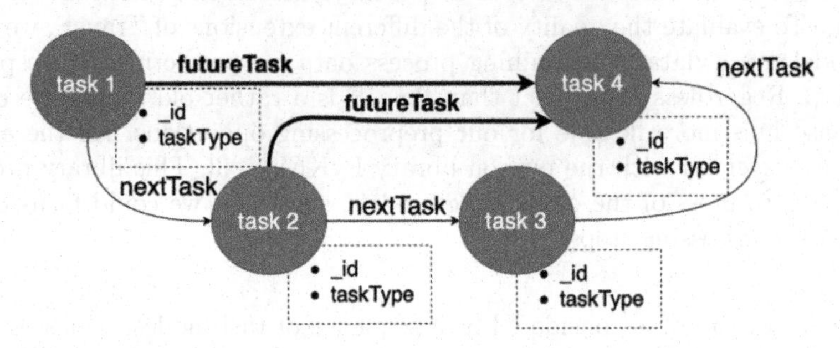

Fig. 2. Design of the first base model extension. Each task is enriched with artificial relations denoting its previous steps.

It uses not only directly following tasks but also the order of the previous steps and is thus also capable to deal with parallel execution of tasks. This parallelism is described by the "task(N)step" relationships, whereas the direct connections are still defined by the "nextTask" edges.

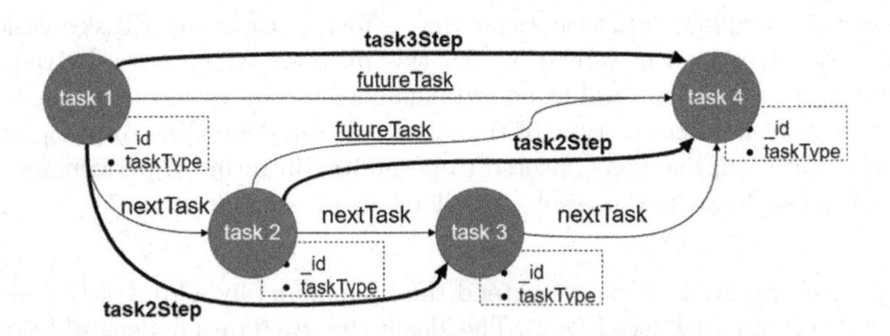

Fig. 3. Design of the second base model extension. Each task is further enriched with the order of previously executed tasks.

3 Experiments and Results

In this section, we discuss the results of applying the base model and its extensions to the described dataset. We use mean rank ($MR = \frac{1}{|\mathcal{I}|}\sum_{r\in\mathcal{I}} r$) and Hits@k ($H_k = \frac{1}{|\mathcal{I}|}\sum_{r\in\mathcal{I}} \mathbb{I}[r \leq k]$) as the metrics for the evaluation. The non-parametric MR is the arithmetic mean over all individual ranks of true triples and better reflects average performance, whereas H_k describes the fraction of true entities that appear in the first k entities of the sorted rank list [7].

Table 1. Experimental results. Mean rank yields values in $[1, \infty)$, whereby lower values are better, and Hits@k in $[0, 1]$, whereby higher values are better.

Dataset	Mean rank	Hit@1	Hit@3	Hit@5	Hit@10
Base model	3.0	0	1.0	1.0	1.0
Extension 1	2.75	0.25	0.75	0.833	1.0
Extension 2	3.63	0.22	0.57	0.8	0.98
Base model (2000 epochs)	2.0	0.5	0.83	**1.0**	1.0
Extension 2 (2000 epochs)	**1.76**	**0.61**	**0.91**	0.99	**1.0**

In the experiments we are interested in the right-side predictions (tail). All experiments were performed with different numbers of epochs (i.e., 10, 50, 150, 500, 750, 1000, and 2000) to evaluate the variation of the prediction quality. The results in Table 1 show that the best performance is achieved with model extension 2, which considers more information about previous paths in the graph than just the basic triples describing the direct pre- and successor nodes. The upper limit of 2000 epochs was chosen as the most optimal setting after a series of hyper-parametrization tests of the model. A further increase of the epochs (tested until 10,000 epochs with a step size of 500) is leading to a significant performance drop but little quality changes. The results were not promising regarding the runtime.

4 Conclusion and Outlook

In this paper, we propose an enhanced way to predict process behavior in temporal knowledge graphs using the TransE embedding model, which exploits its core features without changing the behavior by adapting the model training pipeline and expanding data preprocessing. We created additional relations between the graph nodes that allow to preserve the order of the previously traversed entities. The model was validated on a real-world trademark registration process. The main advantage of our method in comparison to existing graph embeddings is the ability to efficiently work with non-static lengths of the graph paths. In future, we will further optimize the model performance by determining the maximum depth of relations between the nodes (which is specifically critical with long paths). We additionally plan to compare our approach with other extensions of TransE on an open dataset. Since we did not find an open dataset for dynamic events so far, we will do a more comprehensive search for such a dataset, or alternatively, creating a new public, anonymized dataset for future evaluations.

Acknowledgements. The research reported in this paper has been funded by BMK, BMDW, and the State of Upper Austria in the frame of the COMET Programme managed by FFG, and in particular by the FFG BRIDGE project KnoP-2D (grant no. 871299) and the COMET module S3AI (grant no. 872172).

References

1. Aalst, W.M.P.: Object-centric process mining: dealing with divergence and convergence in event data. In: Ölveczky, P.C., Salaün, G. (eds.) SEFM 2019. LNCS, vol. 11724, pp. 3–25. Springer, Cham (2019). https://doi.org/10.1007/978-3-030-30446-1_1
2. Weidlich, M., et al.: Process mining manifesto. In: Daniel, F., Barkaoui, K., Dustdar, S. (eds.) BPM 2011. LNBIP, vol. 99, pp. 169–194. Springer, Heidelberg (2012). https://doi.org/10.1007/978-3-642-28108-2_19
3. Ali, M., et al.: Pykeen 1.0: A python library for training and evaluating knowledge graph embeddings (2020)
4. Breuker, D., Matzner, M., Delfmann, P., Becker, J.: Comprehensible predictive models for business processes. MIS Q. **40**(4), 1009–1034 (2016)
5. Evermann, J., Rehse, J.R., Fettke, P.: Predicting process behaviour using deep learning. Decis. Support Syst. **100**, 129–140 (2017)
6. García-Durán, A., Bordes, A., Usunier, N.: Composing Relationships with Translations. Technical report, CNRS, Heudiasyc (2015)
7. Hoyt, C.T., Berrendorf, M., Gaklin, M., Tresp, V., Gyori, B.M.: A unified framework for rank-based evaluation metrics for link prediction in knowledge graphs (2022)
8. Hübscher, G., et al.: Graph-based managing and mining of processes and data in the domain of intellectual property. Inf. Syst. **106**, 101844 (2022)
9. Hübscher, G., Geist, V., Auer, D., Hübscher, N., Küng, J.: Representation and presentation of knowledge and processes-an integrated approach for a dynamic communication-intensive environment. In: IJWIS (2021)
10. Li, Y., Qin, D., Yang, X.: Path modeling based on entity-connectivity for knowledge base completion. In: ICISCE, pp. 984–989. IEEE (2020)
11. Lin, Y., Liu, Z., Luan, H., Sun, M., Rao, S., Liu, S.: Modeling relation paths for representation learning of knowledge bases (2015). arXiv preprint arXiv:1506.00379
12. Liu, F., Shen, Y., Zhang, T., Gao, H.: Entity-related paths modeling for knowledge base completion. Front. Comp. Sci. **14**(5), 1–10 (2020). https://doi.org/10.1007/s11704-019-8264-4
13. Márquez-Chamorro, A.E., Resinas, M., Ruiz-Cortés, A.: Predictive monitoring of business processes: a survey. IEEE Trans. Serv. Comput. **11**(6), 962–977 (2017)
14. Trivedi, R., Dai, H., Wang, Y., Song, L.: Know-evolve: deep temporal reasoning for dynamic knowledge graphs. In: ICML, pp. 3462–3471. PMLR (2017)
15. Zhang, M., Wang, Q., Xu, W., Li, W., Sun, S.: Discriminative path-based knowledge graph embedding for precise link prediction. In: Pasi, G., Piwowarski, B., Azzopardi, L., Hanbury, A. (eds.) ECIR 2018. LNCS, vol. 10772, pp. 276–288. Springer, Cham (2018). https://doi.org/10.1007/978-3-319-76941-7_21

An Explainable Multimodal Fusion Approach for Mass Casualty Incidents

Zoe Vasileiou[1,2(✉)], Georgios Meditskos[1], Stefanos Vrochidis[2], and Nick Bassiliades[1]

[1] School of Informatics, Aristotle University of Thessaloniki, Thessaloniki, Greece
{zvasileiou,gmeditsk,nbassili}@csd.auth.gr

[2] Centre for Research and Technology Hellas, Information Technologies Institute, Thessaloniki, Greece
{zvasilei,stefanos}@iti.gr

Abstract. During a Mass Casualty Incident, it is essential to make effective decisions to save lives and nursing the injured. This paper presents a work in progress on the design and development of an explainable decision support system, intended for the medical personnel and care givers, that capitalises on multiple modalities to achieve situational awareness and pre-hospital life support. Our novelty is two-fold: first, we use state-of-the-art techniques for combining static and time-series data in deep recurrent neural networks, and second we increase the trustworthiness of the system by enriching it with neurosymbolic explainable capabilities.

Keywords: Pre-hospital life support · Explainable AI · Deep neural network · Neurosymbolic XAI · Mass casualty incident management

1 Introduction

Mass Casualty Incidents (MCIs) are defined as "any event resulting in a number of victims large enough to disrupt the normal course of emergency and health care services" [8]. In MCIs, time is critical and the medical personnel should be aware of potential anomalies and any hazardous situations so as to assign degrees of urgency and decide the order of treatment of a large number of patients.

Deep Learning (DL) has been used to address different challenges in different domains, such as in Healthcare. Healthcare data includes various types of data, such as electronic health records (EHR) and raw signal values collected by ambient and wearable sensors as time-series. By integrating and fusing distributed and heterogeneous data, the complementarity of the multimodality is leveraged to acquire a consistent and accurate understanding of the situation. However, few studies have attempted to combine static and dynamic data, and most of them are focusing on the prediction of a specific disease.

Humans need to understand AI capability and effectively calibrate their trust. For building trustworthy clinical decision support systems, the model should be explainable and transparent through justifications. The main input of the clinical

G. Kotsis et al. (Eds.): DEXA 2022 Workshops, CCIS 1633, pp. 375–379, 2022.
https://doi.org/10.1007/978-3-031-14343-4_35

models are temporal data, but temporal explanations [10] is an unexplored and challenging task. A limited number of explainable clinical early warning systems that provide formal explanations have been developed and primarily focus on feature importance [5,12]. Also, to the best of our knowledge, limited work has been done to combine static and dynamic information of the casualties for early warning systems, let alone to be enriched with explanations.

This paper presents work in progress from ongoing research to address the aforementioned challenges. Concretely, a Long short-term memory (LSTM) architecture, capturing the temporal dependencies in long time series, is followed. This RNN-based solution uses both the static and dynamic information of a casualty for predicting hazardous situations during a MCI, assisting the medical personnel in determining the need of medical treatment and transportation to hospitals. Additionally, the system is enhanced with explainable capabilities by providing indications how the model reached the final prediction by adopting neurosymbolic Explainability Artificial Intelligence (XAI) techniques.

The rest of the paper is structured as follows. A background and related work is provided regarding the LSTM models and existing works. Next, our methodology is presented by means of the architecture and the explainability aspects. Finally, we conclude the paper.

2 Background and Related Work

Recurrent Neural Networks (RNNs) are widely adopted in time-series classification or prediction in various kind of signals, as they allow the information to persist. RNNs are using loops for having a sort of memory. This kind of DL models enable the sequential and time-series to be represented such as the EHR and the long raw signal time sequences. The main disadvantage of RNNs is the problem of vanishing gradients that hinders the knowledge to be retained for long data sequences. LSTM models, a gated variant of RNN, can keep long-term memory by remembering long sequences of data since. Existing works were focused on the use of DL for the temporal data representation in EHR, facing various challenges, such as the data irregularity and data heterogeneity [11]. Mainly, RNN, LSTM and Gated Recurrent Units (GRUs) have been proposed for their suitability in representing temporal sequences.

Regarding the Early Warning Systems in the healthcare domain, various AI-powered solutions have been developed for predicting clinical deterioration [7]. An early detection system of heart failure onset [2] adopted a GRU model using EHR data as input. Other LSTM-based fusion approaches detect Alzheimer's progression [1] and predict early tachycardia [6]. Although temporal data entail several challenges, the opacity of the model is equally important as DL models are black-box models. Neurosymbolic XAI [3] can make black-box models transparent leveraging the inherent self-explainability of symbolic knowledge.

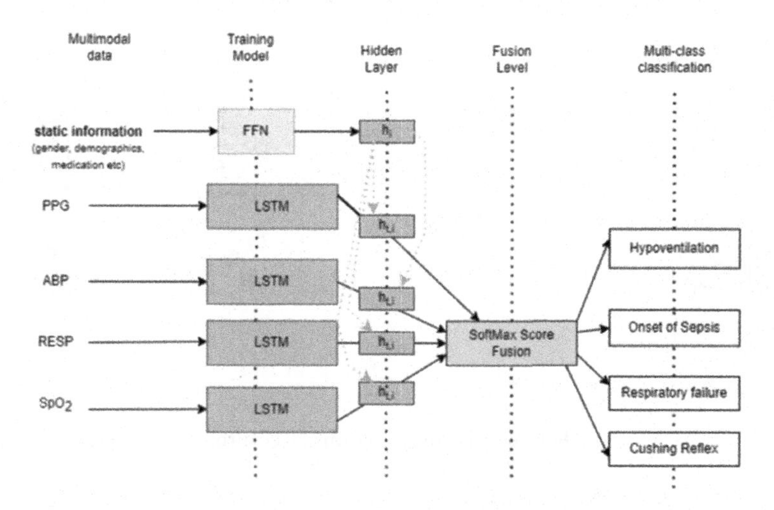

Fig. 1. Multimodal LSTM for decision support in pre-hospital life support in MCIs

3 Methodology

The proposed method provides situational awareness through explainable early warnings and decisions to the medical personnel. A DL-based approach is presented that weaves information from disparate modalities aiming at supporting medical personnel in the important decision for hospitalizing an MCI victim. The main modalities are Photoplethysmography (PPG), Blood Pressure (ABP), Respiratory Rate (RR), and Oxygen Saturation (SpO2).

A Multi-Source Information Fusion Engine integrates multiple sources and fuse data in an abstract way, aiming at detecting complex situations. We propose a LSTM-based multimodal DL algorithm for classifying the sensor data and the EHR. Our methodology combines dynamic and static data. About the dynamic data, the victims, e.g. first responders, are wearing equipment with sensors that are measuring time series. Static data include demographics, comorbidities and medication. For combining static and dynamic data, the LSTM is amalgamated with a feed-forward network. The static information is processed on a separate feed-forward network whose hidden state is concatenated with the hidden layers of the LSTMs. The late fusion scheme is applied as all the modalities are processed in different pipelines. Each modality is a physiological signal, thus processed by a LSTM model for capturing the temporal dependencies and extract patterns. The outputs of the LSTMs are concatenated and used by a softmax layer, a probability distribution over the classes, for making the final prediction. The prediction could be hazardous situations such as (i) respiratory failure, (ii) need for hypoventilation (iii) an onset of a medical emergency such as sepsis (iv) Cushing reflex, a serious situation usually seen in acute head injuries. The flow of our methodology is depicted in Fig. 1.

The overall architecture is depicted in Fig. 2. A pre-processing is performed for cleaning noisy and undesired signals, the data are segmented into fixed-size

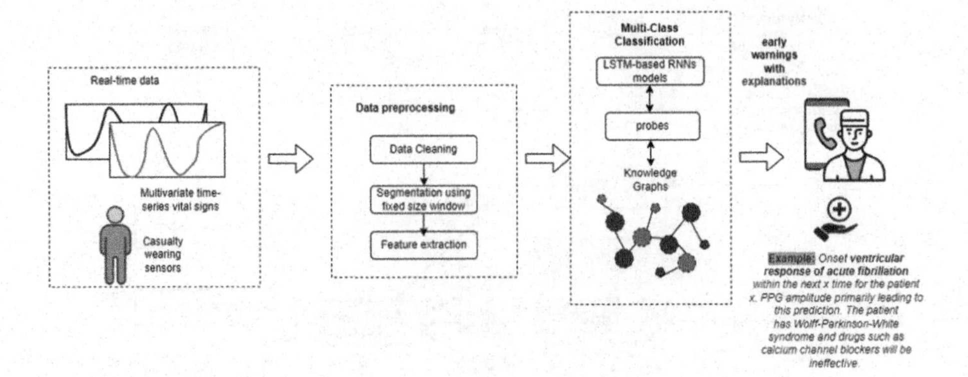

Fig. 2. The proposed architecture.

sliding windows, and then a feature generation is applied. Then, the features are sent as input to a multi-class classification algorithm that classifies the casualties into multiple classes by predicting risky situations. If a risky situation for a casualty is predicted as imminent, an early warning is sent to the medical personnel to incorporate it into their decisions about hospitalization and the level of hospital care that the casualty should receive.

One of the key features of our decision support system is that it is imbued with explainable capabilities which are prominent aspects in a multimodal environment. In order to foster an interpretable decision system, our neural system is endowed with symbolic functionalities forming a neuro-symbolic system. For providing symbolic justifications, a hidden layer analysis is performed for comprehending the concepts extracted by each activation node. A similar approach with the work of [4] is followed by using linear classifiers, known as probes, to be mapped to the intermediate layers running independently from the main model. Those probes are predicting whether a given concept was recognized by the model. The concepts entailed by the probes are mapped to ontology concepts through semantic annotation by populating knowledge graphs in a similar manner with the a mapping network approach [9] aligning artificial neural networks with ontologies. The knowledge graphs represent personalized patient data, information about diseases and clinical terminology leveraging ontologies, such as the SNOMED-CT[1] and ICD10[2].

4 Conclusion and Future Work

This paper presented a preliminary work on the development of an explainable Early Warning System that captures heterogeneous sources of a patient during a MCI. A LSTM-based architecture has been designed that amalgamates the static

[1] http://bioportal.bioontology.org/ontologies/SNOMEDCT.
[2] https://bioportal.bioontology.org/ontologies/ICD10.

and dynamic data of the casualty in a novel way and yields early warnings. By integrating the neural system with knowledge graphs, those early warnings are accompanied with explanations since symbolic approaches are inherently white-boxes. As next steps, we are working on finalising the implementation and testing the framework on real-world use cases[3].

Acknowledgements. This work has received funding from the European Union's H2020 RIA projects NIGHTINGALE (101021957) and INGENIOUS (833435). Content reflects only the authors' view and the Research Executive Agency (REA) and the European Commission are not responsible for any use that may be made of the information it contains.

References

1. Abuhmed, T., El-Sappagh, S., Alonso, J.M.: Robust hybrid deep learning models for Alzheimer's progression detection. Knowl.-Based Syst. **213**, 106688 (2021)
2. Choi, E., Schuetz, A., Stewart, W., Sun, J.: Using recurrent neural network models for early detection of heart failure onset. Am. Med. Inform. Assoc. **24**, 361–370 (2016)
3. Futia, G., Vetró, A.: On the integration of knowledge graphs into deep learning models for a more comprehensible AI-three challenges for future research. Information **11**(2), 122 (2020)
4. Guillaume, A., Bengio, Y.: Understanding intermediate layers using linear classifier probes. In: ICLR (Workshop) (2017)
5. Lauritsen, S., et al.: Explainable artificial intelligence model to predict acute critical illness from electronic health records. Nature Commun. **11**, 1–11 (2020)
6. Liu, X., et al.: Top-Net: tachycardia onset early prediction using bi-directional LSTM in a medical-grade wearable multi-sensor system. Med. Inform. **9**, e18803 (2020)
7. Muralitharan, S., et al.: Machine learning-based early warning systems for clinical deterioration: a systematic scoping review. J. Med. Internet Res. **23**, e25187 (2020)
8. World Health Organization: Mass casualty management systems. Strategies and guidelines for building health sector capacity, Geneva, Switzerland (2007)
9. de Sousa Ribeiro, M., Leite, J.: Aligning artificial neural networks and ontologies towards explainable AI. Artif. Intell. **35**(6), 4932–4940 (2021)
10. Tonekaboni, S., Joshi, S., McCradden, M., Goldenberg, A.: What clinicians want: contextualizing explainable machine learning for clinical end use. In: Machine Learning for Healthcare Conference, MLHC, vol. 106, pp. 359–380. PMLR (2019)
11. Xie, F., et al.: Deep learning for temporal data representation in electronic health records: a systematic review of challenges and methodologies. J. Biomed. Inform. **126**, 103980 (2022)
12. Yang, M., et al.: An explainable artificial intelligence predictor for early detection of sepsis. Crit. Care Med. **48**(11), e1096 (2020)

[3] https://www.nightingale-triage.eu/.

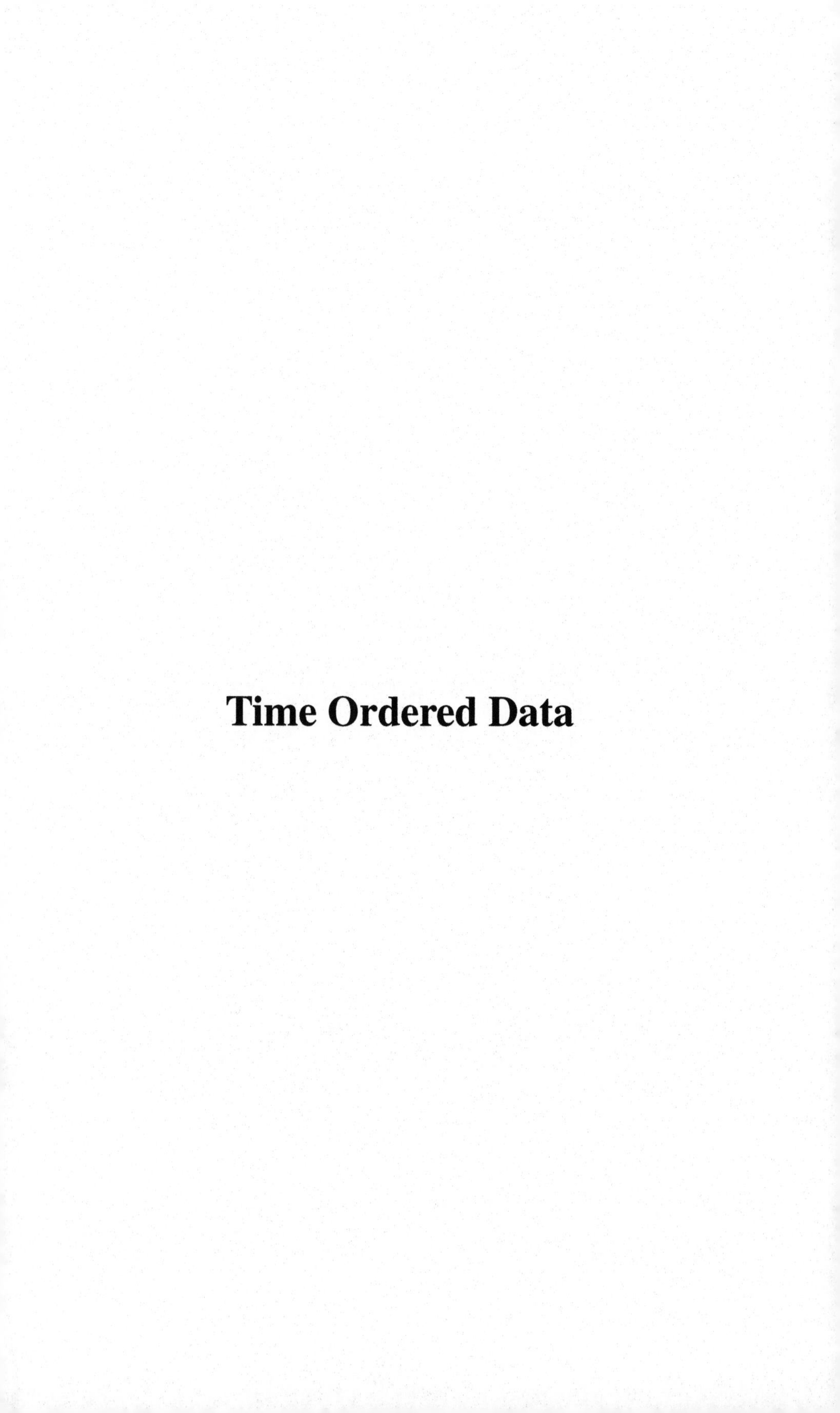

Time Ordered Data

Log File Anomaly Detection Based on Process Mining Graphs

Sabrina Luftensteiner(✉) and Patrick Praher

Software Competence Center Hagenberg, Hagenberg im Muehlkreis, Austria
sabrina.luftensteiner@scch.at
http://www.scch.at

Abstract. In process industry, it is quite common that manufacturing machines repeat a certain order of process steps consistently. These process steps are often declared in programs and have a linear workflow. During the production process, logs are generated for monitoring and further analysis, where not only production steps but also incidents and other deviations are logged. As the manual monitoring of such processes is quite time consuming and tedious, the demand for automatic anomaly and deviation detection is rising. A potential approach is the usage of Process Mining, whereat not all requirements are met. In this paper, we propose a new approach based on spectral gap analysis for the detection of anomalies in log files using the adjacency matrix generated by Process Mining techniques. Furthermore, the experiments section covers their application on a linear process and non-linear processes with deviating paths.

Keywords: Log files · Anomaly detection · Process mining

1 Introduction

Log files are created by devices or systems to provide information about processes and their behavior in a textual or granular view [4]. They do not only provide expected behavior, but also important information about possible incidents [7], e.g. via error logs, which may reveal system or process weaknesses [4]. The manual inspection of log files is often very time consuming and tedious, which demands for an automated approach [7]. Process Mining [14] aims at an automatic extraction of process knowledge, whereat event-logs are utilized to grasp the complex nature of industrial processes [5]. Especially in the field of manufacturing and programming of machines, the usage of Process Mining is not highly present although the area of application fits outstandingly well [13].

In manufacturing, it is common to program machines to repeat certain program workflows consistently. These programs usually have a linear workflow, which consists of a specific order of non-repeating steps. During the execution the

The research reported in this paper has been supported by the Austrian Ministry for Transport, Innovation and Technology, the Federal Ministry for Digital and Economic Affairs, and the Province of Upper Austria in the frame of the COMET center SCCH.

G. Kotsis et al. (Eds.): DEXA 2022 Workshops, CCIS 1633, pp. 383–391, 2022.
https://doi.org/10.1007/978-3-031-14343-4_36

steps and incidents are logged constantly. These logs are evaluated using Process Mining to determine visually, if a process workflow is as expected using certain techniques for graph generation and conformance checking. As the monitoring may be very time consuming and tedious, new approaches for the comparison of such Process Mining graphs are required. It is in demand to provide computationally efficient, yet accurate comparison possibilities, to match workflows and their deviations. In the course of this paper, linear processes represent optimal processes, where anomalies have to be detected. Slightly non-linear process executions may still reflect a good execution, e.g. due to error messages with no direct influence on the execution or process, therefore a comparison to linear workflows may not be sufficient enough. The detection of strong anomalies or deviations, which is reflected by a changing process execution behavior, is significant as they may indicate wearing of machine parts or wrong usage/implementation of programs as well as adapted programs.

This paper proposes an approach to detect anomalies in linear processes using Process Mining techniques and event-logs generated using log files. The approach is based on the spectral gap calculation and uses the adjacency matrix created by Process Mining techniques. Furthermore, an insight into related work is provided, presenting the current state of the art as well as disadvantages of existing approaches. The experimental section covers the effectiveness of the approach in differing scenarios, whereat different deviation types are tackled.

1.1 Discussion on Existing Approaches and Related Work

This subsection provides insight into related work for the comparison of event-logs to find behavior changes or anomalies in linear processes. Various approaches for anomaly detection in log files, conformance checking and process deviance are presented, highlighting their drawbacks regarding our use-cases.

Anomaly Detection in Logs. Frei et al. propose a log file visualization called Histogram Matrix (HMAT), which visualizes the content of textual log files to enable the detection of anomalies using a combination of graphical and statistical techniques [7]. An approach for the detection of abnormal behavior in event-logs of social network websites is provided by Sahlabadi et al. [12]. In a first step the normal user behavior pattern is defined using a genetic Process Mining technique, the second step covers the detection of abnormal behavior via a fitness function, which is time as well as resource consuming. Breier et al. base their anomaly detection approach on data mining techniques for log analysis and the creation of dynamic rules [4].

The presented methods are quite sensitive to deviating process executions and are likely to not work with frequently deviating executions with no stable base, e.g. due to error messages.

Conformance Checking. Conformance checking is used to detect inconsistencies between a process model and its corresponding execution log [6]. Various approaches for conformance checking are available, one of which is provided by Dunzer et al. [6] covering modelling languages, algorithmic types, quality metrics as well as perspectives. An incremental approach is proposed by Rozinat et al. [11] to check the conformance of a process model and an event log by measuring their fitness and analyzing the appropriateness from a structural as well as behavioral perspective. Adriansyah et al. [1] present a robust method for conformance checking using various metrics, which do not cover unsatisfied or unhandled events. Furthermore, a robust replay analysis technique for measuring the conformance is provided Adriansyah et al. [2], which also provides intuitive diagnostics such as skipped or inserted activities.

The conformance checking approaches seem quite interesting, but they require a stable process model as a base to provide valid results. In manufacturing, this may often be not the case as processes are adapted or unpredictable errors are logged, which complicates the automatic creation of a good process model without human intervention.

Process Deviance and Variants. Process deviance and variants provide a third possibility to inspect process workflows and their conformance. The usage of process model variants is rising due to the increased monitoring of dynamic processes. As they are rather difficult to maintain, various methods for generalizing and the deviance from this generalization have been proposed by researchers. Li et al. proposed papers dealing with the goals [9] and issues [8] related to the mining of process model variants with a focus on the integration of process changes into generic process models. Another approach by Li et al. [10] covers the learning from process model adaptations and the discovery of reference models, which are used to configure variants with minimum effort. Process variability and the handling of variants in such varying processes was dealt with by Bolt et al. [3]. They developed an unsupervised technique to detect significant variants in event-logs to detect if variants represent the same process or differ.

Process deviance and variants provide suitable approaches to handling variants and the generation of a reference model. Still, human intervention is likely needed to generate a meaningful base for further usage and to define normal behavior of a process. Furthermore, human input is needed to define positive and negative deviations of the process.

2 Methods

This section describes the preparation of event-logs and our method used for the detection of changing behavior. The subsection covering the data preparation will explain the transformation of event-logs to adjacency matrices, which are used as basis for the spectral gap approach.

2.1 Preparation of Event-Logs

Table 1 contains a simplified event-log of a process. The relevant information covers CaseID, Activity and Timestamp. The CaseID resembles the unique identifier of one process execution, the activity covers a textual representation of a process step and the timestamp is used to keep the data sorted. The data is extracted from log files, which were gathered in industrial environments, e.g. machines or systems. In our use-cases, the events do not include flags to indicate the state of an activity, e.g. start or completed, but rather independent activities, which have to be interpreted correctly. CaseIDs may have overlapping timestamps as machines may already start a new production cycle while the other one is still running.

Table 1. Simplified event-log including unique identifier, activity and timestamp.

CaseID	Activity	Timestamp
1	T1 - Insert workpiece	2022-04-01 00:00:00
1	T2 - Apply heat	2022-04-01 00:00:10
1	T3 - Bend	2022-04-01 00:00:20
1	T2 - Apply heat	2022-04-01 00:00:30
1	T3 - Bend	2022-04-01 00:00:40
1	T2 - Apply heat	2022-04-01 00:00:50
1	T2 - Apply heat	2022-04-01 00:01:00
1	T3 - Bend	2022-04-01 00:01:10
2	T1 - Insert workpiece	2022-04-01 00:01:20

The first step of the data preparation is the extraction of an adjacency matrix, which is extracted from the event-logs using Process Mining techniques to create a directed graph. The values within the adjacency matrix have to be relative in our use-case, referring to the relative amount of edge transitions between nodes. Absolute values complicate further comparisons and evaluations as the processes would have to consist of an identical amount of executions for each comparison timespan. Table 2 contains the matching relative adjacency matrix of the event-log presented in Table 1.

2.2 Spectral Gap Calculation Using Event-Logs

Prior to calculating the spectral gap, the adjacency matrix has to be made doubly stochastic. These matrices fulfill the requirements of being a square matrix

Table 2. Adjacency matrix.

	T1	T2	T3
T1	0	1	0
T2	0	0.25	0.75
T3	0.33	0.66	0

with non-negative real values, having row and column sums of 1, having matrix elements between 0 and 1 as well as the largest eigenvalue being always 1. Equation 1 and Eq. 2 cover the steps to create a doubly stochastic matrix (DSM) from a square matrix M:

$$SM = M + M^T \tag{1}$$

$$DSM_{i,} = \frac{SM_{i,}}{\sum SM_{i,}} \tag{2}$$

At first a transposed matrix is added to the original matrix to generate a symmetric matrix (SM). The symmetric matrix is then normalised and produces a doubly stochastic matrix (DSM), which is used to calculate eigenvalues. The eigenvalues result in the subtraction of the second largest eigenvalue from the largest eigenvalue to receive the spectral gap, see Eq. 3 where λ_1 refers to the largest and λ_2 to the second largest eigenvalue of DSM. The process of calculating the spectral gap can be summarized as follows:

1. Transformation of event-logs to relative adjacency matrices
2. Transformation of adjacency matrices to doubly stochastic matrices
3. Calculation of eigenvalues for each matrix
4. Subtraction of second largest from largest eigenvalue to calculate spectral gap for each matrix
5. Comparison of spectral gaps

$$SG = \lambda_1 - \lambda_2 \tag{3}$$

The calculation of the spectral gap does not include further knowledge of the basic structure or additional information of prior process workflows. It only considers what is available at that point in time and how linear the workflow is. The more deviations from the originally linear process path, the higher the spectral gap. Using this approach on usually linear workflows, it is more straightforward to detect anomalies or deviations in the process flow.

3 Experiments and Discussion

This section covers the experimental setup as well as different types of process graph changes and their influence on our spectral gap approach. The overall goal is to demonstrate that using these methods on event-logs, we are able to detect behavior changes or anomalies in linear process executions.

3.1 Event-Logs Structure

A program consists of a sequence of routines or process steps, whereat the execution is logged in log files for the creation of event-logs. Steps within a program refer to activities of a machine to manufacture a product, ranging from laser

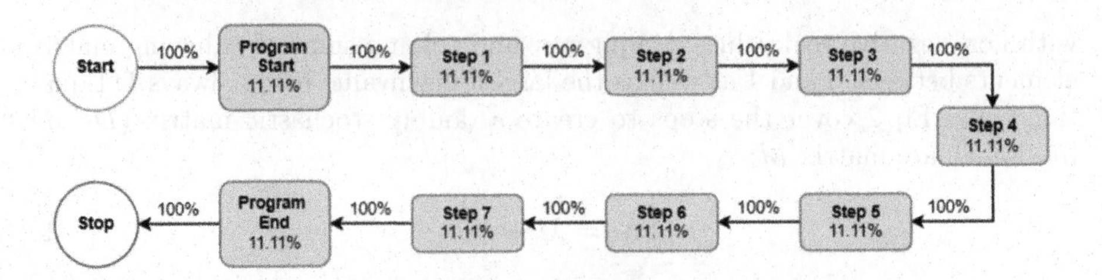

Fig. 1. Workflow example of a flawless production cycle with no repeating activities or additional suspicious activities, e.g. error messages.

cutting to coating to bending and various other activities. Additional errors or warnings may be logged during the process, e.g. occurring due to faulty executions, wrong program usage or wearing of machine parts. The stored event-logs are analyzed on a daily basis, using the execution start of a program as ID for each process cycle. The following event-log activities are covered in our scenario:

- Start Program - Start of a new program cycle
- Stop Program - Stop of a program cycle
- Step x - Process step within a program
- Error x - Error appearing during a program execution

Table 3 covers the simplified event-log of one program execution.

Table 3. Simplified event-log of one program execution.

CaseID	Activity	Timestamp
1	Start program	2021-07-01 00:00:00
1	Step 1	2021-07-01 00:00:10
1	Step 2	2021-07-01 00:00:20
1	Step 3	2021-07-01 00:00:30
1	Step 4	2021-07-01 00:00:40
1	Step 5	2021-07-01 00:00:50
1	Step 6	2021-07-01 00:01:00
1	Step 7	2021-07-01 00:01:10
1	Stop program	2021-07-01 00:01:20

Figure 1 provides a visualization of a manufacturing process using Process Mining on event-logs, whereas Table 3 contains the basic event-log. The event-log contains an optimal and repeating workflow, which covers a start event, a stop event and seven intermediate process steps for each execution. As no deviations are available, the graph represents a linear process without any branching and is therefore used as the basis for further experiments with deviating workflows.

The following list covers deviation types in the program execution, which occur in industrial environments:

- *Repeating Activities* occur due to unsatisfactory outcomes from one step execution or the reduction of program code, e.g. if one step has to be executed twice.
- *Additional Activities* appear either planned, e.g. due to the adaption of existing programs, or unplanned, e.g. due to error or warning messages.
- *Combination of Additional & Repeating Activities* indicate problems within the program or machine if the combination appears unplanned.
- *Program Interruptions* occur often in combination with (unplanned) additional activities, such as errors and lead to an early termination of the execution process.

3.2 Experimental Setup and Evaluation

The experimental scenario consists of a linear program, which was artificially created and covers 12 executions a day for a specific amount of days. A normal process execution has a linear workflow consisting of 7 steps and a likelihood of 10% for an error message to appear. Those additional error messages reflect a typical behavior in production as it is very likely that some incidents or disturbances are happening during a process execution, e.g. wrong alignment of material. A slightly non-linear workflow may still reflect a good execution, e.g. due to error messages with no direct influence, therefore a comparison to linear workflows may not be sufficient enough. The timespan of our scenario covers 7 months to simulate a more realistic area of application. Anomalous days have different characteristics regarding occurring anomalies, with some days having more anomalous behavior and others less, e.g. the occurrence of additional errors.

The scenario covers a timespan of 7 months containing anomalous and non-anomalous data. The individual days cover process executions with weaker and stronger deviations, which indicate problems with either the machine or program as the usual behavior is negatively influenced and changed. The anomalous days may vary regarding their anomalous activity, some having more and others having less different behavior. The sample adjacency matrix is generated for each day and our spectral gap approach is applied for each these matrices. Figure 2 visualizes the results using our approach, whereat the days are colored according to their dedicated category. The results are presented as standard deviations. The visualization shows that it is possible to detect the anomalous days due to their differing behavior using the spectral gap method and a fitting threshold, which is set to the mean and $2 * \sigma$ of the first 20 days. The anomalous days are clearly visible as they have higher sigmas due to their strong deviations in comparison to normal behaving executions.

3.3 Discussion

The strength of the spectral gap approach is its uncomplicated and fast computation. In contrast to other approaches, it is not necessary to know all process

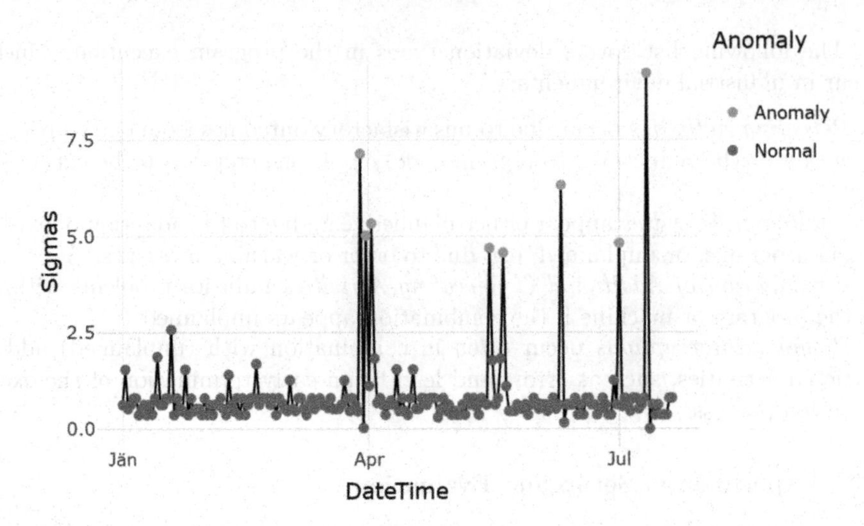

Fig. 2. Visualization of results regarding a repeating process over 8 months, using the standard deviation σ as indicator.

parts in advance and therefore the used adjacency matrices may have different dimensions. Our approach is able to indicate if programs have a linear execution workflow, meaning the more deviating branches are available in the workflow the higher the spectral gap will be. In case of our linear program example, the spectral gap approach enables us to evaluate if the process operates well or if problems are occurring and the program workflow is deviating from its original course. Another advantage of the spectral gap approach is that no common base has to be calculated prior to the actual calculation, increasing the computational efficiency and reducing the overall computation time.

On the downside, the usage of the spectral gap approach is not suitable for non-linear programs, which contain e.g. cycles or elemental deviations within the optimal process path. As only one matrix is focused on, no relations or similarities of multiple matrices, e.g. spanning over a few days, are extracted. Considering this use-case, deviations in processes may not be detected, e.g. if a program is changed at the beginning of a day.

4 Discussion and Conclusion

In our work, we addressed the problem of detecting anomalies and workflow deviations within linear programs. At first, a description of existing approaches and state of the art was presented, highlighting their purpose and disadvantages regarding our use-case. Building up on Process Mining techniques and event-logs, a spectral gap approach was developed to detect anomalies within process executions of industrial machines based on a linear workflow. The experiments section provided a specific setup consisting of a basic linear program and its possible deviations to demonstrate the impact on the methods and their effectiveness. Our method allows the unsupervised identification of flawless and anomalous linear machine executions without the need for any hyper parameter optimization.

As our method only works in combination with linear combinations, non-linear approaches should also be addressed in future work. Further possibilities are experiments with other anomaly detection methods in combination with adjacency matrices as well as the generation and identification of a training phase for supervised approaches.

References

1. Adriansyah, A., van Dongen, B.F., van der Aalst, W.M.P.: Towards robust conformance checking. In: zur Muehlen, M., Su, J. (eds.) BPM 2010. LNBIP, vol. 66, pp. 122–133. Springer, Heidelberg (2011). https://doi.org/10.1007/978-3-642-20511-8_11
2. Adriansyah, A., van Dongen, B.F., van der Aalst, W.M.: Conformance checking using cost-based fitness analysis. In: 2011 IEEE 15th International Enterprise Distributed Object Computing Conference, pp. 55–64. IEEE (2011)
3. Bolt, A., van der Aalst, W.M.P., de Leoni, M.: Finding process variants in event logs. In: Panetto, H., et al. (eds.) OTM 2017. LNCS, vol. 10573, pp. 45–52. Springer, Cham (2017). https://doi.org/10.1007/978-3-319-69462-7_4
4. Breier, J., Branišová, J.: Anomaly detection from log files using data mining techniques. In: Kim, K.J. (ed.) Information Science and Applications. LNEE, vol. 339, pp. 449–457. Springer, Heidelberg (2015). https://doi.org/10.1007/978-3-662-46578-3_53
5. Corallo, A., Lazoi, M., Striani, F.: Process mining and industrial applications: a systematic literature review. Knowl. Process. Manag. **27**(3), 225–233 (2020)
6. Dunzer, S., Stierle, M., Matzner, M., Baier, S.: Conformance checking: a state-of-the-art literature review. In: Proceedings of the 11th International Conference on Subject-Oriented Business Process Management, pp. 1–10 (2019)
7. Frei, A., Rennhard, M.: Histogram matrix: log file visualization for anomaly detection. In: 2008 Third International Conference on Availability, Reliability and Security, pp. 610–617. IEEE (2008)
8. Li, C., Reichert, M., Wombacher, A.: Issues in process variants mining (2008)
9. Li, C., Reichert, M., Wombacher, A.: Mining process variants: goals and issues. In: 2008 IEEE International Conference on Services Computing, vol. 2, pp. 573–576. IEEE (2008)
10. Li, C., Reichert, M., Wombacher, A.: Mining business process variants: challenges, scenarios, algorithms. Data Knowl. Eng. **70**(5), 409–434 (2011)
11. Rozinat, A., Van der Aalst, W.M.: Conformance checking of processes based on monitoring real behavior. Inf. Syst. **33**(1), 64–95 (2008)
12. Sahlabadi, M., Muniyandi, R.C., Shukur, Z.: Detecting abnormal behavior in social network websites by using a process mining technique. J. Comput. Sci. **10**(3), 393 (2014)
13. Son, S., et al.: Process mining for manufacturing process analysis: a case study. In: Proceeding of 2nd Asia Pacific Conference on Business Process Management, Brisbane, Australia (2014)
14. Van Der Aalst, W.: Process mining: overview and opportunities. ACM Trans. Manage. Inf. Syst. (TMIS) **3**(2), 1–17 (2012)

A Scalable Microservice Infrastructure for Fleet Data Management

Rainer Meindl[1,2](✉), Konstantin Papesh[1], David Baumgartner[3], and Emmanuel Helm[1]

[1] Research Group Advanced Information Systems and Technology, University of Applied Sciences Upper Austria, Softwarepark 11, Hagenberg, Austria
[2] Software Competence Center Hagenberg, Softwarepark 32a, Hagenberg, Austria
rainer.meindl@scch.at
[3] Norwegian University of Science and Technology, Høgskoleringen 1, Trondheim, Norway

Abstract. Modern Internet of Things solutions using edge devices produce large amounts of raw data. In order to utilize this data, it needs to be processed, aggregated, and categorized to enable decision making for management and end-users. This data management is a non-trivial task, as the computational load is directly proportional to the amount of data. In order to tackle this issue, we provide an extensible and scalable microservice architecture that can receive, normalize, and filter the raw data and persist it in different levels of aggregation, as well as for time series analysis.

Keywords: Big data · Time series · Data management · Data processing

1 Introduction

Current Internet of Things (IoT) solutions generate large amounts of data that need efficient processing before any information can be extracted. This process is cumbersome and prone to error, as IoT devices provide their information over the internet and wireless networks, where data might be lost or corrupted. There might be on-device data aggregation that fails, and the observed context, i.e., what the IoT device monitors, might be inherently error-prone [1]. We suggest a distributed architecture for accepting and processing this raw data that can handle large amounts of data by horizontal scaling and is error resistant due to preliminary integrity checks and high level data aggregation to extract the needed information. Figure 1 shows an overview of the proposed architecture.

The remainder of this work will elaborate on this. Section 2 explains the components of the service-oriented architecture based on the use case of vehicle fleet monitoring. Section 3 shows the performance of the proposed architecture in a real-world scenario and Sect. 4 elaborates on the strengths and weaknesses. Finally, Sect. 5 briefly gives an outlook on future research directions.

G. Kotsis et al. (Eds.): DEXA 2022 Workshops, CCIS 1633, pp. 392–401, 2022.
https://doi.org/10.1007/978-3-031-14343-4_37

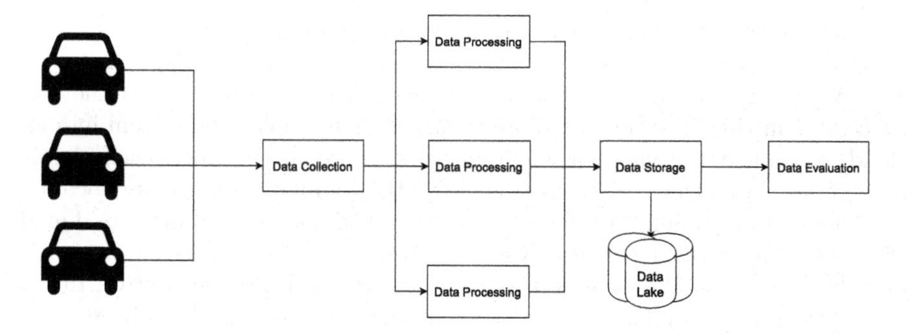

Fig. 1. Overview of the suggested solution.

1.1 Problem Statement

Fleet monitoring systems use a wide range of sensors, such as GPS and accelerometer, to keep track of their active vehicles. Additionally, we also employ additional sensors encapsulated in an edge device, that is mounted in these vehicles. This generates a large amount of data that is not trivial to process. Based on this use case we were able to identify following problems: (i) unreliable data frequency, (ii) overall inconsistent data quality based on the vehicle model, specifically in completeness, correctness and timeliness, and (iii) the correct reassignment of singular data samples to their time series.

1.2 Related Work

Hounsell et al. [2] discuss Automatic Vehicle Location Systems based on Transport for Londons *iBus*. It aims to improve efficiency of road-based passenger transport systems, such as busses, by introducing GPS information and real-time-passenger information. They use a singular relational database to store the real-time, second-to-second basis data of busses. Hounsell et al. provide plenty of interesting use cases, that need highly available and scalable data management systems. They do explicitly mention the need for proper management systems as "'*this vast amount of data also presents challenges for data storage and processing requirements*". We aim to provide a solution to these challenges, as described in Sect. 2.

Wittman et al. [5] present a holistic framework for acquiring and processing of vehicle fleet test data. They use protocol buffers for transmitting GPS and sensor data to integrate different devices. The presented framework offers a multitude of visualisations for both mobile and web applications. The data acquisition approach is similar to ours, as Wittman et al. also use the OBD II interface, yet we miss a description of the used dataset, as it is only described as test data. We aim to elaborate on this, by also focusing on the question how large datasets can be processed in an acceptable amount of time.

A similar approach is presented by Falco et al. [3], whereas they too suggest a microservice architecture to ensure availability and throughput of their system.

The data acquisition is also comparable to our approach, they too access the OBD II interface and extract similar metrics from the vehicle. But they again do not provide any insight into the data itself and the problems of the data processing in this scale. Furthermore their goal is to deploy the system into the cloud, which is not feasible for our framework, as we aim to process confidential data of fleet providers, which must not leave the companies on-premise location.

Killeen et al. [4] builds on top of the COSMO (Consensus self-organized models) algorithm, resulting in the ICOSMO (Improved Consensus self-organized models) algorithm. We deemed this algorithm to be highly restrictive, due to assumptions made by Killeen et at such as the algorithm needs to have access to a database of labeled repair data and all vehicles have to be of the same type. We also lack information how the data is acquired and processed in detail, the test setup states that data is accessed from a hybrid bus, via a WiFi on an LTE station. As we focus on possible real-time processing of data we deem this approach unfeasible for fleet management.

2 Method

Our solution relies on multiple small applications working together to achieve a larger, more complex goal. Such an architecture is referred to as a microservice architecture. Each of these services only serve one purpose, such as retrieving data, triggering alerts, etc. [6]. This allows us to separate the data processing into multiple small problems, that are connected to each other, in order to reliably process large amounts of data with minimum complexity. In accordance to the issues defined in Sect. 1 we first give a short introduction to the data in question, followed by the presentation of the microservice definitions, each solving the stated problem definitions.

2.1 Dataset

We use data extracted from multiple, different vehicle types and brands. Each vehicle has been fitted with a custom edge device that collects data from the vehicle itself, as well as generates information based on its own sensors. The data of the vehicle is accessed using the OnBoard Diagnostic Two (OBD-II) interface. Table 1 shows a small slice of the raw data. The id is the unix timestamp when the data has been generated, date is the instant the data has been retrieved, the Vehicle Identification Number (VIN) uniquely identifies the vehicle that provided the data, lat and long are the GPS coordinates of the vehicle and speed describes the current vehicle speed as measured by the vehicle data-bus. The dataset is neither cleaned nor complete, i.e., it contains null, invalid and false values. Further data values have been omitted in Table 1 due to readability. Further interesting entries also include, but are not limited to:

Engine_rpm represents the current revolutions per minute (RPM) of the combustion engine. These numeric values may contain null values or, depending on the vehicle model, also may contain *INT.MAX*.

Acceleration describes the accelerometer values of the on-board edge device. It is a triplet containing X, Y and Z acceleration from -1 to 1, but may not be correctly normalized depending on different vehicles. Each edge device can be fitted differently to each vehicle resulting in a possibly invalid configuration for the edge device and thus, drifting accelerometer values.

Engine_load is a percentage value how utilized the engine currently is, depending on fuel injection and air intake. This value is calculated on the edge device.

Clutch is a flag on the vehicle data-bus, which is true whenever the clutch is engaged. It is highly fickle and, depending on the speed of the gear shifts, might not be set. Some vehicle models also don't provide this information, causing this flag to be unused.

Table 1. Excerpt of the collected dataset. VIN and GPS coordinates have been anonymized due to privacy concerns.

Id	Date	VIN	Lat	Long	Speed
1617509885	2021-04-04 06:16:04	5N1BV28U16N181386	48.390546	−36.041071	32.33
1617509886	2021-04-04 06:16:05	5N1BV28U16N181386	48.390546	−36.041071	31.95
1617509795	2021-04-04 06:15:35	3VWTG29M11M070886	25.138978	−35.948990	78.01
1617509826	2021-04-04 06:17:05	5N1BV28U16N181386	48.390546	−27.041071	0

The order of the dataset is defined as *First Come First Serve*. Whenever a new sample has been retrieved, it's directly added to the dataset. When multiple vehicles are active at the same time, samples with the same id but different date and VIN might be generated. The id timestamp corresponds to the creation of the time sample on the edge device, while the date corresponds to the time the sample has been received by the system.

2.2 Infrastructure

We define infrastructure as all software, that is used by our microservices. This includes databases, message oriented middleware, external REST endpoints, etc. Within our system, everything relies on message queues for communication. Explicitly, we use RabbitMQ[1] as message oriented middleware for any communication between the services. This allows us to completely decouple the inputs and outputs of the microservices, which further enables us to independently scale the existing services or introduce new ones.

Per default, no service connects to a database. We aim to keep all services as stateless as possible, otherwise scaling and throughput of the system as a whole might be compromised. Instead, we define specific services that only focus on persistence. In Fig. 1 these services are called *Data Storage* and in Fig. 2 they

[1] https://www.rabbitmq.com/.

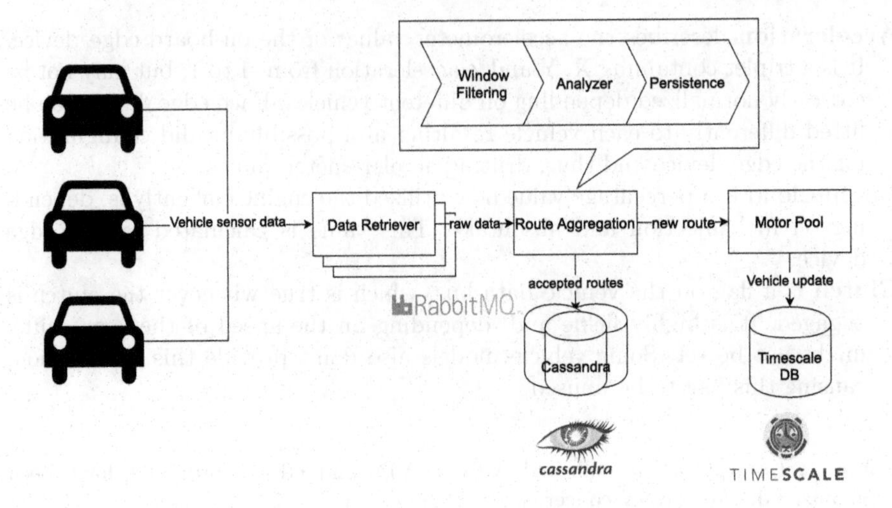

Fig. 2. Detailed overview of the system architecture.

are part of the defined *Data Lakes.* We use an *Apache Cassandra*[2] cluster for the aggregated data and a *TimescaleDB*[3] for the generated timeseries data.

2.3 RouteAggregator

The RouteAggregator accepts raw data from services such as the DataRetriever (see Subsect. 2.4) and transforms it into an aggregated sub-dataset. We call this sub-dataset a *route*, as it is defined as a coherent series of data that describes one trip of a vehicle. In one route, there is a set of ordered samples, that have been filtered and normalized. This means each route is complete and accurate. We achieve this by obtaining one sample, comparing assigning it to an active route, using a sliding window filter with the previous route data samples, and finally determining if the samples quality is sufficient. A sample is deemed sufficient, if the time difference between this and the previous sample is under a certain, externally defined threshold, contains all relevant data, such as GPS, vehicle speed, acceleration, etc., and is coherent, i.e. no sudden jumps in specific data values (going from 50 km/h to 130 km/h within a split-second). We detect the sufficiency of the last sample by applying rule based analyzers on the whole route, including the new sample. Each analyzer checks if their rule applies and thus accepts or rejects the new sample. These analyzers include, but are not limited to:

AccelerationAnalyzer checks each vale of a sample with the previous samples of a route for realistic changes in their value, for example a vehicles GPS coordinates must not change more than 0.0001 per sample. The allowed difference has to be defined per dataset column.

[2] https://cassandra.apache.org/.
[3] https://www.timescale.com/.

DataPresentAnalyzer only ensures the data for the route is present or has been interpolated.

CorrelationAnalyzer determines if the correlation of two values of the new sample is acceptable. Such correlations include the difference in GPS coordinates and vehicle speed, engine RPM and vehicle speed, etc. The values to verify need to be configured externally.

The analyzers allow us to enforce a certain minimum/maximum data frequency. If any of these analyzer rejects a sample, it is not added to the route. It is instead logged with the reason of rejection. If all analyzers accept the sample and the route is deemed complete it is sent to persistence services, allowing other services to use the new route for analysis. A route is complete, when no new sample is processed within a predefined time period, e.g. 30 min.

2.4 DataRetriever

An instance of this service retrieves the raw data of a set of vehicles and transforms it into our defined dataset format, as seen in Table 1. The received data contains both the information of the vehicle data-bus as well as the edge device. We can both process individual data samples and groups of data samples. The former is the default behaviour, the latter is only designed to post process information when no connection to the vehicle is possible, which can happen in remote areas or tunnels. In this subsection we will mainly focus on the default behaviour.

Each accepted data sample from a vehicle undergoes the same fundamental data processing, which is configured on a vehicle model basis. First, we check if all necessary identifiers are present and if missing ones can be assumed. For example if we miss the VIN, but we receive the edge device serial number we can assume the VIN based on previous samples, given the edge device serial number matches the other VIN entries. Second, we ensure all present values adhere to the predefined data types and ranges and convert or prune if possible. If these transformations are not possible the sample is discarded. Third and lastly, we annotate the applied changes to the sample, so other services can differentiate between original or assumed values. As the computational load of this service is assumed to be quite high, due to the possible high load of generated data, no data is persisted in this step, but only shared along in the processing pipeline using Rabbit MQ.

2.5 Motorpool

The Motorpool is a service definition that builds on the generated routes of the RouteAggregator, as defined in Subsect. 2.3. It manages a view on the routes based on the different vehicles and their driven routes in relation to the changes over time. Each new route is added to the vehicle view, where the vehicle itself is updated. We track statistical information of the vehicle, their changes over time, and all routes that have been generated by that vehicle. This information is saved in a TimescaleDB.

Each vehicle is identified by its VIN. This VIN consists of the World Manufacturer Index (WMI) the vehicle descriptor, a check digit and a vehicle identification. Using the WMI and the vehicle descriptor it is possible to identify similar vehicle models and configurations. Small deviations in the vehicle descriptor indicate deviations between the same vehicle model, such as engine type, or gearbox [7]. We use this information to identify similar and comparable vehicles.

A vehicle is defined as their VIN, the sum of their driven routes, their statistical values and a given display name. A route is, when added to a vehicle, annotated with a surrogate key, to allow fast and safe detection of its corresponding vehicle. Using the VIN and the data sample id (see Table 1) and this surrogate key, it is possible to assign each singular sample to a vehicle, and furthermore, to a route.

Applying descriptive statistics allows us to define a basic profile for a vehicle. We calculate these statistics over all driven routes of the vehicle and weight them relative to their recentness. Strong deviations of new routes can indicate changes of driving behaviours, environments or possible maintenance requirements [8].

3 Evaluation

We evaluate our design based on four different vehicles, each of different make and model, that provided 35.617 individual data samples. The data has been aggregated into 28 individual routes, which have been correctly assigned to the corresponding vehicles.

In Table 2 we show the resulting descriptive statistics on vehicle basis. Vehicle four has a relative high standard deviation in both vehicle speed as well as engine rpm, which may indicate rapid acceleration and deceleration of the vehicle, compared to vehicle three, which seems to be driven more moderately. Furthermore, vehicle four seems to have the longest idle time compared to all other vehicles, as the standard deviation of the engine rpm is much higher. We could identify vehicle two as an electric vehicle, as no data sample ever contained engine rpm, or fuel level values. Vehicle one is only used inside cities, which reflects the max vehicle speed values and low mean vehicles speed.

4 Discussion

A major issue we discovered while developing this architecture was the fickle data definition per vehicle model. We get vehicle specific information from the Onboard Diagnostic (OBD) interface, which has (limited) access to the CAN data-bus. Although standardized, the CAN Bus is highly manufacturer dependent, resulting in different data schemas for each vehicle. Whenever a new vehicle model is introduced a new DataRetriever and corresponding config has to be defined.

Table 2. descriptive statistic excerpt of four different vehicle models and makes. The second vehicle is an electric vehicle, hence the missing values in engine rpm.

Vehicle		Vehicle speed	Engine rpm
1	Count	9328	9328
	Mean	10.699	1072.197
	Std	20	599.8
	Min	0	0
	Max	77.36	3032
2	Count	15505	15505
	Mean	12.94	0
	Std	20	0
	Min	0	0
	Max	112.8	
3	Count	10312	10312
	Mean	11.51	1137.32
	Std	16.66	589.39
	Min	0	0
	Max	94.966	4162
4	Count	472	472
	Mean	25.76	949.93
	Std	31.3	1022.15
	Min	0	0
	Max	100.966	3426

Furthermore, many fleet providers started moving their focus from combustion engine vehicles to electric vehicles. This domain shift is difficult, as the data schema is massively different to the traditional combustion engine vehicles. Instead of tracking the engine RPM, fuel levels and gearbox information we need to cover information on the high voltage battery, charging and discharging of said battery, etc. These vehicles can be introduced into our architecture by defining additional DataRetrievers and Analyzers in the RouteAggregator, but in order to properly allow analysis and management in the Motorpool fundamental changes in the vehicle and route definition are necessary.

We tested our solution on one physical machine, where all microservices as well as the infrastructure run in parallel. We identified the bottleneck in the persistence of the routes into the Cassandra cluster. An evaluation on runtime and computational load on a proper distributed environment is still pending.

For evaluation we provide a Grafana dashboard. It visualises the collected information based on vehicles as well as routes in realtime. Figure 3 shows a screenshot of the vehicle visualisation

Fig. 3. Grafana dashboard giving information on one vehicle based on all its routes

5 Outlook

As electric vehicles are slowly becoming the focus for fleet providers, further research should include adaptions of the data preparation and management for electric vehicles. This may include battery management and recharge scheduling.

Independent of vehicle propulsion type, enhancements in data quality analysis or time series analysis can be integrated fairly easily, as the logic can be implemented as its own microservice. It can then receive data using the message queues.

Furthermore, the data prepared by our architecture can be used for a multitude of applications, such as predictive maintenance, driving style analysis, etc.

Acknowledgements. This research was funded by the Austrian Research Promotion Agency (FFG) and the implementation of the presented framework is part of a research project with nexopt (https://www.nexopt.com/) in Austria.

The dissemination of the research reported in this paper has been funded by the Federal Ministry for Climate Action, Environment, Energy, Mobility, Innovation and Technology (BMK), the Federal Ministry for Digital and Economic Affairs (BMDW), and the Province of Upper Austria in the frame of the COMET-Competence Centers for Excellent Technologies Programme and the COMET Module S3AI managed by Austrian Research Promotion Agency FFG.

References

1. Teh, H.Y., Kempa-Liehr, A.W., Wang, K.I.-K.: Sensor data quality: a systematic review. J. Big Data **7**(1), 1–49 (2020). https://doi.org/10.1186/s40537-020-0285-1

2. Hounsell, N.B., Shrestha, B.P., Wong, A.: Data management and applications in a world-leading bus fleet. Transp. Res. Part C Emerg. Technol. **22**, 76–87 (2012)
3. Falco, M., Núñez, I., Tanzi, F.: Improving the fleet monitoring management, through a software platform with IoT. In: IEEE International Conference on Internet of Things and Intelligence System (IoTaIS), vol. 2019, pp. 238–243 (2019). https://doi.org/10.1109/IoTaIS47347.2019.8980429
4. Killeen, P., Ding, B., Kiringa, I., Yeap, T.: IoT-based predictive maintenance for fleet management. Procedia Comput. Sci. **151**, 607–613 (2019)
5. Wittmann, M., et al.: A holistic framework for acquisition, processing and evaluation of vehicle fleet test data. In: 2017 IEEE 20th International Conference on Intelligent Transportation Systems (ITSC), pp. 1–7 (2017). https://doi.org/10.1109/ITSC.2017.8317637
6. Alshuqayran, N., Ali, N., Evans, R.: A systematic mapping study in microservice architecture. In: 2016 IEEE 9th International Conference on Service-Oriented Computing and Applications (SOCA), pp. 44–51 (2016)
7. Dion, F., Rakha, H.: Estimating spatial travel times using automatic vehicle identification data (2001)
8. Mobley, R.K.: An Introduction to Predictive Maintenance. Elsevier, Amsterdam (2002)

Learning Entropy: On Shannon vs. Machine-Learning-Based Information in Time Series

Ivo Bukovsky(✉) and Ondrej Budik

Department of Computer Science, Faculty of Science, University of South Bohemia in České Budějovice, České Budějovice, Czech Republic
ibuk@prf.jcu.cz

Abstract. The paper discusses the *L*earning-based information (***L***) and *L*earning *E*ntropy (***LE***) in contrast to classical Shannon probabilistic *I*nformation (***I***) and probabilistic entropy (***H***). It is shown that ***L*** corresponds to the recently introduced *A*pproximate *I*ndividual *S*ample-point *L*earning *E*ntropy (***AISLE***). For data series, then, the *LE* should be defined as the mean value of *L* that is finally in proper accordance with Shannon's concept of entropy ***H***. The distinction of ***L*** against ***I*** is explained by the real-time anomaly detection of individual time series data points (states). First, the principal distinction of the information concept of ***Ivs.L*** is demonstrated in respect to data governing law that ***L*** considers explicitly (while ***I*** does not). Second, it is shown that ***L*** has the potential to be applied on much shorter datasets than ***I*** because of the learning system being pre-trained and being able to generalize from a smaller dataset. Then, floating window trajectories of the covariance matrix norm, the trajectory of approximate variance fractal dimension, and especially the windowed Shannon Entropy trajectory are compared to ***LE*** on multichannel EEG featuring epileptic seizure. The results on real time series show that ***L***, i.e., ***AISLE***, can be a useful counterpart to Shannon entropy allowing us also for more detailed search of anomaly onsets (change points).

Keywords: Novelty detection · Learning Entropy · Adaptive filter · Least mean square · Multichannel EEG

1 Introduction

The current information theory apparatus is mainly based on the probabilistic Shannon's entropy [1], which is also fundamental in the theory and applications of anomaly detection (novelty detection, change-point detection). These probabilistic information measures, i.e. the Shannon Information *I* and Shannon Entropy *H* are based on quantifying statistical uniqueness of data in the existing dataset. Thus, this estimated information quantity does not explicitly consider the deterministic governing law that can be learned from data by learning systems from smaller datasets and where generalization of a learning system can play its significant role.

G. Kotsis et al. (Eds.): DEXA 2022 Workshops, CCIS 1633, pp. 402–415, 2022.
https://doi.org/10.1007/978-3-031-14343-4_38

Computational learning systems, such as neural networks, can approximate at least the current dynamics of data behavior. Novelty detection in dynamical systems can be approached either by probabilistic approaches, e.g. [2], or by using learning systems, e.g., [3]. As the signal complexity evaluation measures can be recalled the Sample Entropy and Approximate Entropy [4, 5] and fractal measures [6–9] that are based on the concept of multiscale power-law evaluation [10], and these may be seen as some alternatives to the probabilistic concept for novelty detection techniques or purely machine-learning-based ones. Additionally, compensated transfer entropy [14] is another probabilistic technique to assess entropy using conditional mutual information between present and past states. The probabilistic entropy approach for fault detection was published in [15] and the probabilistic technique for sensor data concept drift (also concept shift) appeared in [16].

Among the probabilistic novelty approaches, the currently popular concepts are generalized entropies, especially the extensively studied Tsallis and Rényi entropies and their potentials, e.g., [17] and references therein. The example of their application, e.g., to probabilistic anomaly detection in cybersecurity, can be found in [18].

Another direction of non-probabilistic novelty measures is based on learning systems that employ machine learning excluding probabilistic and Bayesian computations, and *LE* belongs to this research direction. A special deep learning related change point detection technique was presented in [19].

In order to be in accordance to terms Information and Entropy introduced by Shannon, we show in this paper that:

i. the Approximate Individual Sample-point Learning Entropy (*AISLE* [6]), corresponds to the proposed Learning Information L [7] of the data point, i.e. *AISLE*, which is here newly linked as the counterpart to the Shannon Information I of individual data points, and that
ii. the Learning Entropy *LE* for a data point interval is the mean value of L on the data interval, which is here newly linked as the counterpart to Shannon Entropy H,

and further we demonstrate the L on the multichannel time series that is also application extension of *AISLE* in contrast to our previous works.

2 Probabilistic vs. Learning-Based Information

At this point, a brief discourse about quantifying information, in the sense of novelty, the data display to their observer, is drawn as follows. The probabilistic (Shannon based) approaches of information quantification do not explicitly reflect the governing law of data and rely on statistical uniqueness or recurrences. The learning-system-based methods of novelty detection implicitly utilize the governing laws they found in the data, but their novelty quantification of data points is not considered as a potential information concept that might be an alternative to the probabilistic concept of information.

Classically, the Shannon information content I of data sample point $\mathbf{x}$ (k) at sample time k is defined via its reciprocal probability $p(\cdot)$ as follows

$$I(k) = log_2\left(\frac{1}{p(\mathbf{x}(k))}\right), \quad (1)$$

and the Shannon Entropy H of the time series would be the mean value of I over the data.

Although I in (1) is fundamental and well known, we should recall its following aspect, which might not be so obvious and that is important for clarifying the Learning-based information concept L and the Learning Entropy H. It is recalled as follows.

For the given example of time series in Fig. 1, the data sample represents a 2-dimensional data point (state) as follows.

$$\mathbf{x}(k) = \left[y(k-1), y(k)\right], \quad (2)$$

where k is the discrete time index. The natural property of Shannon Information I is its time invariance, because the data points that belong to the same i-th histogram bin is of course independent of time k, so it is usual to use notation as follows

$$\text{if for any } k : \mathbf{x}(k) = \mathbf{x}_i \Rightarrow p(\mathbf{x}(k)) = p(\mathbf{x}_i) = p_i, \quad (3)$$

where i represents the histogram bin index, where similar data points at different times k belongs to. This just recalls that Shannon information of individual data samples is time invariant for equal and very similar same data samples. Therefore, the Shannon Entropy H in a probability space, i.e., via a histogram, depends only on histogram implementation, and it is classically the mean value of information over the dataset as

$$H = \sum_{i=1}^{n} p_i \cdot log_2(I_i), \quad (4)$$

where I_i is the information of data samples belonging to i-th histogram bin, and n is the number of all histograms bins (simplified notation).

The need for machine-learning-based Information concept as the counterpart to the existing Shannon probabilistic concept of information quantity is demonstrated via Fig. 1, Fig. 2, and Fig. 3, where the probabilistic information (second axes from top) becomes high even for deterministically generated data (with low noise) during time $k < 200$. On the contrary, the third axes of the top show magnitudes of increments in real-time learning weights increments $\Delta\mathbf{w}$ (2nd-order HONU [8] as a polynomial class of IPLNAs [9]) and the learning behavior clearly indicates novelty in data samples due to the change in the governing law at $k = 200$.

The concept of Learning Entropy, in particular the Individual Sample Learning Entropy, was originally proposed [6] as a multiscale measure that was aggregated over various detection sensitivities for all weight update magnitudes at their original scales. Here, we propose a formula for AISLE estimation, which is currently discussed in more detail in the work [10]; however, here it is extended to multichannel data as follows

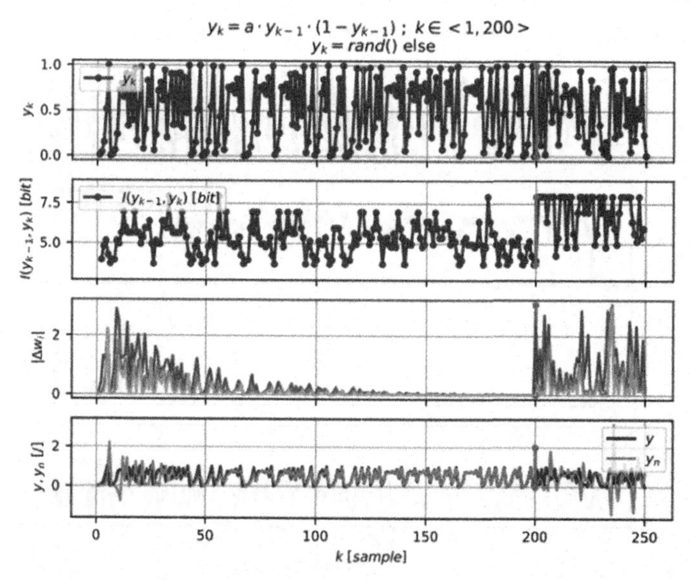

Fig. 1. (top axes) Time series with deterministic part ($k < 200$) and random part ($k \geq 200$), (second top axes) Shannon probabilistic information of 2-D datapoints I(y(k − 1), y(k)), (third axes) the IPLNA's learning weight-update magnitudes $|\Delta \mathbf{w}|$ indicate the anomalous behavior at $k = 200$, bottom axes show the time series data y (blue) and the neural output y_n (green) during learning. (Color figure online)

$$E_A(k) = \frac{1}{n \cdot n_{w_i} \cdot n_\beta} \sum_{i=1}^{n} \sum_{\forall w_i} \sum_{\forall \beta} \{z(|\Delta \mathbf{w}_i(k)|) > \beta\} \;\; E_A \in \langle 0, 1 \rangle \tag{5}$$

where the inequality in (5) results in n_{w_i}-sample long set of zeros or units according to the true or false state of the conditions for i^{th} channel, β is a vector of detection sensitivities to overcome the otherwise single-scale issue of detection sensitivity.

If we drop the lower indexing of weight position and of channel indexing for simplicity, then the z-score for each single weight of each adaptive filter and for each channel writes as follows

$$z(|\Delta w(k)|) = \frac{|\Delta w(k)| - \overline{|\Delta \mathbf{w}^M(k)|}}{\sigma(|\Delta \mathbf{w}^M(k)|)}, \tag{6}$$

where the floating window of M values of recent magnitudes of learning increments is defined as follows

$$\left|\Delta \mathbf{w}^M(k)\right| = \begin{bmatrix} |\Delta w(k-1)| \\ |\Delta w(k-2)| \\ \vdots \\ |\Delta w(k-M)| \end{bmatrix}. \tag{7}$$

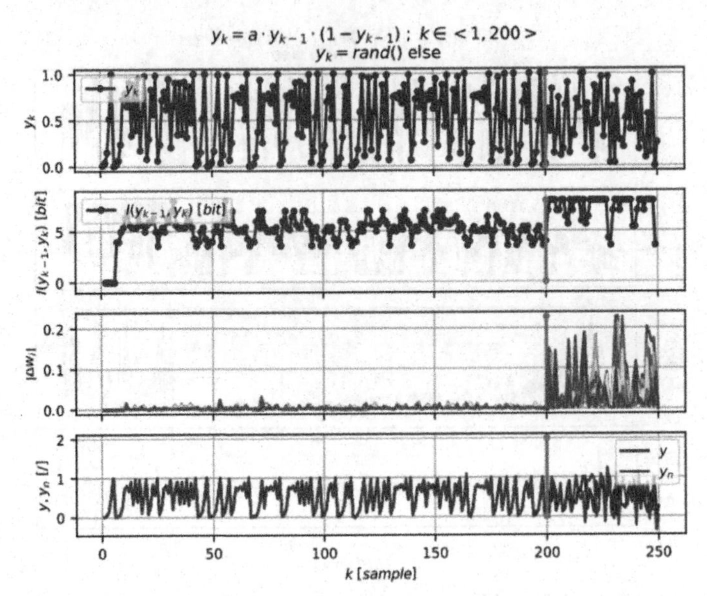

Fig. 2. (top axes) Time series with deterministic part ($k < 200$) and random part ($k \geq 200$), (second top axes) Shannon probabilistic information of 2-D datapoints $I(y(k-1), y(k))$, (third axes) the IPLNA's learning weight-update magnitudes $|\Delta \mathbf{w}|$ indicate the anomalous behavior at $k = 200$, bottom axes show the time series data y (blue) and the neural output y_n (green) during learning. (Color figure online)

It should be emphasized that the vector of detection sensitivities β has now clearly defined statistical meaning in this newly proposed *AISLE* computation via (5)–(7) contrary to the similarly proposed multiple parameter α in the original *LE* estimation algorithm [6]. Due to the of newly proposed z-scoring for the learning increment magnitudes in (6), the values of β are directly related both to the mean and standard deviation of learning increments $\Delta \mathbf{w}_i$ as follows:

- for $\beta = 0$, the weight-update magnitudes larger than their recent mean are summed in (5), i. e., the detection of unusual learning effort is very sensitive,
- for $\beta = 1$, only the weight-update magnitudes larger than their recent mean plus one standard deviation are summed in (5), i. e., the detection of unusual learning effort is less sensitive,
- for $\beta = 2$, only the weight-update magnitudes larger than their recent mean plus two standard deviations are summed in (5), i. e., the detection of unusual learning effort is even less sensitive,
- and similarly so on, so the detection of unusually large learning effort is the less sensitive, the larger parameter β while β shall not necessarily be a discrete number.

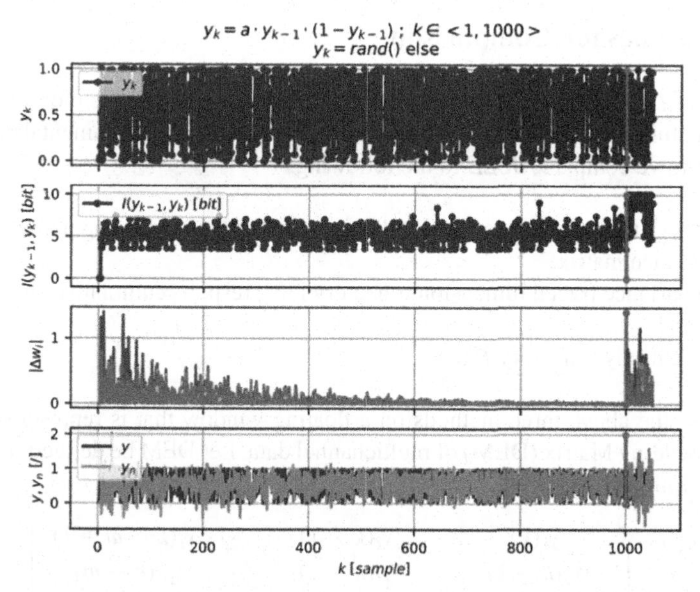

Fig. 3. (top axes) Time series with deterministic part ($k < 1000$) and random part ($k \geq 1000$), (second top axes) Shannon probabilistic information of 2-D datapoints $I(y(k-1), y(k))$, (third axes) the IPLNA's learning weight-update magnitudes $|\Delta \mathbf{w}|$ indicate the anomalous behavior at $k = 1000$, bottom axes show the time series data y (blue) and the neural output y_n (green) during learning. (Color figure online)

Similarly to the originally proposed *AISLE* [6], also this new algorithm for estimation of individual sample learning entropy via (5)–(7) is limited to $\langle 0, 1 \rangle$ meaning that:

- $E_A = 0$ if none of the learning increment magnitudes is unusually large for all β,
- $E_A = 1$ if all of the learning increment magnitudes are unusually large for all β,

3 LE for Multichannel Data

Novelty detection can be based on probabilistic principles [2] or on learning systems [3]. Adaptive filter techniques [11] are an option before more complicated learning systems would also be used for multichannel data, where EEG data signal processing is a challenging topic [12]. We can also employ the linear adaptive filters as learning systems, where the learning rule for each channel is the pure gradient descent (LMS) sample by sample adaptation. When this learning system behavior is monitored as in (5), E_A is aggregated also over all signal channels. In the experimental analysis below, we also evaluate the higher orders of *LE*. The concept of orders of *LE*, i.e. *AISLE*, was proposed in work [6] and it is defined as the order of weight difference that is used for *LE* evaluation. Therefore, the above introduced *AISLE* is the first order *LE*, i.e. $r = 1$, as (5)–(7) are calculated with the first difference of weights.

4 Approaches for Comparison

We compare *LE* (*AISLE*) performance on multichannel data to three more signal processing (floating-window based) methods that are based on more fundamental principles. In particular, we compare *AISLE* to the following:

- $\|\mathbf{covX}^{\Delta}(\mathrm{k})\|$ the trajectory of the Euclidean norm of the upper diagonal elements of the covariance matrix,
- $VFD(k)$ variance fractal dimension trajectory [13], resp. its cumulative sum approximation, and
- Shannon entropy trajectory $H(k)$.

We base the above three methods on a floating window that is represented by the Delay Embedding Matrix (DEM) of multichannel data. Let DEM be defined for a single channel as follows

$$\mathbf{X}_i = \begin{bmatrix} y_i(k) & y_i(k-1) & \cdots & y_i(k-m+1) \\ y_i(k-1) & y_i(k-2) & \cdots & y_i(k-m) \\ \vdots & \vdots & \cdots & \vdots \\ y_i(k-M+m) & y_i(k-M+m-1) & \cdots & y_i(k-M+1) \end{bmatrix}, \tag{8}$$

where m is the embedding. Then, the full multichannel DEM would be as follows

$$\mathbf{X} = [\mathbf{X}_1\ \mathbf{X}_2\ \ldots\ \mathbf{X}_i\ \ldots\ \mathbf{X}_n]. \tag{9}$$

For fundamental comparison of *LE* performance to other methods, we calculate the following floating-window-based measures, i.e., DEM-based measures, as shown below. First, the Euclidean norm trajectory of the upper diagonal values of the covariance matrix is calculated by using the covariance matrix of DEM (9), i.e.

$$\mathbf{covX} = \mathbf{X}^T \cdot \mathbf{X}, \tag{10}$$

where each column of X was z-scored as follows

$$\mathbf{X}_i \leftarrow \frac{\mathbf{X}_i - \overline{\mathbf{X}_i}}{\sigma(\mathbf{X}_i)};\ \forall i, \tag{11}$$

so the upper triangular norm (trajectory) is calculated at every sample time as follows

$$VFDT(k) = \frac{1}{n_\alpha} \sum_{i=1}^{n_\alpha} \sigma^2(\mathbf{X}(k, \alpha_i)), \tag{12}$$

Second, an approximation of variance fractal measure, i.e., the trajectory of cumulative approximation of Variance Fractal Dimension (VFDT) was calculated as follows

$$VFDT(k) = \frac{1}{n_\alpha} \sum_{i=1}^{n_\alpha} \sigma^2(\mathbf{X}(k, \alpha_i)), \tag{13}$$

where $\mathbf{X}(k, \alpha)$ represents vertically re-sampled DEM (9) with re-samplings $\alpha_{\mathrm{i}} \in \alpha = \left[\alpha_1 \alpha_2 \ldots \alpha_{\mathrm{n}_\alpha}\right]$ (for more details on the principle of the approximate concept of characterizing exponents of fractal dimension via a cumulative sum, please see works [6, 14, 15]). Third, the trajectory of Shannon Entropy was calculated as

$$H(k) = -\sum_{\forall i,j} p_{i,j} \cdot log_2(p_{i,j}), \tag{14}$$

where $p_{i,j}$ is the corresponding histogram bin probability of $X_{i,j}$ value, i.e., of the value at i, j position in DEM matrix.

5 Experimental Analysis

The above methods were tested on two data sets. The first dataset is artificial multichannel data generated as

$$\mathbf{Y} = \begin{bmatrix} \mathbf{y}_1 = sin(2\pi \cdot f_1 \cdot k + \Phi_1) + b_1 \\ \vdots \\ \mathbf{y}_{20} = sin(2\pi \cdot f_{20} \cdot k + \Phi_{20}) + b_{20} \end{bmatrix}; \quad k = 0, 1, \ldots, 1000, \tag{15}$$

where the individual signal parameters f_i, Φ_i and b_i were set as random constants. To demonstrate the functionality of the detection methods, all signals in each channel in (15) are simultaneously replaced by pure noise at intervals $k = 400 \ldots 500$ and $k = 700 \ldots 800$. The amplitudes of the sinus signals and the noisy intervals were aligned for similar extrema.

The second dataset is anonymous multichannel EEG recordings with known epileptic seizure. The results are shown and discussed in 0-0 in Appendix. The results confirm the intriguing capability of *AISLE* to detect anomalies with individual samples of data.

6 Conclusions

The main theoretical contribution of this paper is linking the machine-learning-based concept of Information and Learning Entropy to its classical Shannon probabilistic information concepts.

Also, the newly proposed estimation of Learning Information implemented as Approximate Individual Sample Learning Entropy was extended and tested for novelty detection in multichannel EEG data with linear adaptive filters. The method showed potentials for the instant detection of epileptic seizures in multichannel EEG.

Acknowledgment. The research reported in this paper has been funded by the European Interreg Austria-Czech Republic project "PredMAIn (ATCZ279)".

Anonymous real EEG dataset courtesy of Department of Neurology, Faculty of Medicine in Hradec Kralove, Charles University, Czech Republic.

Nomenclature

AISLE	Approximate Individual Sample Learning Entropy
α	Vector of re-sampling setups for estimation of *VFD*
β	Vector of detection sensitivities for *LE*
H	Shannon Entropy
k	Discrete index of time
LE	Learning Entropy (*AISLE*)
n	Number of channels
n_{w_i}	Length of vector $\mathbf{w}_i$
r	Order of *LE*
$\sigma(.)$	Standard deviation
VFD	Variance Fractal Dimension
$\mathbf{w}_i$	Vector of all adaptive parameters (neural weights) of ith channel
x, x, X	Scalar, vector, matrix (multidim. array)
$y_i(k)$	Measured data sample of ith channel at time k
$\tilde{y}_i(k)$	Filter output (neural predictor output) of ith channel at time k

Appendix

(See Figs. 5, 6 and 7)

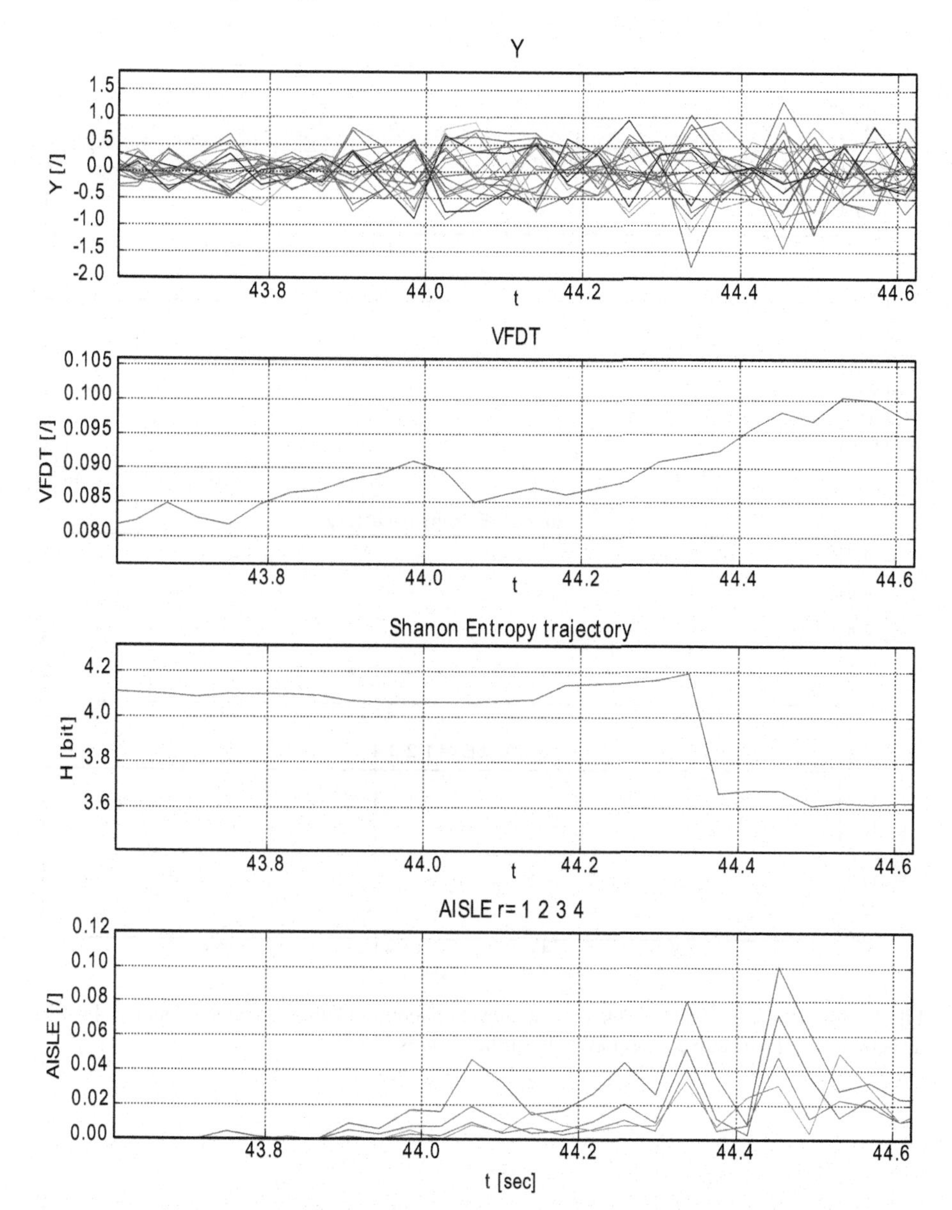

Fig. 4. The comparison of *AISLE* with the other proposed methods on artificial multichannel data (15) with noisy intervals at $k = 400 \ldots 500$ and $k = 700 \ldots 800$; $covDEMT = \left\| \mathbf{covX}^{\Delta} \right\|$ does not detect anything (due to various frequencies and phases of sinusoidal channels), *VFDT* and *H* gradually detects the changes in data, and *AISLE* instantly detects the increased learning effort when dynamical behavior is presented at $k > 400$ and $k > 700$.

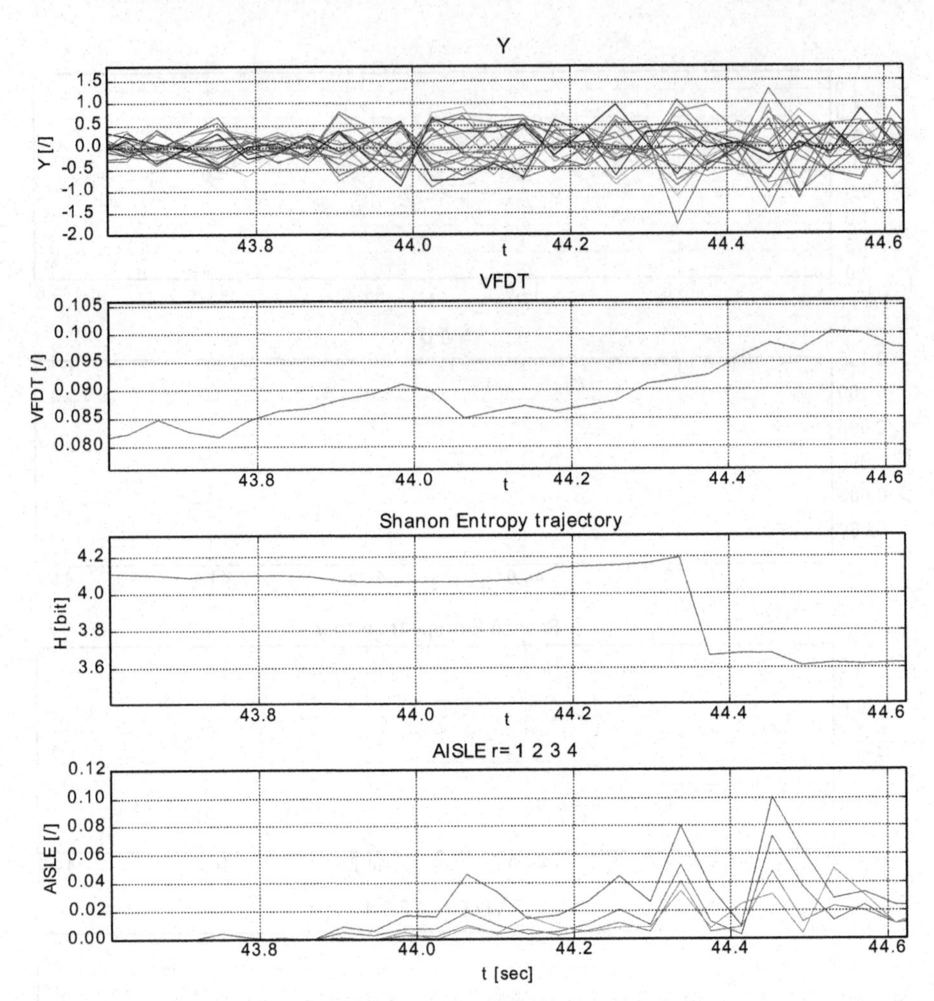

Fig. 5. The data **Y** are EEG channels with known epileptic seizure starting at about t = 44 s, its detection with the discussed methods is in further figures.

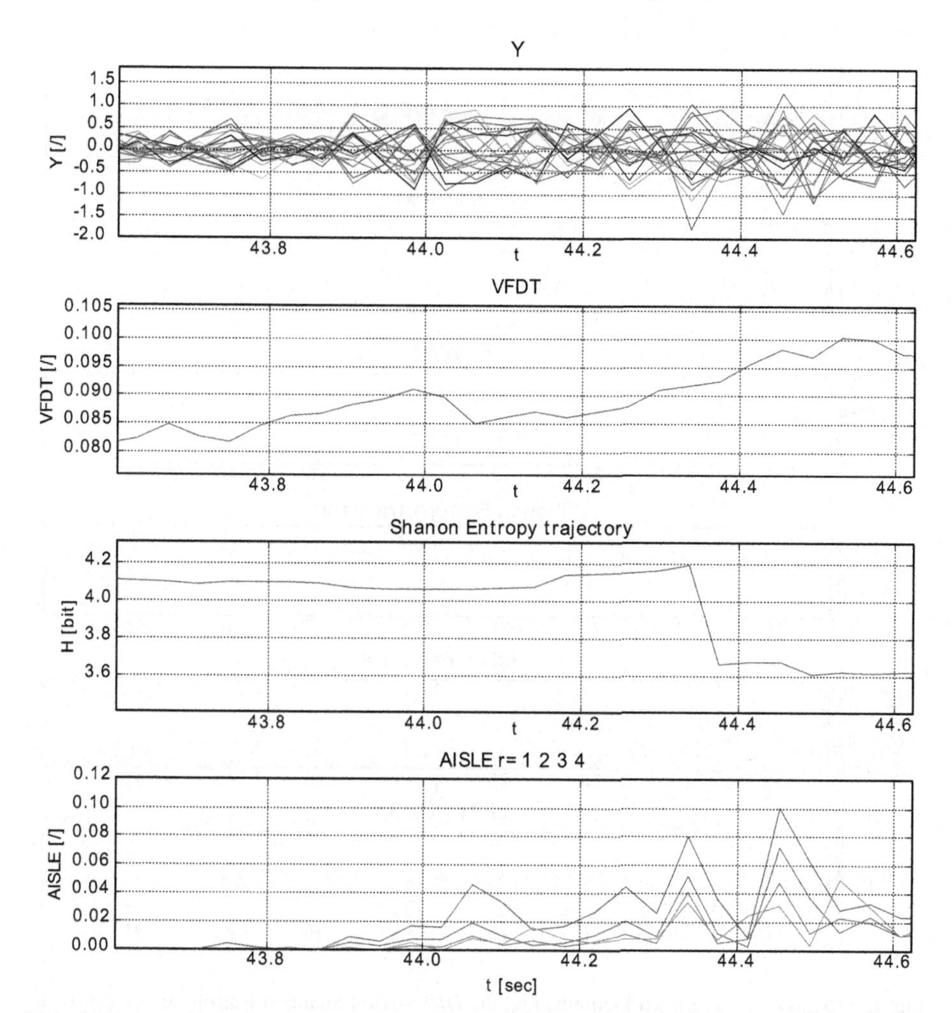

Fig. 6. The evaluation of multichannel EEG data (Fig. 2), where *covDEMT* = does not indicate the seizure onset at t = 44s, *VFDT* gradually increases, and both *H* and *AISLE* indicate the seizure onset rather instantly. The detail of how early *H* vs. *AISLE* detect the seizure is shown in Fig. 4.

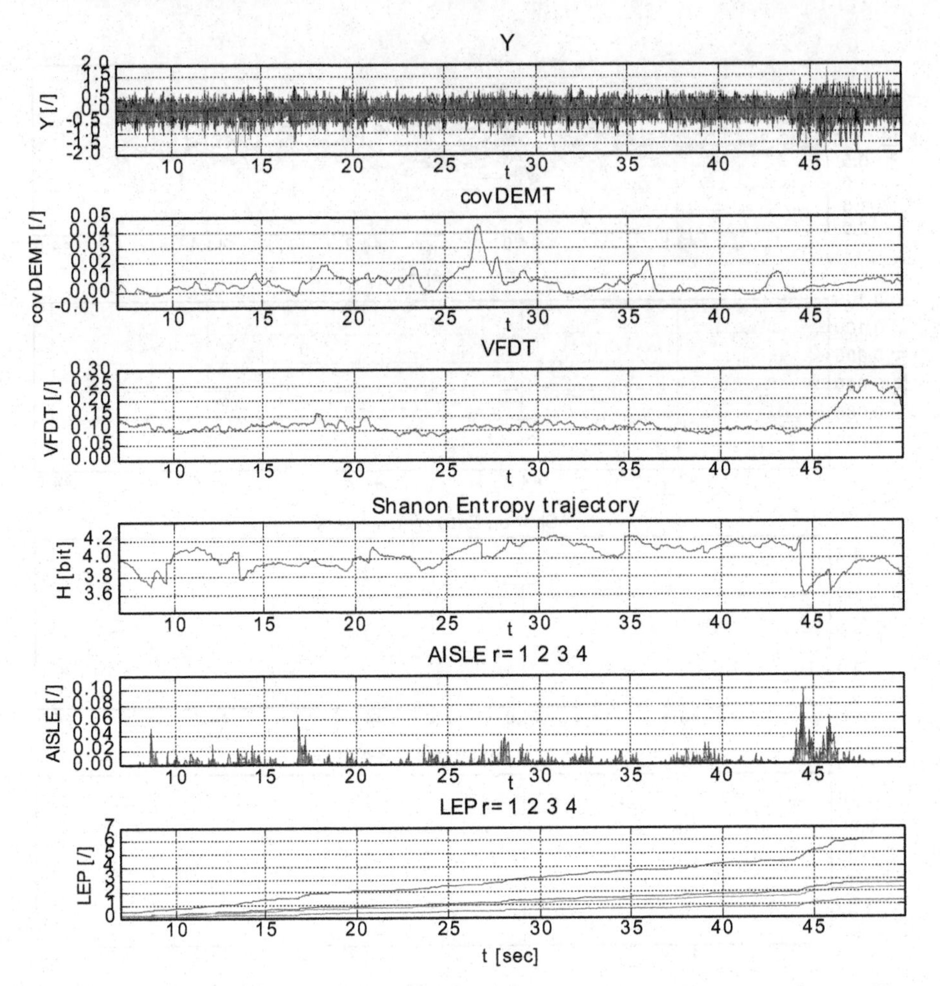

Fig. 7. (Detail of Fig. 3) Both methods, i.e., the *DEM*-based Shannon Entropy trajectory H as well as the Approximate Individual Sample Learning Entropy (*AISLE)* were found practically feasible for early seizure indication while the *AISLE* has potential of earlier detection (bottom axis) because it is in principle more sensitive to individual samples of data.

References

1. Shannon, C.E.: A mathematical theory of communication. Bell Syst. Tech. J. **27**, 379–423 (1948). https://doi.org/10.1002/j.1538-7305.1948.tb01338.x
2. Markou, M., Singh, S.: Novelty detection: a review—part 1: statistical approaches. Sig. Process. **83**, 2481–2497 (2003). https://doi.org/10.1016/j.sigpro.2003.07.018
3. Markou, M., Singh, S.: Novelty detection: a review—part 2: neural network based approaches. Sig. Process. **83**, 2499–2521 (2003). https://doi.org/10.1016/j.sigpro.2003.07.019
4. Pincus, S.M.: Approximate entropy as a measure of system complexity. Proc. Natl. Acad. Sci. U. S. A. **88**, 2297–2301 (1991)
5. Richman, J.S., Moorman, J.R.: Physiological time-series analysis using approximate entropy and sample entropy. Am. J. Physiol. Heart Circ. Physiol. **278**, H2039–2049 (2000)

6. Bukovsky, I.: Learning entropy: multiscale measure for incremental learning. Entropy **15**, 4159–4187 (2013). https://doi.org/10.3390/e15104159
7. Bukovsky, I., Kinsner, W., Homma, N.: Learning entropy as a learning-based information concept. Entropy **21**, 166 (2019). https://doi.org/10.3390/e21020166
8. Bukovsky, I., Homma, N.: An approach to stable gradient-descent adaptation of higher order neural units. IEEE Trans. Neural Netw. Learn. Syst. **28**, 2022–2034 (2017). https://doi.org/10.1109/TNNLS.2016.2572310
9. Bukovsky, I., Dohnal, G., Benes, P.M., Ichiji, K., Homma, N.: Letter on convergence of in-parameter-linear nonlinear neural architectures with gradient learnings. IEEE Trans. Neural Netw. Learn. Syst. **1–4**, 2016 (2021). https://doi.org/10.1109/TNNLS.2021.3123533
10. Bukovsky, I., Vrba, J., Cejnek, M.: Learning entropy: a direct approach. In: IEEE International Joint Conference on Neural Networks. IEEE, Vancouver (2016)
11. Mandic, D.P., Goh, V.S.L.: Complex Valued Nonlinear Adaptive Filters: Noncircularity, Widely Linear and Neural Models. Wiley (2009)
12. Sanei, S., Chambers, J.: EEG Signal Processing. Wiley, Chichester, England; Hoboken, NJ (2007)
13. Kinsner, W., Grieder, W.: Amplification of signal features using variance fractal dimension trajectory. In: 2009 8th IEEE International Conference on Cognitive Informatics, ICCI 2009, pp. 201–209 (2009). https://doi.org/10.1109/COGINF.2009.5250750
14. Bukovsky, I., Kinsner, W., Maly, V., Krehlik, K.: Multiscale Analysis of False Neighbors for state space reconstruction of complicated systems. In: 2011 IEEE Workshop on Merging Fields of Computational Intelligence and Sensor Technology (CompSens), pp. 65–72 (2011). https://doi.org/10.1109/MFCIST.2011.5949517
15. Bukovsky, I., Kinsner, W., Bila, J.: Multiscale analysis approach for novelty detection in adaptation plot. In: Sensor Signal Processing for Defence, SSPD 2012, pp. 1–6 (2012). https://doi.org/10.1049/ic.2012.0114
16. Vorburger, P., Bernstein, A.: Entropy-based concept shift detection. In: 2006 6th International Conference on Data Mining, ICDM 2006, pp. 1113–1118 (2006). https://doi.org/10.1109/ICDM.2006.66
17. Amigó, J., Balogh, S., Hernández, S.: A brief review of generalized entropies. Entropy **20**, 813 (2018). https://doi.org/10.3390/e20110813
18. Bereziński, P., Jasiul, B., Szpyrka, M.: An entropy-based network anomaly detection method. Entropy **17**, 2367–2408 (2015). https://doi.org/10.3390/e17042367
19. Mahmoud, S., Martinez-Gil, J., Praher, P., Freudenthaler, B., Girkinger, A.: Deep learning rule for efficient changepoint detection in the presence of non-linear trends. In: Kotsis, G., et al. (eds.) DEXA 2021. CCIS, vol. 1479, pp. 184–191. Springer, Cham (2021). https://doi.org/10.1007/978-3-030-87101-7_18

Using Property Graphs to Segment Time-Series Data

Aleksei Karetnikov[1(✉)], Tobias Rehberger[1], Christian Lettner[1], Johannes Himmelbauer[1], Ramin Nikzad-Langerodi[1], Günter Gsellmann[3], Susanne Nestelberger[3], and Stefan Schützeneder[2]

[1] Software Competence Center Hagenberg GmbH, Hagenberg im Mühlkreis, Austria
{aleksei.karetnikov,tobias.rehberger,christian.lettner, johannes.himmelbauer,ramin.nikzad-langerodi}@scch.at

[2] Borealis Polyolefine GmbH, Schwechat, Austria

[3] Borealis Polyolefine GmbH, Linz, Austria
{guenter.gsellmann,susanne.nestelberger, stefan.schuetzeneder}@borealisgroup.com

Abstract. Digitization of industrial processes requires an ever increasing amount of resources to store and process data. However, integration of the business process including expert knowledge and (real-time) process data remains a largely open challenge. Our study is a first step towards better integration of these aspects by means of knowledge graphs and machine learning. In particular we describe the framework that we use to operate with both: conceptual representation of the business process, and the sensor data measured in the process. Considering the existing limitations of graph data storage in processing large time-series data volumes, we suggest an approach that creates a bridge between a graph database, that models the processes as concepts, and a time-series database, that contains the sensor data. The main difficulty of this approach is the creation and maintenance of the vast number of links between these databases. We introduce the method of smart data segmentation that i) reduces the number of links between the databases, ii) minimizes data pre-processing overhead and iii) integrates graph and time-series databases efficiently.

Keywords: Property graph · Time-series data · Graph data · Process data · Data integration

1 Introduction

Nowadays more and more industries are using IoT solutions to control their business processes. They offer prominent possibilities in analysis, optimization and prediction of the production. Currently, the vast amounts of data that are obtained from the sensors are stored in data management systems. Some of the solutions can efficiently deal with graph data that helps to model the business process, others are able to store time-series data, but in many projects the bridge

G. Kotsis et al. (Eds.): DEXA 2022 Workshops, CCIS 1633, pp. 416–423, 2022.
https://doi.org/10.1007/978-3-031-14343-4_39

between these two parts is not a trivial one. The main idea of our paper is the introduction of an approach to integrate graph and time-series data. We propose a framework that is capable to solve a modelling problem of an industrial process and show its application on a business case. The paper is organized as following: in Sect. 2 we provide a short overview of existing solutions and their limitations. In Sect. 3 we present our approach. Then, in Sect. 4 we demonstrate a use case that exploits our approach. Finally, in Sect. 5 we give a discussion on the proposed idea and summarise it in Sect. 6.

2 Background Study

The ongoing development of graph storage solutions has made it possible to efficiently manage colossal amounts of data in graph databases. Nowadays, property graph storages are extensively used to model business processes. In [12] the authors use the Neo4j database to model a power transmission grid as a graph. While storing time-series data in a graph database is indeed viable, there are some issues concerning the data model and performance as well as the complexity and expressiveness of queries, which is pointed out in [4,15]. Therefore, utilizing a database designed for handling time-series data should be considered. In [16] the authors are using 3 Database Management Systems (Neo4j, Cassandra and InfluxDB) to build a hybrid storage for industrial IoT data. The attempt to join different types of DBMS was attempted by John Roijackers in [13] by introducing a virtual level between an SQL and a NoSQL system to reduce the complexity of maintaining multiple databases. The authors of [3] are introducing the idea of building a single SQL query engine for the NoSQL solutions but it leads to reduced performance of the latter. This concept of trying to abstract multiple data stores is not new, as is pointed out in [14], where this shift to enabling query processing across heterogeneous data models is analyzed. The ReGraph proposed by Cavotol & Santanchè in [1] is attempting to integrate a relational database and a graph storage, but the architecture of the approach makes limitations on dynamic updates of the graph data.

Considering all this previous experience in building of hybrid DB systems we could not find a working solution that provides such an orchestration of a graph and a time-series database. Considering our requirements:

- Dynamic updates
- Perform predictive modelling
- Possibility to connect the time-series data between the process steps
- Manage time-series segmentation

It creates a need of a solution that is able to deal with the process concepts, retrieving of the data related to them and the opportunities to consolidate the industrial behavioural patterns.

Our idea was to build a hybrid data storage system that is aimed to solve the particular business need. Our system is aimed to create a data integration adapter for Neo4j that allows to query the time-series data from InfluxDB with a main focus on the modelled business process. The querying system should be encapsulated into the adapter and user predictive should take place on the graph model.

3 Our Approach

In this section we will introduce the architecture of the proposed framework. In general, we offer a multi-layer approach (Fig. 1) to model the business case and store its data. First of all, the business process is modelled as a property graph. In our case study we use Neo4j as storage engine.

Then, by using a segmentation model we split the time-series data into little chunks according to the segmentation parameters, explored patterns and static expert knowledge (Fig. 3). A segmentation model is only applied, when specifically requested but not continuously when new sensor data is stored in the time-series database. The first time a specific segmentation is requested, its model parameters and metadata are found in the graph data storage. A machine learning (ML) model uses these inputs to perform the data segmentation on the time-series data. The architecture can be seen in Fig. 4. Because the execution of operations such as segmentation, aggregation and statistical exploration of the time-series data can be very time and resource intensive, the obtained results are also stored. This happens partially as metadata in the graph database (Fig. 5) and partially as timestamped data in the time-series database (Fig. 6). Obviously, the industrial processes are executed in a particular order that leads to correlation between the process steps. That information is also stored in the relations between the concepts in the graph database. They might represent a time-shift or a functional dependency (e.g. coefficient or function that determines the difference between two time-series).

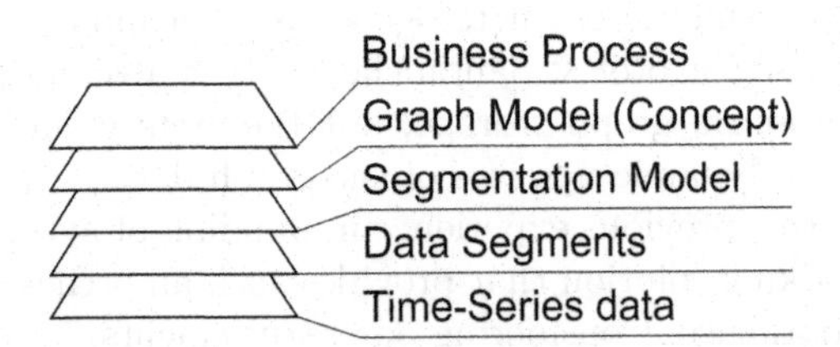

Fig. 1. Multi-layered architecture of the framework. The business process is represented as a graph concept that is linked to the process time-series data through the segmentation model and the data segments

The concepts and the segments' metadata stored in the graph database includes at least a tag and a specific starting and ending time point, which are used to map to data in the time-series storage. For instance, in Fig. 3 there is a chart of two time-series (solid and dotted lines). Firstly, the model identifies the segments where the values are corresponding to a particular rule (this could be a simple threshold or a more sophisticated segmentation approach) and

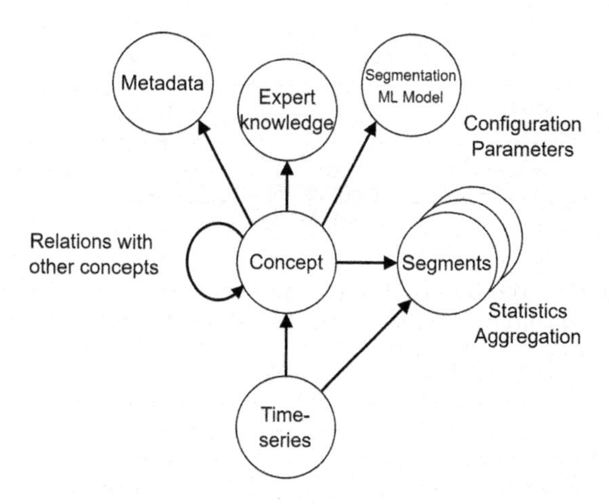

Fig. 2. Overview of the data structure. The concept (e.g. process step) is connected to the data segments that are created by applying ML models with the given parameters on the time-series data. It also includes pre-calculated segment statistics and aggregations, and the concepts' relation with other concepts.

specifies the starting and the ending point of this interval to create the first segment. Secondly, it is obvious that the second (dotted) time-series has a positive correlation with the first one (line), but it is executed with some time-delay. All that data, including those dependencies, starting and ending timestamps and the precalculated statistics and aggregations of the segments are saved in the property graph.

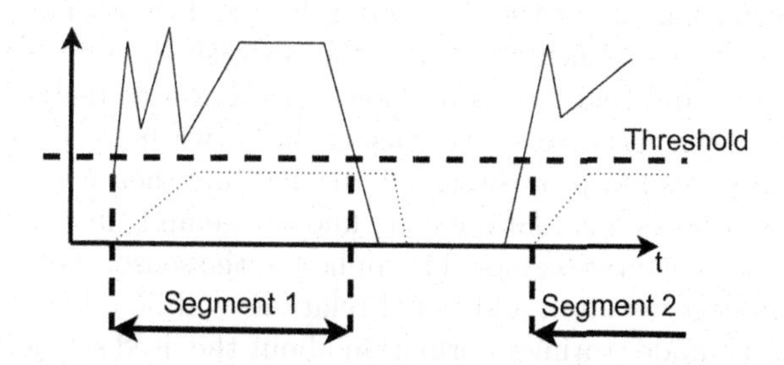

Fig. 3. Example of the time-series data segmentation. The time-interval that starts when the values are higher than the threshold and ends when they are lower is considered to be a time-series segment. Another time-series is shown as a dotted line. This time-series represents another process that is connected to the first one. So one can observe positive correlation between these time-series.

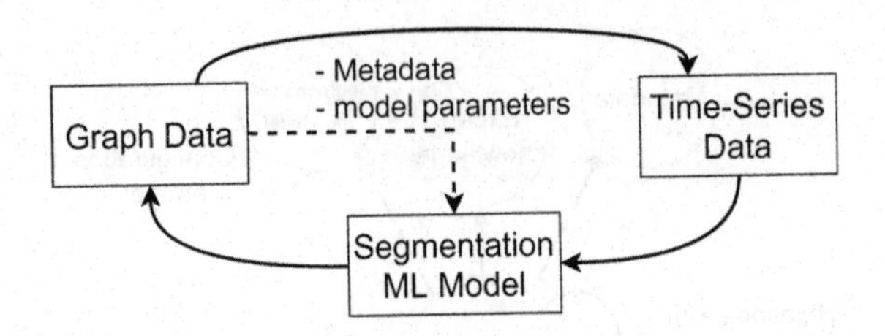

Fig. 4. Architecture of the data connector. The time-series data could be queried from the data graph. Segmentation is done with the ML model based on the metadata and the model parameters

4 Use Case

In this section we describe an industrial use case that applies the framework introduced in the previous section. The framework was tested on an exemplary carwash facility. It consists of the three main business processes: soaping, washing and drying. In terms of our framework they represent three concepts in the graph database. Each of the business processes incorporates two sensors that produce data. One sensor measures the power consumption of the process step, the other sensor measures a process step specific value. For soaping and washing this is the water consumption. For drying it records the air pressure.

To reliably store the data produced in this use case and to show the bridge between the two, a graph database (Neo4j) and a time-series database (InfluxDB) is necessary. Both of these systems are number one in their respective field in the popularity ranking conducted by db-engines.com [8,9]. In this example Neo4j is used to store instances of the modeled business processes, metadata, expert knowledge, information about the time-series data and the segmentation models as well as the segments. When a car uses the carwash the sequence of business processes instances and their relations shown in Fig. 2 are created as nodes and edges in Neo4j. In this use case, one car using the carwash creates three business process instance nodes (soap, washing and drying) and their edges to nodes containing business process metadata, expert knowledge and time-series metadata. Furthermore, if a segmentation model is applied to the sensor data corresponding to a business process instance, additional relations will be added. These consist of a relation to the node storing information about the used segmentation model and a relation to the node of the resulting segment. An example of the nodes and edges created in Neo4j when a car uses the carwash can be seen in Fig. 5.

The timestamped data created by the sensors and segmentation models, is saved in InfluxDB. To accurately map time-series data in InfluxDB to Neo4j, a tag, a start and an end timestamp is stored in all business process instances and segment nodes. In this example the available business process tags are "Soap", "Washing" and "Drying". When a new measurement is inserted into the time-series database, it is also assigned the tag corresponding to the business process it belongs to. For the tag in the segment a combination of the business process tag and an identifier of the segmentation model is used. Tags in InfluxDB are

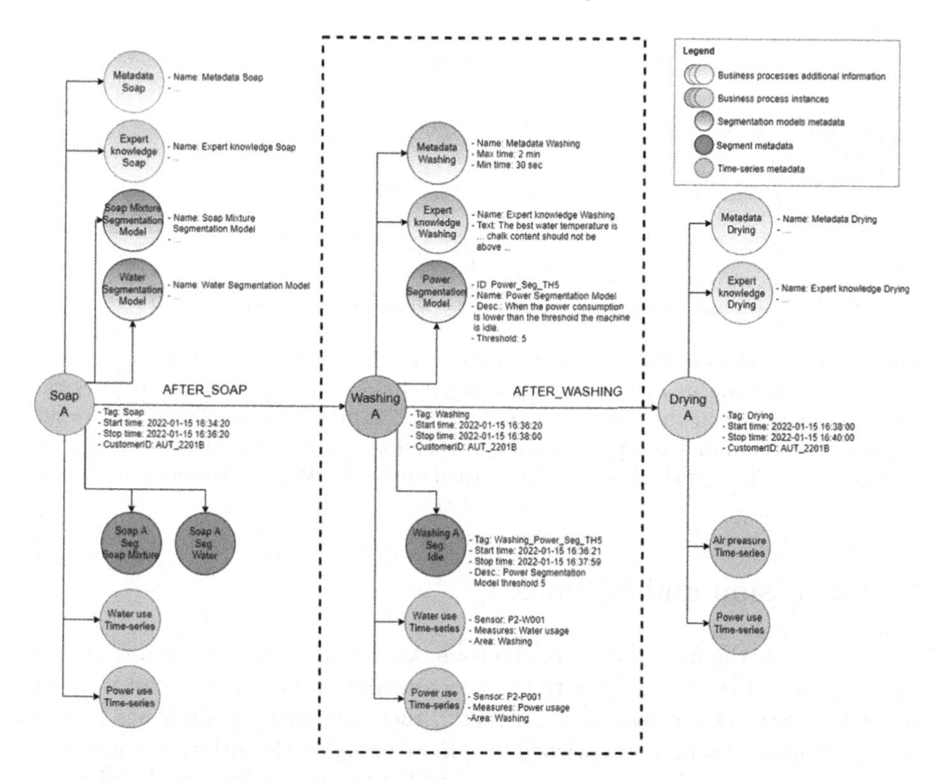

Fig. 5. Nodes and edges that are created in the graph database, when a car "A" uses the carwash. Additionally some segmentations were requested. The nodes "Soap A", "Washing A" and "Drying A" are instances of the business processes. They have a relation to other business processes, their respective expert knowledge and metadata, sensor data, segments and the corresponding segmentation models. For "Washing A" the "Power Segmentation Model" was applied, resulting in the "Washing A Seg. Idle" segment.

indexed, which means that they can be queried more efficiently [5]. In Fig. 6 an example of the power consumption of the washing business process enriched with data generated from a segmentation model can be seen.

With these preparations done, the foundation to accurately map time-series data in InfluxDB to a specific business process instance or segment node in the graph database is completed. The mapping itself can simply be done with a query where the tag, the start and end timestamp of a node are used as input and the result of the query is the time-series data corresponding to that specific tagged time interval. This approach takes advantage of the strengths of both databases: They have a powerful API that can easily be integrated in existing infrastructure and provide easy-to-learn querying languages, InfluxDB excels at writing huge amounts of time-series data and reading data with time-windowed queries [7,11] and Neo4j can store relationships between data in a flexible model and allows for both scalability and querying speed [2,6].

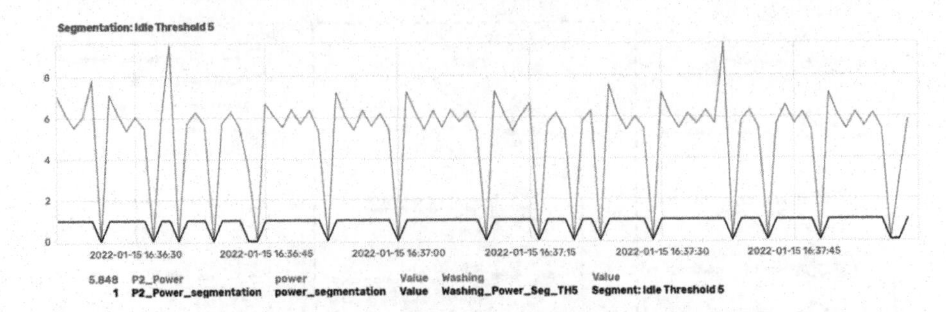

Fig. 6. Visualization of the results created by the "Power Segmentation Model" that can be seen as node in Fig. 5. The segment results are stored in InfluxDB after the model was applied on the time-series data. The "P2_Power_segmentation" is 0 when the power consumption is lower than 5 and therefore the machine is idle. Otherwise, the value is 1. This approach allows for a visual analysis solely on the time-series data.

5 Discussion and Outlook

Currently, it is the first version of the framework that uses only its core principles. The possible directions of the further research are in more detail work on the data segmentation modules. Currently, there are many approaches how to do that. Some of them were enriched by a more complex algorithm. For instance, in [10]. Our current framework is aimed to utilise the possibilities of the Machine Learning techniques to reduces manual operations. The framework also requires a more deep native integration into Neo4j which implies that further investigations on the best approach to reach this goal must be conducted. Additionally, considering the already observed in the work of Cavotol & Santanchè [1] limitation of the dynamic graph data updates we should create an approach that allows to efficiently run our segmentation algorithm.

6 Conclusion

Finally, the paper has presented a concept of the approach that we use for better integration of the graph and the time-series data. The proposed framework allows to map process data to time-series sensor. Furthermore, it is also capable of managing time-series segmentation to support the data analysis tasks on the lower level. The main architectural and conceptual ideas were explained, but the particular implementation is currently limited by the particular use case. At the moment it is technically tested using Neo4j and InfluxDB, but it could be adapted to use other database management systems. The working prototype has shown us the possibilities in data segmentation to increase capabilities of the graph database to easily map to timestamped data in a time-series database. We can confirm that the core goals of this paper were reached and that further steps in developing of this framework are reasonable.

Acknowledgements. The research reported in this paper has been funded by BMK, BMDW, and the State of Upper Austria in the frame of the COMET Programme managed by FFG.

References

1. Cavoto, P., Santanchè, A.: ReGraph: bridging relational and graph databases. In: Proceedings of the 30th Brazilian Symposium on Databases (2015)
2. Fernandes, D., Bernardino, J.: Graph databases comparison: AllegroGraph, ArangoDB, InfiniteGraph, Neo4j, and OrientDB. In: Data, pp. 373–380 (2018)
3. Ferreira, L.P.Z.d.P.M.: Bridging the gap between SQL and NoSQL. Ph.D. thesis, Universidade do Minho (2012)
4. Hofer, D., Jäger, M., Mohamed, A., Küng, J.: On applying graph database time models for security log analysis. In: Dang, T.K., Küng, J., Takizawa, M., Chung, T.M. (eds.) FDSE 2020. LNCS, vol. 12466, pp. 87–107. Springer, Cham (2020). https://doi.org/10.1007/978-3-030-63924-2_5
5. InfluxData: InfluxDB schema design (2022). https://docs.influxdata.com/influxdb/v2.1/write-data/best-practices/schema-design/. Accessed 25 Mar 2022
6. InfluxData: Neo4j graph database (2022). https://neo4j.com/product/. Accessed 25 Mar 2022
7. InfluxData: Optimize flux queries (2022). https://docs.influxdata.com/influxdb/v2.1/query-data/optimize-queries/. Accessed 25 Mar 2022
8. Solid IT gmbh: DB-engines ranking of graph DBMS (2022). https://db-engines.com/en/ranking/graph+dbms. Accessed 25 Mar 2022
9. Solid IT gmbh: DB-engines ranking of time series DBMS (2022). https://db-engines.com/en/ranking/time+series+dbms. Accessed 25 Mar 2022
10. Keogh, E., Chu, S., Hart, D., Pazzani, M.: Segmenting time series: a survey and novel approach. In: Data Mining in Time Series Databases, pp. 1–21. World Scientific (2004)
11. Naqvi, S.N.Z., Yfantidou, S., Zimányi, E.: Time series databases and InfluxDB. Studienarbeit, Université Libre de Bruxelles 12 (2017)
12. Perçuku, A., Minkovska, D., Stoyanova, L.: Modeling and processing big data of power transmission grid substation using neo4j. Procedia Comput. Sci. **113**, 9–16 (2017). https://doi.org/10.1016/j.procs.2017.08.276. https://www.sciencedirect.com/science/article/pii/S187705091731685X. The 8th International Conference on Emerging Ubiquitous Systems and Pervasive Networks (EUSPN 2017) / The 7th International Conference on Current and Future Trends of Information and Communication Technologies in Healthcare (ICTH-2017) / Affiliated Workshops
13. Roijackers, J., Fletcher, G.: Bridging SQL and NoSQL. Eindhoven University of Technology Department of Mathematics and Computer Science Master's thesis (2012)
14. Tan, R., Chirkova, R., Gadepally, V., Mattson, T.G.: Enabling query processing across heterogeneous data models: a survey. In: 2017 IEEE International Conference on Big Data (Big Data), pp. 3211–3220. IEEE (2017)
15. Villalobos, K., Ramírez-Durán, V.J., Diez, B., Blanco, J.M., Goñi, A., Illarramendi, A.: A three level hierarchical architecture for an efficient storage of industry 4.0 data. Comput. Ind. **121**, 103257 (2020)
16. Villalobos, K., Ramírez, V.J., Diez, B., Blanco, J.M., Goñi, A., Illarramendi, A.: A hierarchical storage system for industrial time-series data. In: 2019 IEEE 17th International Conference on Industrial Informatics (INDIN), vol. 1, pp. 699–705 (2019). https://doi.org/10.1109/INDIN41052.2019.8972120

A Synthetic Dataset for Anomaly Detection of Machine Behavior

Sabrina Luftensteiner(✉) and Patrick Praher

Software Competence Center Hagenberg GmbH, Hagenberg im Muehlkreis, Austria
{sabrina.luftensteiner,patrick.praher}@scch.at
http://www.scch.at/

Abstract. Logs have been used in modern software solutions for development and maintenance purposes as they are able to represent a rich source of information for subsequent analysis. A line of research focuses on the application of artificial intelligence techniques on logs to predict system behavior and to perform anomaly detection. Successful industrial applications are rather sparse due to the lack of publicly available log datasets. To fill this gap, we developed a method to synthetically generate a log dataset, which resembles a linear program execution log file. In this paper, the method is described as well as existing datasets are discussed. The generated dataset should enable a possibility for researcher to have a common base for new approaches.

Keywords: Log dataset · Log analysis · Simulated industrial processes · Process Mining

1 Introduction

Logs have been widely adopted in the development and maintenance of software systems enabling developers and support engineers to track system behaviors and perform subsequent analysis [6]. In general, logs are unstructured texts, which record the rich information of the systems behavior used for subsequent anomaly detection [3–5]. Using these logs, it is possible to determine and predict if a system behaves as expected or deviates from given paths. These deviations may indicate problems within the system, e.g. problems in automated machine sequences. The structure of logs get more complex, which requires problem specific solutions for the recognition of anomalies in system behaviors resulting in a labor-intensive and time-consuming analysis [6]. In general, researchers in this field are often working with their own log data, which is not published and therefore not available for others as companies have privacy concerns [6].

The research reported in this paper has been funded by the Federal Ministry for Climate Action, Environment, Energy, Mobility, Innovation and Technology (BMK), the Federal Ministry for Digital and Economic Affairs (BMDW), and the Province of Upper Austria in the frame of the COMET - Competence Centers for Excellent Technologies Programme managed by Austrian Research Promotion Agency FFG.

G. Kotsis et al. (Eds.): DEXA 2022 Workshops, CCIS 1633, pp. 424–431, 2022.
https://doi.org/10.1007/978-3-031-14343-4_40

Due to the lack of available and structured log data, this paper proposes a new method to generate such a dataset synthetically. Using structured log data, it is possible to create structure-independent approaches for anomaly detection in machine behavior. The overall goal is the creation of a labeled linear execution log dataset of a machine program, which covers both anomalous and non-anomalous behavior over multiple days. Through the course of this paper, existing datasets are presented and discussed. Furthermore, possible deviations, indicating anomalous activities, are covered as well as the workflow of the dataset generation itself.

2 Existing Datasets

He et al. [6] provide a large overview of currently existing log datasets, which are covering labeled and unlabeled data. The main problem using these datasets is the non-linear behavior of the dataflow, undefined segments as well as a complex structure of the logs themselves. The following paragraphs cover an overview of interesting datasets and their disadvantages for anomaly detection using Process Mining techniques.

The datasets HDFS_1 and HDFS_2 are based on log files gathered in Hadoop Distributed File System applications [2], whereat HDFS_1 is labeled. The logs contain block IDs, which are used to extract traces, and the labeled data additionally contains a label whether the behavior is normal or anomalous [6]. Due to the structure of the log, overlapping blocks and non available start and end points of blocks as well as highly varying behavior between blocks, it is rather difficult to identify anomalous blocks using Process Mining approaches.

The logs generated by an Hadoop [8] application covers labeled data, whereat at first no anomalies are injected and afterwards logs for events such as machine down, disc full and network disconnections are injected [6]. Due to the nature of injected anomalies, it is complicated to identify anomalies using the process flow on its own.

A dataset consisting of logs provided by the cloud operating system OpenStack [7] contains information about anomalous cases with failure injection for anomaly detection purposes. Due to their complicated structure and non-linear behavior, it is too difficult to apply Process Mining approaches for the identification of anomalies which produce reliable results.

BGL is an open dataset of logs collected from a BlueGene/L supercomputer system at Lawrence Livermore National Labs [1] and contains alert and non-alert messages identified by tags. The structure of log entries is inconsistent between different log types, which makes uniform processing complex. Furthermore, it is one dataflow, which can not be split into multiple smaller batches resulting in one big non-linear behavior tracker.

3 Method

This section covers an overview of the setup, the execution workflow and its deviations as well as the overall workflow of the data generation.

3.1 Setup

The entries of the dataset consist of replicated machine log-files, which simulate multiple program executions of the same linear program. One program execution covers a sequence of routines or process events, which refer to activities of a machine in order to manufacture a product, ranging from laser cutting to coating, to bending, and various other activities. Next to the basic workflow of the execution, additional errors and warnings are also inserted to provide a more realistic workflow. Errors are likely to occur due to faulty executions, wrong program usage as well as wear of machine parts, and indicate changes in the behavior of the program execution. The following event-log activities are covered in our scenario:

- Start Program - Start of a new program cycle
- Stop Program - Stop of a program cycle
- Step x - Process step within a program
- Error x - Error appearing during a program execution

Table 1 covers the simplified event-log of one program execution. The program workflow is started by a *Start Program* entry, followed by a sequence of steps and error entries to describe the program execution, and ended with a *Stop Program* entry. The dataset comprises multiple program executions, whereat each program execution might have some deviations from the optimal/basic workflow. To simulate the routine of manufacturing machines, the dataset covers multiple days with several executions per day.

Table 1. Simplified event-log of one program execution.

CaseID	Activity	Timestamp
1	Start program	2021-07-01 00:00:00
1	Step 1	2021-07-01 00:00:10
1	Step 2	2021-07-01 00:00:20
1	Step 3	2021-07-01 00:00:30
1	Error 1	2021-07-01 00:00:40
1	Step 4	2021-07-01 00:00:50
1	Step 5	2021-07-01 00:01:00
1	Stop program	2021-07-01 00:01:10

3.2 Workflow Deviations

The following four paragraphs cover possible deviations in the program execution, which are likely to occur in industrial environments. Next to a textual explanation, the scenarios are also visualized to enhance the comprehension of the underlying problems. The deviations are based upon the optimal/basic workflow visualized in Fig. 1, covering multiple steps in a linear manner without repeating or deviating events.

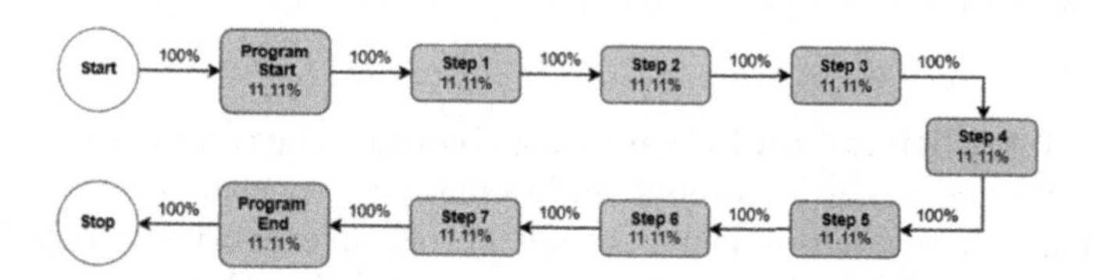

Fig. 1. Workflow of a typical production cycle visualized using relative paths.

Repeating Events. Repeating events are the first workflow deviation covered in our work. They refer to the immediate repeating of an event, either step or error/warning, in the workflow. They occur due to unsatisfactory outcomes from one step execution and its subsequent repeating to enhance the outcome or the reduction of program code, e.g. if one step has to be executed twice. Figure 2 visualizes the special behavior, where the second step is repeated with a chance of 43% having 5 out of 12 executions with repetitions.

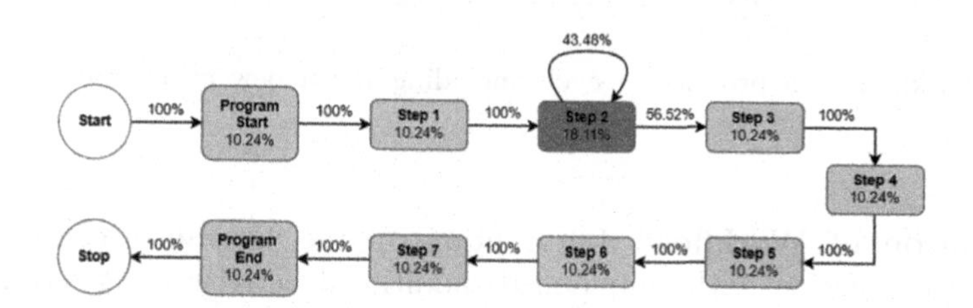

Fig. 2. Workflow of a production cycle including the event of a repeating activity.

Additional Events. During the production process, additional activities may appear within the program execution. The additional events are either planned, e.g. adaptation of existing programs, or unplanned, e.g. unscheduled error or warning messages. Figure 3 visualizes the appearance of an error chain, consisting of Error 1 to Error 3, after the second program step within the workflow. Compared with the basic workflow, see Fig. 1, the errors are present in 9 out of 12 program executions. The presence of an error or error chain could indicate problems with the program itself, machine wearing or operator issues.

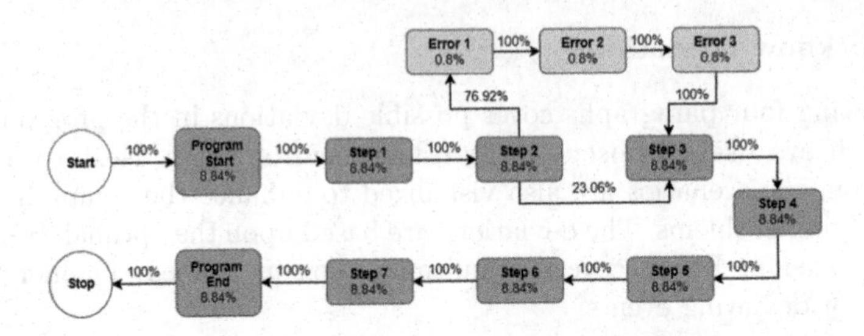

Fig. 3. Workflow of a production cycle containing a chain of error events.

Mixture of Additional and Repeating Events. The combination of the previously mentioned abnormal events, additional and repeating events, is likely to happen in industrial environments. During a program execution, errors/warnings may arise, leading to possible temporary stops of the machine or unsatisfactory outcomes, which demand the repeating of the current step to achieve satisfactory product results. Figure 4 visualizes such a situation, whereat the mixture arises in 8 out of 12 executions based on the basic workflow presented in Fig. 1.

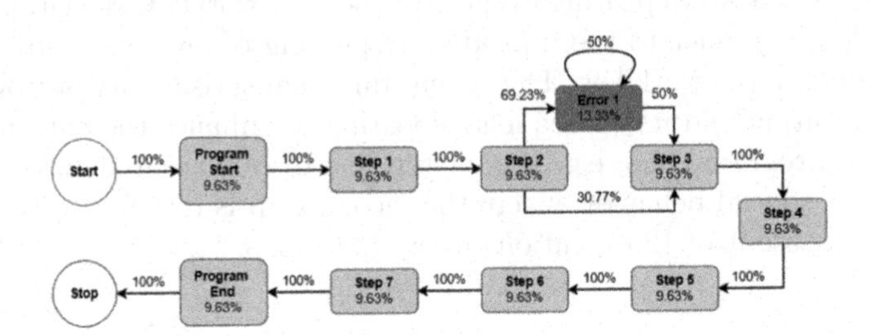

Fig. 4. Workflow of a production cycle including the a new event and a repeating activity.

Interruption of Workflow. Interruptions of the program execution occur often in combination with (unplanned) additional activities, e.g. errors. The occurrence of such an event indicates a major problem in the process as the overall program is stopped completely and not able to be continued. The production has to be started from the beginning. Figure 5 visualizes this scenario, whereat 8 out of 12 program executions include interruptions of the process after the second step.

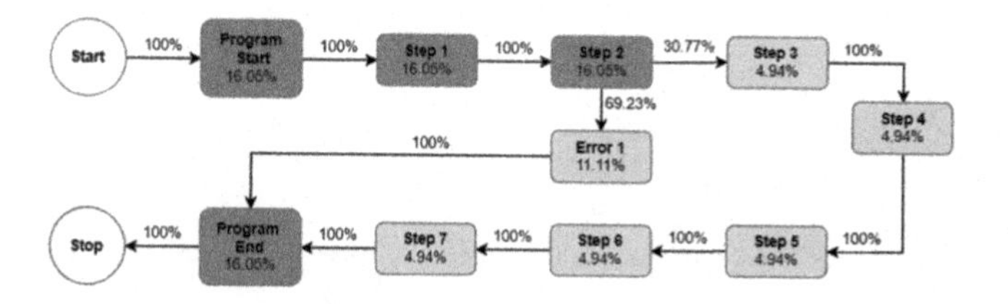

Fig. 5. Workflow of a production cycle including additional activities (e.g. error messages) and a missing activities.

3.3 Dataset Generation Workflow

The steps covering the workflow of the generation of the dataset is visualized in Fig. 6. Regarding our implementation, the programming language R was used. The general process can be mapped to various other programming languages, e.g. Python or Perl.

Prior to the dataset generation, the following parameters have to be set:

- Start date & time
- End date & time
- Time interval as step size.

These parameters define the timespan of the overall process and the breaks between each program execution to simulate an industrial process. After setting the temporal parameters, the timespan is partitioned into a sequence of items, which cover one time interval each, and the iteration over the sequence items starts. At first, it is checked if the item represents the start of a new day. If it is a new day, a flag is set to indicate if the day is anomalous or not by getting a random number and checking if it is above/below a certain pre-defined boundary, in our case 10% of the days are anomalous. Non-anomalous days and non-start items have the same subsequent workflow as items marked anomalous.

The next part of the workflow covers the generation of a new dataset entry for the current item. Start and Stop of a simulated program execution are defined with a timestamp as well as the steps covering the process. The timestamps are set with 10 s difference for each part of the entry, so the temporal flow is visible and available for further usage and analysis. To simulate a proper industrial process, which is not always running perfectly, additional error messages are inserted with a certain predefined probability within the program execution. The error insertions may also occur on normal days and indicate simpler problems, e.g. due to wrong operator settings.

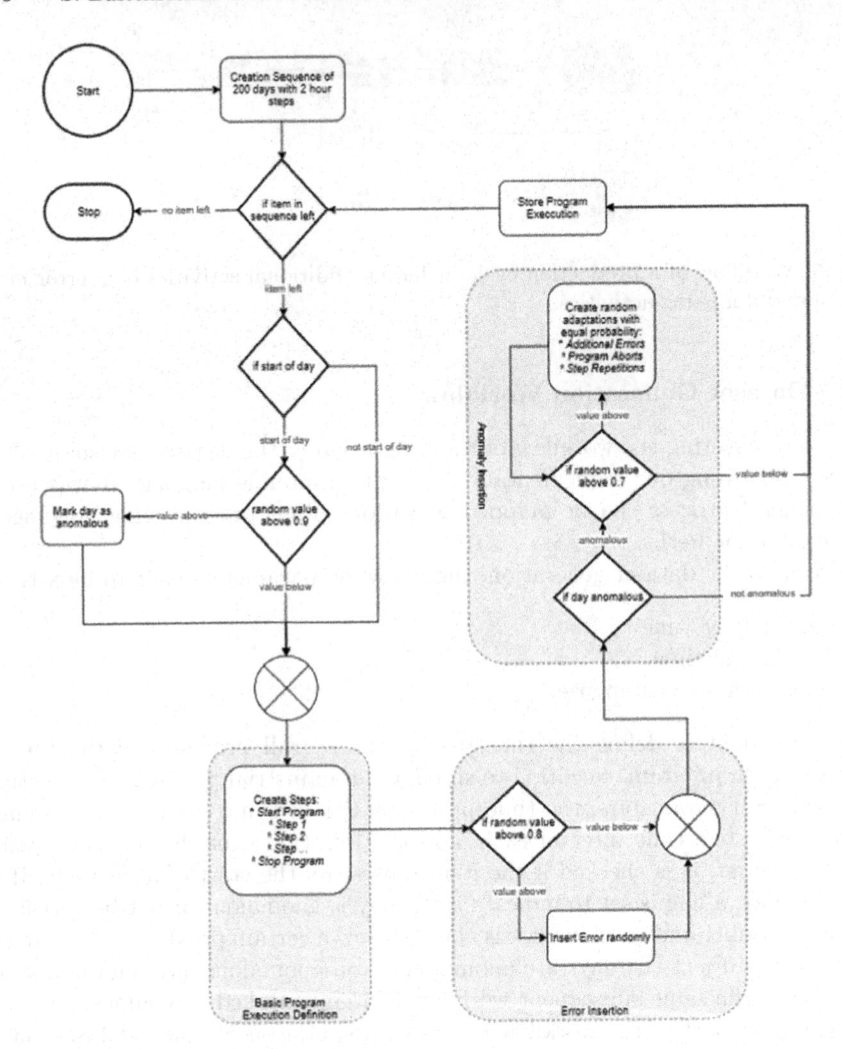

Fig. 6. Dataset generation workflow.

Following the creation of a normal program execution, it is checked if a day is anomalous and further adaptations have to be made. If a day is anomalous, the deviations mentioned earlier are incorporated into the simulated program execution. Each of the deviations has a certain probability to appear and is inserted at a random point in the workflow. Repeating events are inserted right after their counterpart, error message chains are inserted between two steps and program interruptions cut of the remaining workflow. To simulate a more realistic workflow, such deviations are inserted multiple times for an entry as such anomalous events occur often in combinations on problematic days, e.g. due to machine wearing or problematic program definitions.

After the insertion of anomalous events or if a day is non-anomalous, the program execution is stored together with an ID into the dataset. The ID is necessary to identify entries belonging to one program execution and allows grouping, filtering and other operations later on in the analysis process. Next to the final dataset, the anomalous days should also be returned or incorporated into the dataset to provide labeled data.

4 Conclusion

This paper describes the generation workflow of a synthetic log dataset for anomaly detection based on machine behavior. The dataset covers multiple linear program executions, whereat deviations are possible. These deviations range from common error messages to anomalous deviations of the execution path. Next to the generation of the dataset, existing datasets are presented and briefly discussed. The resulting dataset is labeled and is usable for various anomaly detection approaches.

References

1. Adiga, N.R., et al.: An overview of the BlueGene/L supercomputer. In: SC 2002: Proceedings of the 2002 ACM/IEEE Conference on Supercomputing, p. 60. IEEE (2002)
2. Borthakur, D., et al.: HDFS architecture guide. Hadoop Apache Project **53**(1–13), 2 (2008)
3. Du, M., Li, F., Zheng, G., Srikumar, V.: DeepLog: anomaly detection and diagnosis from system logs through deep learning. In: Proceedings of the 2017 ACM SIGSAC Conference on Computer and Communications Security, pp. 1285–1298 (2017)
4. Fu, Q., Lou, J.G., Wang, Y., Li, J.: Execution anomaly detection in distributed systems through unstructured log analysis. In: 2009 Ninth IEEE International Conference on Data Mining, pp. 149–158. IEEE (2009)
5. He, S., Zhu, J., He, P., Lyu, M.R.: Experience report: system log analysis for anomaly detection. In: 2016 IEEE 27th International Symposium on Software Reliability Engineering (ISSRE), pp. 207–218. IEEE (2016)
6. He, S., Zhu, J., He, P., Lyu, M.R.: Loghub: a large collection of system log datasets towards automated log analytics. arXiv preprint arXiv:2008.06448 (2020)
7. Sefraoui, O., Aissaoui, M., Eleuldj, M.: OpenStack: toward an open-source solution for cloud computing. Int. J. Comput. Appl. **55**(3), 38–42 (2012)
8. White, T.: Hadoop: The Definitive Guide. O'Reilly Media, Inc., Sebastopol (2012)

Author Index

Printed in the United States
by Baker & Taylor Publisher Services